DUCKWORTH OVERLOOK
Cosmosapiens

"A 700-page description of the state of current scientific knowledge about the origin of matter, life and humanity, plus a bold attempt to assess its limits. It is lucid and intelligible to the non-specialists—Hands was originally trained as a chemist, but has also published three novels. This is a book of astonishing ambition and scope, more like the work of a great Victorian polymath than most popular science books"
> —TIM CRANE, *Times Literary Supplement* Book of the Year

"A substantial, sceptical survey of the current state of scientific knowledge of about the most basic questions... [Hands] is always fair, and he writes from what must be the best point of view for a scientist—he is a dispassionate agnostic about everything, except those few things which can fulfil Popper's falsification principle... An invaluable, encyclopedic achievement"
> —A. N. WILSON, *Times Literary Supplement* Book of the Year

"Argues that mind and matter evolved in unison and, one day, human consciousness and the star-filled universe will be revealed as part of the same cosmic whole. Such ideas were lapped up by the 19th-century followers of Hegel and it is both shocking and invigorating to hear them stated again... in the context of a quantum universe"
> —NICHOLAS BLINCOE, *Daily Telegraph* Books of the Year

"It would scarcely be possible, without Hands's own overarching perspective, to grasp the central reality of contemporary science... an irresoluble tension between the awesome achievement of science in the recent past and the compelling signs that for all that we might learn there is, after all, more than we can know. *Cosmosapiens* is, as Cambridge Professor of Philosophy Tim Crane has commented, 'a truly remarkable work'... dispassionate, even-handed and, informed by the author's extensive correspondence with more than sixty specialists across the entire range of scientific disciplines, immensely persuasive"
> —JAMES LE FANU, *The Tablet*

"Audacious, ambitious, and philosophically completist... a pearl of dialectical reasoning between Hands and the most celebrated experts he can find. In today's age of specialization, readers will welcome this throwback to the days of the well-informed layperson, conversant and opinionated in a variety of topics... Hands grounds his musings in logic and scientific fact to produce a thoughtful treatise for the eternally curious"
> —*Publishers Weekly* (starred review)

D1638419

"From dark energy to the selfish gene, Hands looks at how we know what we know—and what we don't. An overview of thought on this ever-fascinating subject"
—*Observer*

"An exhaustive, fascinating look at humankind's role in the story of the universe"
—ADAM MORGAN, *Chicago Review of Books*

"A triumph of detailed conceptual analysis covering the fields of fundamental cosmology, physics, biology and the evolution of philosophical and religious ideas… a truly exceptional piece of work"
—DAVID LORIMER, *Science & Medical Network Review*

"A thoughtful, well-written volume"
—*Library Journal*

"A compendious work that will intrigue serious readers"
—*Kirkus*

"*A* stunningly ambitious book that aims to draw together the best of modern scientific knowledge to answer fundamental philosophical questions: How did the universe begin? How did life start? And how did our species evolve from simple organisms?… while the resulting work has won widespread praise… it has also met with hostility from some parts of the scientific estab-lishment for daring to challenge some orthodoxies"
—CRAIG KENNY, *Camden New Journal*

"A game-changer. In the tradition of Thomas Kuhn's *The Structure of Scientific Revolutions*, this lucidly written, penetrating analysis challenges us to rethink many things we take for granted about ourselves, our society, and our universe. It will become a classic"
—PETER DREIER, E P Clapp Distinguished Professor of Politics, Occidental College

"John Hands has attempted a remarkable thing: nothing less than an exhaustive account of the current state of scientific knowledge about the origins and evolution of the cosmos, life and humanity. His driving questions are those that have inspired all of science, religion and philosophy: What are we? Where do we come from? What is the source of consciousness, value and meaning? Hands painstakingly summarises the current state of knowledge in a huge variety of fields, from cosmology to evolutionary psychology, in enviably lucid prose. His conclusions are measured and sceptical, and his conception of the limits of science is well-argued: he gives an extremely clear view of what science has established, what it has not established, and what it never will. This is a truly exceptional piece of work"

—TIM CRANE, Knightbridge Professor of Philosophy, University of Cambridge

"John Hands is an astute observer of recent trends in scientific ideas bold enough to point out what he sees as sense and nonsense and intelligently explain why"

—PAUL J. STEINHARDT, Albert Einstein Professor in Science, Princeton University

"There have been numerous books seeking to tell a tale that opens with the beginning of our universe and... concludes with 'the ultimate' event: the emergence of our species, *Homo sapiens*. None has done this better, more clearly, and with greater thought and documentation than John Hands... a work as bold, broad, and challenging as this will no doubt tweak the bias any one of us may have regarding a particular event, but, then, so did Darwin's *On the Origin of Species*"

—JEFFREY SCHWARTZ, Professor of Physical Anthropology and of The History & Philosophy of Science at the University of Pittsburgh

"An intellectual tour de force: a review of about all the major scientific theories that purport to explain the origins of the universe, matter, life and mind. The approach is refreshingly agnostic, as the author systematically points out how much we still don't know and perhaps never will know. He critically dissects the relevant observations, theories and hypotheses, both mainstream and alternative, pointing out both their strengths, and—alas much more numerous—weaknesses. As such, he provides a welcome counterpoint to the sensationalism that typically accompanies the latest 'discoveries' [and]... manages to explain all these intrinsically very difficult concepts and theories in a clear and readable language... highly recommended for anyone wishing to get a deeper insight into the fundamental but typically arcane theories that purport to explain where we and the universe that surrounds us are coming from"

—PROFESSOR FRANCIS HEYLIGHEN, Evolution, Complexity and Cognition Group, Free University of Brussels

"It often takes an outsider to see the limitations of conventional science. As far as biological evolution is concerned, John Hands has done a remarkable job of disentangling the many topics that are long overdue for reinterpretation. The enormous effort he has made to cover so many evolutionary questions is heroic. That is the first step to making progress. A major accomplishment"

—PROFESSOR JAMES SHAPIRO, author of *Evolution: A View From The 21st Century*

John Hands has devoted more than ten years to evaluating scientific theories about human evolution from the origin of the universe. He graduated in chemistry from the University of London and co-authored two research studies and published one book in the social sciences. He was the founding Director of the Government's Co-operative Housing Agency and served on three Government committees. He has tutored in both Physics and Management Studies for the Open University and was Royal Literary Fund Fellow at University College London. He has written three novels and has been published in eight countries. See www.johnhands.com for more.

Also by John Hands

Non-fiction
Housing Co-operatives

Novels
Perestroika Christi
Darkness at Dawn
Brutal Fantasies

COSMOSAPIENS

Human Evolution from the Origin of the Universe

JOHN HANDS

Duckworth Overlook

First published in the UK in 2015 by
Duckworth Overlook

This edition first published in 2016

LONDON
30 Calvin Street, London E1 6NW
T: 020 7490 7300
E: info@duckworth-publishers.co.uk
www.ducknet.co.uk
For bulk and special sales please contact sales@duckworth-publishers.co.uk,
or write to us at the above address.

978-0-7156-5121-6

Typeset by Fakenham Prepress Solutions, Fakenham, Norfolk NR21 8NN

Printed and bound in Great Britain

In loving memory of my wife, Paddy Valerie Hands

Contents

acceleration of the universe's expansion rate. Redshift. Ripples in cosmic microwave background. Exaggerated claims. WMAP data. Planck telescope's confirmation of contradictory evidence. Data selectivity. Law of Data Interpretation.

Fossil record of animals and plants. Evolution of mammals. Tracing human evolution from the fossil record.

Chapter 21 Causes of Biological Evolution: the Current Orthodox Account

altruism". Game theory. Empirical proof. Selfish gene. Genial gene. Multilevel selection.

Ego-anthropocentricism. Genetic identity. Behavioural difference in degree only.

Reduction in aggression. Increase in cooperation. Rate of change. Globalization. Complexification. Leading edge trend. Convergence. Hominization. Changing duality of human nature. Integration of patterns in the evidence.

Observation and measurement. Data. Subjectivity. Method. Theory. Defective science.

Subjective experiences. Social concepts and values. Untestable ideas. Metaphysical questions.

Acknowledgements

It is impossible to name everyone who helped bring this book to fruition. They include authors whose papers and books I consulted. The Notes section towards the end of the book includes full references to most academic papers and references to books, plus long and significant papers, by author and year. Complete details of the books and significant papers cited in the Notes appear in the Bibliography that follows. The distinction is arbitrary because some of the academic papers are extensive and I have included significant ones in the Bibliography, but in general I want to avoid a huge Bibliography.

Of the very many specialists who generously shared their expertise with me, I'm particularly indebted to those who responded to my request to check draft sections for errors of fact or omission or unreasonable conclusions and to make any other comments. I list them here within broad areas of study, each of which encompasses many specialized fields relevant to the book. Their description is the post held when consulted.

Myths: Charles Stewart and Mukulika Banerjee, Senior Lecturers in the Department of Anthropology at University College London (UCL).

Cosmology and astrophysics: George Ellis, Distinguished Professor of Complex Systems in the Department of Mathematics and Applied Mathematics at the University of Cape Town; Paul Steinhardt, Albert Einstein Professor of Science at Princeton University; Ofer Lahav, Perren Chair of Astronomy and Head of Astrophysics at UCL; Bernard Carr, Professor of Astronomy at Queen Mary, University of London; the late Geoffrey Burbidge, Professor of Astronomy at the University of California, San Diego; Javant Narlikar, Emeritus Professor at the Inter-University Centre for Astronomy and Astrophysics, Pune, India; Jon Butterworth, Professor of Physics and Head of the Department of Physics and Astronomy at UCL; Serena Viti, Reader in the Department of Physics and Astronomy at UCL; Eric J Lerner, President of Lawrenceville Plasma Physics, Inc.

Philosophy: Tim Crane, Professor and Head of the Department of Philosophy at UCL and Director of the Institute of Philosophy, and also later when Knightbridge Professor of Philosophy at the University of Cambridge; Hasok Chang, Professor in the Philosophy of Science at UCL.

Planetary and atmospheric science: Jim Kasting, Distinguished Professor at Pennsylvania State University.

Geology: John Boardman, Reader in Geomorphology and Land Degradation at the University of Oxford.

History of science: Adrian Desmond, biographer and Honorary Research Fellow in the Biology Department at UCL; Charles Smith, Professor and Science Librarian at Western Kentucky University; John van Whye, Founder and Director of *The Complete Works of Charles Darwin* Online; James Moore, biographer and Professor in the Department of History at the Open University; James Le Fanu, physician and historian of science and medicine.

The emergence and evolution of life: Professor Adrian Lister, Merit Researcher in the Department of Palaeontology at the Natural History Museum; Jim Mallet, Professor in the Department of Genetics, Evolution and Environment at UCL; Johnjoe McFadden, Professor of Molecular Genetics at the University of Surrey; Mark Pallen, Professor of Microbial Genomics at the University of Birmingham; Chris Orengo, Professor of Bioinformatics at UCL; Jerry Coyne, Professor in the Department of Ecology and Evolution at the University of Chicago; the late Lynn Margulis, Distinguished Professor of the University of Massachusetts; Jim Valentine, Professor Emeritus in the Department of Integrative Biology at the University of California, Berkeley; Jeffrey H Schwartz, Professor of Physical Anthropology and of The History & Philosophy of Science at the University of Pittsburgh; Hans Thewissen, Professor of Anatomy in the Department of Anatomy and Neurobiology at Northeastern Ohio Universities College of Medicine; Rupert Sheldrake, cell biologist and Director of the Perrott-Warrick Project funded from Trinity College, Cambridge; Simon Conway Morris, Professor of Evolutionary Palaeobiology at the University of Cambridge; Francis Heylighen, Research Professor at the Free University of Brussels; Jonathan Fry, Senior Lecturer in the Department of Neuroscience, Physiology & Pharmacology at UCL; Thomas Lentz, Professor Emeritus of Cell Biology at Yale University School of Medicine; Richard Goldstein of the Mathematical Biology Division at the National Institute for Medical Research, London; Avrion Mitchison, Professor Emeritus of Zoological and Comparative Anatomy at UCL.

Animal behaviour: Volker Sommer, Professor of Evolutionary Anthropology at UCL; Alex Thornton, Drapers' Company Research Fellow at Pembroke College, Cambridge; Heikki Helanterä, Academy Research Fellow at the University of Helsinki; Simon Reader, associate professor in the Biology Department at McGill University.

The emergence of humans: Robin Derricourt, archaeological historian at the University of New South Wales; C Owen Lovejoy, Professor in the Department of Anthropology at Kent State University; Tim White, Professor in the Department of Integrative Biology at the University of California, Berkeley.

The evolution of humans: Steven LeBlanc, Professor of Archaeology at Harvard University; John Lagerwey, Professor in the Centre for East Asian Studies at the Chinese University of Hong Kong; Liz Graham, Professor of Mesoamerican Archaeology at UCL; Subhash Kak, Regents Professor of Computer Science at

Oklahoma State University; Fiona Coward, lecturer in Archaeological Science at Bournemouth University; Dorian Fuller, Reader in Archaeobotany at UCL; Pat Rice, Professor Emerita, Department of Sociology and Anthropology at West Virginia University; Damien Keown, Professor of Buddhist Ethics at Goldsmiths, University of London; Stephen Batchelor, Buddhist teacher and author; Naomi Appleton, Chancellor's Fellow in Religious Studies at the University of Edinburgh; Simon Brodbeck, Lecturer in the School of History, Archaeology & Religion at Cardiff University; Chad Hansen, Professor of Philosophy at the University of Hong Kong; Gavin White, author of *Babylonian Star-Lore*; Magnus Widell, Lecturer in Assyriology at the University of Liverpool; Stephen Conway, Professor and Head of the History Department at UCL; Bruce Kent, founder and vice-president of the Movement for the Abolition of War; Dean Radin, Chief Scientist of the Institute of Noetic Sciences.

Charles Palliser, novelist, commented on selected chapters from the viewpoint of a non-specialist.

Any remaining errors are entirely my responsibility. Not all these specialists agree with the conclusions I drew from the evidence. Indeed, I deliberately sought comments from many whose publications showed they had contrary views. Some engaged in lengthy correspondence, drawing my attention to evidence of which I had been unaware, or giving a different interpretation of the evidence, or questioning my arguments. I greatly appreciate such interchanges, which improved the manuscript and the conclusions. Others were open-minded enough to say that on reflection they agreed with my conclusions. Several specialists disagreed with others in their field.

Although the questions of what we are, where we came from, and why we are here have intrigued me since I was a science undergraduate, the idea to research and write a book on the subject emerged in 2002. My appointment as Royal Literary Fund Fellow at University College London in 2004 provided the income, the ambience, and the library to enable me to develop and shape my ideas, and the book took its current focus in 2006. I'm immensely grateful to UCL and my colleagues there, to the Graduate School for their support, to the graduate students I tutored and thereby learned from, and to the RLF for financing the fellowship. I'm similarly grateful to the Arts Council of England for the award of a literature grant in 2009 that enabled me to work almost full-time on researching and writing. Katie Aspinall very kindly provided her cottage in Oxfordshire for periods of undisturbed reflection.

When my friends wanted to be supportive they said the project was ambitious. When they wanted to be realistic they said it was mad. In my saner moments I agreed with the latter. I owe an enormous debt of gratitude to my agent, Caspian Dennis of Abner Stein, who had faith in the project and the skill to place it with the right publisher at the right time. Andrew Lockett, the newly appointed publishing director at Duckworth, was intrigued by the proposal and in his first week invited us to meet him. The project needed the approval of Duckworth's owner, Peter Mayer, founder of the Overlook Press in New York. The extremely polite but extremely thorough grilling Peter gave me on a visit to London made me understand why for more than 20 years he had been the world's leading and most innovative publisher, including 18 years as Chief Executive Officer of Penguin Books. Andrew championed the book and provided invaluable editorial comments on the manuscript. His team of Melissa Tricoire, Claire Eastham, Jane Rogers, and David Marshall were enthusiastic in their support and ideas for producing an attractive, accessible book and bringing it to the attention of potential readers, as was Nikki Griffiths, Andrew's successor as publishing director, and Deborah Blake, the proofreader. Likewise I'm most grateful to Tracy Carns and Erik Hane at the Overlook Press.

London, 2015

CHAPTER ONE

The Quest

...if we do discover a complete theory, it should in time be understandable in broad principle by everyone, not just a few scientists. Then we shall all, philosophers, scientists, and just ordinary people, be able to take part in the discussion of the question why is it that we and the universe exist. If we find the answer to that, it would be the ultimate triumph of human reason—for then we would know the mind of God.

—STEPHEN HAWKING, 1988

When we have unified enough certain knowledge, we will understand who we are and why we are here.

—EDWARD O WILSON, 1988

What are we? and why are we here? are questions that have fascinated humans for at least 25,000 years. For the vast majority of that time we have sought the answer to these questions through supernatural beliefs. Roughly 3,000 years ago we began to seek the answer through philosophical insight and reasoning. Just over 150 years ago Charles Darwin's *On the Origin of Species* marked a fundamentally different approach. It adopted the empirical method of science and led eventually to the view that we are the product of biological evolution. Around 50 years ago cosmologists concluded that the matter and energy of which we ultimately consist originated in a Big Bang that created the universe. And then some 30 years ago neuroscientists began to show that what we see, hear, feel, and think correlate with the activity of neurons in different parts of our brain.

These towering achievements in science were made possible by advances in technology that generated an exponential increase in data. This in turn drove the ramification of science into ever narrowing and deepening foci of investigation. In recent times nobody has stepped back from examining the leaf on one branch to see what the whole evolutionary tree is showing us about what we are, where we came from, and why we exist.

This quest is an attempt to do just that: to ascertain what science can reliably tell us from systematic observation or experiment about how and why we evolved from the origin of the universe and whether what we are makes us different from all other animals.

I shall approach this task in four parts. Part 1 will examine science's explanation for the emergence and evolution of the matter and energy of which we ultimately consist; Part 2 will do likewise for the emergence and evolution of life, because we are living matter, and Part 3 for the emergence and evolution of humans. In Part 4 I will see if there are any consistent patterns in the evidence that enable overall conclusions to be drawn.

In each part I shall break down the pivotal question What are we? into constituent questions that relevant specialist fields investigate, try to find from academically recognized publications in each field answers that are validated by empirical evidence rather than derive from speculation or belief, and see whether or not there is a pattern in the evidence that enables conclusions to be drawn. Only if such an approach fails to provide a satisfactory explanation shall I consider the reasonableness of hypotheses and conjectures and other possible ways of knowing, like insight.

I shall then ask specialists in each field (listed in the Acknowledgements) to check the draft results for errors of fact or omission and unreasonable conclusions.

At the end of each chapter I shall list any conclusions so that the reader who wishes to skip any of the more technical sections can see my findings.

The question of what we are has intrigued me ever since I was a science under-graduate. Apart from co-authoring two research studies plus writing one book in the social sciences, and four years part-time tutoring physics for the Open University, I have not practised as a scientist, and so in that sense I am unqualified for the task. On the other hand few researchers today possess the relevant knowledge outside the specialized field in which they were trained and now practise.

I anticipate that many such specialists will feel that I have not written in sufficient detail in their field. If so I plead guilty in advance. I am attempting to write a book, not a library, and that necessarily requires summarizing if the goal of revealing the overall picture of human evolution is to be achieved: a vision of what we are and why we are here.

Despite efforts to correct errors, in such an enterprise some details may prove to be flawed, for which I take full responsibility. Or they may be overtaken by the results of new research between writing and publication, but that is how science, as distinct from belief, advances. What I hope is that the book will provide an overarching framework that others can refine and build upon.

A majority of the world's population, however, do not accept that we are the product of an evolutionary process. They believe in various myths to explain our origins. I shall begin, therefore, with a chapter that examines what these origin myths are, why they have endured for nearly 500 years after the scientific revolution began, and whether they have influenced scientific thinking.

Much disagreement arises because different people use the same word to mean different things: meanings change over time and in different cultural contexts. To minimize misunderstanding I shall define the precise meaning I intend for each significant and potentially ambiguous word when I first use it, and compile a list of such terms in the Glossary at the end of the book, which also includes definitions of unavoidable technical terms.

The first word to define is "science". It derives from the Latin *scientia*, which means knowledge. Different kinds of knowledge may be acquired, or claimed to be acquired, in different ways. From about the sixteenth century it came to mean knowledge about the natural world—inanimate and animate—acquired by observation and experiment, as distinct from knowledge acquired by reasoning alone, insight, or revelation. Consequently a definition of science must include the means by which its knowledge is acquired. Our current understanding of science may be summarized as

> **science**   The attempt to understand and explain natural phenomena by using systematic, preferably measurable, observation or experiment, and to apply reason to the knowledge thereby obtained in order to infer testable laws and make predictions or retrodictions.

> **retrodiction**   A result that has occurred in the past and is deduced or predicted from a later scientific law or theory.

Science aims to formulate a law, or a more general theory, to explain the invariable behaviour of a system of phenomena. We use such a law or theory to predict future outcomes by applying it to specific phenomena within the system. For instance, within the system of moving objects we apply Newton's laws of motion to predict the result of firing a specific rocket in specific circumstances.

Science may also inform us about outcomes in the past. An example of a retrodiction is that from the theory of plate tectonics we can deduce that similar fossils dating from before the breakup of the supercontinent Pangaea around 200 million years ago will be found near the complementary western coastline of South America and the eastern coastline of South Africa.

From the eighteenth century the study of natural phenomena encompassed humans and their social relationships. Use of the scientific method in such studies developed by the nineteenth century into the social sciences, an umbrella term covering such disciplines as archaeology, anthropology, sociology, psychology, political science, and even history. I shall evaluate relevant findings in these disciplines in Part 3.

Some refer to mathematics as a science, but its field of study extends way beyond natural phenomena and its theories cannot be tested empirically. In the context of this investigation I think it better to classify mathematics as a language in which some science, particularly its laws, can be expressed.

"Theory" in science has a more specific meaning than in general use, and even in science "theory" and "hypothesis" are often used loosely. It helps to distinguish between the two.

> **hypothesis**   A provisional theory put forward to explain a phenomenon or set of phenomena and used as a basis for further investigation; it is usually arrived at either by insight or by inductive reasoning after examining incomplete evidence and must be capable of being proven false.

The criterion of falsifiability was proposed by philosopher of science Karl Popper. In practice this may not be straightforward, but most scientists today accept the principle that to distinguish a scientific hypothesis from a conjecture or belief it must be subject to empirical tests that can falsify it.

> **theory**   An explanation of a set of phenomena that has been confirmed by a number of independent experiments or observations and is used to make accurate predictions or retrodictions about such phenomena.

The wider the range of phenomena explained the more useful is the scientific theory. Because science advances by the discovery of new evidence and the application of new thinking, a scientific theory may be modified or disproved as a result of contradictory evidence but it can never be proved absolutely. Some scientific theories, though, are generally recognized as well-established. For example, while the theory that the Earth is the centre of the universe and the Sun and other stars revolve around it has been disproved, the theory that the Earth orbits the Sun has been validated by so many observations and accurate predictions that it is accepted as established fact. However, even this may not always be so. Indeed very probably it will not be so in some 5 billion years when the Sun is predicted by most studies of its evolution to turn into a red giant star that will expand to envelop and burn up the Earth.

Any investigation is heavily influenced by prior beliefs. I was born and educated a Catholic, became an atheist, and am now an agnostic. I have no prior beliefs in theism, deism, or materialism. I genuinely do not know. And that is part of the excitement of embarking on this quest to discover from scientific evidence just what we are and may become. I invite readers with an open mind to join me on this quest.

PART ONE

The Emergence and Evolution of Matter

CHAPTER TWO

Origin Myths

I want to know how God created this world.

—ALBERT EINSTEIN, 1955

The world and time had both one beginning. The world was made not in time but simultaneously with time.

—ST AUGUSTINE OF HIPPO, 417

Since 11 February 2003* science's orthodox account, usually presented as fact, is that 13.7 billion years ago the universe, including space and time as well as matter and energy, exploded into existence as a point-like fireball of infinite density and incredibly high temperature that expanded and cooled into the universe we see today. This was the Big Bang from which we evolved.

Before investigating whether science can explain our evolution from the origin of matter and energy, I shall briefly consider the origin myths believed by a large majority of the world's population. It is instructive to examine the principal ideas in the different myths, the various explanations advanced for them by social scientists and whether these explanations meet the tests of evidence or reason, why the myths have endured, and to what extent they have influenced scientific thinking.

Principal themes

Every culture throughout recorded history has one or more stories about how the universe and we humans originated: understanding where we came from is part of an inherent human desire to understand what we are. The Rig Veda, the

---

* The day NASA scientists announced that data from their satellite-based Wilkinson Microwave Anisotropy Probe (WMAP) had confirmed the Big Bang model and had enabled them to determine the age of the universe with an unprecedented accuracy of one per cent margin of error. On 21 March 2013 European Space Agency scientists announced that data from their Planck space telescope meant that the age should be revised to 13.82 billion years.

oldest sacred text in the world and the most important scripture of what is now called Hinduism, has three such myths in its tenth book of hymns to the gods. The Brahmanas, the second part of each veda largely devoted to ritual, have others, while most of the Upanishads, accounts of the mystical insights of seers that tradition attaches to the end of the vedas,* express in various ways a single insight into the origin of the universe.[1] Judaeo-Christian and Islamic cultures broadly share a creation explanation, while other cultures have their own. The Chinese have at least four origin myths that exist in several versions. Although every myth is different,[2] nine principal themes recur; some overlap.

Primordial chaos or water

Many myths tell of a pre-existent chaos, often depicted as water, from which a god emerges to create the world or parts of it. The Pelasgians, who entered the Greek peninsula from Asia Minor in about 3500 BCE, brought with them the story of the creator goddess Eurynome who arose naked from Chaos.[3] The myths of Heliopolis in Egypt dating from the fourth millennium BCE speak of Nu, the primordial watery abyss, from which Atum arose to masturbate the world into existence. By around 2400 BCE Atum became identified with the Sun god Ra and his emergence was associated with the rising of the Sun and the dispelling of chaotic darkness.

Earth diver

Other myths, widespread in Siberia, Asia, and some native American tribes, have a pre-existent animal—often a turtle or bird—that dives into the primordial waters to bring up a piece of earth that later expands into the world.

Cosmogonic egg

In parts of India, Asia, Europe, and the Pacific an egg is the source of creation. The Satapatha Brahmana says the primordial waters produced the creator god Prajapati in the form of a golden egg. After a year he breaks open the egg and tries to speak. His first word becomes the earth, his second the air, and so on. Similarly, one version of the Chinese P'an Ku myth begins with a great cosmic egg inside of which the embryonic P'an Ku floats in Chaos. In the Greek Orphic creation myth, deriving from the seventh or sixth century BCE and contrasting with Homer's Olympian myths, it is time that creates the silver egg of the cosmos out of which bursts the bisexual Phanes-Dionysus who bears with him the seeds of all gods and all men and who creates Heaven and Earth.

World parents

A widespread theme has the world father—usually the sky—mating with the world mother—usually the Earth—to give birth to the elements of the world.

---

*   See Glossary at the end of the book for a more detailed explanation of these terms.

Often they lie locked in sexual embrace, indifferent to their children, as in one Maori creation myth.

Rebellion by children
In several myths the progeny rebel against the world parents. The children in the Maori myth—forests, food plants, oceans, and man—battle with their parents for space. Perhaps the best-known myth of this type is the *Theogony* composed by the Greek Hesiod in the eighth century BCE. This records the rebellion of successive generations of gods against their parents, the first of whom were Chaos, Earth, Tartarus (the underworld), and Eros (love); it leads eventually to the triumph of Zeus.

Sacrifice
The idea of creation through sacrifice often occurs. The Chinese myth of P'an Ku says "The world was never finished until P'an Ku died. Only his death could perfect the universe. From his skull was shaped the dome of the sky, and from his flesh was formed the soil of the fields....And [lastly] from the vermin which covered his body came forth mankind."[4]

Primordial battle
The great Babylonian epic, the *Enûma Elish*, describes warfare between the Sumerian gods and the local deity of Babylonia, Marduk, and his followers. Marduk slays the surviving original goddess, Tiamat, and her monsters of Chaos, establishes order, and becomes the supreme, universal creator god: all nature, including humans, owes its existence to him. Similar myths appear all over the world, for example the Olympian victory of the masculine sky gods of the invading Aryans over the fertile earth goddesses of the Pelasgians and Cretans.

Creation out of nothing
Only a few myths have the theme of creation out of nothing. However, its belief is not only one of the most widespread but also the currently favoured scientific explanation.

The oldest version comes from the Rig Veda. Recent archaeoastronomical investigations challenge Max Müller's nineteenth century dating and claim support for Indian tradition; they conclude that it was compiled over a period of two thousand years, beginning around 4000 BCE.[5] In the tenth and last book, Hymn 129 says "Then was not non-existent nor existent: there was no realm of air, no sky beyond it....That One Thing, breathless, breathed by its own nature: apart from it was nothing whatsoever."

This idea is developed in the Upanishads, the principal ones of which were probably written down between 1000 and 500 BCE. The Chandogya Upanishad epitomizes their central insight that "The universe comes forth from Brahman and will return to Brahman. Verily all is Brahman." Various Upanishads employ metaphor, allegory, parable, dialogue, and anecdote to portray Brahman as ultimate reality existing out of space and time, from which everything springs, and of which everything consists; it is generally interpreted as the Cosmic Consciousness or Spirit or Supreme Godhead beyond all form.

Daoism expresses a similar idea. The principal Daoist text, known in China as *Lao-Tzu* and in the West as *Tao Te Ching*, was probably compiled from the sixth to the third century BCE. It emphasizes the oneness and eternity of the Dao, the Way. The Dao is "nothing" in that it is "no thing": it is without name or form; it is the ground of all being and the form of all being. The Way, or nothingness, gives rise to existence, existence gives rise to the opposites of yin and yang, and yin and yang give rise to everything: female and male, Earth and Heaven, and so on.

The first book of the Hebrew scriptures, written no earlier than the late seventh century BCE,[6] begins with the words "In the beginning God created the heavens and the earth."[7] The next verse describes the Earth in terms reminiscent of the primordial watery chaos myths, after which God says let there be light and light is created, and then God separates light from darkness on this first day of creation. Over the next five days he likewise creates by command everything else in the universe.

In the Qur'an, written from the seventh century CE, God similarly creates the heavens and Earth by command.[8]

Eternal cycle

Several myths originating in India deny that the universe was created and maintain that the universe has always existed, but this eternal universe undergoes cycles.

The Buddha in the fifth century BCE said that to conjecture about the origin of the universe brings madness to those who try.[9] This did not, however, prevent his followers from trying. They applied his insight that all things are impermanent, constantly arising, becoming, changing, and fading, with the result that most Buddhist schools now teach that the universe expands and contracts, dissolves into non-being, and re-evolves into being in an eternal rhythm.

Possibly they were influenced by the Jainists, whose latest Tirthankara (literally Ford-Maker, one who shows how to cross the river of rebirths to the state of eternal liberation of the soul), began teaching before the Buddha in eastern India. The Jainists hold that the universe is uncreated and eternal. Time is like a wheel with twelve spokes that measure out *yugas*, or world ages, each with a fixed duration of thousands of years. Six *yugas* form an ascending arc in which human knowledge and happiness increase, while these attributes decrease in the descending arc of six *yugas*. When the cycle reaches its lowest level even Jainism will be lost. Then, in the course of the next upswing, Jainist knowledge will be rediscovered and reintroduced by new Tirthankaras, only to be lost again at the end of the next downswing in the endlessly rotating wheel of time.

This is similar to most Yogic beliefs, which derive from Vedic philosophy. Typically they posit only four *yugas*. The first, Satya Yuga or Krita Yuga, endures 1,728,000 years, while the fourth, Kali, lasts 432,000 years. The descent from Satya to Kali is associated with a progressive deterioration of *dharma*, or righteousness, manifested as a decrease in the length of human life and quality of human moral standards. Unfortunately we are now in the age of Kali.

Explanations

The many explanations for these origin myths may be grouped into five categories.

Literal truth

Because every origin myth is different, they all cannot be literally true. However, some cultures claim that *their* myth is literally true. 63 per cent of Americans believe the Bible is the word of God and literally true,[10] while the overwhelming majority* of the world's 1.6 billion Muslims believe in the literal truth of the Qur'an because it is the eternal word of God written on a tablet in Heaven and dictated to Muhammad by the Archangel Gabriel.

Many believers in the literal truth of the Bible endorse James Ussher's calculation from Genesis that the six-day creation of the universe was completed on Saturday 22 October 4004 BCE at 6pm.†[11] However, overwhelming geological, palaeontological, and biological evidence, using radiometric dating of rocks, fossils, and the ice core puts the age of the Earth as at least 4.3 billion years. Astronomical data indicate that the universe is 10–20 billion years old. The evidence against the literal truth of the creationist belief is conclusive.[12] Moreover, to believe in the literal truth of the Bible is to believe in at least two contradictory accounts of creation. In Genesis 1:26–1 God creates plants and trees on the third day, fishes and birds on the fifth day, animals early on the sixth day, and male and female humans in his own image only at the end of the sixth day. In Genesis 2, by contrast, God first creates a male human from dust; only after that does he create a garden and cause plants and trees to grow, and then from the earth creates all the animals and birds—the fish don't get a mention—and finally he creates a woman from the man's rib.

It is illogical, too, for believers in the literal truth of the Qur'an to believe that God created the Earth and the heavens in eight days (Sura 41:9–12) and that he created the Earth and the heavens in six days (Sura 7:54).

---

* The mystic and the modernist strands of Islam are now marginalized, see Ahmed (2007).
† Since Ussher (1581–1656) was Archbishop of Armagh in Ireland, presumably this is Greenwich Mean Time.

Metaphor

Barbara Sproul, one of the leading scholars of origin myths, argues that, while they may not be literally true, all myths use metaphors to express their truths. The only evidence she quotes is the ethnologist Marcel Griaule's interpretation of a Dogon wise man's explanation that his people's myth has to be spoken in words of the lower world. For the rest she explains what different origin myths really mean. Thus in the Heliopolis myth the creator god masturbating the world into existence represents the internalised duality manifesting all duality "and becomes sacred and revealing about the nature of reality if only we understand what is meant by it".[13] She supplies no evidence that the Heliopolis myth-makers, still less the population of Heliopolis, of five thousand years ago shared her understanding.

For the other examples she cites it is difficult to avoid the impression that she is simply projecting onto those myths her own late-twentieth-century interpretations. If 63 per cent of the most technologically sophisticated nation on Earth believe that a Genesis creation myth is literally true, is it reasonable to suppose that nomadic tribes of four thousand years ago, or even King Josiah's scribes of two and a half thousand years ago, believed it to be a metaphor?

While it is reasonable to conclude from their context that *some* origin accounts, like those in the Upanishads, deliberately employ metaphor, Sproul offers no evidence to demonstrate that most such myths were either intended or recognized as other than literal accounts.

Aspect of absolute reality

Sproul maintains that all religions declare an absolute reality that is both transcendent (true for all time and places) and immanent (true in the here and now), and that "Only creation myths have as their primary task the proclamation of this absolute reality".[14] Moreover, her collection of creation myths "does not show any essential disparity in understanding; rather it reveals a similarity of views from a rich variety of viewpoints".[15]

Thus many origin myths speak of polar opposites: light and dark, spirit and matter, male and female, good and bad, and so on. The more profound trace these back to Being and Non-being, with some, like the Chandogya Upanishad, saying that Non-being was produced from Being, while others, like a Maori myth, assert that Not-Being-Itself is the source of all Being and Non-being. Some see the origin of all polarities as Chaos, in which all distinctions are potentially there; creation occurs when Chaos coalesces into form and acts on the rest of the unformed to produce further distinctions and thereby create the world. "Which is the absolute reality here? The Chaos itself? Or the child of Chaos that acts on it? *Both*. They are one."[16]

Apparent differences arise only because the myths speak about the unknowable in terms of the known, commonly by anthropomorphizing or using relative words to try to describe the absolute. According to Sproul, even the Buddhist, Jainist, and Yogic rejections of a creation event do not set their eternal universe apart from one that is created; myths that tell of creation events are simply temporalizing: they speak of the absolute in terms of the first.

The claim that all origin myths reveal aspects of the same absolute reality is a fascinating one. It is not, however, supported by any evidence. It is equally explained by Sproul's interpreting these myths to accord with her own belief as to what constitutes absolute reality.

Archetypal truth

According to Sproul, who was a student of Joseph Campbell, creation myths are important not for their historical values alone but also because they reveal archetypal values that help us understand our own personal growth "physically, mentally, and spiritually, in the context of the cyclical flow of being and not-being and ultimately in the absolute union of the two".[17]

Her use of Campbell's Jungian-derived psychology fails to present a convincing explanation.

Foetal experience

Molecular biologist Darryl Reanney suggests that the common theme of pre-existent dark, formless waters into which light appears and the birth of the universe begins may be explained by subliminal memories of the foetus's birthing experience from the dark, formless, nourishing waters of the womb. "Pre-natal brain imprints of experience of birth predispose myths to evolve particular configurations of symbolic imagery which strike deeply responsive cords in psychology."[18] In support he says that electrical activity can be recorded in the cerebral cortex of foetuses from about the seventh month onwards (more recent data suggests this occurs before six months).

This is an interesting conjecture, but it is difficult to see how it can be either validated or disproved.

I suggest three other explanations.

Limited comprehension of natural phenomena

At the stage of human evolution when these myths developed, most cultures had a mistaken or limited understanding of natural forces and, except for eastern India and parts of China, philosophical inquiry had not begun.

The primordial water element of many myths may stem from the reason many late Neolithic peoples developed their settlements by the banks of a river. They used water to drink and sustain their own lives and to irrigate their crops. Water was the source of life and fertility; before the growth of cities it was commonly associated with the spirit or goddess of life.

Most myths stem from Bronze Age cultures in which science, apart from astronomy, was unknown. When asked where the world came from, the wise men drew upon their own experiences of creation. Humans and animals were created by the sexual union of their fathers and mothers, and so the world itself was created by the union of a father and mother. To fertilize the world this father must be all-powerful, and the most powerful force they knew was the sky, whence came heat from the Sun, thunder, lightning, and rain to fertilize everything that grows. To gestate the world this mother must be all-fecund, and the most fecund thing they knew was the Earth, whence came all trees, vegetation, and crops. Hence the sky-god father and earth-god mother.

Sages of different peoples saw the egg as the thing from which life emerges. Hence the cosmos, or the god that creates it, must have emerged from an egg. Other sages noted the cycles of the Sun, the Moon, the seasons, and the crops. Each of these wanes, dies, re-emerges, and develops in an apparently endless series. This is how the essential elements of the universe function, and so must the universe itself.

Political and cultural need
By the Bronze Age the spirits of nature invoked by the hunter-gatherer and primitive agricultural cultures had evolved into gods, whose functional hierarchy reflects that of the evolving city-states, while their origin myths often meet a political or cultural need.

Atum, the self-sufficient creator god worshipped in Heliopolis in the fourth millennium BCE, was downgraded by the theologians of the Pharaoh Menes to an offspring and functionary of Ptah, hitherto god only of destiny, whom they wished to elevate to creator god because he was a local Memphite deity and Menes had built a new capital at Memphis.

The creation through primordial battle myths typically conforms to this explanation. Thus in the Babylonian *Enûma Elish* myth, Marduk's slaying of Tiamat and her monsters of Chaos and his elevation to supreme creator god sanctifies and legitimizes the Babylonians' triumph over the old Sumerian powers and the creation of their order throughout Sumeria.

Late twentieth century archaeological evidence[19] suggests that the written biblical account of creation by the word of God is most probably explained by political and cultural need. In the late seventh century BCE King Josiah charged his scribes to compile from the region's myths and legends a canonical text that sanctified and legitimized the union of his kingdom of Judah with the now-fallen kingdom of Israel under one absolute patriarchal ruler with one code of laws. Yahweh, the local god of Judah, who originally had the goddess Asherah as his spouse, became not only the chief god but also the only god. Yahweh is the name for God in the Genesis 2 creation account. But to persuade the people of Israel to accept the union, he is seen to be the same as their gods. Elohim, the name given to God in Genesis 1, is the generic term for a divine being and was used by the Canaanites, whose territory and culture the Israelites had taken over, to denote their entire pantheon of gods; in Genesis 1 they are subsumed into a single deity. Reflecting the role of absolute ruler of the United Kingdom of Judah and Israel that Josiah wanted sanctified, this one God had only to say a thing and it was done; thus was the world created.

Such a change of myth is not the prerogative of the conqueror. The creation story of the Chiricahua Apache is a tragi-comical fusion of the Old Testament and their pre-conquest mythology. The biblical Flood drowns those who worship the mountain gods Lightning and Wind. After the waters subside a bow and arrow and a gun are put before two men. The first takes the gun and becomes the white man, while the second has to take the bow and arrow and becomes the Indian.

Insight

Some cultures in India and China valued training the mind to focus within and gain direct knowledge by becoming one with the object of inquiry. From such meditating seers in India came the insight that *atman*, the essential Self, was identical with the universe, which itself was identical with Brahman, the ineffable self-existent entity from which it came forth. This mystical insight is very similar to that of the early Daoists and of later seers in other countries. The essence of these common insights should be distinguished from their culturally refracted interpretations by disciples, which often showed a lack of understanding of natural phenomena or a social or political need.

Tests of evidence and reason

We have no evidence to validate in a scientific sense any origin myth or the explanations for them. We do have sufficient evidence, however, to disprove the literal truth of most such myths, including those claimed to be revealed by an external, transcendent God.

A limited, if not false, understanding, of natural phenomena, plus cultural and political need, and culturally refracted interpretations of mystical insights may be more prosaic explanations than those advanced by most mythologists, ethnologists, psychologists, or other scholars, and I cannot cite conclusive evidence in support. However, they have the advantage of being in accord with such facts as we know, and are arrived at by applying Ockham's Razor, or the scientific rule of parsimony: they are the simplest explanations.

The origin accounts that rest their claim for truth not on material evidence or reasoning or revelation by a transcendent God, but on insight, can neither be validated nor disproved by science or reasoning. I shall return to insight in more depth when considering the development of philosophical thinking. However, purely from a scientific and rational perspective, most origin myths fall into the category of superstition, which I define as

> **superstition**   A belief that conflicts with evidence or for which there is no reasonable basis and which usually arises from a lack of understanding of natural phenomena or fear of the unknown.

Reasons for endurance

One explanation for the endurance of creation myths, even in our most scientifically advanced cultures, is that science examines only the physical world, but there exists an ultimate reality that transcends the physical world; all the different creation myths express this ultimate reality in terms—frequently anthropomorphic—that reflect the different cultures.

While this could be true in some cases, too many myths are mutually contradictory for this proposition to be valid generally. A simpler explanation is that the endurance of such conflicting beliefs is not testimony to their truth but rather to the power of inculcation by two hundred generations of human societies over five thousand years.

Influence on scientific thinking

Their enduring power not only withstood the first scientific revolution, but most of the architects of that revolution saw their role as discovering the laws by which the Judaeo-Christian God governed the universe he had created. Isaac Newton, the consummator of that revolution, believed that the universe "could only proceed from the counsel and dominion of an intelligent and powerful being".[20]

Their enduring power also withstood the second scientific revolution that began in the mid-nineteenth century with Darwin's arguments for biological evolution and culminated in the first third of the twentieth century with the transformation of physics by relativity and quantum theories. Darwin himself abandoned his Christian beliefs and ended life an agnostic,[21] but Albert Einstein, originator of the Special and General Theories of Relativity, shared Newton's belief that a supreme intelligence had created the universe, although he denied that such a God intervened in human affairs.[22]

Many pioneers of quantum theory espoused the belief that matter does not exist independently but only as the construction of the mind. Some, like Erwin Schrödinger, had a life-long fascination with the Upanishadic insight that everything, including the universe, sprang from the consciousness of Brahman, the ultimate reality existing out of space and time;[23] to what extent this influenced his work is an open question. David Bohm's scientific thinking was certainly influenced by such a belief.[24]

Today a minority of scientists openly profess their religious faith. They include John D Barrow, cosmologist and member of the Christian Emmanuel United Reformed Church, Francis Collins, former director of the Human Genome Project and evangelical Christian, who sees "DNA, the information molecule of all living things, as God's language, and the elegance and complexity of our own bodies and the rest of nature as a reflection of God's plan,"[25] and Ahmed Zewail, Muslim and the 1999 Nobel laureate in chemistry. Generally such scientists hold that science and religious belief operate in different domains, although some, like John Polkinghorne, theoretical physicist and Anglican priest, actively promote debate on the intersection of science and theology.

Moving on from myth, science gives us a clearer understanding of the origin of the universe, and hence of the matter and energy from which we evolved. Or does it?

CHAPTER THREE

The Emergence of Matter: Science's Orthodox Theory

Now this Big Bang idea seemed to me to be unsatisfactory.
—FRED HOYLE, 1950

[The Big Bang] is a witness to that primordial Fiat lux.
—POPE PIUS XII, 1951

We are matter. We may be more than matter. We may be manifestations of a cosmic consciousness, as mystical insights maintain, or three-dimensional simulations generated by a super-intelligent computer, as one philosophical conjecture proposes. But this quest seeks to establish what currently we know, or can reasonably infer, from experiment or observation of the world that we perceive: in other words, what science tells us we are and where we came from.

The starting point, therefore, is what we know from science of the origin of matter, and science's current orthodox theory is that matter and energy emerged in the Big Bang 13.8 billion years ago.

I stress the word "current" because too often the media and popular science books present a scientific theory or even conjecture as though it were an indisputable fact. Scientific theories change. To emphasize this I shall describe the theory in the early part of the twentieth century, show why and how it changed to produce the Big Bang model, examine problems with this model, and consider cosmologists' current solutions to these problems.

First half of the twentieth century

If I had been writing this book in 1928 I would have said science's current orthodox theory is that the universe is eternal and unchanging.

So firmly was this theory held that Einstein made what he later admitted was the biggest blunder of his life. In 1915 he completed his General Theory of Relativity, which incorporated gravity in its description of all known matter and forces. When he applied it to the universe as a whole, however, he found that it predicted a changing universe—gravity had the effect of pulling all the matter in the universe together—and so two years later he introduced an arbitrary constant, Lambda (Λ), into his field equations. By fine-tuning the value of Lambda he was able to make the extra term in his equations exactly balance the inward pull of gravity, thus producing a static universe.

For the next fifteen years or so nearly all theoretical physicists accepted this because it was supported by evidence: the stars moved very little. This static universe view remained after the astronomer Edwin Hubble demonstrated in 1924 that some blobs of light were not clouds of gas in the then only known galaxy—the Milky Way—but were very distant galaxies of stars.

Between 1929 and 1931, however, Hubble showed that the light emitted from these distant galaxies was redshifted, and the redshift increased with the increase in distance from us. White light consists of a mixture of colours, revealed when a prism splits it into a spectrum of wavelengths, with the shorter wavelengths appearing blue and the longer wavelengths appearing red. When a source of light is moving away from an observer its wavelength appears to increase and is shifted towards the red end of the spectrum. Hubble's observations were interpreted as indicating that the galaxies are moving away from us and the further away they are the faster they are moving.

Only then did theoretical physicists take seriously the work of those who had produced other solutions to Einstein's field equations of general relativity that resulted in an expanding universe. One of these was the Belgian Jesuit and scientist Georges Lemaître who incorporated Hubble's data into his own 1927 ideas and ran the expansion of the universe backwards to produce his hypothesis of the primeval atom. This said that at time zero everything in the universe—all light and all the galaxies, stars and planets—was compressed into a single, super-dense atom that exploded outwards to form an expanding universe.

The astronomer Fred Hoyle dismissively referred to this as the Big Bang after he, together with Thomas Gold and Herman Bondi, had developed the steady state theory in 1948. According to this hypothesis the universe is expanding, but not from a point: matter is continually created in the expanding space to produce an overall uniform density in an infinitely large universe.

For a decade or so after the Second World War several theoretical physicists turned their attention to the puzzle of cosmogenesis, or how the universe began. Enrico Fermi, Edward Teller, Maria Mayer, Rudolf Peierls, George Gamow, Ralph Alpher, and Robert Herman were among those who examined the Big Bang idea.

Gamow, Alpher, and Herman tried to work out how all the different kinds of atoms we see in the universe today could have been made from the hypothesized incredibly small, dense, hot plasma of protons, neutrons, electrons, and photons.*

---

* See Glossary for an explanation of these terms.

They showed how the nuclei of helium and the isotopes* of hydrogen could be produced by protons and neutrons combining in the first three minutes after a Big Bang as this plasma expanded and cooled to below a billion Kelvin.† Alpher and Herman's calculation of the ratio of hydrogen to helium produced this way approximately matched that observed in the universe, adding support to the Big Bang hypothesis, but neither they nor anyone else could show how heavier elements were made because of the instability of nuclei comprising five or eight protons and neutrons in combination. This cast doubt on the Big Bang, and Fermi and his colleagues dropped it as a model of cosmogenesis.[1]

According to the orthodox account, Gamow and Alpher calculated that, after 300,000 years of expansion from a Big Bang, the plasma cooled to 4,000K‡ when the negatively charged electrons would be captured by the positively charged atomic nuclei to form electrically neutral, stable diatomic molecules of hydrogen, its isotopes, and atoms of helium. The photons—neutrally charged particles of electromagnetic radiation—would no longer be bound to the plasma but would decouple and travel freely through the expanding space. As they did so they cooled and their wavelength increased. By the time the universe reached its present size their wavelength would be in the microwave region and they would fill all of space, producing a cosmic microwave background. In 1948 they estimated the temperature of this cosmic microwave background to be about 5K. In 1952 Gamow estimated this temperature to be 50K.[2]

In the meantime Fred Hoyle and colleagues had shown how the heavier elements could be produced by nuclear fusion inside stars.

This post-war work thus left the steady state and the Big Bang as two competing hypotheses to explain the origin of the universe, the first maintaining that the universe was eternal and so had no beginning, the second saying that the universe began as an explosion of light and plasma from a point.

Without waiting for the further evidence that the scientific community needed in order to choose between the two, the Roman Catholic Church gave its verdict. In 1951 Pope Pius XII told the Pontifical Academy of Sciences that the Big Bang was a witness to the creation account given in Genesis when God said let there be light. The alacrity with which the Church responded to this scientific hypothesis contrasts with the two hundred years it took to accept that Galileo was right when his observations supported the Copernican theory that the Earth is not the centre of the universe but the Earth and the other planets orbit the Sun.

---

*   Atoms with the same number of protons but a different number of neutrons in their nucleus are called isotopes. The nucleus of a hydrogen atom consists of one proton, that of deuterium consists of one proton and one neutron, while that of tritium consists of one proton and two neutrons.

†   Kelvin is a degree of temperature measured on the Kelvin temperature scale that begins at 0K, the absolute zero below which molecular energy cannot fall. A Kelvin is the same size as a Celsius degree, and 0K corresponds to –273.15°C.

‡   K is the abbreviation of Kelvin.

Unlike the Catholic Church, the scientific community remained divided between Big Bang and steady state supporters until 1965 when, according to the orthodox version of history, a chance discovery produced the decisive evidence.

Astronomers Arno Penzias and Robert Woodrow Wilson had failed to eliminate background "noise" that came from every region of the sky when surveyed by their radio telescope at the Bell Laboratories in New Jersey. They turned for advice to Robert Dicke of Princeton who, unknown to them, had been trying to find Gamow's predicted cosmic background radiation. Dicke realized that this uniform "noise" in the microwave region was this radiation which had cooled to a temperature of 2.7K.[3]

It is rarely, if ever, reported that Geoffrey Burbidge, astrophysics professor at the University of California, San Diego, asserted that this orthodox account is distorted. He maintained it was Alpher and Herman's choice of parameters in their equations that led to the production of hydrogen and the other light elements in a ratio that roughly corresponds to the observed ratio. Moreover, he pointed out that Andrew McKellar had discovered the cosmic microwave background radiation, estimating its temperature as between 1.8 and 3.4K, and had published his findings in 1941; he alleged that Gamow, at least, knew of these results and so did not predict the cosmic microwave background radiation that subsequent observation confirmed.[4]

However, the orthodox account prevailed, and Penzias and Wilson received a Nobel Prize for their discovery. The vast majority of the scientific community adopted the Big Bang as the model for the origin of the universe, and those who didn't agree had a difficult time. According to John Maddox, Hoyle's continued espousal of the steady state theory "led to his being ostracized by his academic colleagues and to his almost unprecedented resignation of his Cambridge professorship".[5]

The latter was probably due also to Hoyle's characteristically blunt criticisms of evidence against the steady state theory advanced by his fellow Cambridge academic, Martin Ryle, which resulted in a feud between the two. While Hoyle never held another academic post, Ryle went on to become Britain's Astronomer Royal and Nobel laureate. Unaccountably the Nobel Prize in 1983 was awarded for work on stellar nucleosynthesis solely to William Fowler and ignored Hoyle and Geoffrey and Margaret Burbidge, the other three authors of the seminal 1957 paper that describes in detail how all the naturally occurring elements other than hydrogen and helium are formed in the interior of stars. Fowler himself freely acknowledged that Hoyle was the first to establish the concept of stellar nucleosynthesis, and he had come to Cambridge as a Fulbright scholar in order to work with Hoyle.[6]

The orthodox account exemplifies the scientific method in that a firmly held theory—the eternal universe—was discarded when new data confirmed predictions made by a different hypothesis—the Big Bang—which then became the orthodox theory. It also exemplifies in its treatment of Hoyle how the scientific establishment can behave towards those who dissent from orthodoxy.

Since the mid-1960s the Big Bang model has been held with at least as great a conviction as was the eternal and unchanging universe theory in 1928. But does evidence continue to validate the model and, if not, how has the scientific community responded?

Current theory: the Big Bang

In order to see if the Big Bang model provides a satisfactory explanation of the origin of the universe, we need to examine its theoretical basis.

Theoretical basis

Unlike the commonly accepted scientific method,* the Big Bang theory didn't derive from observation; it arose from solutions to the equations of Einstein's General Theory of Relativity, one of which was selected because it best fitted observations.

Einstein's insight produced laws of motion that do not depend on the observer moving in a particular way with respect to what he or she is observing. Einstein assumed that the speed of light (c) is constant, that it is the same for all observers at all times and in all regions of the universe, and that nothing can travel faster. His Special Theory of Relativity of 1905 dispenses with the idea that space and time are independent and absolute: it provides a four-dimensional space-time matrix in which space or time may expand or shrink depending on the motion of the observer, but *space-time* is the same for everyone.

A consequence of his Special Theory of Relativity is that mass (m) and energy (E) are equivalent, related by the famous equation $E = mc^2$.

When Einstein incorporated gravity into these laws of motion to produce a General Theory of Relativity, his insight was that gravity is not a force that acts instantaneously between masses as Newton's law defined it, but that gravity is a distortion in the space-time fabric caused by mass, and the greater the mass the greater the distortion. Such distortions then dictate how other masses move through space-time. To paraphrase John Archibald Wheeler, matter tells space-time how to curve and space-time tells matter how to move.

In order to produce an equation that quantifies this concept and enables predictions to be made, Einstein used a difficult branch of mathematics called differential geometry that deals with curved surfaces. He arrived at what is now known as Einstein's field equations. The plural is used because the single equation contains tensors† that have ten possibilities, creating, in effect, ten equations. The very many possible solutions to these equations produce very many theoretical universes, and the challenge was to find a solution that best fitted the observational data.

---

* See Glossary for a full description.
† A tensor is an array of components that change in a transformation from one coordinate system of space to another.

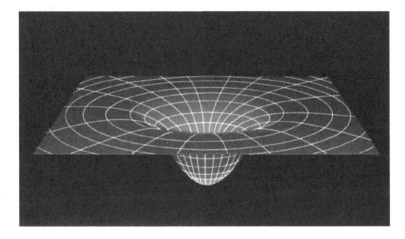

Figure 3.1 Two-dimensional representation of the curvature of space-time round a spherical mass like a star

Solving these equations is extremely difficult. Four men led the attempts. Apart from Einstein and Lemaître, the other two were the Dutch astronomer Willem de Sitter and the Russian meteorologist Alexander Friedmann.

Simplifying assumptions: isotropy and omnicentrism
All of them made two simplifying assumptions: at any given time the universe appears the same in whatever direction we look (it is isotropic) and this is true if we observe the universe from anywhere else (it is omnicentric). These two assumptions necessarily mean that the universe is the same at every point (it is homogeneous).*

The isotropic assumption clearly is not totally valid: the stars in our own galaxy form a distinct band of light across the night sky, which we call the Milky Way. However, the assumptions were made for three reasons: (a) intuition that this was a good approximation on the scale of the universe; (b) a belief that we do not occupy a special, or privileged, place in the universe, just as Copernicus had demonstrated that we do not occupy a unique place in the solar system; and (c)

---

* *Isotropic* means that to an observer the universe appears the same in every direction. *Homogeneous* means that the universe is the same at every point. These are not necessarily the same. For example, a universe with a uniform magnetic field is homogeneous since all points are the same, but it is not isotropic because an observer sees different magnetic field lines in different directions. Conversely, a spherically symmetric distribution of material is isotropic when viewed from its central point, but it is not necessarily homogeneous: the actual material at one point may not be the same as that at a different point in the same direction. However, if we assume that the distribution of material is isotropic when viewed from *every* point, then the universe is necessarily homogeneous.

mathematical convenience—it drastically reduces the number of possible geometries, or space-times, that describe the shape of the universe because, if matter produces curvature and if the universe is homogeneous, then the curvature of the universe is the same everywhere.

Friedmann showed that one consequence of these assumptions is that the universe can have only three possible geometries: closed (spherical), open (hyperbolic) or flat, each of which changes with time according to the scale, or expansion, factor of the universe. It is the whole thing—three-dimensional geometry plus time-dependent scale factor—that defines the space-time that is curved by matter according to Einstein's field equations.

These changing four-dimensional mathematical matrices are difficult to visualise. Figure 3.2 gives only two-dimensional representations of the three space dimensions that change with time.

Friedmann showed that universes having these three possible geometries had three different fates. The closed (or spherical) universe expands out of a Big Bang, but the gravitational effect of its matter is strong enough to slow down, halt, and then reverse the expansion until the contracting universe ends in a Big Crunch. The open (or hyperbolic) universe expands out of a Big Bang, but the gravitational effect of its matter is too weak to stop this expansion, which continues indefinitely at a steady rate until its elements no longer have contact with each other, leading to an empty universe. The flat universe expands out of a Big Bang, but the gravitational effect pulling all its matter together exactly balances the kinetic energy of expansion, with the result that the expansion rate decreases but not enough to halt it, so that the universe expands forever at a continuously slowing rate.

As a consequence of the simplifying assumptions both the flat and the open universes are necessarily infinite in extent: if either came to a definite edge it would contradict the assumption that the universe looks the same from all points. The same does not follow for a spherical universe: a perfect sphere looks the same from all points on its surface.

Unlike Einstein, Friedmann did not add an arbitrary constant Lambda to achieve the result he wanted. In his mathematical model the contracting gravitational pull of matter compared with its expanding kinetic energy may be expressed as the density parameter, Omega (Ω). In a closed universe Omega is greater than 1; in an open universe Omega is less than 1; in a flat universe Omega is exactly 1.

After Hubble published his data, most scientists concluded that a flat universe that had started from a very hot Big Bang most closely matched observation, and so the Friedmann-Lemaître mathematical model became the orthodox one.[7]

Problems with the Big Bang theory

The universe as a subject of scientific inquiry was no longer the exclusive domain of observational and theoretical astronomy. A new science of cosmology had emerged, which I define as

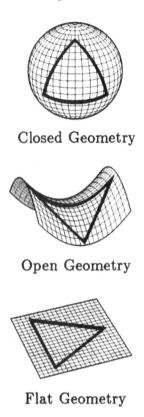

Closed Geometry

Open Geometry

Flat Geometry

Figure 3.2 Two-dimensional representation of the geometry of Friedmann universes (no arbitrary cosmological constant)

A closed geometry is the three-dimensional analogue of the surface of a sphere: a triangle in this geometry contains more than 180°, and the circumference of a circle is shorter than π times the diameter. An open geometry is the analogue of a hyperbolic, or saddle-shaped, surface: a triangle in this case contains less than 180°, and the circumference of a circle is longer than π times the diameter. A flat geometry is the Euclidean geometry with which we are familiar: each triangle contains exactly 180°, and the circumference of a circle is exactly π times its diameter. Each spatial geometry changes with time according to the scale, or expansion, factor of the universe. If we introduce a nonzero cosmological constant, however, then any type of geometry can occur with any type of time evolution.

cosmology The study of the origin, nature, and large-scale structure of the physical universe, which includes the distribution and interrelation of all the galaxies, clusters of galaxies, and quasi-stellar objects.

Relativity theory had played a crucial role in investigating the universe as a whole compared with astronomy's traditional focus on individual stars and galaxies. Theoretical and experimental particle physics, plasma physics, and quantum physics were now used to examine what happened at, and immediately after, the

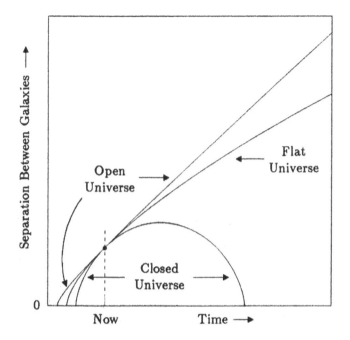

Figure 3.3 The evolution of Friedmann's universes

With no cosmological constant a mass density high enough to reverse the expansion leads to a closed universe; a low mass density is insufficient to reverse the expansion, which continues at a constant rate leading to an open universe; a critical mass density bordering the two has a flat geometry: it will expand forever, but at an ever decreasing rate.

Big Bang when the universe was incredibly tiny and hot. When scientists applied these different disciplines to the Big Bang model they found four problems.

Magnetic monopole

Particle and plasma physicists theorized that the extremely high temperature and energy of the plasma immediately after the Big Bang should have created magnetic monopoles, which are particles with only one pole of magnetic charge instead of the usual two.* Indeed, using relativity theory, they calculated that the Big Bang should have produced enough to make one hundred times the observed energy density of the universe.[8]

Not a single magnetic monopole has been detected in the universe.

Homogeneity

The two assumptions in the orthodox model produce a universe that is homogeneous, or completely uniform, whereas all other solutions of Einstein's field equations without these assumptions produce irregular universes.

---

*   See Glossary for full definition.

By contrast to this model, when we observe the universe we see that it is not completely uniform. It has solar systems, galaxies, clusters of galaxies, and super-clusters, all separated by vast voids containing little or no matter. The Earth has a mass density roughly 10^{30} times higher than the universe average, the air we breathe has a mass density 10^{26} times higher, the average mass density of our Milky Way galaxy is 10^6 times higher, our local group of galaxies has an average mass density about 200 times higher, while voids between sheets of superclusters are typically 150 million light years across.[9]

If the universe were perfectly homogeneous we wouldn't be here to observe it.

However, cosmologists thought that on a universe-size scale the departure from homogeneity is only about 1 part in 100,000.

The Big Bang model fails to explain how and why the universe should be so extremely close to, but fall short of, perfect homogeneity in such a way as to allow the formation of structures like our solar system with planets like the Earth on which we evolved.

Isotropy of cosmic microwave background (Horizon problem)
The Big Bang model says that the cosmic microwave background (CMB) is the radiation that decoupled from the matter plasma about 380,000 years (according to revised estimates) after the Big Bang and that lost energy and cooled to its current temperature of 2.73K as the universe expanded.

Detectors on the Cosmic Background Explorer (COBE) and Wilkinson Microwave Anisotropy Probe (WMAP) satellites showed that this temperature is very nearly isotropic, that is, the same in all directions. In order to achieve such a near-uniformity of temperature all the particles of radiation (photons) must have mixed through repeated collisions just after they decoupled from the plasma.

Relativity theory says that nothing can travel faster than the speed of light. Hence, in order to mix, each photon cannot have been separated from the others by a distance greater than it can travel at the speed of light. This distance is termed the photon's contact horizon.

However, the Big Bang model says that the rate of expansion of the universe slowed. Accordingly, a photon's contact horizon was very much smaller in the very much younger universe than it is now. Hence it would have been impossible for each photon to be in contact with all the others just after they decoupled from the plasma. Therefore the energies of the photons then would have been different, and this should be shown now in different temperatures of the cosmic microwave background measured in different directions.

The Big Bang theory cannot explain this conflict with observational evidence.

Flatness (Omega)
A flat universe is inherently unstable. It relies on the precarious balance between the kinetic energy of expansion and the attractive gravitational energy of its

matter to produce Omega = 1. But the mathematics of the Big Bang model show that Omega is extremely sensitive, especially in the early universe. Minute deviations from unity either way very rapidly magnify, tipping the universe into either a closed or an open configuration. Dicke calculated that when the universe was one second old the value of Omega must have been somewhere between 0.99999999999999999 and 1.00000000000000001, that is, a sensitivity of $\pm\ 1^{-17}$. If Omega had deviated from 1 by more than this, then the universe would have either collapsed into a Big Crunch or else expanded to emptiness long before solar systems and planets could form, and we wouldn't be here to speculate about the Big Bang.

Most cosmologists infer from theoretical, rather than observational, grounds that the universe has been expanding since it was 10^{-43} seconds old* (for no other reason than quantum theory† breaks down before this time). If this is so, then the value of Omega could not have departed from unity by more than 10^{-64}, a sensitivity that is unimaginable.[10]

But Omega couldn't have been exactly one from the very beginning, otherwise the universe would never have expanded in the first place.

The Big Bang model fails to explain how or why the value of Omega should be so incredibly close, but not equal, to one, so as to allow the universe to expand stably.

There is a fifth, even more fundamental, problem that the vast majority of cosmologists did not address. I shall consider this elephant in the room in the following chapter when examing what the revised model fails to explain.

Inflation theory solution

One idea claims to solve these first four problems at a stroke.

Who first came up with the idea was, and still is, a matter of dispute. Alan Guth of the Massachusetts Institute of Technology says "I view the official debut of inflation as the seminar that I gave at SLAC on January 23, 1980."[11] The Russian Andrei Linde, now of California's Stanford University, claims that the essential ideas for inflation had been developed prior to that by Alexei Starobinsky, David Kirzhnits, and himself in the then Soviet Union.[12]

According to the version published by Guth in 1981, very shortly after the Big Bang the universe underwent a huge and near-instantaneous expansion, inflating by trillions of times in what may have been less than a trillion trillionth of a second. To achieve this phenomenon the universe was in an unstable supercooled state; this decayed, inflation stopped, and the now vast universe began the decelerating expansion predicted by the basic Big Bang model.[13]

---

*   This is known as *Planck time*; see Glossary for an explanation.
†   See Glossary for a definition of *quantum theory* and page 87 for a fuller description.

After inflation the universe is so immense that all we can see now is just a minute fraction of it. That is why, just like a tiny area on the surface of an enormous balloon, our part appears flat. Put another way, any imbalances between the explosive energy of expansion and the attractive gravitational energy of matter in the Big Bang are diluted to such a degree by inflation that they now achieve the otherwise highly improbable balance that enables the post-inflation universe to undergo stable decelerating expansion. After inflation Omega has become, in effect, 1, and the universe is not fated to undergo rapid acceleration to emptiness or rapid contraction to collapse: it follows the mathematical model of the flat universe. That solves the flatness problem.

Any irregularities arising from the explosive Big Bang have similarly been diluted by the huge inflationary expansion. That solves the homogeneity problem.

Likewise, all those magnetic monopoles exist somewhere out there in the vast universe, but our region is so incredibly tiny that it doesn't contain any. That solves the magnetic monopole problem.

The universe that we are able to see now, whose furthest distances are limited by the speed of light and the age of the universe, is just the post-inflation, normally expanded region of what was an incredibly tiny part of the inflated universe. In this incredibly tiny part all the photons were in contact and achieved a uniform temperature. That solves the isotropy of the cosmic microwave background problem.

Guth's inflation conjecture, however, was fatally flawed. His mechanism for ending the runaway exponential expansion so that the universe could then begin its decelerating expansion according to the basic Friedmann-Lemaître flat universe model produced gross inhomogeneities in the universe that are disproved by observation. Guth withdrew it after a year.

Andreas Albrecht and Paul Steinhardt and, separately, Linde put forward modified versions.

According to Linde these didn't work either. He claims that the inflation conjecture's problems were solved only when he devised a new, simpler version in 1983 in which he dispensed with supercooling, quantum gravity effects, and even the standard assumption that the universe originally was hot; instead he just relied on scalar fields. A scalar field is a concept in mathematics in which every point in space is associated with a scalar, a quantity such as mass, length, or speed that is completely specified by its magnitude.* Linde simply assumed that the universe had all possible scalar fields, each with all possible values. This assumption gave his mathematical model limitless possibilities and theoretically produced domains of the universe that remain small and others that inflate exponentially. He called this "chaotic inflation" because of its arbitrary nature. It became a popular version of inflation, but many different versions have since been proposed.[14]

---

*   See Glossary for a full definition.

The inflation conjecture doesn't solve the elephant in the room—it actually makes that elephant even larger—but cosmologists were so relieved that it solved the four problems they had identified that they eagerly embraced it. Indeed, they accorded it the title of Inflation Theory, and the Inflationary Big Bang model became scientific orthodoxy.

Validity of Inflationary Big Bang theory

Whether this revised cosmological orthodoxy provides a scientific explanation of the origin of the universe depends on whether both the basic theory and the inflationary add-on are valid on two counts: (a) is the theory reliable? (b) is it supported by observation or experiment?

Reliability of basic theory

The basic Hot Big Bang theory comprises two parts. The first is the solution to Einstein's field equations that assumes the universe is both omnicentric and isotropic (and hence is homogeneous) and the choice of the geometrically flat universe. The second part is the Standard Model of Particle Physics.

The omnicentric assumption is not testable. Even if an advanced civilization in a distant galaxy sent us their view of the universe, it would be long out of date when it arrived.

Although the isotropic and homogeneous assumptions are clearly not totally valid because the universe comprises solar systems, galaxies, local groups of galaxies, galaxy clusters, and superclusters separated by huge voids, cosmologists believe that on the scale of the universe these assumptions are valid. However, every time astronomers have surveyed larger sections of the universe at greater distances with more sophisticated instruments they have found structures as large as the survey size. In 1989 Geller and Huchra identified a nearly two-dimensional structure approximately 650 million light-years long that they dubbed the Great Wall. In 2005 Gott and colleagues detected the Sloan Great Wall at more than twice that length, 1.3 billion light-years, located approximately a billion light-years away. In 2013 Roger Clowes and colleagues identified a high-membership quasar group that is 4 billion light-years long at a distance of 8 to 9 billion light-years.[15] In 2014 István Horváth and colleagues reported their 2013 discovery of an object more than six times the size of the Sloan Great Wall, around 7 to 10 billion light-years long, at a distance of approximately 10 billion light-years.[16] The sizes of these objects contradict the assumptions of isotropy and homogeneity.

As for the choice of the flat geometry, we cannot test the consequence that the universe is infinite in extent.

Furthermore, the idea that the universe, including space-time, erupts into existence from nothing in a Hot Big Bang came from extrapolating the expansion of the universe back to time zero. But quantum theory breaks down at this time because its Uncertainty Principle says that nothing can be specified within a period of less than 10^{-43} seconds, known as Planck time.* Moreover, this extrapolation compresses the universe into a point of infinite density where the curvature of space-time is infinite, causing relativity theory to break down.[17] As Guth says, "the extrapolation to arbitrarily high temperatures takes us far beyond the physics that we understand, so there is no good reason to trust it. The true history of the universe going back to 't = 0' remains a mystery."[18]

A theory based on a mystery in which the underpinning theories break down and on simplifying assumptions, one of which cannot be tested while the others are contradicted by astronomical observations, falls somewhat short of total reliability.

The second part of the basic Big Bang theory is the Standard Model of Particle Physics, which uses quantum field theory to explain how subatomic particles form from the energy released at the Big Bang by a mechanism called symmetry breaking.

> **Standard Model of Particle Physics**   Aims to explain the existence of, and interactions between, everything except gravity we observe in the universe in terms of fundamental particles and their movements. Currently it describes 17 such types of fundamental particles, grouped as quarks, leptons, or bosons. When corresponding antiparticles and boson variations are taken into account, the total of fundamental particles is 61.

According to this model, different kinds of quarks combine to make up the proton and neutron (whose different combinations make up the nuclei of all atoms). The interactions between 12 of these types of fundamental particle are the movements of 5 other fundamental particles, bosons that are force carriers, such as gluons that provide the binding force for quarks.†

The model has successfully predicted the existence of particles that have subsequently been detected, directly or by inference in the case of quarks, by experiment or observation. A key prediction is the existence of a particle named the Higgs boson, essential to explaining why the other 16 types of fundamental particle, except the photon and the gluon, have mass. In 2012 two experiments using the Large Hadron Collider (LHC), built by the European Organization for Nuclear Research (CERN) beneath the Franco-Swiss border, identified the very short-lived existence of the Higgs boson, or possibly a family of Higgs bosons in which case the Standard Model would need revising.

Even if only one Higgs boson is confirmed after the LHC reopens in 2015 with much higher energy levels, significant problems remain. The Standard Model contained 19 parameters, revised after 1998 to 29 to allow for neutrinos to have

---

*   See Glossary for an explanation of *Planck time* and the *Uncertainty Principle*.
†   See Glossary for a definition of *quark*, *lepton*, and *boson*.

mass, which hadn't been predicted by the model. These parameters are freely adjustable constants whose values must be chosen: the theory is mathematically consistent no matter what values are inserted. These constants specify the properties of matter, like the charge of an electron, the mass of a proton, and coupling constants—numbers that determine the strength of interactions between particles. The constants are measured experimentally and then inserted into the model "by hand". As Guth concedes "[according to the Standard Model] the masses of the W^+ particle and the electron arise in essentially the same way, so the fact that the mass of the electron is 160,000 times smaller is built into the theory only by rigging the parameters to make it happen."[19] Such a theory is inherently less reliable than one whose predictions are subsequently confirmed by experiment or observation.

When corresponding antiparticles and boson variations are taken into account the number of particles becomes 61,[20] which seems a large number of particles defined as fundamental or irreducible. Furthermore, the current Standard Model is necessarily incomplete because it does not take gravity into account. If it were to do so, then more conjectured fundamental particles such as gravitons would be needed.

The reliability of the basic Big Bang theory depends, too, on its correspondence with reality. Cosmologists have adopted Friedmann's interpretation of his mathematical solution to Einstein's field equations. This says that stars (later revised to galaxies, and then to galactic clusters) do not move. They are encrusted in space, and it is the space between galaxies that expands. The mathematical logic of this may be fine, but to many non-cosmologists the interpretation appears Jesuitical: in the real world, if the distance between two galaxies increases over time then they move apart during this time. Indeed, cosmologists refer to the observed redshift of a galaxy as the measure of the speed with which that galaxy is moving away from our galaxy.

Claims of evidential support for basic theory

The vast majority of cosmologists claim that three planks of evidence provide powerful support for the basic Big Bang theory: (a) the observed redshifts of galaxies that show the universe is expanding; (b) the existence and the blackbody form of the cosmic microwave background; and (c) the observed relative abundances of the light elements.

Cosmological redshift

Interpreting the observed redshift of celestial objects as their movement away from us and the further away the faster their movement led to the Hubble constant, the ratio of recessional speed to distance. Its calculation proved notoriously difficult, not least because measuring distance is immensely challenging.* Nonetheless

---

*   See page 75.

orthodox cosmologists assumed that all redshifts from objects further away than our Local Group of galaxies are due to the expansion of the universe and adopted redshift as a *measure* of distance.

Several reputable astronomers have challenged this assumption, claiming the evidence shows that many redshifts have a different cause. I shall examine these conflicting claims in Chapter 6 when considering the problems facing cosmology as an explanatory means. But *if* their interpretation of the data* is correct, then redshift by itself—and especially very high redshift without evidence that it arises from emission or absorption spectra of starlight—is not a reliable indicator of either cosmological distance or recessional velocity, and consequently age. And that would undermine one of the three key planks of evidence supporting orthodox cosmology's Big Bang model.

Cosmic microwave background
The 2.73K temperature of the cosmic microwave background (CMB) is consistent with radiation decoupling from matter in the early stages of the Hot Big Bang and cooling as it spreads through the expanding universe. Moreover, such radiation would have what is known as a Planck blackbody spectrum. The COBE satellite, launched in 1989, detected such a spectrum, providing powerful support for the orthodox model.[21]

However, as we shall see when I consider data interpretation in Chapter 6, advocates of other cosmological models also claim that the existence and characteristics of the CMB are consistent with their hypotheses.

Relative abundances of the light elements
Gamow, Alpher, and Herman showed how the nuclei of helium, deuterium, and lithium could have been made by the nuclear fusion of protons and neutrons in the intensely hot plasma existing for the first couple of minutes after the Big Bang.† The relative abundance of these light elements produced before expansion and cooling shut down the process of nucleosynthesis should remain largely unchanged in the universe today. Alpher and Herman's prediction of the ratio of approximately 75 per cent hydrogen to 25 per cent helium by mass is the same as the observed ratio and is cited as compelling evidence for the Hot Big Bang.

As noted previously, Burbidge claimed that Alpher and Herman chose for their equations a value for the ratio of the density of baryons (normal, visible matter) to the density of radiation that was calculated to produce the then observed ratio of hydrogen to helium, and so it wasn't a prediction.‡ He conceded that this chosen parameter also produces the observed ratio of hydrogen to deuterium and this does lend support to the Big Bang hypothesis.

---

* See page 78 for further discussion of this data interpretation.
† See page 20.
‡ See page 21.

However, in 2004 Michael Rowan-Robinson, then Professor of Astrophysics at Imperial College, London and President of the Royal Astronomical Society, noted that more recent estimates by Tytler and colleagues of the deuterium abundance from absorption lines in the line of sight to high-redshift quasars requires a revision of the estimate of baryon density. The new value is supported by analysis of fluctuations in the cosmic microwave background. This then produces a poor agreement with the helium abundance.[22]

All this suggests that this plank of evidence for the Big Bang may not be as firm as most cosmologists claim.

An alternative hypothesis put forward by Hoyle and Burbidge maintains that all the elements are produced by nucleosynthesis inside stars. In support they claim that if the known helium abundance is made this way from hydrogen, then the energy released when thermalized generates a blackbody CMB with a temperature of 2.76K, almost exactly the observed one. They argue that the other light elements can be made either in flaring activity on the surfaces of stars, as is known to occur with the Sun and other stars, or in incomplete hydrogen burning in the interiors.[23]

Ancient objects in young galaxies?

Reputable astronomers claim that galaxies with very high redshifts, which the orthodox model says are therefore very young, contain very old objects such as red giant stars plus iron and other metals. Because galaxies cannot contain objects older than themselves, these astronomers argue that the orthodox Big Bang theory is therefore wrong. I shall consider these claims in more detail when I examine the evolution of matter in the universe in Chapter 8.

Less controversial conflicts with evidence are three of the five problems mentioned previously: the absence of magnetic monopoles, the departure from homogeneity of about 1 part in 100,000, and the isotropy of the cosmic microwave background. It was these three inconsistencies with the evidence, plus the flatness problem in the Big Bang model, that prompted most cosmologists to embrace inflation theory's solution as part of the orthodox model of the universe. Hence we need to examine the validity of this addition to the basic Big Bang theory.

Reliability of inflation theory

Linde says that "if the universe at the beginning of inflation was as small as 10^{-33} centimetre,* after 10^{-35} second of inflation this domain acquires an unbelievable size. According to some inflationary models, this size in centimetres can equal $(10^{10})^{12}$," that is, $10^{100000000000}$ centimetres.[24] What one of the originators of inflation theory is saying is that, in one hundred thousand million million million million millionth of a second, a universe with an assumed diameter of one thousand million million million million millionth of a centimetre could have expanded

---

*   An assumption based on the length below which quantum theory breaks down.

to more than 10 billion orders of magnitude greater than the size of the universe we observe today. It would be presumptuous to demur from Linde's description of "unbelievable".

To be unbelievable, however, does not mean that a conjecture is scientifically invalid. But to achieve such an increase in size in such a short time does mean that the universe expanded very many magnitudes faster than the speed of light. Inflationists maintain that this does not violate relativity theory. By appealing to Friedmann's interpretation they argue that it wasn't the stuff of the universe that travelled faster than light speed, but the space between the stuff, and relativity theory prohibits only matter and information, not space, from travelling faster than light.

A majority of inflationists now take the view that the Hot Big Bang occurred after inflation, that is, there was no stuff—matter and radiation—but only a bubble of vacuum that underwent inflation, and at the end of inflation this converted into energy and matter. However, inflationists also argue that the space or vacuum that expanded has a ground state energy, and since energy and mass are equivalent, then consequently the mass-energy of the space or vacuum has travelled very many magnitudes faster than light, thus conflicting with relativity theory.

How and when the proposed inflation of the universe began is an open question. The version Guth first put forward was based on grand unified theories (GUTs) of particle physics, and inflation began about 10^{-35} seconds after the Big Bang. Since then theorists have developed more than a hundred versions of inflation based on various mechanisms, which usually incorporate some form of scalar field referred to generically as an inflation field; these include chaotic, double, triple, and hybrid inflation, and inflation using gravity, spin, vector fields, and branes from string theory.* Each has a different starting time, a different period of faster-than-light exponential inflation, and a different ending time, producing a vast range of sizes of the universe. Yet each version claims that its inflationary period ends with the universe having the highly improbable critical mass density in which Omega = 1 so that a flat universe subsequently undergoes stable, decelerating expansion.

If the Big Bang occurs after the universe undergoes inflation, this begs the questions of what came before inflation, and why, how, and when inflation began. Guth seems unconcerned: "inflationary theory allows a wide variety of assumptions concerning the state of the universe before inflation,"[25] and "No matter how unlikely it is for inflation to start, the external exponential growth can easily make up for it."[26] Such vague responses are not the hallmark of a reliable theory.

Another question raised by these inflationary theories is what drives these different mechanisms to inflate exponentially a superdense primordial universe against its immense gravitational field that would be expected to crush it into

---

* The principal versions are considered in more detail on page 116 and the following pages.

a black hole* from which nothing can escape? To answer this most theorists reintroduced into their equations the arbitrary constant Lambda that Einstein had discarded as a mistake. By assigning this arbitrary constant a positive value very much larger than the one Einstein had given it they are able to provide their notional inflation fields with a huge negative (or repulsive) gravitational energy that renders insignificant the immense normal gravitational field.

As to what Lambda is in physical reality, as distinct from an arbitrary mathematical constant that gives a desired solution to an equation, I shall consider the various views in Chapter 4 because, more than fifteen years after the inflation hypothesis, cosmologists invoked it again—albeit with yet another very different value—in order to account for another astronomical observation that contradicted the Big Bang model.

Self-evidently all the very different versions of inflation cannot be right, but Guth comments "From the many versions of inflation that have been developed, a few conclusions can be drawn....Inflation requires only that there is *some* state to play the role of false vacuum, and there is *some* mechanism to produce the baryons (e.g. the protons and neutrons) of the universe after inflation has ended. Thus inflation can survive even if grand unified theories are wrong [my italics]."[27] This does not define a theory as understood in science but a collection of different conjectures so abstract and generalized as to lack meaning in the physical world.

This problem of definition arises because cosmology has been led by theorists whose principal instrument is mathematics. Mathematicians use the word "theory" to describe a collection of propositions in a subject that are demonstrable by deductive reasoning from a set of assumptions or axioms and that are expressed by symbols and formulae. A mathematical theory need bear no relation to physical phenomena, as Einstein freely conceded: "The skeptic will say: 'It may well be true that this system of equations is reasonable from a logical standpoint. But that this does not prove that it corresponds to nature.' You are right, dear skeptic. Experience alone can decide on truth."

While most cosmologists claim their discipline is a science, many conflate mathematical theory with scientific theory, which is something quite different. Science is an empirical discipline, and a scientific theory is an explanation of a set of phenomena that has been validated by a number of independent experiments or observations and is used to make accurate predictions or retrodictions about phenomena.

Claims of evidential support for inflation theory

Do any of the inflation hypotheses make unique predictions that are confirmed by evidence? In 1997 Guth wrote "It is fair to say that inflation is not proven, but I feel that it is rapidly making its way from working hypothesis to accepted fact."[28] By 2004

---

*   See Glossary for a fuller description of *black hole*.

he claimed that "The predictions of inflation agree beautifully with what has been measured in cosmic microwave background."[29] Indeed, the scientific team responsible for the space-based Wilkinson Microwave Anisotropy Probe (WMAP) announced in 2006 that the ripples detected in the cosmic microwave background (CMB) favour the simplest versions of inflation, and this confirmed inflation as an essential element of orthodox cosmology's explanation of the emergence of the universe.

Moreover, in 2014 the team surveying one to five degrees of the sky (two to ten times the width of a full Moon) from near the South Pole, known as the BICEP2 project, announced that they had found direct evidence for inflation.[30]

What in fact they had found was a B-mode polarization signal in the cosmic microwave background (CMB). The BICEP2 team concluded that this signal was caused by primordial gravitational waves generated by the inflationary expansion of the universe. After the initial excitement and talk of Nobel prizes, two independent studies of the BICEP2 data subsequently claimed the signal could just as easily be accounted for by dust and galactic magnetic fields in our own Milky Way galaxy.[31] Moreover the signals were much stronger than expected and were inconsistent with data from the WMAP and Planck telescopes.

Whether versions of inflation actually make predictions or whether the parameters of their field equations are contrived to produce outcomes consistent with observation is a question I shall examine in Chapter 8. Furthermore, cosmology literature rarely mentions that the CMB ripples are claimed to be consistent with other hypotheses, such as a spherically symmetric inhomogeneous model of the universe, the cyclic ekpyrotic universe model, quasi-steady state cosmology, and plasma cosmology's eternal universe model.

I shall examine these claims and the WMAP data in more detail when I consider the question of data interpretation in Chapter 6. Suffice it to say here that Peter Coles, Professor of Astronomy at the University of Nottingham (UK), highlights discrepancies between WMAP data and inflation, citing the inexplicable alignment of certain components of hot and cold spots, which theoretically should be structureless. He concludes

> There is little direct evidence that inflation actually took place. Observations of the cosmic microwave background…are consistent with the idea that inflation took place, but that doesn't mean it actually happened. What's more, we still don't even know what would have caused it if it did.[32]

This echoes the conclusion of Rowan-Robinson:

> Several different versions of how inflation occurred have now been proposed. The essential common feature is the period of exponential expansion in the very early universe, which solves the horizon and flatness problems. However there is no evidence that any such phase ever occurred and it is in fact quite hard to see how such evidence can be obtained.[33]

Ellis draws attention to the weakness in explanatory and predictive power of inflation theory.

> If the hypothesis solves only the specific issues it was designed to solve in the early universe and nothing else, then in fact it has little explanatory power, rather it is just an alternative (perhaps theoretically preferable) description of the known situation...the supposed inflaton field underlying an inflationary era of rapid expansion in the early universe has not been identified, much less shown to exist by any laboratory experiment. Because this field φ is unknown, one can assign it an arbitrary potential $V_{(\varphi)}$....It can be shown that virtually any desired scale evolution $S_{(t)}$ of the universe can be attained by suitable choice of this potential; and also almost any desired perturbation spectrum [which is claimed to give rise to the CMB ripples] can be obtained by a (possibly different) suitable choice. Indeed in each case one can run the mathematics backwards to determine the required potential $V_{(\varphi)}$ from the desired outcome.[34]

Finally, if a hypothesis is to become a scientific theory it must be capable of being tested. The central claim of the various inflation hypotheses is that the universe we observe is only an incredibly minute part of the whole universe. If no information can travel faster than the speed of light, then we cannot communicate with, or obtain any information about, any other part of this universe. Until the proponents of the various inflation models devise a method of unambiguously testing the existence of something with which we cannot communicate or obtain information about, their central claim is not only untested but also untestable. Henceforth I shall refer to it as the inflation conjecture.

As John Maddox, editor-in-chief of *Nature* for twenty-three years, observed "It is a telling comment on the habits of the research community that its perpetual and healthy scepticism has not been lavished with the accustomed generosity on this daring and ingenious theory."[35]

Conclusions

It is difficult to avoid the conclusion that cosmology's orthodox theory is unreliable because many of its underlying assumptions lack validity and depend on inserting and changing the values of arbitrary parameters in order to accord with observations. Moreover, the various inflation models introduced in order to explain admitted contradictions of the basic Big Bang model with observational evidence not only lack reliability but also their central claim is untestable.

Furthermore, it fails to address or adequately explain several key issues that I will consider in the next chapter.

CHAPTER FOUR

What Science's Orthodox Theory Fails to Explain

As a general scientific principle, it is undesirable to depend crucially on what is unobservable to explain what is observable.

—HALTON ARP, 1990

Not only can we not see what most of the universe is made of, we aren't even made of what most of the universe is made of.

—BERNARD SADOULET, ~1993

If science's orthodox theory, the Inflationary Big Bang model, is to furnish us with a convincing account of the origin of the matter of which we consist then it must provide a satisfactory answer to six key questions.

Singularity

According to the Big Bang model, if you run back the clock of the universe's expansion it produces a singularity. Theoretical physicists developed the idea of a singularity when considering black holes. It may be defined as

> **singularity** A hypothetical region in space-time where gravitational forces cause a finite mass to be compressed into an infinitely small volume and therefore to have infinite density, and where space-time becomes infinitely distorted.

In 1970 Stephen Hawking and Roger Penrose published a mathematical proof that there must have been a Big Bang singularity provided only that the General Theory of Relativity is correct and that the universe contains as much matter as we observe. This became the orthodox theory.

Hawking, however, has since changed his mind and maintains that the singularity disappears once quantum effects are taken into account (see Hartle-Hawking no-boundary universe in the next chapter).

Hence, was there a singularity at the Big Bang, and if there was, what do we know about the universe at this point?

Orthodox theory is equivocal on the first part of the question. On the second part, if there was a Big Bang singularity, then it tells us nothing because, as noted in the previous chapter,* its underpinning relavity and quantum theories break down. While a period of 10^{-43} seconds may seem an absurdly small period of time in which nothing can be specified, various inflation models speculate that highly significant events occurred at or within such a period before or after a Big Bang.

Observed ratio of matter to radiation

Orthodox theory about the origin of matter invokes the Standard Theory of Particle Physics to explain how matter was created from the energy explosively released at the Big Bang.

According to the Standard Theory an elementary particle of matter can spontaneously materialize from an energy field together with its symmetrically opposite particle of antimatter, which has the same mass and spin but opposite charge. Thus an electron (negatively charged) appears with a positron (positively charged), and a proton (positively charged) with an antiproton (negatively charged). In laboratory conditions these particles and antiparticles can be separated and "bottled" by electromagnetic fields. However, without externally applied fields the lifetime of such elementary particles and anti-particles is minute, typically 10^{-21} seconds, after which they annihilate each other in an explosive burst of energy—a reverse of the process by which they are made.

Hence the Inflationary Big Bang model needed to explain the following: (a) since any particle-antiparticle pairs produced were pressed next to each other in the extremely high density following the Big Bang, why did all the particles and antiparticles not annihilate each other; and (b) since we know that an enormous amount of matter exists in the universe, where is the corresponding amount of antimatter?

Speculations about anti-galaxies gave way to observational estimates of the ratio of photons to protons in the universe, which is roughly two billion to one. Hence, theorists concluded, for every one billion antiparticles—antiprotons and positrons—that materialized from the Big Bang energy release, one billion and one corresponding particles—protons and electrons—must have materialized. Each billion particles and antiparticles annihilated one another in an explosive burst of energy producing two billion photons, which are quanta of electromagnetic energy. According to the Big Bang model it is this energy, now expanded and cooled, that forms the cosmic microwave background radiation energy we see today. The billionth-and-one orphan protons and electrons survived and subsequently combined to make all the matter—all the planets, solar systems, galaxies, and galactic clusters—of the universe.

---

*   See page 30.

However, this conflicted with the Standard Theory of Particle Physics, which said that, according to the law of symmetry, only pairs of particles and antiparticles could materialize.

The reason for this conflict remained a problem for theoretical physicists until the mid 1970s, when they conjectured that at the extremely high temperatures of the Big Bang three fundamental forces of nature—electromagnetic, weak nuclear, and strong nuclear*—are just different aspects of the same force. The theorists devised different mathematical models, which they called grand unified theories (GUTs), although experimental data has disproved the original GUT and so far has failed to validate any others. These conjectures allow every type of elementary particle to interact with, and transmute into, every other particle. As a consequence, theoretical physicists believed that the symmetry between matter and antimatter need not be conserved. They adjusted the Standard Model to provide for asymmetry. This adjustment doesn't predict the amount of asymmetry but, like the charge of an electron, requires observational measurements that are then inserted into the model so that the model becomes consistent with observation.

Despite the hopes expressed in the 1970s, matter-antimatter asymmetry in the laboratory wasn't detected and measured until 2001, when B mesons—particles hypothesized to contain a bottom quark and an anti-down quark—and anti-B mesons were produced and survived for 10^{-12} seconds. However, the asymmetry observed was not large enough to explain the estimated ratio of energy to matter in the universe.[1]

Hence science's current orthodox theory of the origin of matter needs to supply a convincing answer to the question of how matter was formed from the energy released by the Big Bang in such a way as to produce the ratio of matter to energy that we observe in the universe today.

Dark matter and Omega

Two problems arise here.

First, if you estimate the mass of a galaxy by the conventional method of measuring its luminosity, the gravitational attraction of the mass is only about one tenth of that needed to keep all its stars in orbit round its centre. Likewise, the gravitational attraction of the mass of a cluster of galaxies measured by its luminosity is only about one tenth of that needed to keep the galaxies together in a cluster.

This isn't so surprising because the conventional method is simply measuring light emitted. Stars and galaxies of different masses vary in their luminosity and their distance from us, and more distant ones may be obscured by intervening gas and dust or concealed by the light of nearer stars and galaxies. Accordingly, mass

---

*   See Glossary for an explanation of these different forces.

measurement is only a crude estimate based on averaging estimated "known" values.

More significantly, the conventional method doesn't measure the mass of anything that doesn't emit or reflect light. If the General Theory of Relativity is valid, then there must be about ten times the amount of non-luminous matter— dark matter—extending beyond the radius of each visible galaxy in order to prevent the galaxy flying apart. Likewise, there must be about ten times the amount of dark matter in what had been thought of as the void surrounding visible galaxies in a cluster.

The many speculations as to what constitutes this dark matter may be grouped into two types:

MACHOs   Massive Compact Halo Objects are forms of dense matter, like black holes, brown dwarfs, and other dim stars—favoured by astrophysicists to explain dark matter.

WIMPs   Weakly Interacting Massive Particles are particles left over from the Big Bang, like neutrinos with a hundred times the mass of a proton, and so on—favoured by particle physicists to explain dark matter.

Although dark matter has been inferred from its gravitational effects, investigations for more than 30 years have failed to identify the nature of dark matter or experimentally confirm the existence of WIMPs. Many particle physicists are hoping that the Large Hadron Collider will produce evidence of WIMPs after its reopening in 2015 with almost double its previous collision energy levels.

The second problem is that, even if all the dark matter estimated to keep stars orbiting in galaxies and galaxies grouped into clusters is added to all the estimated visible and known matter, this is far too small a total mass to provide enough gravitational attraction to balance the kinetic energy of the universe's expansion according to the orthodox Inflationary Big Bang model. The estimates produce a value of the density parameter Omega of about 0.3,[2] which is significantly less than the 1.0 of the flat universe assumed in the orthodox Friedmann-Lemaître model and rationalized by inflation conjectures.

Hence, science's orthodox theory of the origin of matter currently fails to answer: (a) what is the dark matter that apparently keeps the stars in orbits and the galaxies in clusters; and (b) what and where is the additional dark matter necessary for consistency with the orthodox model of the universe?

Dark energy

If this were not bad enough for the orthodox model, in 1998 astronomers announced a discovery of even greater significance.

With the development of technology and astrophysical theory two international teams of astronomers were able to collect data from highly redshifted Type 1a supernovae; they considered that these violent explosions of white dwarf stars produce standard luminosities. According to orthodox cosmology the degree of redshift meant that they were distant and therefore young, exploding when the universe was around 9 to 10 billion years old. However, they were dimmer than expected. Cosmologists concluded that they must be further away than predicted by their Friedmann-Lemaître flat geometry model, which says that the expansion rate of the universe is slowing down. Hence they decided that something must have caused the expansion of the universe to accelerate. They called this unknown ingredient "dark energy".[3]

Based on the orthodox theory's assumptions and interpretations of astronomical data, the scientists conducting the space-based Wilkinson Microwave Anisotropy Probe announced in 2003 that the universe consists of 4 per cent known matter, 23 per cent of an unknown type of dark matter, and 73 per cent of this mysterious dark energy.* In other words, the unknown dark matter that dwarfs the matter we know about is itself dwarfed by an unknown dark energy that accounts for more than two-thirds of the universe. Perhaps Bernard Sadoulet's quote at the beginning of the chapter should read "Not only are we not made of what most of the universe consists of, we don't even know what most the universe consists of."

Figure 4.1 portrays this revised orthodox version of the universe's history.

Scientists immediately began trying to identify this dark energy. Theoretical cosmologists thought they could account for this relatively recent increase in the rate of expansion of the universe if they re-introduced Lambda into Einstein's field equations.

As we saw in Chapter 3, Einstein had ditched this cosmological constant, describing it as his biggest blunder. It doesn't appear in the Friedmann equations used in the basic Big Bang model, but most inflation models reintroduced it, albeit with an immensely larger value than Einstein's and only for an incredibly brief period, in order to account for an inflation of the primordial universe so that it achieved the critical mass density with Omega = 1 for a stable decelerating expansion. (Since it is an arbitrary constant, cosmologists can give it any value, positive, negative, or zero.)

Now theorists introduced the cosmological constant again—but with a very different value than either Einstein or inflationists had given it—to try to account for the implied acceleration, which was very much less than inflationary acceleration. Not without reason does University of Chicago cosmologist Rocky Kolbe jokingly refer to it as the cosmo-illogical constant.

---

* Data from the European Space Agency's Planck space telescope in 2013 resulted in a revision of these proportions to 4.9 per cent known matter, 26.8 per cent dark matter, and 68.3 per cent dark energy.

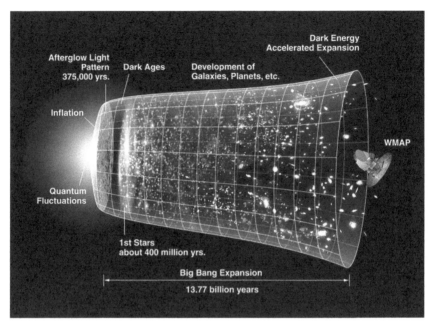

Figure 4.1 Orthodox cosmology's history of the universe

But if it is to be more than a mathematical constant that theorists can arbitrarily insert into their equations and fiddle about with its value until their solutions match observation, it must actually mean something in the real world. Particle physicists thought it represented the zero point, or quantum mechanical ground state, energy of the universe: that is, the lowest possible energy of the universe, which is the energy associated with the vacuum of empty space. But when they calculated its value this way it turned out to be an enormous 120 orders of magnitude greater than that observed by astronomers.[4]

Theoretical physicist Martin Kunz and his colleagues threw a spanner in the Lambda works by pointing out that, first, there is a large spread in the astronomical data and, second, interpretation of the data is implicitly sensitive to assumptions about the nature of the dark energy. They suggested that by comparing data on a range of astrophysical phenomena it might be possible to rule out a cosmological constant as the origin of dark energy.[5]

Another particle physicist, Syksy Rasanen of CERN, threw a further spanner by proposing that the accelerating expansion of the universe is driven not by a mysterious dark energy but, paradoxically, by a continuously greater *decrease* in the expansion rate of those small regions of space dominated by matter. As these regions suck in more matter by their gravitational attraction, they condense and become a progressively smaller—and less important—percentage of the universe's volume. The expansion of the voids continues unchecked as they become a progressively larger percentage of the universe's volume. The overall effect,

Rasanen suggests, is that the average expansion rate of the universe increases without the need for dark energy.[6]

In 2011 theoretical physicist Ruth Durrer of Geneva University pointed out that all the different kinds of evidence claimed to date for dark energy depend on distances calculated by redshift that are larger than expected from the orthodox model.[7] Richard Lieu, astrophysics professor at the University of Alabama, goes further, arguing that much of the orthodox model, which now includes dark matter and dark energy, "has been propped by a paralyzing amount of propaganda which suppress counter evidence and subdue competing models". He claims two competing models, one eliminating dark matter and the other eliminating both dark matter and dark energy, fare no worse in matching or failing to match the evidence, and concludes that the more the orthodox camp, which dominates the funding agencies, fails to find these unknown dark ingredients, the more they invest taxpayers' money in trying to find them to the point of totally choking alternative approaches.[8]

Ellis also maintains that alternative explanations of the astronomical data are possible. They can fit a spherically symmetric inhomogeneous universe model, or could be due partly to the back reaction of inhomogeneities on the cosmic expansion, or the effect of inhomogeneities on the effective area distance.[9]

Cosmologist Lawrence Krauss concludes that "the nature of 'dark energy' that is causing the apparent accelerated expansion of the universe is, without doubt, the biggest mystery in physics and astronomy".[10]

Hence, if science's orthodox theory of the origin of matter is to be convincing, it needs to answer: (a) whether or not the expansion of the universe is accelerating; (b) if it is, when did the decelerating expansion change to an accelerating one; and (c) what verifiable thing is causing this acceleration?

Fine-tuning of cosmological parameters

When examining the flatness problem of the basic Big Bang model in Chapter 3, we saw how extremely tiny differences in the value of Omega—a measure of the gravitational attraction of the matter in the universe compared with its expansion energy—produce very different types of universe.

In 2000 Britain's Astronomer Royal, Martin Rees, argued that our universe wouldn't have evolved to the stage it is now, where it contains humans like us who think about the origin of the universe, if the value not only of Omega but also of five other cosmological parameters had been different by only minuscule amounts.

In fact the fine-tuning of many parameters additional to Rees's six is claimed to be necessary for human evolution, and I shall examine the "anthropic universe" question more fully in later chapters. Suffice it to say here that cosmology's orthodox theory fails to answer the question of how and why the universe that emerged from the Big Bang took the form that it did when very many other forms were possible.

Creation from nothing

This is the elephant in the room. It is the biggest question that comology's orthodox theory about the origin of matter must answer. Put simply, where did everything come from?

Specifically, from where did the energy come not only to produce the entire universe but also to counteract the immense gravitational attraction of the super-dense matter thereby created—infinitely dense if it began as a singularity—and expand the universe to its present size?

Many cosmologists espouse the idea that it comes from the net zero energy of the universe. As a consequence of Einstein's Special Theory of Relativity, every mass m has an energy equivalence measured by $E = mc^2$, and this rest mass-energy of matter conventionally has a positive value. Guth argues that the energy of a gravitational field is negative. Based on an idea that appears to have been advanced first by Richard Tolman in 1934,[11] Guth's argument for creation from nothing using the net zero energy idea may be summarized as follows:

1. if the Principle of Conservation of Energy applies to the universe, then the universe must have the same energy as that from which it was created;
2. if the universe was created from nothing, then the total energy of the universe must be zero;
3. because the observable universe is clearly filled with an unfathomable mass-energy of a hundred billion galaxies that are expanding, then this must be offset by some other energy;
4. because the gravitational field has negative energy, then the immense energy we observe can be cancelled by a negative contribution of equal magnitude from the gravitational field;
5. because there is no limit to the magnitude of energy in the gravitational field, there is no limit to the amount of mass-energy it can cancel;
6. therefore the universe could have evolved from absolutely nothing in a manner consistent with all known conservation laws.[12]

Proposition (2) depends on the condition that the universe *was* created from nothing. This is far from a self-evident truth. Hence the validity of the proposition is questionable.

Proposition (5) rests on a questionable assumption; I shall discuss terms like "limitless" and "boundless" in Chapter 6 when examining infinities in a physical cosmos.

Even if we accept these questionable propositions, however, the argument shows how the universe *could in theory* have evolved from nothing, but it doesn't tell us *how in practice* it evolved.

Edward Tryon proposed the "vacuum quantum fluctuation" answer in 1973. According to quantum theory's Uncertainty Principle, we cannot measure the precise energy of a system at a precise time. Quantum theory accordingly allows the conjecture that even a vacuum, a space from which everything has been removed, has a zero point, or ground state, of fluctuatating energy from which a pair of matter and antimatter particles can spontaneously materialize, exist for an incredibly brief period, and then disappear. Tryon suggested that the universe spontaneously materialized from a vacuum by such a quantum fluctuation.[13]

However, the probability in quantum theory of an object materializing from a vacuum decreases dramatically with its mass and complexity, and so the probability that a 14-billion-year-old complex universe of some ten thousand billion billion times the mass of the Sun could arise this way is so remote as to be practically impossible. Tyron's suggestion was not taken seriously until inflation rode to the rescue.

Guth and others speculate that, during the infinitesimally short lifetime of such vacuum quantum fluctuations, a proto-universe inflated by fifty or so orders of magnitude in almost an instant, despite the gravitational field due to its matter acting to crush it out of existence.

This prompts two challenges. First, as we saw in Chapter 3, despite the attempt by inflation theorists to argue otherwise, this mass-energy cannot travel faster than the speed of light without conflicting with relativity theory.*

Second, as we also saw in the last chapter, there are a hundred or so different versions of inflation, but most follow Linde and assume the mechanism is some kind of scalar field, generically called an inflation field. However, unlike an electromagnetic field that can be detected and measured, no one has yet found a way of detecting, still less measuring, an inflation field. This crucial conjecture lacks empirical support.

The energy to power this speculative inflation field is assumed to come from the net zero energy of the universe. Whereas Rees is careful to use the word "conjecture" about this idea,[14] Hawking has no such reservations. He asserts that, in the case of a universe that is approximately uniform in space, negative gravitation energy exactly cancels the positive energy represented by matter. Hence the total energy of the universe is zero. "Now twice zero is also zero. Thus the universe can double the amount of positive matter energy and also double the gravitational energy without violation of the conservation of energy....During the inflationary phase, the universe increases in size by a very a large amount. Thus the total amount of energy available to make particles becomes very large. As Guth has remarked, 'It is said that there's no such thing as a free lunch. But the universe is the ultimate free lunch.'"[15]

I'm not aware of many scientists other than cosmologists who believe in free lunches. But even if the universe were a free lunch, this idea still doesn't tell us where the ingredients came from. Specifically, a vacuum with a ground state energy that undergoes random quantum fluctuations is not nothing. Where did this vacuum come from? Furthermore, how can this conjecture be tested?

---

*   See page 35.

These are the questions orthodox cosmology must answer if its creation from nothing conjecture is to be treated as a scientific theory.

Conclusions

Chapter 3 concluded that cosmology's orthodox theory is unreliable and the central claim of inflation, added to explain admitted contradictions with observational evidence, is almost certainly untestable.

This chapter has concluded that, even with two more major add-ons—dark matter and dark energy—the current theory fails to provide convincing answers to six key questions: whether or not and, if so, how the universe originated as a singularity; how did matter form from the energy released by the Big Bang so as to produce the ratio of matter to energy observed today; what is the dark matter apparently needed to explain why galaxies and galactic clusters don't fly apart, and what and where is the far greater additional dark matter needed to explain why the rate of expansion of the universe is that predicted by the theory; how and when did this slowing rate of expansion apparently change to an accelerating one and what is the dark energy invoked as the cause; how and why did the universe take the form that it did when many other forms were possible; and, crucially, how was everything created out of nothing when the originating vacuum bubble has a ground state energy and so is not nothing?

In 1989 *Nature* published an editorial describing the Big Bang model as "unacceptable" and predicting that "it is unlikely to survive the decade ahead".[16] It has survived longer than that: the Inflationary Hot Big Bang is still cosmology's orthodox explanation for the origin of the universe. But for how much longer?

Other hypotheses are competing to either modify the orthodox model or else supersede it. Do they provide a more scientifically rigorous account of the origin of the universe?

CHAPTER FIVE

Other Cosmological Conjectures

Seeking an alternative [to the Orthodox Model] is just plain good science. Science proceeds most rapidly when there are two or more competing ideas.

—PAUL STEINHARDT, 2004

Many of today's practising theorists seem to be unconcerned that their hypotheses should eventually confront objective, real-world observations.

—MICHAEL RIORDAN, 2003

Some cosmologists, unlike some popes, are unhappy with the idea that the universe exploded into existence from nothing. For them the idea that the universe is eternal is far more appealing. One difficulty in evaluating their ideas and other alternatives to the orthodox Inflationary Big Bang model is that, just as most of the literature on religion is written by believers in particular religions, so too most of the literature on cosmological speculation is written by believers in particular speculations; like their religious counterparts they are sometimes less than objective when presenting their hypotheses and selecting and interpreting claimed supporting evidence.

I have chosen what I think are the most significant ideas.

Hartle-Hawking no-boundary universe

To address the first question raised in the previous chapter—whether or not the universe, including time and space, began as a singularity in which the known laws of physics break down—Stephen Hawking examined ways in which quantum theory can be applied to the initial state of the universe. Making no distinction between the three dimensions of space and the one dimension of time in his equations allowed him to introduce imaginary time. This was equivalent to the long-accepted notion in mathematics of imaginary numbers. If we take a real number, like 2, and multiply it by itself the result is a positive number, 4. The same is true if we multiply a negative number by itself: -2 times -2 equals 4.

An imaginary number is one which, when multiplied by itself, gives a negative number. For example, i multiplied by itself equals -1, while $2i$ multiplied by itself equals -4.

The result of this work, which he developed with Jim Hartle in 1983, was to produce a universe in which time and space are finite but without any boundaries.[1] A simplified two-dimensional analogue of the Hartle-Hawking four-dimensional space-time universe is the surface of the Earth, which is finite but has no boundaries, as shown in Figure 5.1.

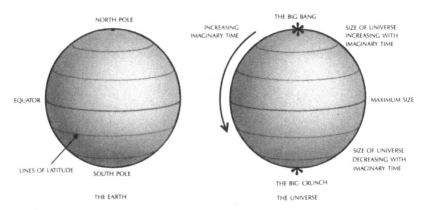

Figure 5.1 Simplified two-dimensional representation of the surface of the Hartle-Hawking no-boundary universe compared with the surface of the Earth

Here the universe starts in the Big Bang with zero size at the equivalent of the North Pole, expands in imaginary time to reach maximum size at the equivalent of the equator, and contracts in continuing imaginary time to end in a Big Crunch with zero size at the equivalent of the South Pole. Just as the laws of physics hold at the real North Pole of the Earth's surface, so too they hold at imaginary time zero.

This solution describes a universe in which, according to Hawking, "There would be no singularities at which the laws of science break down and no edge of space-time....The boundary condition of the universe is that it has no boundaries. The universe...would neither be created nor destroyed. It would just BE."

Hawking concedes that when we go back to the real time in which we live, there will still appear to be singularities, but suggests that what we call imaginary time might actually be the real time while what we call real time is just a figment of our imagination.

This ingenious proposal solves many of the problems of cosmology's orthodox model, not least of which is creation from nothing. However, Roger Penrose, who co-authored with Hawking the mathematical proof that there must have been a Big Bang singularity provided that the General Theory of Relativity is correct,*

---

*   See page 39.

describes the model as a "clever trick" for producing consistent quantum field theories but has "severe difficulties" with it when used in conjunction with the approximations needed to solve the equations.[2]

To see if it describes the world we experience, it must be tested. Hawking claims that it makes two predictions that agree with observations: the amplitude and spectrum of fluctuations of the cosmic microwave background. However, like those of inflation, the "predictions" derive from an arbitrary choice of scalar fields rather than from actual predictions of the no-boundary model itself.

Conceptually appealing though the model is, after more than 30 years Hawking has failed to convince many theoretical cosmologists that its mathematical constructions are viable and that his imaginary time is actually real time. Moreover the model has not made any unique predictions that are supported by observation.

Eternal chaotic inflation

If the Big Bang followed a period of inflation, as most cosmologists now favour, this raises the question of what preceded it. Accordingly Linde developed in 1986 what he termed the "eternally existing self-reproducing chaotic inflationary universe" model.[3]

Although cosmologists educated in the post-1965 culture of the Big Bang as the only cosmogony rarely, if ever, recognize it, this model shares much in common with the updated version of the steady state theory known as quasi-steady state cosmology.* Linde proposes that chaotic inflation continues eternally as an ongoing process, creating new regions of space with different properties. Some of these regions could be as large as our entire observable universe. "With eternal inflation, it would really all be the same universe, but its parts would be so far away from each other that for all practical purposes you can call them different universes."[4]

The model suggests that once these regions have inflated they will necessarily contain within themselves tiny parts that will inflate, and once these parts inflate they will contain tiny parts that in turn will inflate. And so the inflationary process is eternally self-reproducing.

Guth gives this his enthusiastic endorsement because "eternal inflation lays to rest the difficult question of deciding how plausible it is for inflation to have started". Moreover

> if the ideas of eternal inflation are correct, then the big bang was not a singular act of creation, but was more like the biological process of cell division....Given the plausibility of eternal inflation, I believe that soon any cosmological theory that does not lead to the eternal reproduction of

*   See page 57.

universes will be considered as unimaginable as a species of bacteria that cannot reproduce.[5]

Guth is right to couch this view as a statement of belief rather than a scientific conclusion.

In principle the conjecture answers the question about the origin of our region, or bubble, of the universe: it had a beginning and may or may not have an end, but the whole process will never end. As for how the whole process began, Linde is less certain. In 2001 he wrote "There is a chance that all parts of the universe were created simultaneously in an initial big bang singularity. The necessity of this assumption, however, is no longer obvious."[6]

Seven years previously, cosmologists Arvind Borde and Alexander Vilenkin had reached a more definite conclusion. They argue that, provided some technical assumptions are accepted, a physically reasonable space-time that is eternally inflating to the future must have started from a singularity.[7]

The weight of argument clearly supports the conclusion that the "eternal" chaotic inflation conjecture is not eternal: while chaotic inflation may continue indefinitely into the future, it had a beginning. Accordingly, it doesn't answer the fundamental question of where everything came from, nor does it lay to rest the difficult question of deciding how plausible it is for inflation to have started. Moreover, it has the same problems of untestability of its principal claim that put other versions of the inflation idea outside the realm of science and into the realm of philosophical conjecture.

Variable speed of light

A young cosmologist, João Magueijo, then holder of a prestigious Royal Society Research Fellowship, proposed an alternative to the inflation conjecture which, he maintains, has become sacrosanct in the American cosmological community. His central idea is that, in the very early universe, the speed of light was very many times faster than it is now. This conjecture solves all the problems that the inflation conjecture solves and, whereas there is no evidence for an inflaton particle or its corresponding inflationary field, Magueijo claims that some observational evidence of very young stars supports the variable speed of light hypothesis developed by him and Andreas Albrecht (who had jointly developed one of the early, modified versions of the inflation conjecture, see p. 29). This idea does, of course, break one of the tenets of Einstein's relativity theory that nothing can travel faster than the speed of light, and Magueijo is attempting to rework Einstein's equations accordingly.

To speculate that, in a period very shortly after the Big Bang (in which Einstein's relativity theory breaks down), the speed of light was thousands of times faster than it is now seems no more unreasonable than to speculate that the mass-energy of a vacuum inflated thousands of times faster than the speed of light.

Nonetheless, Magueijo and Albrecht found it extremely difficult to get their paper published, just as did Hoyle and others whose hypotheses differed from cosmological orthodoxy.

Magueijo's account of their attempts and the responses of referees to their paper paints a picture of science's vaunted peer review process as one of review by bishops of orthodoxy instinctively dismissing heretical views that challenge the conjectures on which their own reputations are based. It is a sad reflection on *Nature* that I had to buy the American edition of Magueijo's book in order to read the full version: *Nature* threatened the UK publisher with legal action if it did not pulp the first edition and publish an expurgated version. Among other things, Magueijo alleged that the consensus in his research field is that *Nature's* cosmology editor is not up to the job but his colleagues dare not say so for fear of their careers. He uses intemperate language, referring to the editor as "a first class moron" and "a failed scientist" with penis envy. Such language detracts from his case. Surely it would have been more in keeping with the ideals of science for *Nature* to desist from acting like the Holy Office and allow readers to decide to what extent Magueijo's account is that of an immature, erroneous egomaniac or else a reasonable man driven to frustration.[8]

Magueijo and Albrecht's paper was eventually published in 1999 by *Physical Review*. Their conjecture could lead to a more robust theory if it can make predictions that can be tested by observation, but judgement on it must wait until more theoretical work and more observational evidence are produced.

Cyclic bouncing universe

The hypothesis that the Big Bang arose from the collapse of a previous universe was suggested by Richard Tolman at Caltech as early as 1934. He based this on another solution to Einstein's equations of general relativity for the universe, which he also assumed to be isotropic and omnicentric, but which was closed rather than flat. The particular solution shows an oscillating universe, which expands and then contracts to a Big Crunch and then expands again, repeating the process *ad infinitum*. When Tolman applied the Second Law of Thermodynamics to the model he found that each cycle of the oscillating universe became bigger and lasted longer than the previous one.[9]

It fell out of favour for several reasons, particularly when cosmologists concluded that observational evidence supported the flat universe model. As we have seen, however, the flat universe model encountered major conflicts with observational evidence that required the incorporation of the inflation conjecture plus vast amounts of unknown dark matter and unknown dark energy.

Tolman's hypothesis appears to avoid the singularity problem. However, if each cycle of the oscillating universe becomes bigger and longer, then each preceding cycle was smaller and lasted less time. Running the clock back leads to a point when the cycle approaches infinite smallness and infinite density in zero time, the conditions conjectured for the basic Big Bang singularity. Hence the model is not a true eternal one and does not avoid the basic origin problem: how and why did such an infinitely small and dense singularity come into existence?

Furthermore, no one has yet devised any means of observing or otherwise testing the physical existence of any previous universe in the cycle. Until they do, this conjecture, too, must remain outside the realm of science as currently understood.

Natural selection of universes

Lee Smolin is a theorist who is not only prepared to think outside the box of orthodoxy, but also believes it is necessary to do so if physics is to progress. His conjecture of an evolution of universes by natural selection has been taken seriously in several quarters of the scientific community and so is worth examining in some detail.

In 1974 John Wheeler speculated that the collapse of the universe in a Big Crunch might lead to a Big Bounce in which the universe is reprocessed to produce a new universe where the laws of physics would be the same but the physical parameters, such as the mass of a proton and the charge on an electron, which are not predicted by the laws, would be different.

As we saw in the last chapter when considering the fine-tuning of cosmological parameters, very small changes can produce very different universes. For example, if the only change were that protons are 0.2 per cent heavier, then no stable atoms would form and the universe would remain a plasma: complex matter like humans could never evolve.

Taking this notion further, Smolin conjectures that it is not only the collapse of a universe in a Big Crunch that produces another universe with different parameters through a Big Bounce, but also the collapse of a star into a black hole produces another universe with different parameters on the other side of the black hole. Generations of universes produced this way from a progenitor universe with random parameters will lead by a process of natural selection like that in biology* to universes best fitted to survive and allow the evolution of intelligent life.[10]

Eight assumptions underpin this speculation:

1. Quantum effects prevent the formation of a singularity in which time starts or stops when a universe collapses into a Big Crunch or a star collapses into a black hole, and so time continues into some new region of space-time connected to the parent universe only in its first moment.

---

*   The cumulative effect of small genetic changes occurring in successive generations of the members of a species that leads to the dominance of those members whose mutations make them best fitted to compete and survive; the mutations eventually produce a new species whose members do not breed with those of the original species.

2. Such a new region of space-time where time continues after the collapse of a star into a black hole is necessarily inaccessible to us but "could be as large and varied as the universe we can see".

3. Because our own visible universe contains an enormous number of black holes, then "there must be an enormous number of these other universes... at least as many as there are black holes in our universe...[moreover] there are many more than that, for why should not each of these universes also have stars that collapse to black holes and thus spawn other universes?"

4. The parameters of the first universe are such that it produces at least one descendant universe.

5. Each successive descendant universe produces at least one progeny.

6. The parameters of a new universe formed by the collapse of a universe or a star are slightly different from those of its parent.

7. The rules of natural selection apply: the cumulative effect of small random mutations in the parameters of the daughter universes eventually generates universes whose parameters are best fitted to produce many black holes—and hence many progeny—until we end up with universes like our own that creates around 10^{18} black holes.

8. The parameters of universes like our own with such large numbers of black holes are fine-tuned for the evolution of intelligent life.

These assumptions are by no means self-evident.

Smolin shares assumption (1) with many other theorists, but acknowledges that whether this corresponds to reality depends on the details of the quantum theory of gravity, which is not complete.

Assumption (2) appears unreasonable. If the current theory of black holes is correct, while black holes may possess a huge, if not infinite, density, they possess limited mass. For instance, a black hole should be formed by the gravitational collapse of a star with a mass greater than three times that of our Sun. The mass of the known luminous and hypothesized dark matter of the universe is estimated as roughly ten thousand billion billion Suns. Even without taking into account the energy needed to drive the expansion of the new universe, to assume that a mass of, say, five Suns collapses into a black hole and explodes out the other side as a mass of ten thousand billion billion Suns seems illogical. Presumably Smolin, like Guth and other theorists,* is following Tolman but is being somewhat less ambitious and creating a new universe out of five solar masses rather than out of nothing.

On assumption (4) Smolin concedes that if the parameters of the first universe were random then it is "most likely" (overwhelmingly probable is, I think, more appropriate) that in microseconds this first universe either inflates into emptiness or collapses. That is, the evolutionary process never starts. To avoid this, Smolin

---

* See page 46 for Guth's argument for the net zero energy of the universe.

assumes that the parameters of the first and subsequent universes are finely tuned to allow them to undergo at least one bounce. However, he provides no justification for this assumption other than that it is necessary for his speculation to work. Moreover, the speculation doesn't explain how such a progenitor universe came into existence in the first place. As such it cannot reasonably claim to possess greater explanatory power than the orthodox Inflationary Big Bang model.

Assumption (8) implies that the physics that maximises black hole production is the physics that enables the evolution of life, but Smolin offers no basis for such an assumption.

As for its empirical basis, this speculation posits the existence of an enormous number of other universes that have either vanished or, if they still exist, with which we cannot communicate. Smolin nonetheless maintains that his conjecture is testable by arguing that it predicts the parameters in the laws of elementary particle physics are close to a value that maximises the number of black holes in our universe. This argument is circular. The speculation is untestable by any known means and hence is philosophical conjecture rather than science.

Loop quantum gravity

Like Smolin's conjecture, a majority of alternative hypotheses to the Big Bang model deal with the singularity problem by speculating that, if the expansion of the universe is run backwards, quantum effects prevent the formation of a singularity at which time starts or stops. They attempt to unify quantum and relativity theories and propose a collapsing universe on the other side of the Big Bang to which our universe is connected by a quantum tunnel. If this were so, then it would demolish the key claim of the Big Bang model, namely that space and time erupted into existence from nothing at the Big Bang.

A major problem, as Smolin and other cosmologists acknowledge, is that we don't yet have an adequate theory of quantum gravity.

> **quantum gravity**   The hoped-for quantum theory of gravity that would enable gravitational energy to be unified with other forms of energy in a single quantum theoretical framework.

> **quantum theory**   The theory that energy is emitted and absorbed by matter in tiny, discrete amounts, each of which is called a quantum that is related to the radiation frequency of the energy, and thus possesses properties of both a particle and a wave. It gave rise to quantum mechanics. The term is now used in a general sense to refer to all subsequent theoretical developments.

Abhay Ashtekar, however, makes the bold declaration that he and his colleagues at the Penn State Institute of Gravitational Physics and Geometry are the first to provide a robust mathematical description that systematically establishes the existence of a prior collapsed universe and deduces properties of space-time geometry in that universe. Ashtekar and his team use an approach called loop quantum gravity in which they claim to show that in place of a classical Big Bang there is a quantum bounce, on the other side of which is a classical universe like ours.

Ashtekar does admit one limitation, namely the assumptions that the universe is homogeneous and isotropic. "It is an approximation done in cosmology, even though we know that the universe is not exactly like that. So the question is how to make the model more and more realistic. And this is an ongoing work."[11]

The jury is out on the mathematical model, but even if the verdict were "proven", science would require physical evidence to support any mathematical proof, and no one has yet suggested how this can be obtained.

Quasi-steady state cosmology

In 1993 Fred Hoyle, Geoffrey Burbidge, and Jayant Narlikar modified the steady state theory in the light of observations to produce what they termed quasi-steady state cosmology (QSSC). This says that in the longer term of 1,000 billion years the universe expands in a steady state, but it does so in 50-billion-year cycles of expansion and contraction where the contraction never reaches zero, or a singularity.

This team of astronomers and astrophysicists postulates that a universal creation field (C-field) is responsible for the continual creation of matter and for the expansion of the universe. The C-field has negative energy and creates matter in the form of Planck particles, the most massive elementary particle possible: with a greater mass than this an elementary particle would be overwhelmed by its own gravitational force and collapse into a black hole.*

The C-field is only strong enough to create Planck particles near very massive, compact, and dense objects they call near black holes (NBHs) that lie at the centre of galaxies. These are formed when the growing strength of the C-field prevents a contracting celestial object reaching a radius of $2GM/c^2$, after which it would become a black hole. Instead, the shrinking object slows down to a halt and begins to bounce back at the NBH radius. Here the strength of the C-field is high enough to create Planck particles, of about one hundred thousandths of a gram, in a non-singular mini-bang, or matter creation event (MCE). These then decay into many smaller particles, including baryons (like protons and neutrons) and leptons (like electrons and neutrinos), with the production of radiation, to form the matter from which galaxies evolve.

The C-field increases in strength as it creates matter, thus increasing its production of matter and radiation. However, the negative energy of the C-field

---

* See *Planck mass* in Glossary for a fuller explanation.

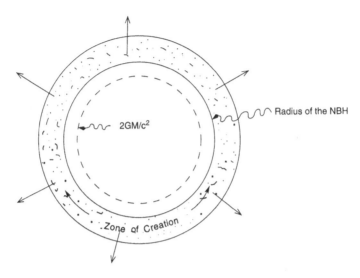

Figure 5.2 Matter creation at a near black hole (NBH)
where the growing strength of the C-Field prevents a contracting celestial object reaching a
radius of $2GM/c^2$, after which it would become a black hole.

acts as a repulsive force that ejects this newly created matter and radiation from the NBH, which may be thought of as a "white hole".

In the QSSC model the cosmological constant is negative, whereas in the orthodox model it is positive. Thus the newly created matter and radiation is subject to the two attractive forces of gravitation and the cosmological constant and to the repulsive force of the creation field. Initially the latter dominates, and matter and radiation are ejected at very high speed from the NBH causing expansion of the universe.

As matter expands density falls, and with it the strength of the C-field to the point that it can no longer create Planck particles. The attractive forces of gravity and the cosmological constant come to dominate and cause the universe to contract. Once it contracts to a sufficiently high density, the C-field becomes strong enough to create new matter and the next cycle begins.[12]

When the theoretical basis of this was challenged, Hoyle wryly pointed out at a meeting of the Royal Astronomical Society in London in December 1994 that the relevant equations of QSSC are the same as the corresponding equations of inflation if you replace the Greek letter "Φ" by the letter "C".

Proponents of QSSC claim that it requires only one assumption, the C-field; the rest is derived from observation and explained by normal physics, unlike the orthodox model that has to invoke ideas like a quantum gravity era, an inflation field, grand unified theories, unknown dark matter, and unknown dark energy in order to preserve the Big Bang idea and maintain consistency with observations.

Moreover, whereas the Big Bang occurs only once and cannot be observed, MCEs occur constantly and can be observed in the jets of plasma from radio sources and bursts of energy in the form of radio, infrared, visible, ultraviolet, and gamma rays from near the centre of galaxies. (Orthodox cosmology interprets these observations as emissions from quasars as matter is sucked into a black hole at the centre of very distant, and therefore very young, galaxies, see page 79.)

Its proponents further claim that QSSC provides a better explanation than the orthodox model for other observations. For example, orthodox cosmologists accept the stellar nucleosynthesis theory advanced by Hoyle and colleagues to explain how all the elements beyond lithium are made by stars, but claim that only the orthodox model shows how helium created in the Big Bang fireball is produced in the amount observed. Burbidge and Hoyle counterclaimed that it is illogical to think that helium is made by a different method: *all* the elements are made in stars, and the amount of helium observed today includes helium produced in the very much longer timescale of the QSSC universe than orthodox cosmology's (then)13.7 billion years.

Moreover, calculations show that thermalization of the energy radiated by the stellar production of helium produces almost exactly the black body temperature of the cosmic microwave background (CMB) of 2.73K observed today.* Thermalization is achieved by the absorption of this radiated energy and emission at millimetre wavelengths by whiskers of iron in the intergalactic medium. These whiskers are produced by supernova explosions that eject vaporous atoms of iron which cool and condense as tiny, fine whiskers rather than balls, as demonstrated experimentally; this whisker dust is an effective absorber and emitter of radiation at millimetre wavelengths.

Evidence for the whisker dust is provided by radiation from the Crab Pulsar, which shows a gap in its spectrum at millimetre wavelengths; QSSC ascribes this to absorption by iron whisker dust produced by the Crab Supernova that resulted in the Crab Pulsar. A similar gap appears in the emission spectrum at the centre of our galaxy where supernova activity is believed to be intense.

QSSC explains that the small-scale inhomogeneities in the cosmic microwave background (CMB) reflect the inhomogeneities in galactic cluster distribution at the minimum contraction phase of the previous cycle.

Whereas orthodox cosmology has to invoke an arbitrary change in its positive cosmological constant to account for the redshifted Type 1a supernovae, in QSSC the negative cosmological constant stays constant; these supernovae appear fainter than expected because their light is partially absorbed by iron whisker dust.

Narlikar and Burbidge argued that astronomical observation will be able to show which model reflects reality. QSSC predicts very faint objects with

* See page 33.

blueshifts, which are sources of light in the previous cycle when the universe was larger than it is now; they are not predicted by the orthodox model.

QSSC further predicts the existence of very young galaxies formed from matter ejected by the C-field at MCEs that are relatively close, whereas the Big Bang model says that very young galaxies are necessarily very distant because they formed in the early epoch of the universe. Narlikar and Burbidge contended that astronomical observations support the QSSC prediction. However, this depends on the interpretation of redshifts.*

QSSC also predicts the existence of very old stars formed in the previous cycle. For example, a star of half a solar mass formed 40–50 billion years ago should now be a red giant. Hence if low-mass red giants are detected, this will confirm QSSC. The orthodox model, by contrast, says that no matter can predate the Big Bang 13.8 billion years ago.

Narlikar and Burbidge complained that proponents of QSSC are denied time on telescopes, like the Hubble telescope, in order to test their claims; orthodox Big Bang astronomers monopolize such instruments and this is not compatible with open scientific inquiry.

Proponents of cosmology's orthodox model dismiss QSSC. UCLA astronomy professor and COBE and WMAP project scientist Ned Wright claims that the QSSC model is incompatible with observational data, in particular counts of bright radio sources, and maintains that the claim that the model fits the CMB data is false. Moreover he asserts that the paper written in 2002 claiming that the QSSC model better accounts for the Type 1a supernovae data requires a large optical opacity for the universe, whereas another paper in the same year claiming conformity with the CMB anisotropy requires a low opacity. "These articles were submitted to different journals, and refer to each other as successful calculations of the QSSC model, but they in fact contradict each other. Presumably this is a deliberate attempt to deceive the casual reader."[13]

The unnecessary addition of an attack on an assumed deceitfulness of Narlikar and Burbidge sadly reflects attitudes too often struck by believers in particular cosmological models. Undoubtedly these claims and counterclaims will continue as long as believers survive. But this may not be long. Fred Hoyle died in 2001 and Geoffrey Burbidge in 2010, while Halton Arp was born in 1927 and Jayant Narlikar in 1938. Younger generations of cosmologists have been taught the orthodox beliefs and, for the vast majority of them, researching alternatives is not an option for an academic career in cosmology.

Plasma cosmology

In his 1991 book *The Big Bang Never Happened* plasma physicist Eric Lerner compiled observational evidence that contradicts orthodox cosmology's model.

---

*   See page 78.

This, he maintained, is a repackaged myth of a universe created from nothing that violates one of the best-tested laws in physics, the Principle of Conservation of Energy. Moreover, in order to reconcile the mathematical model with observation, it requires three major conjectures—an inflation field, dark matter, and dark energy—that have no empirical support.

Based on the work of plasma physicist and Nobel laureate Hannes Alfven, Lerner proposed a cosmology that he claims accounts for astronomical observations by empirically established plasma physics and gravity. He assumes, like the proponents of the orthodox model, a Euclidean, or flat, geometry—the kind with which we are familiar (see Figure 3.2)—for the universe, but proposes that it does not have a beginning or end and is not expanding.

According to Lerner, within this non-expanding universe plasma filamentation theory predicts the formation of intermediate mass stars during the formation of galaxies and the observed abundance of the light elements. Large-scale structures such as galaxies, clusters, and superclusters form from magnetically confined vortex filaments; since the conjecture proposes no beginning for the universe, there is no limit to the time it takes for the observed large-scale structures to evolve from an initial disordered plasma.

Radiation emitted by early generations of stars provides the energy for the CMB. This energy is thermalized and isotropized by a thicket of dense, magnetically confined plasma filaments that pervade the intergalactic medium. This accurately matches the CMB spectrum and predicts the observed absorption of radio waves. Furthermore, the alignment of the CMB anisotropy, which contradicts orthodox cosmology, is explained by the density of absorbing filaments being higher locally along the axis of the local supercluster and lower at right angles to this axis.[14]

Many orthodox cosmologists dismiss Lerner as not a serious scientist largely, I suspect, because he is not an academic. He is president of Lawrenceville Plasma Physics Inc. (which undertakes research in fusion energy), a member of the Institute of Electrical and Electronic Engineers, the American Physical Society, and the American Astronomical Society, and has published more than 600 articles. The evidence he cites includes many papers from academic astronomers, and in 2014 he co-authored a paper with two such astronomers that compares the size and brightness of about a thousand nearby and extremely distant galaxies.[15] He claims these contradict the surface brightness predicted by an expanding universe but are consistent with a non-expanding universe.

What may be termed plasma cosmology's static evolutionary universe model proposes an eternal universe and, unlike QSSC, does not require the creation of matter from nothing. It explains the observed evolution of the universe as an interaction of known physical forces: electromagnetism, gravity, and nuclear reactions within stars and by cosmic rays. However, as currently developed, it doesn't explain what caused the initial disordered plasma of an eternal universe to exist, and what caused these known physical forces to exist and interact in such a way as to produce successively more ordered, complex states of matter.

Quintessence

Paul Steinhardt, Albert Einstein Professor in Science at Princeton University, is another cosmologist who is prepared to think outside the orthodox box. Instead of using an arbitrary constant, Lambda, with a value very different from either the one Einstein had discarded or the one inflationists introduced for an incredibly brief period, Steinhardt proposed that the dark energy invoked to account for the apparent increase in the expansion rate of the universe is actually a new component of the universe.

Since cosmologists had previously thought the evolution of the universe was determined by four components—baryons,* leptons,† photons,‡ and dark matter§—Steinhardt called this fifth component quintessence, after the ancient Greek belief in the fifth and highest essence, from which the celestial sphere was made, as distinct from the four basic elements of earth, air, fire, and water.

Its principal difference from Lambda is that, whereas the cosmological constant has the same value everywhere in space and is inert, the density of quintessence decreases slowly with time and its distribution in space is not uniform.

Orthodox cosmologists criticized quintessence, pointing out that observations to date show no evidence of time or spatial variation in dark energy. This rules out some quintessence models but, according to Steinhardt, still allows a significant range of possibilities.[16]

A new model that eliminates an arbitrary Lambda with a value fifty orders of magnitude less than the inflationary arbitrary Lambda would be more elegant. However, like the orthodox model, quintessence models fail to explain where this variable dark energy comes from.

Steinhardt and others subsequently developed an alternative model of the universe that claims to provide such an explanation. I examine this next.

Cyclic ekpyrotic universe¶

This alternative to the orthodox cosmological model was based on M-theory, the latest version of string theory, which claims that everything in the universe reduces to infinitesimally small strings of energy. The different masses and other properties of both the fundamental particles—electrons, neutrinos, quarks, and so

---

*   Heavy subatomic particles, like protons and neutrons.
†   Light or almost massless elementary particles that do not interact through the strong nuclear force, like electrons.
‡   Massless quanta of electromagnetic energy.
§   The unknown form or forms of non-radiating matter invoked to make theory consistent with observation.
¶   Steinhardt calls this model the cyclic universe, but I use the label cyclic ekpyrotic universe to differentiate it from other cyclic models, like Tolman's cyclic bouncing universe and QSSC's cycles.

on—and the force particles associated with the four forces of nature—the strong and weak forces, electromagnetism, and gravity—are simply reflections of the various ways in which these tiny one-dimensional strings vibrate.

M-theory allows strings to expand, and an expanded string is known as a "brane" (an abbreviation of membrane); such branes can have 0, 1, 2, 3, or any number of dimensions. With enough energy a brane can grow to an enormous size, even as large as our universe.

In 1999 Steinhardt and Neil Turok, then Professor of Mathematical Physics at Cambridge University, attended a Cambridge University cosmology conference at which University of Pennsylvania string theorist Burt Ovrut suggested that our universe consists of three large observable spatial dimensions (height, width, and length) on a brane, plus six curled-up spatial dimensions too small to be observed, plus a tenth spatial dimension—a finite line—that separates this brane from the brane of another universe that also has three large and six tiny curled-up spatial dimensions. Because the other universe occupies different dimensions it is hidden from our perception. This raised the question of how two such universes could interact.

Steinhardt and Turok concluded that if the tenth spatial dimension separating two such universes contracted to zero the interaction would release an enormous energy, like that released at the Big Bang; moreover, a colliding universes scenario might well answer some of the problems of orthodox cosmology's Inflationary Big Bang model. The three of them, plus Justin Khoury, one of Steinhardt's graduate students, subsequently developed the ekpyrotic universe model, named after the Greek word for "out of fire" that describes an ancient Stoic cosmology in which the universe goes through an endless cycle of fiery birth, cooling, and rebirth.

This ekpyrotic model encountered problems, and Steinhardt and Turok developed it to produce a cyclical version that has the ambitious goal of explaining "the entire history of the universe, past and future, in an efficient, unified approach".[17] They based it on three ideas:

a. the Big Bang is not the beginning of time but a transition from an earlier phase of evolution;
b. the evolution of the universe is cyclic;
c. the key events that shaped the structure of the universe occur during a phase of slow contraction of the tenth dimension before the Bang rather than an incredibly brief inflationary period of expansion after the Bang.

In constructing their mathematical model they made three assumptions. The first two are as follows.

1. M-theory is valid. In particular, observable particles in our universe—protons, electrons etc.—lie on our brane; any particles lying on the other universe-brane can interact gravitationally with particles on our brane, but not electromagnetically or any other way.

2. The two branes are attracted to each other by a spring-like force that is very weak when the two branes are thousands of Planck distances apart (still an incredibly small distance), as they are in the current phase of the universe's evolution, but which grows stronger as the branes draw closer.

Figure 5.3 illustrates the cycle.

"You Are Here" shows the current stage of the cycle (the brane on the right is a two-dimensional representation of our universe's observable three dimensions). Dynamic dark energy (quintessence) increases the rate of expansion of the universe so that, over the course of the next trillion years, all the matter and radiation are diluted exponentially until the average matter density is less than a single electron per quadrillion cubic light years of space: in effect each brane is almost a perfect vacuum and is almost perfectly flat.

At this point the interbrane attractive force takes over. As it draws the two branes together its strength increases and it halts the accelerating expansion of each brane. There is no contraction of the three large dimensions of each brane, but only of the extra tenth dimension (a line) between them. Although each brane is a near-perfect vacuum, each possesses an enormous vacuum energy. As they approach, quantum effects cause these flat branes to wrinkle before they make contact and blow themselves apart with an explosive release of energy that is the Big Bang; the two branes bounce apart, reaching maximum separation almost immediately. Since contact occurs first between the wrinkle peaks, the explosion of energy isn't precisely homogeneous: hot spots correspond to the wrinkle peaks and cold spots to the troughs. As the released hot fireball of energy in each brane expands and cools, matter condenses out of the hot spots and evolves into galactic clusters while the cold spots form the voids between.

Each brane expands at a decreasing rate, as with the Big Bang model, until its energy density is sufficiently dilute that the positive interbrane potential energy density dominates. This acts as a source of dark energy that causes the expansion of the branes to accelerate—back to where we started—and so the cycle continues.

Unlike Tolman's cycles, no matter is recycled and entropy is not increased each time; ekpyrotic cycles of trillions of years repeat themselves endlessly.

A third assumption is necessary for the model to work.

3. The branes survive the collision. This collision is a singularity in the sense that one dimension momentarily disappears, but the other dimensions exist before, during, and after the collision.

Steinhardt and Turok claim that their mathematical model has all the advantages of the orthodox Inflationary Big Bang model in that it predicts the production of elements in the ratio found today, an observable universe that is almost homogeneous but with sufficient inhomogeneities for galactic clusters to form by

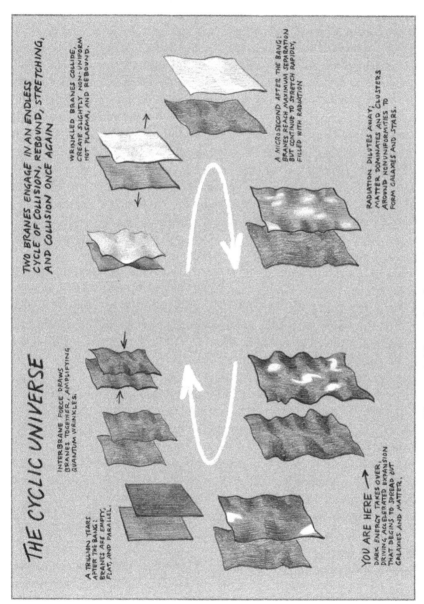

Figure 5.3 The cyclic ekpyrotic universe

the gravitational attraction of matter, and the observed ripples in an otherwise isotropic cosmic microwave background.

They argue that it has the additional advantage of parsimony: it requires the least modification to the basic Big Bang model to be consistent with observation. The ripples in matter and radiation energy don't arise from an inflationary add-on, which invokes an arbitrary cosmological constant, but are imprinted on the Big Bang energy release by the brane rippling as the two cold, empty universe-branes approach each other prior to making contact. Dark energy isn't the unexplained reappearance of the cosmological constant with a considerably smaller value; rather, as a dynamically evolving quintessence, it has an integral role throughout each cycle. There is no singularity problem because density and temperature don't reach infinity at the transition from brane collision to Big Bang.

Moreover, the model answers the question of how time and space can begin at the Big Bang. They don't. From the perspective of our own three space dimensions and one time dimension they *appear* to begin but, in the megaverse of 10 spatial dimensions in which our observable three-dimensional universe brane is embedded, space is infinite and time is continuous: the cycles continue forever.

The response to this apparently problem-solving hypothesis is revealing, particularly since Steinhardt had developed an early inflationary model.* At one conference Andrei Linde, a founder of the inflationary conjecture that the ekpyrotic model challenges, drew a caricature of a key U-shaped graph used by Turok and slashed through the U. At the USA's National Academy of Sciences conference in 2006 Alan Guth, another founder of the inflation conjecture, responded to Turok's presentation by showing a slide of a monkey.[18] This perhaps falls somewhat short of reasoned debate.

Arguing that the cyclic ekpyrotic universe model is flawed, other theorists claim that the third assumption is invalid: when the branes come into contact the extra dimension separating the two branes goes from vanishingly small to zero, a singularity *is* produced, and the laws of physics break down as they do in the basic Big Bang model.

Steinhardt counterclaims that no singularity occurs because of special conditions that obtain when two branes collide, while other theorists speculate that the branes bounce apart before collision, thus avoiding a singularity.[19]

It is not currently possible on mathematical or any other grounds to establish who is right, just as it is not currently possible to establish whether or not there is a singularity in the basic Big Bang model.

The cyclic ekpyrotic universe model needs to answer five questions if it is to meet its ambitious goal of explaining the entire history of the universe, past and future, in an efficient, unified approach.

First, does it conserve energy? The matter and radiation energy produced by the collision of the two branes—one of which makes up the universe we see

*   See p. 28.

now—aren't converted into anything. At the end of the cycle they remain in their brane, albeit stretched to such extreme dilution that their constituents are beyond their mutual contact horizons. In the next cycle the brane collision produces new mass-energy that is exponentially greater than that produced in the previous cycle.

According to Steinhardt and Turok this doesn't violate the Principle of Conservation of Energy. Like other competing conjectures, apart from plasma cosmology, the cyclic ekpyrotic universe model claims to run on gravity: the energy needed to continually produce new matter and radiation energy, plus the positive kinetic energy to expand this almost infinitely dense matter against its constituents' mutual gravitational attraction, comes from the negative gravitational energy field that increases with each cycle. I questioned the validity of this argument when considering the net zero energy of the universe claim in Chapter 4.* It depends on the assumption that the cosmos was created from nothing; it also assumes that gravitation is a source of infinite energy.

Second, does the cyclic ekpyrotic universe model describe an eternal universe? It *appears* to do so provided that two further assumptions are made:

4. the three spatial dimensions we observe can stretch, halt, and stretch further in the next cycle, with no distance limit; and
5. there are no limits to the number of cycles.

The equations allow these assumptions. However, this means extrapolating evidence-based physics theory not only beyond what is testable but also to an infinite extent; they take the model out of the realm of science and into philosophical conjecture.

Moreover, the fourth and fifth assumptions raise the question of what happens if we run the cycles back to the point from which the stretching began. In 2004 Steinhardt and Turok suggested that "the most likely story is that cycling was preceded by some singular beginning",[20] although Steinhardt later said that he would take back the words "most likely" because the arguments he had used were flawed; whether or not the model is truly eternal is an open question.[21] However, it cannot be eternal if the universe increases in size with each cycle because running the clock back reaches the point when the stretching was infinitesimally small, which would produce an initial singularity.

If there was a singular beginning, Steinhardt and Turok are considering an effect called "tunnelling from nothing", a way to quantum create space, time, matter, and energy all in one go.[22] However, this depends on the prior existence of a quantum field, which is not nothing.

The third question is what is the nature of the inter-brane spring-like force that acts along the hypothesized tenth dimension and how can it be tested? We have no information on this.

---

*   See page 46.

A fourth set of questions is why should two different universe-branes be so close to each other and be parallel? This is mathematically required by M-theory, where the tenth spatial dimension is a line joining two branes whose existence is necessary for particles with different spin properties to exist. In effect, the two universe-branes share this tenth dimension. This implies that universes necessarily come in pairs. But why should this be so other than to make M-theory consistent with the evidence that particles with different spin properties exist? In principle there is no limit to the length of this tenth dimensional line; in practice it cannot be more than 1 millimetre otherwise M-theory produces gravitational effects inconsistent with what we observe. Witten and Horava suggested that, for an M-theory consistent with observation, the extra dimension should be about 10,000 Planck lengths (10^{-28} cm), and this is the length Steinhardt and Turok took as a working hypothesis. They maintain that the two universe-branes are parallel because bending into each other would cost much energy and that repeated cycling maintains the alignment.[23]

Hence the answers to this fourth set of questions depends on the fifth question: is the underpinning M-theory valid? I shall consider this in the section after next.

Steinhardt and Turok have since claimed that their model doesn't rely on M-theory. The model works if you remove the six curled-up spatial dimensions too small to be observed and replace the tenth spatial dimension (the distance between the branes in M-theory) by a scalar field that plays the same role. This results in a mathematical model no more exotic than inflation models.[24] However, it is also no more empirically testable than the many inflation models based on a similarly arbitrary scalar field.

String theory's landscape of possibilities

Another conjecture arose from string theory. Leonard Susskind asserts that there is no reason to constrain the fundamental energy strings to vibrate only in ways that produce those particles and forces we observe. In myriad different universes the strings vibrate in different ways to produce myriad different particles and forces and, consequently, myriad different laws of physics and different cosmological constants, and so on. Susskind called this string theory's "landscape of possibilities".[25]*

This answers the problem of the fine-tuning of cosmological parameters: why did our universe emerge the way it is when it could be otherwise? String theory's landscape of possibilities means that there is nothing special about our universe. We just happen to be in a universe in which the strings are vibrating in such a way

---

*   Susskind called it the "megaverse", but since I have used this term to describe superstring theory's larger universe of 10 spatial dimensions, I shall label Susskind's version as a speculation about the cosmos because it invokes a myriad of such megaverses.

as to lead to the laws of physics and the particles that we observe; in myriad other universes they are different.

Problems with string theory

A theory that unifies particles and fundamental forces, including gravity, and quantum and relativity theories is the holy grail of physics. String theory's claim to do just that by replacing the 61 "fundamental" particles of the Standard Model of Particle Physics, and incorporate gravity, by a string of energy has attracted many of the brightest minds in theoretical physics, not only because of its undoubted conceptual appeal but also because of its mathematical elegance. It seemed to herald a new era in physics—one Harvard string conference was called the Postmodern Physics Seminar—in which its proponents are so convinced it must be true that some dispense with the criterion of experimental or observational testability and seek only mathematical proof.

But this raises a question we have encountered before: do its various mathematical models—which can readily employ fewer or more dimensions than the three dimensions of space and one of time that we perceive—constitute a scientific theory? According to many respected theorists—including Nobel physics laureate Sheldon Glashow, mathematical comologist Sir Roger Penrose, and theoretical physicist and former string theorist Lee Smolin—they do not. The grounds for this view are as follows.

Inadequacy of theory

The initial versions of string theory required 25 dimensions of space, a particle that travels faster than light, and particles that cannot be brought to rest.[26] These differences from the world we observe may reasonably be called a significant problem for the theory.

In the face of scepticism from the theoretical physics establishment its pioneers developed string theory through the 1970s until 1984, when John Schwarz and Michael Green convinced Ed Witten, a leading mathematical physicist, that a string theory using nine dimensions of space plus supersymmetry—hence "superstring"—was a good candidate for a unified theory of everything. Superstring theory suddenly became the hottest game in theoretical physics.

However, five mathematically consistent string theories emerged, each of which posited the existence of ten dimensions: one of time, the three spatial dimensions we observe, plus six other curled-up spatial dimensions that are too small to be observed. But certain symmetries meant that the string theories couldn't explain the existence of matter particles, like electrons and neutrinos, with left- or right-handed spin properties (called chirality) as required by quantum theory. Moreover, five different theories meant that something was wrong. String theory went out of fashion.

Witten "solved" this second problem when he proposed in 1995 that the five distinct theories could be united by a deeper M-theory if an extra eleventh dimension were added. This allows strings to expand into a brane, which can have any number of dimensions.* No symmetry condition exists on a boundary, which enables a universe having particles with left- and right-handed spin properties to exist on a brane. String theory came back into fashion.

However, neither Witten nor anyone else has yet formulated such a deeper M-theory. As João Magueijo, then physics professor at Imperial College, London put it with typical bluntness, "M-people say [that all of these string and membrane theories have become unified in a single construction, M-theory] with such religious fervour that it is too often not noted that there is no M-theory. It's just an expression referring to a hypothetical theory that no one actually knows how to set up." [27] Even Nobel laureate, leading string theorist, and Witten's former mentor, David Gross concedes "we're still so far from a true understanding of what string theory is".[28]

It faced a third significant problem in 1998 when most cosmologists concluded on their interpretation of the dimness of highly redshifted Type 1a supernovae that the universe began an accelerating expansion after about 10 billion years, and this required a positive cosmological constant to be introduced into the orthodox mathematical model.† Not only had the revised superstring theories not predicted this, but one of the few things they concluded at the time was that the cosmological constant could only be zero or negative. Witten admitted in 2001, "I don't know any clear-cut way to get de Sitter space [a universe with a positive cosmological constant] from string theory or M-theory."[29]

A group of Stanford theoreticians "solved" this third problem in early 2003 when they produced yet another version in which, among other things, they theoretically wrapped antibranes round the six unobservable curled-up dimensions and chose values of parameters to produce a positive cosmological constant.

However, the result of this and other work implies the existence of 10^{500}, if not an infinity of, string theories. Susskind conceded that "One might argue that the hope that a mathematically unique solution will emerge [from this landscape of string theories] is as faith-based as ID [Intelligent Design]."[30] It places the conjectured underlying M-theory in the realm of belief, not science. As for superstring theory, if there are infinite versions then no one version can ever be falsified, thus failing the generally accepted Popper test of what constitutes a scientific hypothesis.

Writing in 2003, Dan Friedan, a string theorist since 1985, concluded "String theory, as it stands, cannot give any definite explanations of existing knowledge of the real world and cannot make any definite predictions. The reliability of string theory cannot be evaluated, much less established. String theory, as it stands, has no credibility as a candidate theory of physics."[31]

---

*   A point particle is considered a brane of zero dimension, a string as a brane of one dimension, a membrane is a two-dimensional brane, and so on.
†   See page 42.

According to Smolin, string theory's "search for a single, unique, unified theory of nature has led to the conjecture of an infinite number of theories, none of which can be written down in any detail. And if consistent, they lead to an infinite number of possible universes. On top of this, all the versions we can study in any detail disagree with observation....Those who believe the conjectures find themselves in a very different intellectual universe from those who insist on believing only what the actual evidence supports."[32]

These different universes are separated by more than a difference of view over the need for empirical support. According to both Smolin and Peter Woit,[33] string theorists' domination of committees deciding academic appointments and grants in theoretical physics in the USA has made it very difficult for alternative approaches to obtain funds. They paint a picture of a cult that engages in dubious practices to suppress dissenting opinions from other physicists. The bitterness of the dispute between string theorists and its critics is exemplified by Harvard assistant professor Luboš Motl's denigration in an Amazon.com customer review of the book Smolin published in 2006, *The Trouble with Physics*. Woit countered this in his own review, alleging that Motl's review was dishonest and accusing Motl of offering $20 to anyone successfully posting a one-star review of his own book. Amazon deleted both these reviews after a week.

Lack of empirical support
A theory that has been under development by some of the best minds in physics for more than thirty years might be expected to have achieved significant empirical support. However, each string is 100 billion billion times smaller than the protons in the nucleus of an atom. Put another way, if an atom were scaled up to the size of the solar system, a string would be the size of a house. This means that there is no foreseeable way in which we can detect strings.

Nonetheless, proponents of string theory like Brian Greene believe—the word "believe" occurs in their publications and interviews far more frequently than you might expect in scientific literature—that its predictions can be validated. One requirement of superstring theories is supersymmetry, which says that for every subatomic particle we're familiar with, like an electron or a proton, there should also be a much heavier partner called a "sparticle". But supersymmetry does not depend on string theory—other hypotheses like the minimal supersymmetric extension of the Standard Model and loop quantum gravity either require or are compatible with it[34]—and so it is not a unique prediction whose confirmation would validate string theory. Moreover, no one has ever detected a sparticle. Although some researchers are hopeful, it seems unlikely that even the revamped Large Hadron Collider, which began much higher energy operations in 2015, will be able to do so.

Superstring's pivotal prediction concerning the origin of our universe is that there exist other dimensions with which we cannot communicate. Greene believes that this prediction can be demonstrated, if not validated, using another superstring prediction about gravitons, the massless particles hypothesized to transmit the force of gravity. Superstring theories claim that gravity is so weak in our universe compared with the other forces of nature because the strings of which gravitons consist are closed loops that are not confined to the brane in which our universe of three observable spatial dimensions exists: a graviton can move into other dimensions. Hence, if a particle detector observed a graviton that suddenly disappeared, this would provide experimental support for string theory's prediction of extra dimensions. However, no one has yet detected a graviton, still less a graviton that suddenly disappears.

Hence we must conclude that there is no foreseeable way to test superstring theorists' claims, and their principal claim is almost certainly untestable.

While I find the idea that all energy and matter ultimately consists of strings of energy intuitively more appealing than 61 fundmental particles, at this stage it is no more than an idea that has gained various mathematical expressions. In future I will use the term string conjecture or string "theory" to make clear that this idea does not meet the principal criterion for a scientific theory as currently understood.

Universe defined

As we have seen, the word "universe" is used nowadays to mean very different things. To avoid misunderstandings, I shall define what I mean by the term and related terms.

universe All the matter and energy that exists in the one dimension of time and the three dimensions of space perceived by our senses.

observable universe That part of the universe containing matter capable of being detected by astronomical observation. According to current orthodox cosmology it is circumscribed by the speed of light and the time since matter and radiation decoupled some 380,000 years after the universe came into existence in the Big Bang.

megaverse A speculated higher dimensional universe in which our universe of three spatial dimensions is embedded. Some speculations have many megaverses comprising the cosmos.

cosmos All that exists, which includes various speculations of dimensions other than the three dimensions of space and one of time that we perceive and also other universes with which we have no physical contact and about which we cannot obtain observable or experimental information.

multiverse A speculated cosmos that contains our universe plus multiple if not infinite other universes with which we have no physical contact and about which we cannot obtain observable or experimental information. Several different kinds of multiverse, each having different properties, have been proposed.

Conclusions

Neither any modifications to the Inflationary Big Bang model nor any competing conjectures currently provide a satisfactory scientific, as distinct from mathematical, explanation of the origin of the matter of which we consist and why the universe took the very particular form, rather than any other form, that allowed the eventual evolution of humans.

There must be an explanation—and one of these conjectures may eventually supply it—but cosmology has problems meeting the generally accepted tests that differentiate science from speculation or belief. I shall consider these in the next chapter.

CHAPTER SIX

Problems Facing Cosmology as an Explanatory Means

When generating theoretical ideas scientists should be fearlessly radical, but when it comes to interpreting evidence we should all be deeply conservative.

—PETER COLES, 2007

Faith in theory tends to trump evidence.

—GEORGE ELLIS, 2005

If an explanation is to be scientific it must be testable. In particular, generally accepted scientific criteria decree that the validity of an explanation of a thing depends on our ability to detect and collect, and preferably measure, data about it, interpret that data correctly, and draw a provisional conclusion, or hypothesis, from the data as a basis for making predictions or retrodictions that can be tested by observation or experiment and which independent testers can confirm or refute.

Cosmology is different from other branches of science, like chemistry and biology, in three respects: we have only one universe; we are part of it; and it is incomparably large. We can't experiment with it by, say, changing its temperature or pressure or initial conditions, nor can we compare it with other universes because, by definition, the universe is all that can be perceived by our senses; we can't observe it from the outside; and its size presents formidable challenges. These factors play a significant role in four interrelated problems confronting cosmology in its attempts to explain the origin and evolution of matter: practical difficulties, data interpretation, inadequate theory, and intrinsic limitations.

Practical difficulties

The practical difficulties fall into two categories: detection limits and measurement problems.

Detection limits

If relativity theory is valid, then nothing can travel faster than the speed of light. This creates the particle horizon.

> **particle horizon** We cannot be causally influenced by, or obtain information on—and therefore detect—any particles, whether of positive or zero mass, further from us than the distance travelled at light speed since time began.

If cosmology's current orthodox account is valid, then we face a second detection limit.

> **visual horizon** According to the Big Bang model, we can only see back to the time of decoupling of matter and electromagnetic radiation—currently estimated at 380,000 years after the Big Bang—because before that photons were scattered by continual interaction with the initial plasma, thereby making the universe opaque.

This means that we are unable to detect electromagnetic radiation from earlier times.

Measurement problems

The development of technology since the 1960s has enabled a far greater range of the means, and more accurate methods, of detecting cosmic phenomena. Not only can we make visual observations but we can also detect emissions across the whole electromagnetic spectrum of radio waves, microwaves, infrared light, visible light, ultraviolet light, X rays, and gamma rays. Equipment such as charge-coupled devices and fibre optics, plus the ability to place detectors in space above the Earth's atmosphere, make measurements much more refined. All this has produced a wealth of data in the last fifty years. Nonetheless, according to cosmologist George Ellis, "the underlying problem in all astronomy is determining the distance of observed objects".[1]

Crucial cosmological parameters, such as the age of celestial objects and the age and rate of expansion of the universe, depend on the determination of distances. But astronomers can't make direct distance measurements as we do of objects on Earth, nor can they use brightness as a measure of distance because, although stars and galaxies are dimmer the further away they are, such stellar objects have different intrinsic brightnesses, technically called luminosities. Accordingly, astronomers calculate the distance of nearby stars by parallax, basically local trigonometry by which they measure the angles subtended by a star from different positions of the Earth as it orbits the Sun. They measure more distant objects by a series of distance indicators, most commonly "standard candles". These are objects that astronomers consider to have known luminosities.

Hence they compare a standard candle's observed luminosity with its known luminosity in order to calculate its distance from us and, consequently, the distance of a larger object, like a galaxy, of which it is a part. The most commonly used standard candles are Cepheid variables, giant yellow stars that brighten and dim in a very regular manner and whose period of variation depends on their intrinsic brightness. For the furthest objects orthodox cosmologists use the object's redshift, the shift in the wavelength of its radiation towards the red end of the spectrum.

The standard candles, however, are less standard than some announcements of distances might suggest. For example, astronomers discovered in 1956 that the Cepheid variables came in two types and are more variable than previously thought. It would hardly be scientific to assume that, as observational methods improve and the volume of data increases, no other assumptions or interpretations will be found to be false.

Moreover, a distant standard candle's observed luminosity may be dimmed by interstellar gas and dust, or hidden by the brightness of stars or galaxies in the line of sight. To make adjustments for such factors is far from easy and requires assumptions that are, at best, less than watertight.

Astronomers also discovered that not only is the Earth orbiting the Sun at 30km per sec while the Sun is orbiting the centre of our Milky Way galaxy at 220km per sec, but also our galaxy is speeding at 200km per sec towards the centre of a local cluster of galaxies, which in turn is thought to be moving at roughly the same speed, but in a different direction, towards the centre of the local supercluster.[2] Each discovery required adjustments to the assumed velocity of a celestial body away from Earth calculated by its redshift. Orthodox cosmologists use redshift to measure not only the distance of very far—and therefore very young—cosmic objects but also the expansion rate of the universe and, consequently, the age of the universe.

Age of universe estimates

One consequence of these measurement problems has been variations in estimates of the universe's age. Hubble's own first estimate was less than 500 million years.[3] Even in the 1950s astronomers calculated the age of the universe as 2 billion years, whereas radioactive dating of rocks had shown the age of the Earth to be at least 3 billion years. Allan Sandage, Hubble's successor at Mount Wilson, estimated 20 billion years, while Gérard de Vaucouleurs of the University of Texas claimed 10 billion years.

In 1994 an international team of astronomers used the Hubble Space Telescope to make the then most accurate measurement of the distant galaxy M100 and concluded that the universe was between 8 and 10 billion years old.[4]

In 2003, following observation of the whole sky by the space-based Wilkinson Microwave Anisotropy Probe (WMAP), principal investigator Charles L Bennett confidently declared that the age of the universe was 13.7 billion years within a margin of error of one per cent.[5] This unprecedentedly precise figure was accepted by most cosmologists. It was, however, revised in 2013 to 13.82 billion years following data from the Planck space telescope.[6]

A scientist would be less than wise if he or she does not anticipate that future discoveries may well overturn today's estimates when the detection and measurement of data underpinning estimates face such practical difficulties. And, moreover, when such data is subject to differing interpretations.

Data interpretation

Before Galileo used a telescope in 1610 to study the planets, you could interpret your observations of the movements of the Sun and planets as either the planets, including Earth, moving around the Sun or else the Sun and planets moving around the Earth; each theory was consistent with the data. Nearly all observers adopted the latter interpretation. They were, of course, wrong, but their religious beliefs determined their interpretation.

Much cosmology literature is reminiscent of belief-driven interpretations. Advocates of different cosmological hypotheses tend to seize on or interpret evidence to support their beliefs. This applies both to those who hold orthodox interpretations and to those who challenge them.

Furthermore, we naturally pay more attention to the conclusions drawn from observation than the assumptions—often unstated—that underpin the conclusions. In the case of cosmology such assumptions are frequently questionable.

Age of universe

The accepted conclusion by Bennett and the NASA team in 2003 that the WMAP data showed the age of the universe to be 13.7 billion years ± 1 per cent depends on a number of assumptions. One of these is the value of the Hubble constant. According to Rowan-Robinson, estimates of this constant have "continued to be a matter of raging controversy for the past thirty years".[7] Three years after Bennett's announcement, a research team led by Alceste Bonanos at the Carnegie Institution of Washington used what they claim is a more accurate measurement of the distance to the M33 galaxy to propose a 15 per cent decrease in the Hubble constant, which would put the age of the universe at 15.8 billion years,[8] confirming Rowan-Robinson's conclusion that we still need a more accurate estimate of the age of the universe.[9]

While such questioning of the generally accepted universe age comes from orthodox cosmologists, challenges have repeatedly been made to the orthodox interpretation of data and assumptions underlying age-of-the-universe estimates. For example, plasma physicist Eric Lerner claims that the time needed for the structures we see in the universe, such as the gigantic sheet-like superclusters of galaxies like the Great Wall and the huge voids between them, must have taken upwards of 100 billion years to form.[10] His interpretation of data has been challenged in turn by that stout defender of cosmology's current orthodoxy, Ned Wright.[11]

Redshifted Type 1a supernovae

As we saw in Chapter 4, the discovery that highly redshifted Type 1a supernovae—very bright but short-lived objects considered to be exploding white dwarf stars—were dimmer than expected was interpreted by orthodox cosmologists to mean that the decreasing rate of expansion of the universe after the Big Bang had changed to an increasing rate. While other studies claim to support this interpretation, they depend on interpreting all redshift as a measure of distance (see below). Moreover Type 1a supernovae brightness may be dimmed by interstellar dust, as quasi-steady state cosmology claims,* or, like the Cepheid variables, their luminosity may not be as standard as astronomers currently assume.

Apparent acceleration of the universe's expansion rate

Chapter 4 noted too that orthodox cosmologists ascribed their interpretation of this dimness to an acceleration of the universe's expansion rate caused by a mysterious dark energy. They accounted for this mathematically by reintroducing the arbitrary cosmological constant Lambda with a very different value than either the one Einstein had discarded or the one inflationists had used; they interpreted this as the zero point in the quantum mechanical field energy of the universe, notwithstanding that its calculated value is an enormous 10^{120} times greater than that consistent with their interpretation of the universe's expansion rate.

Kunz and colleagues, Rasanen, and Lieu each gave a different interpretation of the data that dispense with dark energy, while Ellis offered other interpretations, adding that "Current orthodox cosmology regards such proposals as very unappealing—but that does not prove they are incorrect."[12]

As discussed in Chapter 5, Steinhardt and Turok argue that dark energy is not an arbitrary constant but is a dynamically changing basic ingredient of the universe, quintessence,† while proponents of quasi-steady state cosmology argue that the supernovae data is equally explained by QSSC without any need for a mysterious dark energy.‡

Redshift

One the most critical interpretations of data is that redshift of itself is *always* a measure of the distance and, combined with the Hubble constant, of the recessional velocity, and therefore the age, of celestial objects. Chapter 3 noted that Halton Arp, Geoffrey Burbidge, and others challenged this interpretation.§

---

* See page 59.
† See page 60.
‡ See page 59.
§ See page 33.

The key to this conflict of interpretation is the nature of quasars, otherwise known as quasi-stellar objects or QSOs. These powerful sources of variable radio emission were detected in 1961 and identified with tiny visible objects first thought to be stars in our galaxy. However, when the spectra of their light were analysed, they showed very large redshifts. Other tiny, highly redshifted objects were then detected; these didn't emit radio waves but emitted visible light, mainly blue, that varied in periods of days, while most of them strongly emitted X-rays that varied in periods of hours compared with years to months for those emitting radio waves.

Orthodox cosmologists interpreted the very large redshifts as meaning that these quasars were extremely distant and were receding from us at speeds of up to 95 per cent of light-speed. Because of the time it takes for their light to reach us, what we are seeing now are these quasars when they—and the universe—were very young. The problem was to explain that such huge distances meant their electromagnetic emissions were the equivalent of those from a thousand galaxies combined, but their small period of variation in emission meant that these sources were very small; moreover, only about one twentieth of them were emitting radio waves, while the majority were emitting visible light and X-rays, and some gamma rays as well.

By the 1980s orthodox cosmologists reached a reasonable consensus interpretation. These huge optical and X-ray emissions were caused by a disc of very hot gas and dust spinning round, and being sucked into, a massive black hole at the centre of a very young galaxy, while the relativistic radio emissions were due to jets ejected along the axis of rotation, as observed in star formation. It is simply our viewing angle that differentiates strong radio sources from strong optical and X-ray sources.[13]

Arp, Burbidge, and others claimed, however, that their studies of high-redshift quasars show very many aligned either side of nearby active galaxies and, in some cases, a physical link with such galaxies; moreover analyses show an increase in brightness and decrease in redshift with increasing distance from the parent galaxy. They interpreted these data to suggest that such quasars are small proto-galaxies emitted at near light-speed from near black holes at the nucleus of existing active galaxies, and these evolve into galaxies, thus becoming brighter, as they increase their distance while decelerating from the parent galaxy.

In 2007 Royal Astronomical Society President Michael Rowan-Robinson dismissed this out of hand: "The redshift anomaly story was over thirty years ago. Some of these associations were chance, others were due to gravitational lensing."[14] But Burbidge maintained that it is the accumulation of data over the last thirty years that reinforces their interpretation.

> Unless this can be explained by arguing that dark matter near the bright galaxies is giving rise through gravitational lensing to the brightening of faint QSOs [quasars] which are far away—and no satisfactory gravitational

model of this kind has been made—most QSOs are not very far away. The only other way out for the conventional people is to conclude that all of the configurations are accidental and/or that the statistics are wrong.... The data continue to accumulate. In 2005 Margaret Burbidge and her colleagues showed that an X-ray-emitting QSO with a redshift of 2.1 lies only 8 arc seconds from a nearby active galaxy NGC 7619. The likelihood that this is accidental is one part in 10,000. And there are many, many other cases in the literature.[15]

But, according to Arp, not as many as there should be. He says he finds it difficult to get his papers published in scientific journals. In 1998 he reproduced some of his exchanges with anonymous referees, whom he says are "manipulative, sly, insulting, arrogant, and above all angry".[16] This view was echoed by Burbidge, who claimed that cosmology is dominated not by observational scientists but by mathematical theorists who give only secondary attention to the data. "Mostly our views on cosmology are ignored, and over the last twenty years there have been many successful attempts to stop us giving invited papers and doing other things to stop us at cosmological conferences. Probably because when we are given a platform we are quite convincing."[17]

These are reputable scientists. Arp is perhaps the most experienced extra-galactic observational astronomer. For 29 years he was on the staff of the Palomar Observatory and then joined the prestigious Max Planck Institute in Germany; his awards include the American Astronomical Society's Helen B. Warner Prize. Burbidge was an astrophysics professor at the University of California San Diego and was awarded the Royal Astronomical Society's Gold Medal in 2005. Their complaints about suppression of alternatives to the orthodox view were echoed by Richard Lieu.*

I found one thing puzzling about this alternative interpretation of some redshifts: if protogalaxies are ejected at near light speed from the centres of existing active galaxies, why are they not ejected randomly, with as many ejected towards us as away from us, producing high blueshifts as well as high redshifts?

Arp pointed to his 2008 analysis of a cluster of 14 high-redshift quasars around the AM 2330-284 galaxy in the Two-degree-Field Galaxy Redshift Survey.[18] He claims that their redshifts are in a narrow range greater and lesser than the cosmic recessional velocity of the galaxy; this is consistent with them being ejected randomly from the parent galaxy at velocities of plus and minus 1,800km per second. This is much less than light speed because the mass of each ejected object increases with time, and to conserve momentum it must reduce velocity.[19] Arp is basing this interpretation on the Hoyle-Narlikar variable mass hypothesis of conformal gravitation theory that says newly created matter starts off with zero mass that grows with time through interaction with the rest of the matter in the universe.[20]

---

*   See page 45.

Arp may be wrong, as may Burbidge and others who also challenged the orthodox redshift interpretation. However, until orthodox cosmologists engage in reasoned debate with Arp and other reputable scientists with different interpretations of the data rather than ignoring or denigrating them, a question mark must hang over the orthodox interpretation of all redshifts and, consequently, of the Big Bang model.

Ripples in cosmic microwave background

The attitude of belief rather than reason is reflected in the language sometimes used to announce results. For instance, when chief investigator George Smoot reported in 1992 that the Cosmic Background Explorer (COBE) space-based detector had found lengthy density ripples of 0.001 per cent in the cosmic microwave background (CMB), he declared it was "like seeing the face of God". Stephen Hawking said that COBE had made "the discovery of the century, if not of all time".[21] "They have found the Holy Grail of Cosmology," claimed Chicago astronomer Michael Turner when the announcement was made at the American Physical Society in Washington.[22]

The reason for the euphoria was that most cosmologists interpreted the COBE data as proof of orthodoxy's Inflationary Big Bang model on the grounds that the ripples reflected inhomogeneities in the plasma at the time photons decoupled from the plasma. They assumed these inhomogeneities had been caused by the inflationary expansion of quantum fluctuations at the Big Bang and were the seeds of the structure of galaxies, galactic clusters, and superclusters separated by vast voids. Smoot and his colleague, John Mather of the Goddard Flight Center, shared the 2006 Nobel Prize in Physics for detecting the ripples.

An editorial in *Nature* the week following the euphoria took a more sober stance:

> ...the simple conclusion, that the data so far authenticated are consistent with the doctrine of the Big Bang, has been amplified in newspapers and broadcasts into proof that "we now know" how the Universe began. This is a cause of some alarm.

The article went on to mention the problems of how such ripples could account for the structures that we see in the universe today and of orthodox cosmology's explanations, commenting that "for neither dark matter nor inflation is there true independent support".[23]

Cosmology literature, moreover, rarely mentions that these ripples in the CMB are also consistent with other models. Ellis maintains that spherically symmetric inhomogeneous models of the universe probably can produce similar ripples. quasi-steady state cosmology says the microwave background radiation arises from thermalization of the energy generated by the production of helium in stars and interprets the ripples as localized effects.[24] Plasma cosmology's eternal universe model gives a similar explanation of the background radiation energy and interprets the ripples as due to the imperfect isotropization of this energy by a thicket of dense, magnetically confined plasma filaments that pervade the inter-galactic medium.* The cyclic ekpyrotic universe model interprets them as being imprinted on a Big Bang energy release by the prior rippling caused as two cold, near-empty universe-branes approach each other.†

One cosmology professor assured me that the CMB ripples made "almost certain" the correctness of the current orthodox model from one second after the Hot Big Bang. When he dismissed other interpretations of the CMB ripples, I asked him for his reasons. He admitted that he hadn't read any of the relevant papers; he simply didn't have time to read everything that was published, and that was true for most cosmologists. In which case, how could they dismiss alternative interpretations if they hadn't read them? There were, he explained, probably no more than half a dozen cosmologists who set the agenda and established the view of what was the best fit for the data.

He is a sincere, honest man and it simply didn't occur to him to ask whether these half dozen cosmologists have a vested interest—consciously or subcon-sciously—in interpreting data to fit only the model on which they have worked for most of their lives and in which they believe deeply. I was left with the impression of a bishop of orthodoxy happily deferring to cosmology's college of cardinals.

Exaggerated claims

Belief rather than reason produces exaggerated claims. "NASA Satellite Glimpses Universe's First Trillionth of a Second" was the heading of the news release issued by NASA on 16 March 2006 to report the three-year findings of the space-based Wilkinson Microwave Anisotropy Probe (WMAP), launched a decade after COBE with much more sensitive instruments. It quoted principal investigator Charles Bennett saying "We can now distinguish between different versions of what happened within the first trillionth of a second of the universe."[25]

Sadly, it is not only politicians who engage in spin. Bennett is an orthodox cosmologist and so presumably believes that we cannot see further back in time than the visual horizon, when radiation decoupled from matter around 380,000 years after the Big Bang.‡

Closer examination of the data shows that WMAP instruments actually recorded temperature variations and polarization of the microwave background radiation, which the investigators *assume* to have originated 380,000 years after the Big Bang. Based on a nest of other assumptions, the WMAP scientists are inferring that this polarization was caused by conjectured events during the first trillionth of a second. Making unwarranted assertions of fact falls somewhat short of best scientific practice.

---

*   See page 61.

†   See page 62.

‡   See page 75.

WMAP data

The WMAP project scientists interpreted their data as further validation of the orthodox Inflationary Big Bang model, saying that the pattern of hot and cold spots was consistent with the simplest inflation conjecture's predictions.

In 2005, however, analyses of the WMAP raw data by other cosmologists raised significant doubts. Richard Lieu and Jonathan Mittaz of the University of Alabama in Huntsville found the WMAP data indicate a slightly "super critical" universe where there is more matter—and hence a stronger gravitational field—than the WMAP project scientists' interpretation, and that this presented serious problems for the inflationary conjecture.[26] They also found a lack of what is called gravitational lensing effects in the cosmic microwave background that the standard Big Bang model predicts. According to them, cool spots in the microwave background are too uniform in size to have travelled across almost 14 billion light years from the edges of the universe to Earth. The consequences suggest several alternative explanations. The most conservative is that the cosmological parameters of the orthodox model, including the Hubble constant, are wrong. The most contentious possibility is that the microwave background radiation itself isn't a remnant of a Big Bang but was created by a different process, a local process so close to Earth that the radiation wouldn't go near any gravitational lenses before reaching our telescopes.

David Larson and Benjamin Wandelt of the University of Illinois found a statistically significant deviation from the Gaussian distribution of hot and cold spots predicted by inflation.[27] Kate Land and João Magueijo of Imperial College London analysed the microwave background radiation by its three components and found that two of them—the quadrupole and octupole—showed the orientation of hot and cold spots were aligned along what Magueijo dubbed "the axis of evil", which contradicts inflation's prediction of random orientation. Magueijo suggests they could be due to the universe being shaped like a slab, or a bagel, or that the universe is rotating, each of which conflicts with the assumptions of isotropy and omnicentricity on which the orthodox model is based.[28]

Most cosmologists, however, interpret these non-random orientations as ordinary statistical deviations from the mean of over a hundred measures.[29]

Whether of not these and other scientists who found problems with the WMAP raw data are right—and the later Planck data strongly suggest they are (see next section)—their willingness to examine the data with an open mind and consider alternative interpretations contrasts with the project's own scientists who seem to draw only conclusions supporting the hypothesis they are investigating.

Planck telescope's confirmation of contradictory evidence

When announcing in March 2013 the initial 15 months of data from the European Space Agency's Planck telescope, project scientist Jan Tauber, following his WMAP counterpart, said "we see an almost perfect fit to the standard model of cosmology", but then contradicted himself by continuing "but with intriguing features that force us to rethink some of our basic assumptions".[30]

The Planck telescope observed the cosmic microwave background (CMB) with greater resolution and sensitivity than ever before. It revealed many features that challenge the orthodox Inflationary Big Bang model. These include not only revisions of the age of the universe and the proportions of the conjectured dark matter and dark energy, but also a cold spot that extends over a much wider patch of sky than expected and fluctuations in the CMB temperature that do not match those predicted. The data confirmed that the orientation of hot and cold spots could not be dismissed as a statistical deviation but were indeed aligned along an axis, refuting the orthodox model's prediction that they should be isotropic, that is, broadly similar in any direction we look.

Data selectivity

Selectivity, not just of interpretation but also of data, occurs when scientists seek to justify a hypothesis rather than objectively examine the evidence.

I have mentioned string theorist Leonard Susskind's landscape of possibilities idea that gives rise to a multiverse.* In reviewing Susskind's *The Cosmic Landscape*, Ellis points out that this multiverse is hypothesized to have started out by quantum tunnelling, resulting in a spatially homogeneous and isotropic universe with negative spatial curvature, and hence with a value of Omega (Ω) less than 1.† The best observationally determined value is $\Omega = 1.02 \pm 0.02$. Given statistical uncertainties, this doesn't absolutely conflict with Susskind's conjecture, but it certainly doesn't support it. However, Susskind doesn't even discuss this unfavourable data. Ellis concludes that this is "a symptom of some present-day cosmology, where faith in theory tends to trump evidence".[31]

Law of Data Interpretation

I have given space to alternative interpretations of astronomical data by reputable scientists because such alternatives are rarely heard outside the cosmological community and, in many cases, even within it. This community is a human institution in which apprentice cosmologists are taught by, and have their publications, grants, and careers decided by, proponents of the current orthodoxy. The pressures to conform in any human institution are considerable.

Moreover, cosmology requires an investment of time over the many years it may take between proposing a particular investigation, through submitting a funding application, persuading funding bodies to support it and, say, NASA to launch a satellite, and to analysing and interpreting the raw data—18 years in the case of the COBE satellite. Similarly, particle physicists may invest decades of their career undertaking an investigation that requires the cooperation of several governments to fund a particle accelerator. Scientists are only human and naturally want to see that such personal investment has been worthwhile.

---

* See p. 68.

† See p. 24 for an explanation of Omega and its relationship to the geometry of the universe.

Comparing announced results with the rarer, balanced views of cosmologists like Ellis and Rowan-Robinson suggests that a law may be operating here.

> **The Law of Data Interpretation** The degree to which a scientist departs from an objective interpretation of the data from his investigation is a function of four factors: his determination to validate a hypothesis or confirm a theory, the time the investigation has occupied his life, the degree of his emotional investment in the project, and his career need to publish a significant paper or safeguard his reputation.

Inadequate theory

Chapters 3 and 4 found serious problems with cosmology's Inflationary Hot Big Bang model, while Chapter 5 concluded that other conjectures competing to either modify the orthodox model or else supersede it do not yet provide a scientific account of the origin of the universe: their main claims are untested and most are almost certainly untestable.

I shall consider here some deeper theoretical problems that underlie the orthodox cosmological model and most alternatives.

Incompleteness of quantum and relativity theories

Quantum theory and relativity theory underpin the Big Bang model and its competing ideas. Each has been extremely successful in making predictions that have been verified by observation and experiment within its own realm: the extremely small—subatomic—for quantum theory and the extremely large—stellar masses and near light-speeds—for relativity. Yet each is necessarily incomplete because each cannot explain phenomena in the other's realm.* This suggests that each is a limiting case of a deeper, more complete theory of nature.

Many attempts have been made to unite quantum and relativity theories— string "theory" and loop quantum gravity, for example—but as we have seen these

---

* For example, Wolfgang Tittel and colleagues reported that two quantum-entangled photons more than 10km apart instantaneously behaved in the same way when confronted by alternative, equally possible pathways, in stark violation of Einstein's Special Theory of Relativity that prohibits information travelling faster than light [Tittel, W, et al. (1998) "Violation of Bell Inequalities by Photons More Than 10km Apart" *Physical Review Letters* 81: 17, 3563–3566]. But it is not only in the subatomic realm that relativity theory does not apply. Rainer Blatt and colleages [Riebe, M, et al. (2004) "Deterministic Quantum Teleportation with Atoms" *Nature* 429: 6993, 734–737] and D J Wineland and colleagues [Barrett, M D, et al. (2004) "Deterministic Quantum Teleportation of Atomic Qubits" *Nature* 429: 6993, 737–739] each reported the near-instantaneous transmission of the quantum states of calcium ions and beryllium ions respectively.

have not yet been scientifically tested. Moreover, each raises its own problems: the suitability of general relativity for predicting or even describing the universe as a whole, and the nature of reality in quantum theory.

Suitability of general relativity

Astronomers and physicists adopted Einstein's General Theory of Relativity because it incorporated gravity into the Special Theory of Relativity, thereby describing all known forces, and it explained the anomalous precession of the closest approach of the planet Mercury to the Sun, which Newtonian mechanics had failed to do. But is the theory fit for purpose to predict or even describe the universe as a whole?

The General Theory isn't an equation that predicts a unique outcome for specific initial conditions; rather it comprises a set of ten field equations in which arbitrary scalar fields, parameters, and values of those parameters can be inserted. It allowed Einstein to choose a parameter, and a value of that parameter, to produce a static universe, various cosmologists to choose a notional scalar field with various values to produce various inflationary universes, Hoyle and colleagues to choose an identical scalar field but with different parameters and values to produce a quasi-steady-state universe, and Steinhardt and Turok to choose a different scalar field, parameters, and values to produce a cyclic universe in which dark energy is not a constant but dynamically evolves.

All claimed that their version of the universe is consistent with observational data, although there is no evidence for the existence of an inflationary, creation, or dark energy scalar field, unlike, for example, the evidence underpinning an electric potential scalar field. In fact, as Ellis points out,* it is perfectly possible to run the mathematics backwards to choose parameters that produce the desired outcome. To paraphrase Humpty Dumpty in *Through the Looking Glass* by Lewis Carroll, when I choose scalar fields, parameters and their values, the equations of general relativity mean just what I choose them to mean—neither more nor less. (Carroll, of course, was a mathematician.)

The set of equations also allowed the proponents of inflation, cyclic, quasi-steady state, and other models to draw on gravity in order to provide their versions of the universe with a limitless source of energy to create matter and power its expansion against the immense gravitational attraction of the super-dense matter thereby created.† While limitless sources of energy—and unicorns and gods—may exist in the conceptual world of which mathematics is a part, there must be some doubt as to whether unicorns or gods or limitless sources of energy exist in the physical world.

---

* See p. 38.
† Newton's gravitational equation similarly allows this provided that the separation between point masses is boundless.

The reality of the quantum world

The question of the extent to which logically consistent mathematical forms represent the real world arises again when we consider quantum theory, one of the most difficult and technical concepts in science.

The equations and principles that constitute quantum theory have been outstandingly successful in making experimentally tested predictions, none more so than the number and atomic structure of the elements that can exist, and how their atoms bond to form molecules, thus providing the theoretical foundation of chemistry. Nonetheless, quantum theory disturbed those of its founders, like Einstein, Erwin Schrödinger, and Louis de Broglie, who subscribed to the philosophical view of realism by which science explains reality as it is in our absence. The reason is that the world that quantum theory describes is paradoxical, intrinsically uncertain, dependent on measurement, and nondeterministic, that is, it allows effects with no cause.

The theory has generated many interpretations of what its equations mean in reality. For example, quantum theory says that an electron behaves as both a particle and a wave. So does light. The wave is not a wave of physical stuff, like an ocean wave, but a wave of information. An analogy is a crime wave, which tells us where there is the greatest probability of a particular crime being committed. A quantum wave tells us the probability of where we can expect the particle to be and the probability of its possessing properties such as rotation or energy. It is non-localized: it is infinite in extent and contains all possible existence states for that quantum entity. Thus an electron potentially can be anywhere.

According to the standard (otherwise known as the Bohr, or Copenhagen) interpretation, what cannot be measured does not have a physical existence. Only when it is measured does the wave function collapse into the probability of a physical particle having a specific position, momentum, and energy. However, we cannot measure both a particle's precise position and its precise momentum at the same time or its precise energy at a precise time.

The standard interpretation says that no independent reality in the ordinary physical sense can be ascribed to either the quantum phenomenon or to the agency of its measurement. Many supporters of the standard interpretation, like Nobel laureate Eugene Wigner, take the view that the measurement requires the presence of a conscious observer. This doesn't present a fundamental problem when measuring, say, the scattering pattern of a beam of light after hitting a plate with slits in it. It does raise problems, however, when considering the reality of the universe or of the electrons, baryons, and photons that materialized from a Big Bang.

The distinguished theoretical physicist John Wheeler, a collaborator with Einstein in his later years, took the conscious-dependent view of physical reality to its logical conclusion. He argued that the universe depends for its existence on the presence of conscious observers to make it real, not only today but also retrospectively to the Big Bang. The universe existed in a kind of indeterminate probabilistic ghost state until conscious beings observed it, thus collapsing the wave function for the entire universe and bringing it into physical existence.

Different interpretations, like Everett's quantum multiverse (which I shall consider in the next chapter) have sought to avoid this kind of problem, but have generated others.

Infinities in a physical cosmos

Quantum theory also has problems with infinities. The quantum mechanical description of a field, like an electromagnetic field, ascribes values to every point in space. This produces an infinite number of variables, even in a finite volume; each of these has a value that, according to quantum theory, can fluctuate uncontrollably. According to Smolin, this leads to the prediction of infinite numbers for the probability of some event happening or the strength of some force.

In addition to allowing a limitless source of energy, the General Theory of Relativity has other problems with infinities. As we saw in Chapter 4,* inside a black hole the density of matter and the strength of the gravitational field become infinite, while the same thing is generally believed to occur when the expansion of the universe is run backwards to the Big Bang. However, when density becomes infinite the equations of general relativity break down.

As a consequence of the simplifying assumptions made to solve the equations of general relativity, both the flat (the orthodox model) and the hyperbolic universe models are necessarily infinite in extent: if either came to an edge it would contradict the assumption that the universe looks the same from all points.† Most cosmologists do not see this as a problem. As Max Tegmark puts it "How could space not be infinite?"[32]

Some proposals claim that, in a multiverse that includes all possible universes, it necessarily follows that there is an infinite number of universes.

Infinity, however, is quite different from a very large number. David Hilbert, who laid much of the foundation of twentieth century mathematics, states

> Our principal result is that the infinite is nowhere to be found in reality. It neither exists in nature nor provides a legitimate basis for rational thought.[33]

If Hilbert is right, hypotheses that employ infinities to describe the physical world are invalid.

If Hilbert is wrong to the degree that the mathematical construct of infinity does have a correspondence in the physical world but that, as finite beings, we are unable to sense it, then we can never validate or disprove any such hypothesis by the scientific method.

---

* See page 39.

† The same does not follow for a spherical universe: a perfect sphere looks the same from all points on its surface.

If Hilbert is entirely wrong, and it is only our current inability to devise empirical tests for hypotheses involving infinities that is limiting us, then such hypotheses are still problematic. For example, several alternative cosmological conjectures considered in Chapter 5 claim that the universe is eternal. In such a case Ellis points out that, if an event happens at any point in time, any such conjecture needs to explain why it did not occur before that time since there was infinite previous time for it to occur.[34]

I shall consider the Principle of Increasing Entropy when examining the evolution of matter. Suffice it to say here that this physical law maintains that, during any process in an isolated system, disorder increases until a state of equilibrium is reached. The universe is the ultimate isolated system since, by definition, it either contains all the matter and energy there is or else is disconnected from other universes in a conjectured multiverse. It follows that, if this physical principle is valid in the universe and if the universe has an infinite existence, then it would have reached its equilibrium end state an infinite time ago and we would not be alive to reflect on this question.[35]

Inadequacy of mathematics

Newton developed a new form of mathematics—calculus—that played a key role when he developed his physical laws. Theoretical physicists and cosmologists from Einstein onwards have borrowed or adapted existing mathematics—the differential geometry of four-dimensional space, gauge theory, scalar fields, and so on—in order to express and quantify their ideas about the origin and evolution of the universe. As we have seen, the mathematics of some of these ideas breaks down when we go back to the origin of the universe or fails to correspond with reality as we sense it. It may well be necessary to develop a new mathematics in order to express and quantify a complete theory of the origin and evolution of the universe. Such a theory would explain all that relativity theory explains on the large scale and all that quantum theory explains on the subatomic scale.

Intrinsic limitations of science

As we have seen, many conjectures advanced to explain the origin of the universe not only are untested but also are untestable: if we cannot detect a phenomenon, or its attributable effect on something we can detect, then we cannot test it. Any such untestable conjectures necessarily lie outside the domain of the empirical discipline that is science.

Conclusions

Despite the bullish, and sometimes triumphalist, announcements made by project scientists after analysing data from expensive and lengthy observations, cosmology faces considerable practical difficulties. It also faces problems in interpreting raw data, which include the questionability of often-unstated underlying assumptions. Consequently we do not know with confidence the value of many key parameters, like the Hubble constant and the density of the universe, and consequently the age of the universe and its expansion rate. Neither the orthodox considerably modified Big Bang model nor competing conjectures provide scientifically robust theories to explain the origin and form of the universe. Moreover, each of the two theories underpinning all cosmological models—relativity and quantum theories—is necessarily incomplete and has problems of its own.

As detection techniques improve, and interpretation and theories develop in response to new data and new thinking, the practical, interpretation, and theory limitations will be pushed back, and cosmology will provide us with a greater understanding of the origin of the universe, and hence the origin of the matter of which we are composed.

However, until cosmologists produce a new definition of science and scientific method that is acceptable to both the scientific community and the wider intellectual community, many cosmological "theories" must be classified as untestable conjectures and so remain outside the domain of science.

It may, of course, be argued that cosmology is different from other branches of science in the three respects that I listed at the beginning of this chapter and, consequently, if cosmology is constrained by conventional scientific methodology then it will have little explanatory power. This argument can be used to justify cosmologists going beyond the conventions of science in order to explain the universe. Hence in the next chapter I shall consider if cosmology's conjectures produce compelling reasoning even if they do not pass the stricter tests of science.

CHAPTER SEVEN

Reasonableness of Cosmological Conjectures

Through space the universe grasps me and swallows me up like a speck; through thought I grasp it.

—BLAISE PASCAL, 1670

Reason is natural revelation.

—JOHN LOCKE, 1690

Two questions to ask when evaluating the reasonableness of cosmological explanations that lie outside the empirical realm of science are what should be the scope of such conjectures? and how do we test their reasonableness? In the light of answers to these questions I shall examine cosmological conjectures in two areas: the origin of the universe and the form of the universe, because each is central to understanding the emergence of the matter of which we consist.

Scope of cosmological conjectures

Should cosmological conjectures be limited to the material element of the cosmos?

Many scientists are materialists, and for them this is a trivial question because they believe the material cosmos *is* all that exists, and that such things as consciousness, mind, and so on will eventually be explained by science in terms of matter and its interactions. However, such a view is itself unscientific according to Popper's criterion because it can never be falsified.

I think it reasonable to extend the scope by examining non-material things that have a direct bearing on the origin and evolution of the material universe. This raises a series of interrelated metaphysical questions; they may be viewed as different aspects of the same question, but it is helpful to separate them into three categorics, even if this division is arbitrary and permeable.

Cause of the laws of physics

Most cosmological explanations say or assume that matter behaves and evolves according to the laws of physics. Hence the fundamental question is what caused these laws to exist?

As we shall see in Chapter 28 when I examine the evolution of philosophical thinking, there is no clear answer. Even the archetypal rationalist Aristotle was compelled by following the chain of causality to conclude that the first cause must be self-causing, eternal, unchanging, without physical attributes, and hence divine.

Nature of the laws of physics

One level up from this fundamental question is what is the nature of the laws of physics? This divides into three subquestions:[1]

1. *Are they descriptive or prescriptive?*
   If they simply describe the way things are, then why does all matter and its interactions (forces) have the same properties everywhere in the observable universe? Why are all electrons identical? Why is electromagnetic force calculated the same way everywhere? If, on the other hand, they determine the way things are, then matter necessarily will be the same everywhere, assuming the laws themselves are invariable. In which case, how can theoretical laws impose themselves on matter in the universe?

2. *Do they pre-exist the universe and control its coming into being, or do they come into being with the universe, or are they co-existent with an eternal universe?*
   If the Big Bang is the start of everything, how can such laws be created in a lawless creation event? If not, how do they pre-exist the universe? If the universe is eternal, are the physical laws unchanging and eternally co-existent with the universe or do they change over the course of infinite time?

3. *Why are they expressed by mathematical relationships that, in most cases, are very simple?*
   With notable exceptions like general relativity, most physical laws are expressed by very simple equations, like the inverse square law of electromagnetic force. Why should this be so? Does mathematics describe or determine physical laws? What is the nature of mathematics?

Nature of mathematics

Cosmologist Max Tegmark of the Massachusetts Institute of Technology proposes that a mathematical structure is "an abstract, immutable entity existing outside of space and time".[2] (This conjecture implicitly denies materialism.)

Roger Penrose similarly follows Plato and argues on "powerful (but incomplete) evidence" that mathematical forms have an objective reality outside the physical world. Moreover, while only a small part of this mathematical world has any relevance to the physical world of which we are a part, the entire physical world is governed according to mathematical laws. If this is so, "then even our own physical actions would be entirely subject to such ultimate mathematical control, where 'control' might still allow for some random behaviour governed by strict probabilistic principles".[3]

The conjecture that mathematical forms exist as a transcendent reality outside the physical universe and that they cause and/or govern the universe must explain how a mathematical form can generate and/or control a material universe.

The Christian Church long ago assimilated this particular Platonic notion and gave a simple explanation: the transcendent reality is God. Consequently the vast majority of leading scientists in the West, from the first scientific revolution of the mid-sixteenth to the mid-eighteenth century up to the early twentieth century—Copernicus, Kepler, Newton, Descartes, through to Einstein—sought to discover the mathematical laws that govern our universe as ways of discovering, in Stephen Hawking's notable phrase, the mind of God.

Keith Ward, Anglican cleric and former Regius Professor of Divinity at Oxford, attempts a rational explanation by arguing that mathematical necessities are fully existent only when they are conceived of by some consciousness. For a mathematical theory of everything, that consciousness necessarily has the attributes of God, the supreme self-existent mind.[4]

The non-religious idea of transcendent mathematical forms that govern a physical universe has been taken to a logical conclusion by mystic and ecologist Duane Elgin. He proposes a "dimensional evolution" in which the universe is a living system held together through the cohering influence of a "sacred geometry of exquisite subtlety, depth of design, and elegance of purpose". This pervades the cosmos and provides the organizing framework for the orderly manifestation of our material universe as well as the organizing context through which life evolves. This sacred geometry is the creation of the "Meta-universe", which is "an unimaginably vast, incomprehensively intelligent, and infinitely creative Life-force that chose to bring our cosmos into manifest existence".[5] If this Meta-universe isn't quite the personally interventionist God of Judaeo-Christian-Islamic beliefs, it sounds remarkably like the ancient insights of Brahman and the Dao.*

Lee Smolin, on the other hand, believes that the second scientific revolution is liberating science from this essentially spiritual view of the world. "What ties together general relativity, quantum theory, natural selection, and the new sciences of complex and self-organized systems is that in different ways they describe a world that is whole unto itself, without any need of an external intelligence to serve as its inventor, organizer or external observer."[6] The new ingredients here are natural selection, complexity, and self-organizing systems theories. Since they are usually invoked to explain the emergence of life, I shall defer consideration of them until Part 2.

---

*   See Glossary for definition or page 9 for an explanation.

Tests for cosmological conjectures

If there is no foreseeable, or even possible, way to test cosmological conjectures by experiment or observation, then what tests should we use to evaluate them and determine their reasonableness compared with, say, the myth that the cosmos emerged from an egg?

The following are commonly used.

Beauty

Theoretical physicists frequently seek aesthetics in their theories and equations. As quantum theorist Paul Dirac put it, "It is more important to have beauty in one's equations than to have them fit experiment." Nobel laureate and theoretical particle physicist Steven Weinberg observed that "Time and again physicists have been guided by their sense of beauty not only in developing new theories but even in judging the validity of physical theories once they are developed."[7]

But beauty is subjective. Is π, which occurs as a constant in many equations and equals 3.141592653… (its exact value cannot be computed), beautiful? Is my hypothesis or my equation more beautiful than yours? Is the creation account given in Chapter 1 of Genesis beautiful?

If the theorists who make this claim actually mean that they have an insight that beautifully explains a set of phenomena, then that is a different thing; elsewhere I shall examine ways of knowing other than by reasoning. But I do not think beauty is a proper test for the reasonableness of a conjecture.

Parsimony

This is variously called economy, or Ockham's razor, or simplicity. Essentially it means that the least complex of several competing explanations of data is preferable.

I think it useful as a rule of thumb, but it must be employed with circumspection and only alongside other tests because it can be disputable whether other explanations better meet this test. For example, it may be argued that the simplest explanation for the laws of physics is that God or a god designed them that way.

Internal consistency

The conjecture should be coherent, that is to say it should have an internal logical consistency such that its separate parts fit together to form a harmonious whole. If it suffers from an internal contradiction, then the conjecture clearly is unreasonable, and so this is an essential test.

External consistency with evidence

Here a conjecture may be shown to be consistent with such evidence as is known, even if it cannot make predictions or retrodictions that can be independently tested.

This is a useful test, albeit one that falls short of scientific validation.

External consistency with other scientific tenets
This is what Edward O Wilson, drawing on the nineteenth century philosopher William Whewell, calls consilience: the conjecture conforms with solidly verified knowledge in other scientific disciplines to form a common basis for explanation.

If it is not possible to show consistency with evidence in a conjecture's subject of study, then consistency with other major tenets of contemporary science is a useful test.

Origin of the universe

The principal cosmological conjectures about the origin of the universe divide between those that posit a beginning for the universe and those that maintain the universe is eternal.

Orthodox model: the Big Bang
The prime example of the first category is the current orthodox model, but as we saw in Chapter 4 it doesn't explain how matter was created from nothing. Attempts to do so resulted in the conjecture of the net zero energy of the universe, whereby negative gravitation energy exactly cancels the positive energy represented by the rest mass and kinetic energy of matter plus radiation.* It prompted Guth to call the universe a free lunch.

However, if the Big Bang is the start of everything, including space and time, then there is no universe with a net zero energy to provide the energy for everything that follows and there is no pre-existing vacuum that obeys the laws of quantum theory. This conjecture fails the essential test of internal consistency.

If, as inflationary theorists now favour, the Big Bang occurred at the end of a period of inflation, then this is not an add-on to the basic model but a contradiction of its fundamental tenet. Not even cosmologists can have their free lunch and eat it: either the Big Bang was the start of everything or it wasn't.

Although cosmologists call the current orthodox model the Standard Cosmological Model or the Concordance Model, it is perhaps more accurately described as the Quantum-Fluctuation-Group-of-Inflationary-Conjectures-Either-Before-Or-After-the-Hot-Big-Bang-Unknown-27-per-cent-Dark-Matter-Unknown-68-per-cent-Dark-Energy Model.

The versions with inflation occurring before a Big Bang are more internally self-consistent than either the ones in which the Big Bang is the start of everything or else the internally inconsistent Judaeo-Christian and Islamic divine creation myths.† However, since they do not offer a convincing account of what dark matter and dark energy are, they can hardly claim parsimony or even great explanatory power since they currently leave 95 per cent of even the claimed minute

---

*   See page 46.
†   See page 11.

observable part of the universe unexplained. Moreover, since these versions don't explain where the quantum vacuum, the laws of quantum mechanics, and the inflation field came from, it is difficult to argue that they are more reasonable than the insight that Brahman or the Dao is ultimate reality that exists out of space and time, from which everything springs and of which everything consists.

Multiverse conjectures

Motivated principally by dissatisfaction with orthodox cosmology's origin explanation, other ideas, such as Linde's chaotic inflationary model, Smolin's natural selection of black holes universes, and string conjecture's landscape of possibilities, speculate that our universe is simply one of many, if not an infinity, of other universes in a multiverse.

While these suggestions may explain from where and how our universe emerged, they simply postpone the origin question: they fail to explain from where and how and why the multiverse, or Smolin's progenitor universe, came into existence. If it held that a multiverse is eternal, then this fails to explain how or why our particular universe should have come into existence at a particular point in time rather than one of the infinite number of possible previous times in eternity.

I shall consider the reasonableness of other multiverse claims in the next section, ***Form of the universe.***

"Eternal" models

Chapter 5 examined several self-proclaimed eternal models, like eternal chaotic inflation, the cyclic bouncing universe, and the cyclic ekpyrotic universe. We saw that, while the mathematics allows each to continue indefinitely into the future, each necessarily has a beginning. Logically, there can be no such thing as a "semi-eternal" universe that has a beginning but no end, and so they fail the test of internal consistency.

Chapter 5 also showed how Hoyle and colleagues modified the original steady state model to produce quasi-steady state cosmology (QSSC). Its central idea is that the universe is eternal and continues expanding indefinitely: both time and space are infinite.

There is no logical contradiction in infinite space continuing to expand, and so such a conjecture is internally consistent.

However, QSSC seeks to achieve external consistency with observational data by claiming that non-singular mini-bangs continually create regions of new matter in cycles that produce a long-term steady-state expansion of the universe. Creation from nothing by an endless series of mini-bangs is no more reasonable than creation from nothing by one Big Bang. Moreover, while QSSC avoids the singularity problem for each cycle, extrapolating back the overall expansion of the universe leads to an infinitely small universe indistinguishable from a singularity. Arguably this constitutes a beginning for the universe, which cannot therefore be eternal.

Form of the universe

How does the universe come into existence with one specific form rather than another when other forms are logically possible? These other forms include universes with different physical constants, or with different physical laws, or with different numbers of dimensions, and so on. This question is an essential part of the wider anthropic question: why is our universe fine-tuned to enable the evolution of humans?

Fine tuning of cosmological parameters

In Chapter 4 we saw that Martin Rees maintained that if any one of six cosmological parameters differed by only a minute amount from its measured value then the universe could not have evolved in such a way as to enable the emergence of carbon-based thinking humans like ourselves. The laws of physics do not predict the values of these parameters, and cosmology's current orthodox model fails to explain how or why these parameters are so finely tuned.

These six parameters are:

1. *Omega (Ω): a measure of the gravitational attraction of the matter in the universe compared with its expansion energy*
   If there was a Big Bang, then when the universe was one second old the value of Omega must have been somewhere between 0.99999999999999999 and 1.00000000000000001, otherwise the universe would have either rapidly collapsed into a Big Crunch or rapidly expanded into emptiness.*
2. *Lambda (Λ): the cosmological constant*
   As we saw in Chapter 4, this conjectured constant representing an unknown anti-gravity dark energy is questionable, as are many of the assumptions underlying its estimation. Nonetheless, it forms a key part of current orthodox cosmology and the value that astronomers have estimated is incredibly small, about 10^{-29} gram per cubic centimetre. If it were not so small, Rees argues, its effect would have stopped galaxies and stars from forming and cosmic evolution would have been stifled before it could even begin.
3. *Nu (N): the ratio of the strength of the electromagnetic force compared with the strength of the gravitational force*
   Its value is approximately 10^{36} (1,000,000,000,000,000,000,000,000,000, 000,000).

   The electromagnetic force provides atoms and molecules with their stability by balancing the attractive and repulsive forces of oppositely charged nuclei and electrons. At this scale the relatively minute gravitational force is negligible. At the size of the almost electrically neutral

* See page 28.

planets and larger, however, gravitational force dominates. If *Nu* were a just few zeroes less, such a relatively stronger gravitational force would produce a short-lived miniature universe, no complex structures would form, and there would be no time for biological evolution.

4. *Q: a measure of how tightly structures like stars, galaxies, clusters of galaxies, and superclusters are held together*
It is the ratio of two energies: the energy needed to break up and disperse these cosmic structures compared with their total rest mass-energy calculated by $E = mc^2$. It has been estimated as approximately 10^{-5} or 0.00001. If Q were even smaller the universe would be inert and structureless. If Q were much larger no stars or solar systems could survive: the universe would be dominated by black holes.

5. *Epsilon (ε): a measure of how firmly helium nuclei bind together*
The key nuclear chain reaction in creating all the elements and in powering stars is the fusion of two protons (hydrogen nuclei) and two neutrons into the nucleus of helium.* The mass of a helium nucleus is 0.7 per cent less than the mass of its constituent parts. This conversion of 0.007 of its mass into energy, mainly heat, according to the equation $E = mc^2$, measures the force that binds together the constituent parts of the helium nucleus, overcoming the mutually repulsive electrical force of the two positively charged protons.

If this conversion factor were less, say 0.006, the first stage of the chain reaction, the binding of one proton with one neutron, would not occur, no helium would therefore be produced, and the universe would consist only of hydrogen. If it were greater, say 0.008, two protons would readily bind together directly to create helium and no hydrogen would remain to provide the fuel for ordinary long-lived stars or to allow the eventual production of molecules essential for human life, like water.

6. *D: the number of spatial dimensions in the universe*
In the universe it is three (plus one time dimension). One consequence, according to Rees, is that forces like gravity and electricity obey an inverse-square law: double the distance between masses or charged particles and the force between them is four times weaker; triple the distance and the force is nine times weaker; and so on. This permits a balanced relationship between, for example, the centrifugal motion of a planet and the centripetal pull of its sun's gravity, thus allowing for a stable orbit. If there were four spatial dimensions the forces would follow an inverse cube law and structures would be unstable: if an orbiting planet were slowed down only slightly it would plunge into its sun; if it speeded up only slightly it would quickly spiral outwards into darkness. If there were less than three spatial dimensions no complex structures could exist.[8]

---

*   See Chapter 9 for a more detailed explanation.

John Barrow and Frank Tipler claim there are parameters additional to Rees's six that are necessary for human evolution and I shall examine the "anthropic universe" question as it emerges in the course of this quest. Here, I am examining the reasonableness of cosmological conjectures to answer the particular question of how and why the universe took the form it did in contrast to the beliefs that God or a god designed it this way.

The multiverse explanation

The conjecture favoured by Rees and most cosmologists is the multiverse explanation. At first sight this seems eminently reasonable. A conjectured multiverse denies the uniqueness of the universe in which we live by applying probability to the cosmos. Its essential claim is that everything is possible, and so in a cosmos consisting of an unimaginably large number, or even an infinity, of universes, each of which has different properties, it is overwhelmingly probable that there is a universe having the properties that our universe has. We just happen to exist in that one.

Closer examination, however, raises several questions. The first of these is what is the nature of the multiverse? There are almost as many different types of conjectured multiverse as there are conjectured universes in any one multiverse. They may be grouped into four principal categories.

1. *Quantum multiverse*

   This interpretation of quantum theory, which conflicts with the standard or Copenhagen interpretation,* was proposed in 1957 by graduate student Hugh Everett. Here, all possible outcomes of every single event at a quantum level give rise to alternative universes that exist in parallel as disconnected alternative versions of reality on another quantum branch in infinite-dimensional space. The initial version of this conjecture proposes that such universes exist with the same number of space-time dimensions as ours and are described by the same laws of physics with the same constants; only the outcomes of each event are different. For example, a whole series of event outcomes at the subatomic quantum level of a woman leads to different outcomes at the macro level to the question asked by a man, will you marry me? This produces, among others, one universe in which she does marry him and one in which she doesn't. Some later versions suggest the laws of physics are different in these alternative quantum branches of reality.

   Whether the quantum multiverse conjecture is internally consistent is questionable. Logically it produces one universe in which Everett devises and believes in the conjecture and another universe in which he doesn't. Moreover, invoking an inconceivably large number of universes in order

---

*   See page 87.

to explain the one we perceive falls somewhat short of parsimony;* this applies to all multiverse conjectures. A claim of external consistency with quantum theory's outstanding empirical success in explaining the subatomic foundations of chemistry is seductive, but what the equations and principles constituting quantum theory mean in reality, and whether they can be extrapolated from the subatomic realm to the large-scale realm of the universe, are questions that the best minds in physics and philosophy have so far failed to resolve.

2. *Weak cosmological multiverse*

These multiverses were introduced by orthodox cosmologists to account for the fine-tuning of key physical parameters needed to create the physico-chemical environment whereby human life can evolve as it has in our universe, the so-called "anthropic universe". Most such conjectures propose that the other universes are either short-lived or exist in the same three-dimensional space as ours but at a remote distance, far beyond our contact horizon. (A quantum multiverse, by contrast, has no physical distance separating universes existing in parallel quantum branches.)

I have labelled these "weak" because only the values of the constants, or parameters, of physics—like the charge of an electron or the value of the gravitational constant—are assumed to vary; their advocates give no good reason why, in a multiverse in which anything is possible, only physical constants vary while the laws of physics are the same. To assume that the laws of physics observed in a minute part of one universe are the same as those in other universes of which we can have no knowledge is unreasonable.

Unlike other versions, Smolin's conjecture of a multiverse that has evolved by cosmological natural selection of black hole universes claims external consistency with other scientific tenets because the mechanism of natural selection operates in biology. Whether natural selection in biological evolution has been proven in the scientific sense is a question I shall examine in Part 2, but this seems a fair claim. However, this conjecture also depends on a series of questionable assumptions, at least three of which are unreasonable, as noted in Chapter 5.†

3. *Moderate cosmological multiverse*

Such ideas allow factors other than physical constants to vary. One example is universes with different dimensions. String conjectures speculate that our perceived universe of three spatial dimensions is part of a megaverse

---

*   Since the number of alternative outcomes of all quantum events since the Big Bang is unimaginably large, Paul Davies described this conjecture as "cheap on assumptions but expensive on universes".

†   See page 54.

having 11 dimensions (this number has differed in the past and may change in the future).*

Another example arises from string conjectures' landscape of possibilities, in which all possible megaverses have different constants and laws of physics as well as different dimensions.†

While allowing more variations than the weak version, these conjectures do not allow universes that are not governed by string "theory", and fail to explain why this is so.

Moreover, as we saw in ***Problems with string theory***,‡ while each string theory is internally consistent, Smolin presents a strong case against external consistency with evidence, concluding that "all the versions we can study in any detail disagree with observation".[9] He also maintains that it is externally inconsistent with the scientific tenet of relativity theory: "Einstein's discovery that the geometry of space and time is dynamical has not been incorporated into string theory."[10]

Without more positive results from these tests of reasonableness for a scientific (as distinct from a mathematical) conjecture, it is difficult to see how the speculated existence of other dimensions is any more tenable than the belief of many Buddhist schools that there are 31 distinct realms of existence.[11] Furthermore, the so far untestable idea that the matter of the universe reduces not to fundamental particles but to strings of energy seems no more or less reasonable than the Upanishadic insight that *prana* (vital energy) is the essential substrate of all forms of energy and, in many interpretations, of all matter.§

4. *Strong cosmological multiverse*

This takes the conjecture to its logical conclusion: universes in which everything is possible.

Tegmark is an enthusiastic proponent of this view, labelling it a Level IV universe, which "eliminates the need to specify anything at all".[12]

Taking a Platonic approach, he asserts that a mathematical structure satisfies a central criterion of objective existence because it is the same no matter who studies it: a "theorem is true regardless of whether it is proved by a human, a computer or an intelligent dolphin".

He further proposes that "all mathematical structures exist physically as well. Every mathematical structure corresponds to a parallel universe. The elements of this multiverse do not reside in the same space but exist outside of space and time." However, he fails to explain or suggest how these mathematical structures originated.

---

* See page 63 and page 69.
† See page 68.
‡ See page 101.
§ See Glossary for a full definition.

He acknowledges that no known mathematical structure exactly matches our universe and concludes that either we will find one or else "we will bump up against a limit to the unreasonable effectiveness of mathematics" and have to abandon this level.

He claims that the multiverse concept does not fail the test of parsimony. To the contrary, he contends that the argument that nature is not so wasteful as to indulge in an infinity of different worlds we can never observe can be turned round to argue *for* a multiverse, because the entire ensemble is often simpler than one of its members. For example, the set of all solutions to Einstein's field equations is simpler than a specific solution. "In this sense the higher-level multiverses are simpler." Since "the Level IV multiverse eliminates the need to specify anything at all...the multiverse could hardly be simpler".

I think it may be argued that if you do not specify anything at all the multiverse could hardly be more meaningless.

Likewise, that same lack of specificity means that it is impossible to demonstrate any meaningful external consistency either with evidence or with other scientific tenets.

Moreover, Tegmark gives no reason to stop at mathematical structures. If everything is possible, then it follows that one possible universe has properties that are determined not by a mathematical structure but by God in such a way that the evolution of humans such as ourselves is a necessary outcome. This is precisely the anthropic universe by divine design that the multiverse advocates seek to counter.

Conclusions

Neither science nor reasoning offers a convincing explanation of the origin and form of the universe, and hence of the origin of the matter and energy of which we consist. I think it most likely that it is beyond their ability to do so. In Ellis's view the ability of science to answer foundational questions is strictly limited. The evidence so far from this quest supports his "profound conclusion that certainty is unattainable at the foundations of understanding in all areas of life, including fundamental physics and cosmology as well as philosophy [and] even the apparently impregnable bastion of mathematics".[13] This is not a counsel of despair or pessimism. If we accept the limitations of science and reasoning, "we can attain satisfying and even profound understandings of the universe and the way it works, at all times regarded as provisional but nevertheless providing a satisfactory worldview and foundation for action".

Hopefully science will have greater explanatory power when I move on from the emergence of matter to its evolution.

CHAPTER EIGHT

Evolution of Matter on a Large Scale

We must explain why the universe is so uniform on large scales and at the same time suggest some mechanism that produces galaxies.

—ANDREI LINDE, 2001

No matter how impressive a cosmological theory is, it has to fit in with what we actually see in the sky.

—MICHAEL ROWAN-ROBINSON, 1991

As we have seen, cosmology's current orthodox and competing explanations of the origin of matter are conjectures rather than scientific theories established by evidence. I shall now examine science's explanation of how matter evolved from its primordial state into more complex forms that eventually included humans.

> **evolution**   A process of change occurring in something, especially from a simple to a more complex state.

I use this meaning because I wish to make clear that evolution is not limited to biological evolution but is a phenomenon we perceive throughout the universe.

Because the evolution of matter depends critically on the way in which different elements of matter interact, first I shall review science's current understanding of the four fundamental interactions to which all known natural forces reduce. Next I shall summarize cosmology's current orthodox account of the evolution of matter on a large scale, distinguishing between conjecture and hypothesis on the one hand and theory supported by firm evidence on the other, and consider reasonable challenges and alternative scientific explanations where necessary. In the following chapter I shall examine the parallel and related evolution of matter on the small scale.

The fundamental forces of nature

Subhash Kak, currently Regents Professor and Head of Computer Science Department at Oklahoma State University, claims that gravity was understood in ancient India.[1] Philosophers from Aristotle in the fourth century BCE have speculated about the force that causes planets to move and also the force that causes objects to fall to the Earth. However, Isaac Newton is generally credited as the first to formulate a law of universal **gravitation** that applies to both the force holding an object to the Earth and also the force holding the Moon and planets in their orbits, which he published in 1687 in *Principia Mathematica*.

Magnetic force was known at least as early as the fifth century BCE, static electricity was apparently referred to by Thales around 600 BCE, while electrical currents were discovered in 1747 by William Watson. Recognition that electrical force and magnetic force were the same began in 1820 when Hans Ørsted discovered that electric currents produce magnetic fields, while Michael Faraday demonstrated in 1831 that a changing magnetic field induces an electrical current. From 1856 to 1873 James Clerk Maxwell developed the theory of the electromagnetic field on a mathematical basis, derived the laws of **electromagnetism**, and discovered the electromagnetic nature of light.

Following the discovery of the neutron in 1932, Hideki Yukawa proposed in 1935 that a force existed between nucleons* which took the form of an exchange of massive particles called bosons. This idea was developed in the late 1970s when particle physicists concluded that quarks, not nucleons, were fundamental particles, and it was their interaction—the **strong force**—that constituted a fundamental force of nature; this relegated the nuclear force to be a "residual" of the strong force.

Henri Becquerel discovered "uranic rays" in 1896, but it was decades before scientists understood the various processes of radioactive decay by which an unstable atomic nucleus loses energy by emitting radiation in the form of particles or electromagnetic waves. The **weak force**, responsible for one type of radioactive decay, was first properly characterized in 1956 by Chen Ning Yang and Tsung Dao Lee when they predicted that the law of conservation of parity†—hitherto considered universal—would break down in weak interactions. Their hypothesis was confirmed experimentally within a year by Chien-Shiung Wu.

The four fundamental forces that act between elementary particles, of which all matter is assumed to be composed, are now referred to as the four fundamental interactions. Science's current understanding of them may be summarized as follows.

---

* Nucleon is the collective name for protons and neutrons, the particles that make up an atomic nucleus.

† This law states that symmetry is conserved at the subatomic level: if a nuclear reaction or decay occurs, then so does its mirror image and with equal frequency.

Gravitational interaction

In Newtonian physics gravitational force is an instantaneous force of interaction between all particles of mass. Among the four fundamental interactions it is the only universal one. Its range is infinite, it is always attractive, and its strength is given by multiplying the masses and dividing the product by the square of the distance between the centres of mass of the particles, and then multiplying the result by a universal constant, G, called Newton's gravitational constant. Mathematically it is expressed as

$$F = G \frac{m_1 m_2}{r^2}$$

where F is the gravitational force, m_1 and m_2 are the masses, r is the distance between the centres of mass, and the constant, G, is an incredibly small number, 6.67×10^{-11} metre^3 (kilogram-second^2)$^{-1}$.

Physicists and engineers still use this equation today because at most masses and speeds that we experience it is in excellent accord with the data. It is used, for instance, in calculating trajectories for space flights. Current science theory, however, regards it as only a good approximation. Einstein's General Theory of Relativity changed the concept of gravitation: it is not a force of interaction between masses but a warp in the space-time fabric caused by mass; it is not instantaneous.*

According to quantum field theory, the gravitational field created by a mass should be quantized, that is, its energy should appear in discrete quanta, called gravitons, just as the energy of light appears in discrete quanta called photons, but this theory breaks down at very high energies (and hence very short wavelengths).

Accelerating masses should emit gravitational waves—which are propagating gravitational fields—just as accelerating charges emit electromagnetic waves. In 2016 scientists at the twin Laser Interferometer Gravitational-wave Observatory (LIGO) detectors, located in Livingston, Louisiana and Hanford, Washington, announced the detection of gravitational waves produced during the final fraction of a second of the merger of two black holes, to produce a single, more massive spinning black hole. Gravitons have no current evidential support.

Electromagnetic interaction

The electromagnetic interaction is associated with electric and magnetic fields, which are manifestations of a single electromagnetic field. It governs the interaction between two electrically charged particles like a proton and an electron, and is responsible for chemical interactions and the propagation of light.

Like the gravitational interaction its range is infinite and its strength is inversely proportional to the square of the distance between the particles; unlike

---

* See illustration on page 23.

the gravitational interaction it can either be attractive, when the two charges are unlike (positive and negative), or else repulsive, when the two charges are like (both positive or both negative). The electromagnetic interaction between atoms is 10^{36} times stronger than their gravitational interaction, i.e. a million, billion, billion, billion, billion times stronger.

According to the Standard Model of Particle Physics it operates by the exchange of a messenger or carrier particle, the massless photon, between charged particles. The photon is a quantum of electromagnetic energy possessing both particle and wave properties and has an indefinite lifetime. Its existence is evidenced by the photoelectric effect, whereby electrons are ejected from metals when irradiated by light in a manner that cannot be explained by classical physics but is explained by Einstein's photon theory.

On the scale of atoms and molecules, the electromagnetic interaction dominates; it is responsible for holding atoms together. A hydrogen atom consists of a positively charged proton around which a negatively charged electron is held in orbit by the electromagnetic force of attraction between them. When two hydrogen atoms are held together in a molecule, the repulsive electrical force of the two protons is balanced by the attractive force of the two orbiting electrons, making the molecule electrically neutral and stable.

According to quantum theory, because electrons exhibit both particle and wave-like qualities they don't orbit the positively charged nucleus in one plane, as the Earth orbits the Sun; rather they smear in a shell-like orbit. This means that the negative charge is spread over the outside of the atom or molecule. Consequently, when two moving molecules collide, the force of repulsion between the two negatively charged shells causes the molecules to bounce off each other. Because this electromagnetic interaction is 10^{36} times stronger than the gravitational interaction between the molecules, the gravitational interaction can be ignored at the atomic and molecular level.

As an illustration, if you jump off the top of the Empire State Building, the gravitational interaction between you and the centre of the Earth will cause you to accelerate towards the centre of the Earth. However, you won't reach there because the negatively charged shell of electrons round the molecules on your outermost layer is repelled by the negatively charged shell of electrons round the molecules of the pavement's outermost layer: the collision splatters you.

Gravity, however, dominates for large, planetary-sized masses. The reason is that gravity is always attractive: if you double the mass then you double the gravitational pull it exerts. But two charges can only exert twice the force of one if they are both positive or both negative. A large body, like the Earth, consists of nearly equal numbers of positive and negative charges. Thus the attractive and repulsive interactions between the individual particles nearly cancel each other out and there is very little net electromagnetic interaction. On the scale of a small planet and above (including our unusually large Moon), gravity takes over from electromagnetism and is responsible for its spherical shape.

Strong interaction

The strong interaction is thought to be the force holding quarks together to form protons, neutrons, and other hadrons, and for binding protons and neutrons together to form the nucleus of an atom, thereby overcoming the electrical repulsion of the positively charged protons. It is thus responsible for the stability of matter.

Its range is approximately that of an atomic nucleus and at such distances its strength is about 100 times that of the electromagnetic interaction. If it were stronger it would be difficult to break nuclei apart and there would be no nuclear chain reactions inside stars and no elements beyond lithium would be produced. If it were weaker no atomic nuclei possessing more than one proton would be stable and so there would be no elements beyond hydrogen. If it acted on electrons it would pull them into the nucleus and no molecules and no chemistry would be possible. If it had infinite range, like gravity and electromagnetism, it would pull all the protons and neutrons in the universe together into one big nucleus.

According to the Standard Model of Particle Physics it operates by the exchange of a massless gluon—a messenger or intermediary particle—between quarks, of which protons and neutrons are thought to be composed. No free gluons, of which eight types are hypothesized, have been observed. Their existence was inferred in 1979 from electron-positron collisions at the DESY particle accelerator in Hamburg.

Weak interaction

The weak interaction is the fundamental force between elementary particles of matter that plays an important role in transforming particles into different ones, for example, through one type of radioactive decay. It is responsible for transforming an electron and a proton into a neutron and a neutrino, an essential stage of nuclear reactions inside stars.

It is the interaction between fundamental particles of spin ½, such as neutrinos, but not between particles of spin 0, 1, or 2, such as photons. It is several orders of magnitude weaker than the electromagnetic interaction and much weaker still than the strong nuclear interaction, while its range is about one thousandth the diameter of an atomic nucleus.

According to the Standard Model of Particle Physics it operates by the exchange of messenger particles, the heavy, charged W^+ and W^- and the neutral Z bosons; these particles were detected in 1983 at the CERN particle accelerator in Geneva.

Since two of these interactions were only discovered and confirmed in the last eighty years, it would be unwise to assume that no other forces or interactions of a different character will be discovered in future. Indeed, some claim to have identified one or more other forces in studies of human consciousness. I shall examine any such reasonable claims at the appropriate point of this quest, but they have no apparent bearing on cosmology's account of the evolution of matter.

Cosmology's current orthodox explanation of the evolution of matter

Drawing together summaries of hypotheses and conjectures discussed in previous chapters and augmenting them from other sources produces a timetable for cosmology's current orthodox explanation of how matter evolved.

Hot Big Bang

> Time: 0; Temperature: Infinite? Universe radius: 0

The universe, including space and time and a single force of nature, erupts into existence from nothing as a point-like fireball of radiation in the Hot Big Bang.

However, extrapolating the expansion of the universe back in time using the General Theory of Relativity produces a singularity, a point of infinite density and infinite temperature, at which relativity theory breaks down; quantum theory's Uncertainty Principle implies that nothing meaningful can be said before 10^{-43} seconds after the beginning of time.* Such extrapolation of known physics to time $t = 0$ is untrustworthy.† This explanation of the origin of matter is conjecture.

> Time: 10^{-43} second; Temperature: 10^{32}K; Universe radius: 10^{-33} centimetre (cm)

The radius of the universe is the shortest distance travelled at the speed of light for which quantum theory can be applied (Planck length‡). The gravitational force separates from the universal force leaving a grand unified force.

The universe is expanding rapidly, but its rate of expansion is slowing. As it expands and cools the radiation created by the Big Bang produces fundamental particles and antiparticles that annihilate each other and convert back into radiation. The expanding universe is thus a seething soup of radiation energy in the form of photons together with a much smaller proportion of electrons, quarks, gluons, and other fundamental particles plus their corresponding antiparticles all predicted to exist by the Standard Model of Particle Physics.

Individual quarks (and the corresponding gluons hypothesized to act as the carrier particles for the strong force between them) have never been detected. Their existence was first inferred in the late 1960s from scattering patterns detected when electrons were fired at atomic nuclei in experiments at the Stanford Linear Accelerator Center. To explain why individual quarks have never been detected, particle physicists subsequently conjectured that they are confined within baryons (three quarks) and mesons (one quark and one antiquark); if

---

*   See page 31.
†   See page 35.
‡   See Glossary for definition.

energy is put in to knock out a quark from a baryon it is transformed into a quark-antiquark pair.[2]

The speculation that gravity separates out from a single universal force is based on conjectures for a Theory of Everything (ToE), which extrapolate back in time by a hundred million times from the symmetry-breaking ideas of grand unified theories (GUTs) of particle physics (see below), which are themselves problematic. It is not yet supported by any evidence and so constitutes a conjecture.

Time: 10^{-35} second; Temperature: 10^{27}K; Universe radius: 10^{-17}cm

---

According to grand unified theories (GUTs), as the expanding universe cools below 10^{27}K (a billion billion billion degrees), conjectured carrier or messenger particles that feel both the strong and weak forces are no longer produced by radiation. These particles decay and so the strong force—which binds together quarks and, ultimately, protons and neutrons—and the electroweak force separate out of the grand unified force. This phase transition separates relatively large fundamental particles—positively charged quarks and negatively charged antiquarks—from the relatively small leptons—particles that include negatively charged electrons and neutral neutrinos. Symmetry-breaking at this event is conjectured to explain the apparent absence of antimatter in the universe.* The Standard Model of Particle Physics hypothesizes that at temperatures above 10^{15}K all these funda-mental matter particles—quarks and leptons and their corresponding antimatter particles—are massless.

The original and simplest GUT, proposed by Howard Georgi and Sheldon Glashow in 1974 and known as SU(5), is mathematically elegant, logical, and gives a precise prediction about proton decay. However, more than twenty-five years of sensitive experiments failed to detect any evidence of proton decay when, statistically, they should have done. SU(5) was thus disproved. Other GUTs were developed, adding more symmetries and particles, and hence more constants to adjust, enabling the rate of proton decay to be changed so that theorists could, in the words of theoretical physicist Lee Smolin, "easily make the theory safe from experimental failure". Such GUTs have "ceased to be explanatory".[3] They are conjectures.

In several models inflation either starts or ends at this conjectured phase transition for the universe. Currently favoured ones speculate that inflation starts and then ends in a Hot Big Bang before the universe cools to 10^{27} K when the strong and electroweak forces separate.† We are still in the realm of conjecture.

Time: sometime between 0 and 10^{-11} second; Temperature: ? Universe radius: inflates to somewhere between 10^{10}cm and $(10^{10})^{12}$cm

---

* See page 40.
† See page 35.

Depending on which of the hundred or so versions of the inflation conjecture you choose, at some indeterminate time in the past, between 0 and 10^{-11} seconds after the beginning of space-time, the universe undergoes exponential inflation for an indeterminate but incredibly brief period that increases the radius of the universe to between 10^{10}cm and $10^{100000000000}$cm.* (The current radius of the observable universe is thought to be about 10^{28}cm.) This huge discrepancy in predicted size stems not only from differences in the assumed period of inflation but also from differences in the assumed initial radius, with some versions supposing that it was considerably less than the Planck length and that inflation began at a time considerably less than Planck time, 10^{-43} second after the beginning of time. These versions raise theoretical problems because quantum theory breaks down at a time less than 10^{-43} second.

Cosmologists now tend to support versions in which the creation of the universe is followed by an extremely rapid exponential inflation of a false vacuum state and ends with the reheating of a vacuum bubble to produce a Hot Big Bang, from which the slowing expansion of the universe proceeds as above.

If this appears confusing, it is because most cosmologists present the Inflationary Big Bang model as the only explanation of the evolution of the universe, but they do not agree when and how inflation began and ended; evidence so far does not confirm the validity of any version, still less supports one rather than another.†

> Time: 10^{-10} second; Temperature: 10^{15}K; Radius of our part of the universe: 3cm

As the universe expands, the average particle energy drops to the typical energy scale of the weak force, which corresponds to a temperature of 10^{15}K, when it separates out from the electromagnetic force.

The theory that above this temperature these two forces are the same—the electroweak force—was developed in the 1960s by Sheldon Glashow, Steven Weinberg, and Abdus Salam, for which they shared the Nobel Prize. It gained support in 1983 by the discovery of three of several of the elementary particles that it predicted and became the foundation of the Standard Model of Particle Physics.

Many of the predictions of the electroweak theory have since been tested to high precision. A key prediction is the existence of the Higgs boson, the messenger particle whose interaction with quarks and leptons provides these fundamental particles—and consequently all particles of mass in the universe—with their mass. In 2012 two experiments at the Large Hadron Collider in Geneva identified the very short-lived existence of the Higgs boson, or possibly a family of Higgs bosons in which case the Standard Model will need revising.‡

---

*   See page 34.
†   See pages 36 to 38.
‡   See Page 31.

> Time: 10^{-4} second; Temperature: 10^{11}K; Radius of our part of the universe: 10^6cm

The universe expands and cools to the point where quark triplets become confined inside a range of particles called hadrons, of which the stable protons and neutrons constitute the basic building blocks of the matter with which we are familiar. The proton has an electrical charge equal in strength to that of an electron, but is positive compared with the electron's negative charge; it has a mass of 1836 times that of an electron. A proton is also called a hydrogen ion.* Initially the numbers of protons and neutrons are equal; however, the mass of a neutron is very slightly greater than the mass of a proton, and hence it requires more energy to create a neutron.

> Time: 1 second; Temperature: 10^{10}K; Radius of our part of the universe: 10^{10}cm

Fewer neutrons are now produced because their greater mass requires greater energy. Protons and neutrons separate out in the ratio of 7:1.

> Time: 100–210 seconds; Temperature: 10^9K → 10^8K; Radius of our part of the universe: ~ 10^{12}cm

At this point colliding neutrons and protons fuse together, bound by the strong interaction, with the release of a photon of energy; the energy of photons that collide with the fused particles is no longer greater than their nuclear binding energy and so photons are unable to break them apart.

A proton-neutron pair is called a nucleus of deuterium, which is an isotope† of hydrogen. Deuterium nuclei fuse with each other and with other fusion products to produce nuclei of helium-3, helium-4, tritium, and lithium-7 (see Figure 8.1). This extremely rapid, multi-stage process of nuclear fusion is called nucleosynthesis.

Rare collisions result in minute amounts of lithium-7 nuclei. Apart from these, because there is no stable nucleus with five particles the Big Bang nucleosynthesis

---

* Atoms are electrically neutral; ions are atoms that have either lost or gained one or more electrons and so have either a positive or a negative electrical charge. A hydrogen atom consists of one proton and one electron, and so a proton is the same as a positively charged hydrogen ion.

† One element is distinguished from another by having a different number of protons in the nucleus of its atoms; this determines its chemical activity. The most commonly occurring form of an element is the one with the most stable nucleus, which consists of protons and neutrons. In this case the most stable form of hydrogen has no neutrons in its nucleus. Other forms of an element that have the same number of protons but a different number of neutrons are called isotopes of the element. Hence deuterium is an isotope of hydrogen. See Glossary for *atom*, *element*, *isotope*, and *atomic number*.

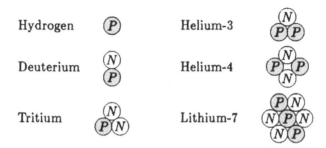

Figure 8.1 Products of the Big Bang nucleosynthesis

The isotopes of hydrogen are given special names: hydrogen-2 is called deuterium and hydrogen-3 is called tritium.

model produces no nucleus larger than helium-4, which consists of two protons and two neutrons. This has by far the highest binding energy of all nuclei with a mass number (number of nucleons) less than 5, and is the principal product of Big Bang nucleosynthesis.[4]

When the universe expands and cools below a hundred million degrees (10^8K), the temperature is not hot enough to cause fusion and so nucleosynthesis shuts down, leaving about 95 per cent of the number of nuclei as stable protons (hydrogen-1), 5 per cent by number as stable helium-4 nuclei, with just traces of deuterium, helium-3, and lithium-7 nuclei.[5]

The relative abundance in the universe today of these elements* is claimed as powerful support for the Big Bang model, but this has been challenged.† We are in the realm of hypothesis.

At this time the average density of matter is the same as the density of water today.

> Time: 3½ minutes to 380,000 years; Temperature: 10^8K → 10^4K;
> Radius of our part of the universe: 10^{13}cm → 10^{23}cm

For the next 380,000 years or so the expanding and cooling universe consists of a plasma of these positively charged nuclei plus negatively charged electrons coupled to neutral photons of radiation. Initially radiation dominates because the energy density of the photons is very much greater than the energy density of matter. As the universe continues to expand and cool, however, the energy density of matter decreases less than that of radiation: the density of both photons and matter particles decreases in proportion to volume but, whereas the

---

*   The relative abundance is usually quoted by mass, not by number: about 75 per cent hydrogen, 25 per cent helium-4 (which is four times more massive than hydrogen), and traces of the rest.

†   See page 33.

matter particles retain their mass-energy (measured by $E = mc^2$), each photon additionally loses energy by being stretched to longer wavelengths.

> Time: 380,000 years; Temperature: 3,000K; Radius of our part of the universe: 10^{23}cm

The universe cools to the point when negatively charged electrons are captured by the positively charged nuclei to form electrically neutral, stable diatomic molecules of hydrogen (H_2) plus traces of deuterium (D_2 and HD) together with atoms of helium (He) and traces of lithium (Li). Electromagnetic radiation decouples from matter and spreads through the expanding and cooling universe to form the cosmic microwave background that we detect today.*

> Time: About 200–500 million years; Temperature: Variable; Radius of the observable universe: 10^{26}cm (1 per cent of current radius) to 10^{27}cm (10 per cent of current radius)

Density differences in the expanding cloud of molecules—mainly hydrogen gas—create gravitational fields that slow down the denser regions until they separate out into different clouds light-years across that continue to contract under their own gravity. Their centres heat up as they do so by the conversion of gravitational potential energy to kinetic energy of the infalling molecules, thus increasing the temperature at the centre, or core; the space between the clouds continues to expand.

About 200–500 million years after the Big Bang some of these clouds have contracted so much, and their cores have become so hot—15 million Kelvin—that they ignite through hydrogen fusion, emitting hot, bright radiation from their cores that counteracts further gravitational collapse. Thus the first generation of stars form and galaxies begin to take shape under the gravitational influence of unknown dark matter.

The only elements in the universe are hydrogen, helium, and traces of lithium.

> Time: About 500 million years to 13.8 billion years; Temperature: Variable; Radius of the observable universe expands to 13.8 billion light-years

The first generation of larger stars consume their hydrogen and undergo gravitational collapse to the point where the increased temperature causes helium fusion to produce carbon. The process continues, producing successively heavier elements through collapse and nuclear fusion. When they have burned up their nuclear fuel so that there is insufficient release of radiation to counteract gravity,

---

*   See page 21.

they implode and then explode into supernovae, ejecting the heavier elements into interstellar space. Second and third generation stars form from clouds of interstellar hydrogen gas mixed with supernovae dust and gas, while galaxies evolve, thus creating the structures that we see today.

This orthodox history of the universe is illustrated in Figure 4.1 (page 44).

Structure of the universe

As we saw in Chapter 3 the simplifying assumptions made to solve the field equations of Einstein's Theory of General Relativity result in a universe that is homogeneous, but observations show this is not the case: it consists of many different structures. I shall now examine these structures in more detail, followed by explanations of how such structures evolved.

A galaxy consists of stars orbiting a centre, like our own Milky Way galaxy that comprises some hundred billion stars and is about 100,000 light years across. From edge on it looks like a fried egg surrounded by more than a hundred bright points, which are globular clusters, tight knots of hundreds of thousands of old stars (see Figure 8.2); the central bulge contains old stars and, viewed from above, the disc appears as a spiral of younger stars plus gas and dust (see Figure 8.3). In addition to spiral galaxies like the Milky Way, other galaxy shapes have been observed, like ellipticals—thought to be spheroidal—and irregulars. Evidence indicates that some such galaxies result from the collision of previously separate galaxies.[6]

Figure 8.2 Milky Way galaxy viewed from side on
The central bulge consists of old stars, with a hypothesized massive black hole at is centre; the disc is filled with younger stars, gas and dust, while globular clusters and hypothesized dark matter surround the disc.

Figure 8.3 Galaxy like the Milky Way viewed from above
In the Milky Way the Sun lies off one of the spiral arms, about half way to the edge of visible matter from the centre around which it revolves at 220km per second, taking 200 million years to complete one revolution.

At the next level galaxies form local groups, like our own Local Group, which is a few million light-years across and consists of the Milky Way, a larger spiral galaxy named Andromeda that we are moving towards, plus 30 or so smaller galaxies. Our Local Group lies near the edge of the Virgo Cluster, which consists of more than a thousand galaxies whose centre is 50 million light-years from us.

As Chapter 3 noted, in 1989 astronomers discovered another level of structure: massive, sheet-like superclusters separated by large bubble-like voids. Subsequent observations of larger and more distant sections of the universe with more sensitive instruments revealed ever larger superclusters—up to 10 billion light-years long—whose size is limited only by the size of the survey.* These contradict the orthodox assumption that the universe is isotropic and homogeneous on a large scale.

---

* See page 30.

Cause of structure in the universe

Cosmologists explain that what caused structure in the universe was gravitational instability. According to this account, small inhomogeneities in the matter (principally hydrogen molecules) of the early universe create regions that are very slightly denser than the rest. The gravitational field of each denser region attracts other matter, which makes it denser still, creating a bigger gravitational field that attracts still more matter, and so on.

This sounds reasonable, but it does raise two questions: (a) how did these initial inhomogeneities arise? and (b) how did such inhomogeneities produce the structures that we observe today?

Cause of initial inhomogeneities

The orthodox explanation was that subatomic quantum fluctuations in the matter created by the Big Bang are stretched by inflation to the size of galaxies or larger. Evidence for this was claimed by the precise patterns of the ripples in the cosmic microwave background revealed in 1992 by COBE.*

If we examine this explanation more closely, however, it turns out to be somewhat less convincing than it first seems.

In Guth's first version inflation ended with what is called a first-order phase transition, when randomly formed vacuum bubbles collided. Guth assumed that this would produce the required inhomogeneities, but when the calculations were made they showed the resultant inhomogeneities were far too large.

In the second version the entire observed universe was assumed to lie deep inside a single bubble, so that any bubble collisions are far too remote to have any observable effects. Inflation would produce a smooth universe, even if matter were extremely lumpy before inflation began. But this couldn't explain how *any* inhomogeneities occurred.

A solution was worked out by collaboration between Guth, Steinhardt, Hawking, and others using grand unified field theories. According to this scenario inflation ends with the spontaneous symmetry-breaking of the universal Higgs field, the conjectured scalar energy field mediated by the Higgs boson that gives mass to fundamental particles. Guth and his collaborators assumed that the spectrum of the density perturbations—fluctuating inhomogeneities—took a simple scale-invariant form, meaning that each wavelength has the same strength. This is what had been found in the cosmic microwave background, interpreted as the relic of the decoupling of radiation from matter then estimated at 300,000 years after the Big Bang. But the calculated magnitude of the perturbations at decoupling was far too large to produce the structure we observe today.

They were convinced that the concept was right but, since the universal Higgs energy field gave the wrong result, their solution was to conjecture the existence of

---

* See page 81.

another universal scalar energy field, an inflation field mediated by a conjectured inflaton particle, that did give the right result. As Guth admits, "a theory of this sort is contrived with the goal of arranging the density perturbations to come out right".[7]

Later versions proposed that the quantum fluctuations occurred in a pre-existing vacuum and were inflated by the inflation field before transforming into slightly inhomogeneous matter in a hot Big Bang.

In 2014 Steinhardt concluded that the inflationary model is so flexible that it "is fundamentally untestable and hence scientifically meaningless".[8] Moreover, as we saw in Chapter 6, several cosmologists claim that analysis of the ripples revealed by WMAP, the space-based detector 45 times more sensitive than COBE and launched a decade later, show significant inconsistencies with the inflation model, and these claims have been confirmed by data in 2013 from the even more sensitive Planck telescope.* Other cosmological conjectures also claim to account for the density ripples in the CMB.†

The only reasonable conclusion is that we really don't know how these initial inhomogeneities arose, and the current orthodox explanation is no more than a contrived mathematical model that nonetheless produces inconsistencies with observation.

Cause of large structures
As for how these initial inhomogeneities produced the large structures observed today, most of the work has been carried out at the level of galaxy, using our own and nearby galaxies as evidence because, until relatively recently, these were the principal structures shown by observation.

Two sets of models competed. Top-down models, such as the one advanced by Eggen, Linden-Bell, and Sandage in 1962, propose that a higher-level structure like a galactic cloud forms first and this collapses over a hundred million years into stellar clouds that produce stars.[9] Bottom-up models, like Searle and Zinn's of 1978, propose that stars form first and these aggregate by gravitational attraction into globular clusters that in turn aggregate into a galaxy.[10]

The COBE data of 1992 showed that both models were inadequate. According to the orthodox interpretation at the time they indicate a degree of inhomogeneity in matter 300,000 years after the Big Bang of one in 100,000, which is far too little density variation for gravitational instability to cause *any* structures to form.

Several conjectures were advanced to explain structure formation, including cosmic strings (long spaghetti-like filaments speculated in some grand unified theories to have formed in the very early universe as topological defects in the fabric of space-time) and shock waves caused by quasars that created regions of much higher matter density. The latter, however, fail to explain how the

---

*   See page 83.
†   See page 81.

high-energy-emitting quasars, or the massive black holes believed to cause them, formed in the first place.

Most cosmologists resurrected the idea first put forward by Fritz Zwicky in 1933 of dark matter.* To generate the observed structures, such dark matter would have to make up more than 90 per cent of all matter in the universe.

Two conjectures developed. The top-down hot dark matter model speculated that the dark matter consists of particles moving close to the speed of light. One candidate was neutrinos. Physicists had thought these particles were massless and moved exactly at light speed, but now they didn't rule out the possibility that neutrinos had a small mass and moved at slightly less than light speed. These would form structures on very large scales that collapse to give pancake-shaped aggregates out of which galaxies subsequently form. However, it proved difficult to match this top-down picture with the actual distribution of galaxies in clusters.

The favoured version was the bottom-up cold dark matter model, in which the dark matter consists of slowly moving—hence cold—WIMPs, or weakly interacting massive particles, left over from the Big Bang. There are no known particles with the necessary properties to fit the model, but particle physicists conjectured several candidates, like a photino, a superheavy version of the massless photon. These weakly interacting superparticles would have decoupled from radiation much earlier than the baryons (protons and neutrons that make up the matter we observe). Because they are slow moving they would have clumped together under the influence of gravity to form large galactic masses. When baryons decoupled from radiation they would be pulled to the centre of a dark galactic mass by its gravitational field where they would form a visible galaxy surrounded by a large halo of invisible cold dark matter. Such supermassive galaxies—ten times the size of what we see—would be drawn together by gravitational attraction to form clusters and superclusters. The model, however, did have to assume that galaxy formation was "biased", so that galaxies form only where there are exceptionally large fluctuations in the density of the cold dark matter.

Even so, when the estimated mass of all galaxies, including their dark haloes, was computed to give an average mass density for the universe, this amounted to no more than 10 per cent of that necessary for the critical density assumed by the orthodox model in which the kinetic energy of expansion is matched by the gravitational attraction of matter.†

Hence, cosmologists conjectured, there must be even more—very much more—dark matter in the universe in order to achieve this critical density. On this assumption, the bottom-up cold dark matter (CDM) conjecture became part of the orthodox model.

However, the large structures and large voids identified by Geller and Huchra in 1989 cast serious doubts on the CDM model. According to Michael Rowan-Robinson

---

* See page 41.
† See page 24.

a paper published in 1991 by *Nature* delivered the *coup de grace*.[11] Will Saunders and nine collaborators, who included Rowan-Robinson, had undertaken an all-sky redshift survey of galaxies detected by the Infrared Astronomical Satellite and demonstrated that there is much more structure on large scales than predicted by the CDM model.[12] It generated an opinion piece, "Cold Dark Matter Makes an Exit", by David Lindley, *Nature* Associate Editor, who pointed out that this disavowal of the CDM model came from a group who included some of the model's long-time supporters. He warned against saving the model by introducing other conjectured parameters, like a cosmological constant, comparing such attempts to Ptolemy's fixes to explain an Earth-centred solar system.[13]

But this is what happened. According to Volker Springel and colleagues in 2005, "During the past two decades, the cold dark matter (CDM) model augmented by a dark energy field (which may take the form of a cosmological constant, Λ) has been developed into the standard theoretical model of galaxy formation."[14]

Evidence for orthodox model
The evidence for this model comes principally from two sources. First, cosmo- logists claim that the Millennium Run of 2005, a detailed computer simulation, gives a good fit to the orthodox model. However, like other computer simulations, it is based on many assumptions, including the densities of visible and dark matter and dark energy that are required to achieve orthodoxy's flat universe consistent with inflation conjectures. It also depends on "post hoc modelling of galaxy formation physics".[15]

Again, by adjusting the model in the light of observations the outcome neces- sarily accords with observation and so is not predictive.

Second, the existence of dark matter—though not of what it consists—is claimed to have been proved by gravitational lensing, by which, according to general relativity, the gravitational field of the inferred dark matter bends light from more distant objects causing multiple images of those objects.[16] However, these effects could be explained by alternative mathematical models, like those for a small universe or a spherically symmetric inhomogeneous universe.[17]

Evidence against orthodox model
According to Riccardo Scarpa, because dark matter cannot emit light or any other form of electromagnetic radiation, it cannot radiate away its internal heat, a process vital for gravitational contraction to the relatively small scale of a globular cluster. Hence dark matter should not be found in these small knots of stars that orbit the Milky Way and many other galaxies. Yet Scarpa and colleagues at the European Southern Observatory in Chile found evidence in 2003 that stars in three globular clusters are moving faster than the gravity of visible matter can explain.

Scarpa concludes that there is no need to conjecture the existence of any dark matter in the universe. The explanation, first put forward more than twenty years previously by Mordehai Milgrom, is that Newton's law of gravity is only valid above a critical acceleration. Jacob Bekenstein developed a relativistic version of Milgrom's modified Newtonian dynamics that Constantinos Skordis of the University of Oxford claimed in 2005 to have used to explain both the cosmic microwave background ripples and the distribution of galaxies throughout the universe.[18]

Furthermore, the Sloan Digital Sky Survey discovered very bright quasars with very large redshifts. The orthodox interpretation of their redshifts places them at such large distances that they existed when the universe was less than a tenth of its present age.* Most cosmologists think that each such enormous emission of radiation is produced by a large quantity of very hot gas just before it is sucked into a massive black hole at the centre of a galaxy. One quasar, for example, has been calculated as emitting the light of ten trillion Suns, corresponding to a black hole of almost a billion solar masses, and is estimated to have formed only 850 million years after the Big Bang. This discovery raised doubts that such a massive structure could have formed by the bottom-up model so soon.

However, Springel and colleagues maintain that the Millennium Run computer simulation shows black holes forming at such early stages of the universe.[19] Nonetheless, I think we should be careful not to equate computer simulations with reality, especially when such simulations depend on the range of assumptions and post hoc modelling noted in the previous section. Such caution is reinforced by the identification in 2013 of a large quasar group that appears to be the largest structure in the early universe. According to its discoverers its size challenges orthodox cosmology's assumptions.[20]

Moreover, NASA's Spitzer space-based infrared telescope, launched in 2003, has detected high-redshift galaxies, estimated to have formed between 600 million to a billion years after the Big Bang. Such young galaxies should consist only of young stars, but they include red giants like those in our own galaxy that astrophysicists consider to have taken billions of years to burn up the hydrogen in their cores, after which they undergo gravitational collapse, heating up their outer layers to the point of fusion that makes them swell out and radiate red light. Interpretation of the Spitzer data is controversial, with some astrophysicists claiming that these red giants are young.

However, these young galaxies also appear to contain iron and other metals. According to the orthodox model such a young universe consists solely of hydrogen and helium plus traces of lithium;† iron is produced only after large first-generation stars have burned up not only hydrogen but also successively helium, carbon, neon, oxygen, and silicon before collapsing and exploding in supernovae that disperse iron and other metals.

---

*   As Chapter 6 noted, this quasar redshift interpretation is disputed. See page 78.
†   See page 113.

Cause of star formation

Evidence for star formation comes from observations of our own solar system, observations of young stars, many of which are surrounded by discs of dust and gas, observations of giant molecular clouds in our own galaxy, and computer modelling.

These studies led to the stellar nebular hypothesis according to which supernova explosions eject nuclei and electrons in various directions into interstellar space where they mix with existing interstellar gas, principally hydrogen. As they cool they form atoms and simple molecules of gas and dust with different velocities and different angular momenta. Gravitational fields separate this turbulent mix of supernova debris and interstellar gas into roughly spherical clouds.

Each such dynamic nebula contracts under its own gravitational field. As it does so, three processes occur: its centre heats up because the gravitational potential energy of the infalling material converts to kinetic (heat) energy; it rotates faster in order to conserve its net angular momentum as its radius decreases; and it flattens as collisions of its gas and dust particles average out motion in favour of the direction of the net angular momentum.

The vast majority of this shrinking, flattening cloud spirals into its increasingly dense and massive centre of gravity, the core of which heats up sufficiently to begin nuclear fusion: a second-generation star is born. Most of the remaining gas is ejected back into the interstellar medium as huge jets along the rotational axis. The now flat rotating disc consists of heavier dust—mainly silicates and ice crystals—and gases relatively close to the star, while the stellar wind forces the lighter hydrogen and helium gases to the outer part of the disc. This disc has an uneven density, creating a variety of gravitational fields. These produce violent collisions and aggregations to form planetesimals which, in a few hundred million years, coalesce into planets that sweep up most of the remaining gas and dust as they orbit the star in the plane of the disc.[21] See Figure 8.4.

Figure 8.4 **Hypothesized star and planetary formation from spiralling nebula**

Closer examination of this orthodox account, however, raises two problems. First, studies show that when the densest part of a molecular cloud—called a clump—is about to collapse to form a star, it contains between ten thousand to a million molecules of hydrogen per cubic centimetre.[22] This compares with the density of the air we breathe of more than ten billion billion molecules per cubic centimetre. How can something that is, at most, ten thousand billion times less dense than air contract under its own gravitational field to form a star?

While observations of giant dusty molecular clouds within our own galaxy show high-density clumps or denser protostars or new-born stars, they don't show what initiated the gravitational collapse of the clump, still less what caused these clumps in the first place. A surrounding massive halo of dark matter cannot reasonably be invoked because the scale is far too small.

Second, these observations show forces acting against the gravitational collapse of the cloud, specifically:

1. great turbulence and;
2. non-uniform magnetic fields in clouds that contain ions (mainly protons, or hydrogen ions) and electrons that cause ionized matter to stream along magnetic field lines.

Astrophysicists devised computer models, many based on hydrostatics, to try to show how these counter-forces are overcome, but with so many variables of unknown values the task is horrendously difficult and too many simplifying assumptions are needed to make the models work.

One reasonable conjecture for the cause of the initial regions of high density and their subsequent gravitational collapse is that they are due to compression created by the supernovae blast waves partly responsible for turbulence within a giant molecular cloud. That cause cannot be invoked, however, for the formation of the first generation of stars whose lives ended as first-generation supernovae.

According to Kashlinsky and colleagues in 2005, "Recent cosmic microwave background polarization measurements indicate that stars started forming early— when the Universe was 200 M[illion] yr old."[23] How first generation stars formed raises once more the question of what were the first structures in the universe if the orthodox Hot Big Bang model is correct. Over the last fifty years different cosmologists have argued that these were globular clusters, supermassive black holes, or low mass stars.

Although contemporary press releases, popular science books, and TV programmes frequently portray star formation as well established, I find it difficult to disagree with Cardiff University astrophysicist Derek Ward-Thompson. After reviewing the literature, he concludes "Stars are among the most fundamental building blocks of the universe, and yet the processes by which they are formed are not understood."[24] In 2002 Britain's then Astronomer Royal, Martin Rees, similarly concluded that "present-day star formation is poorly understood".[25]

This does not mean that science will never be able establish how stars currently form or how the first stars formed. However, in the case of the first stars, it will be difficult to obtain empirical evidence that will distinguish between different claims if the orthodox Hot Big Bang model is correct. According to this model, about half a million years after the Big Bang the temperature fell below 3,000K. Thereafter the primordial blackbody radiation shifted into the infrared, and the universe remained dark for a billion years.[26] In this period, known as the cosmic dark age, individual stars would be far too faint for any foreseeable technology to detect.

Alternative explanation

Burbidge pointed out that the current orthodox model of structure in the universe depends on many assumptions, including:

1. the universe was compressed at one time into a point of infinite or near infinite density;
2. the universe inflated exponentially before resuming decelerating expansion;
3. density fluctuations are present in early matter;
4. most matter of the universe is unknown dark matter;
5. the redshift in the spectra of quasars is caused solely by cosmic expansion, with the absorption spectra being due to intervening gas.

He argued that quasi-steady state cosmology (QSSC) offers a more convincing explanation without the need for such assumptions.*

An impartial evaluation must conclude that not only is cosmology's current orthodox explanation of how initial inhomogeneities arose in primordial matter conjecture rather than empirically supported theory, but also orthodox cosmology doesn't yet provide a scientific explanation of how any such inhomogeneities produced the stars, galaxies, local groups, clusters, and sheet-like superclusters separated by bubble-like voids that we observe today. QSSC's alternative explanation has the benefit of greater parsimony but does not provide a robust scientific theory.

Continuing evolution?

According to cosmology's current orthodox account, matter has evolved from a disordered primordial soup of fundamental particles forming from, and annihilating into, very hot radiation to the complex hierarchy of structures that we observe today. But does it continue to evolve in this antientropic manner?†

---

* See page 57.
† Entropy is a measure of the degree of disorder of the constituent parts of a system. See Chapter 10 for a more detailed consideration of this concept and its relevance to the evolution of matter.

Five speculations on the future of the universe provide contrasting views on the evolution of large-scale matter.

Perpetually self-sustaining galaxies

Borrowing self-organizing systems theory from biology, Lee Smolin conjectures that galaxies like our own are perpetually self-sustaining ecosystems. Criss-crossing waves from different supernovae interact to sweep up interstellar gas and dust into collapsing clouds that generate new stars that, after billions of years of nuclear fusion, undergo explosive death in supernovae that eject gas and dust into interstellar space, and so the cycle continues. Feedback mechanisms regulate the clouds to optimum conditions to produce main-sequence stars and maintain a balance between the density of the clouds and the number of stars and supernovae produced each generation. This self-organizing system is maintained in a state far from thermodynamic equilibrium by energy flows from starlight and supernovae explosions. "In a spiral galaxy such as our own, this process apparently occurs perpetually, causing waves of star formation that sweep continually through the medium of a spiral galaxy,"[27] thus sustaining the current level of complexity.

In self-organizing system theories, however, a system is kept in a dynamic far-from-equilibrium state by a throughput of energy and matter from outside the system, not from within it.*

Fractal universe

Another speculation is that the universe is fractal, that is, the complex shape at each level of the hierarchical structure repeats itself on a larger scale at each higher level and so on *ad infinitum*. According to Rees, however, there is no evidence to indicate that there is any higher level of structure than the sheet-like superclusters first identified by Geller and Huchra.[28] Moreover the complex shapes at the levels of solar system, galaxy, cluster, and supercluster are not identical.

Big Crunch

The conjecture that there is around ten times the amount of as yet undetermined dark matter than visible matter prompted the view that the gravitational attraction of all matter in the universe could be sufficient to slow down and then reverse the expansion of the universe. This contracting universe reverses the increase in complexity and ends in a Big Crunch, recreating the high entropy, or disorder, of the Big Bang.

---

*   See page 145.

Long-term heat death

By contrast, University of Michigan physics professor Fred Adams and NASA staff scientist Greg Laughlin claimed in 1999 that large-scale matter in the universe reaches maximum entropy through a process leading to heat death.

They argue that star formation continues in galaxies only as long as they have dense molecular clouds of interstellar hydrogen in their spiral arms. The supply of star-forming material is finite, and once stars have converted the available supply of hydrogen into heavier elements no new stars form, evidenced by elliptical galaxies, which are largely devoid of interstellar hydrogen gas and no longer produce new stars (thus contradicting Smolin's conjecture).

They estimate that the current age of star-forming will last up to a hundred billion years, and then the Stellar Era of the universe will wind down after ten to a hundred trillion years (10^{13} to 10^{14} years) as the smallest, longest lived stars, red dwarfs, fade leaving no stars to shine.

In the universe's Degenerate Era, lasting a further 10^{25} years, galaxies initially consist of compact objects: brown dwarfs, white dwarfs that are cooling or cold ("black dwarfs"), neutron stars, and black holes. White dwarfs sweep up most of the dark matter, and galaxies degenerate by evaporating these into intergalactic voids. Finally, white dwarfs and neutron stars degenerate by proton and neutron decay.*

This is followed by the Black Hole Era, in which the only stellar-like objects are black holes. These evaporate by a quantum process known as Hawking radiation until the universe is some 10^{100} years old.

Beyond this unimaginably vast age lies the final enveloping desolation of the Dark Era when only the leftover waste products from previous astrophysical processes remain: photons of colossal wavelength, neutrinos, electrons, and positrons plus, perhaps, weakly interacting dark matter particles and other exotica. Low-level annihilation events take place, but eventually the universe undergoes heat death when entropy (disorder) is at a maximum: the universe has reached a state of equilibrium in which no energy is available for use.[29]

Shorter-term heat death

From the late 1990s most cosmologists swung behind the orthodox interpretation of redshifted Type 1a supernovae data; they believed that the universe ceased its decelerating expansion after about 9 billion years and began an accelerating expansion that would produce heat death but more rapidly than that estimated by Adams and Laughlin. As we saw in Chapter 4, to explain this mysterious behaviour they invoked a mysterious anti-gravity dark energy and accounted for it in their mathematical models by re-introducing the cosmological constant Λ with yet another arbitrary value.†

---

* Predicted by grand unified theories, but so far unsupported by any evidence.

† See page 42.

According to John Barrow, this accelerating expansion prevents matter coalescing under the influence of gravity and stopped the formation of galaxies and clusters. He comments that if this unknown dark energy had switched on a little earlier no galaxies and stars would have formed and we wouldn't be here to speculate on the future of the universe.[30]

Lawrence Krauss and Michael Turner, however, caution against using such data in predictions. After re-evaluating standard notions about the connection between geometry and the fate of our universe, they conclude that "there is no set of cosmological observations we can perform that will unambiguously allow us to determine what the ultimate destiny of the Universe will be".[31]

Conclusions

1. Neither cosmology's orthodox Inflationary Hot Big Bang Bottom-Up Cold Dark Matter model nor any alternative model currently provides a scientifically robust explanation of the evolution of matter on a large scale.

2. If primordial matter consisted of an extremely dense, hot, seething, disordered plasma of fundamental particles spontaneously forming from, and annihilating into, radiation energy as the current orthodox model maintains, then it has evolved to form large-scale complex structures in a hierarchy from stellar systems, through galaxies, local groups of galaxies, clusters of galaxies, to superclusters separated by large bubble-like voids.

3. Gravitational fields interacting with matter possessing kinetic energy are conjectured to cause such an antientropic process, but the ultimate cause of the kinetic energy and the gravitational fields is not known. Moreover, although competing conjectures have been advanced, we do not have a satisfactory scientific explanation of how these complex structures—from stars through to superclusters—formed, nor do we know which formed first from the primordial plasma.

4. Each structure, like a galaxy, is not identical to a structure at the same level (in the way, say, that the units of silicon crystals are identical to each other); each higher level of the hierarchy, like a local group, is not a repeat of the lower level (in this case, a galaxy) on a larger scale. The universe is a complex whole.

5. The universe is dynamic: stars burn up their fuel, explode, and new stars form; solar systems, galaxies, local groups, clusters, and probably superclusters too, move with respect to each other; galaxies move away from and also towards other galaxies in the local group when collisions can occur.

6. Cosmology's current orthodox view that the universe's decelerating expansion mysteriously changed to an accelerating expansion after some

9 billion years implies that this antientropic process of complexification of matter on a large scale has ceased.

7. Whether the universe remains at this level of dynamic complexity in perpetuity, or is maintained at this level through perpetual cycles as quasi-steady state cosmology proposes, or whether it has begun a process that will lead eventually to maximum disorder in a heat death, or whether it will contract back to maximum disorder in the singularity of a Big Crunch, are speculations for which cosmological observations are unlikely to provide clear evidence.

In order to learn what science can tell us about the matter of which we are made we need to examine the parallel and interdependent evolution of matter on a small scale. This is the focus of the next chapter.

CHAPTER NINE

Evolution of Matter on a Small Scale

We are stardust.

—JONI MITCHELL, 1970

A lean male weighing 70kg consists of around 7 x 10^{27} atoms, that is, nearly 10 billion billion billion atoms. Approximately 63 per cent of these are hydrogen atoms, 24 per cent oxygen atoms, 12 per cent carbon atoms, 0.6 per cent nitrogen atoms, while atoms of some 37 other elements comprise the remaining 0.4 per cent.[1] How these atoms formed, and how in turn they formed us, constitutes an essential part of this quest.

Evolution of nuclei of the elements[2]

We saw that the Big Bang model could not explain how, apart from minute traces of lithium, the nuclei of elements beyond helium formed from the tiny, hot soup of quarks, electrons, and photons created by the energy release of the Big Bang before expansion and cooling of the universe shut down the fusion process.*

The discovery in 1950 by Martin and Barbara Schwarzchild that older stars contained significantly more heavier elements than did younger stars gave the first clue as to how larger nuclei might form.

In 1957 Hoyle and colleagues published the seminal paper establishing orthodox cosmology's current theory of how all naturally occurring elements heavier than helium are produced by stars. They concluded that

> The basic reason why a theory of stellar origin appears to offer a promising method of synthesising the elements is that the changing structure of stars during their evolution offers a succession of conditions under which many different types of nuclear processes can occur. Thus the internal temperature can range from a few million degrees at which the *pp* [proton-proton]

---

* See page 111.

chain first operates, to temperatures between 10^9 and 10^{10} degrees when supernova explosions occur. The central density can also range over factors of about a million. Also the time-scales range between billions of years, which are the normal lifetimes of stars of solar mass or less on the main sequence, and times of the orders of days, minutes, and seconds, which are characteristic of the rise to explosion.[3]

Although the details have since been refined, no cosmologist has made a serious challenge to this theory, which is supported by spectroscopic evidence. We may reasonably conclude that all naturally occurring elements beyond helium formed by the following stellar nucleosyntheses.

Elements from helium to iron

Stars range from one-tenth of the mass of the Sun to more than 60 times its mass; smaller proto-stars would never become hot enough to begin nuclear fusion, while larger masses would burn up too quickly to become stable stars. The size of the star determines the product of nucleosynthesis.

For small and moderately sized stars up to 8 solar masses, helium is produced in the core of the star by a series of nuclear fusions called the proton-proton chain, in which protons—nuclei of hydrogen—fuse in a chain of reactions to produce helium-4 plus energy in the form of heat and light. This outward release of energy balances the inward gravitational force of contraction so that the star remains stable, as illustrated by Figure 9.1, for billions of years.

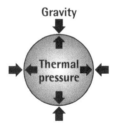

Figure 9.1 **A stable star in which the outward thermal pressure equals the inward pull of gravity**

When all the hydrogen in the core has burned up, no fusion energy is released to counteract gravity and the star begins to collapse. Its gravitational potential energy transforms into kinetic energy, heating up the contracting—and hence denser—star. The hydrogen in the star's middle layers becomes hot enough to fuse into a shell of helium around the helium core. The heat from this reaction expands the star's outer layers, swelling the star far beyond its previous size. The expansion cools these outer layers, increasing the wavelength of the light they emit. The star has become a red giant.

The star's helium core continues to collapse until its temperature rises to 100 million Kelvin. This is hot enough for the helium to fuse into carbon, releasing energy that halts further gravitational collapse and produces another period of stability. Depending on the size of the star—and hence its potential to generate increasingly higher core temperatures through gravitational contraction—this process either halts or continues.

With a star of between 2 and 8 solar masses different temperatures are generated in the core and in various layers, giving rise to different fusion products, while outer layers are blown off as a stellar wind. For a star of around 8 solar masses, thermonuclear reactions progressively fuse carbon into nitrogen, nitrogen into oxygen, and so on through elements of higher atomic number up to iron.

Unlike the elements before it, iron-56 does not release energy when fused because it is the most stable of all the elements. Consequently, when the core consists of iron there is nothing to stop further gravitational contraction. When this stage is reached in larger stars, or when smaller stars have burned up their nuclear fuel and haven't generated sufficient temperature through gravitational contraction to begin further fusion, the star collapses. In a small star this crushes its electrons together to form a white dwarf star; a medium-size star collapses further, crushing its neutrons together to form a neutron star; a larger star collapses still further into a black hole.

The huge energy and shock waves generated by such a gravitational collapse explode into space most of the mass of the star, creating a short-lived increase in the star's luminosity of up to a hundred million times its previous brightness; this is called a supernova.

A star from 8 to more than 60 times the mass of the Sun undergoes similar but much more rapid nucleosyntheses that produce a red supergiant with successive layers of fusion products, somewhat like an onion as shown in Figure 9.2.

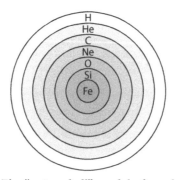

Figure 9.2 The "onion shell" model of a red supergiant
where H is the symbol of hydrogen, He of helium, C of carbon, Ne of neon,
O of oxygen, Si of silicon, and Fe of iron

Once its core consists only of iron, a red supergiant, too, becomes a supernova.

Table 9.1 summarizes the principal and secondary elements produced by nucleosyntheses in stars, the temperature at which the reactions occur, and how long it takes to use up the available nuclear fuel at these temperatures.

Table 9.1 Nucleosyntheses of the elements occurring in successive stages in large stars

Fuel	Main product	Secondary products	Temperature (billion Kelvin)	Duration (years)
H	He	N	0.03	1×10^7
He	C, O	Ne	0.2	1×10^6
C	Ne, Mg	Na	0.8	1×10^3
Ne	O, Mg	Al, P	1.5	0.1
O	Si, S	Cl, Ar, K, Ca	2.0	2.0
Si	Fe	Ti, V, Cr, Mn, Co, Ni	3.3	0.01

Source: Lochner, James C et al. (2005)

Elements heavier than iron

Although a proton cannot fuse with iron-56 to produce the next element, cobalt, iron-56 can capture three neutrons over a period of thousands of years to create the unstable isotope iron-59, whereupon one neutron decays into a proton and an electron to create the more stable cobalt. This process of slow neutron capture inside stars creates many of the stable elements heavier than iron.

The intense heat of a supernova creates a large flux of neutrons that are rapidly captured by nuclei to create heavier, unstable nuclei that decay into other stable elements, like gold, and most of the naturally occurring radioactive elements, like thorium and uranium, which are flung into cold interstellar space by the explosion.

Cosmic ray production of elements

Because the nuclei between helium and carbon are not very stable—hence the five-nucleon gap—very little lithium, beryllium, and boron are produced in stars. They are thought to be created when cosmic rays—considered to be electrons and nuclei ejected at near-light-speed from supernovae—collide with interstellar gas and dust; the collision chips off fragments, creating nuclei of smaller elements.

Second- and third-generation stars

Orthodox cosmology says that clouds of interstellar gas and dust produced by supernovae and stellar winds undergo gravitational collapse, although the mechanism is not understood.* This produces second-generation stars with more complex material than the first-generation stars, which initially consisted principally of hydrogen molecules; consequently the fusion processes are more complex. These, too, end in supernovae, ejecting part of their products into space. Third generation stars are generally thought to have formed in a similar way.

Elements produced

Hence, from the simple nuclei of hydrogen, all these processes generate successively larger and more complex nuclei, leading to about 95 naturally occurring elements found in the universe.† Two of these, hydrogen (75 per cent) and helium (23 per cent), account for most of the mass of the universe.

Of the 95 elements, 8 make up more than 98 per cent of the mass of the Earth's crust, which consists mainly of oxygen (47 per cent), silicon (28 per cent), aluminium (8 per cent), and iron (5 per cent); the oceans comprise mainly oxygen (86 per cent) and hydrogen (11 per cent).[4]

While we consist of 41 elements, 99 per cent of the mass of an average human body comprises just 6 elements: oxygen (65 per cent by mass), carbon (18 per cent by mass), hydrogen (10 per cent by mass), nitrogen (3 per cent by mass), calcium (2 per cent by mass), and phosphorus (1 per cent by mass).[5]

Although they constitute only traces, the heavier elements nonetheless play a significant role in our evolution. For example, through their radioactive decay the elements uranium, thorium, and potassium-40 generate heat in the Earth's crust that causes plate tectonics which, as we shall see later, may be necessary for a biosphere, while molybdenum is needed for nitrogen fixation, essential for plant and animal metabolism.

Fine-tuning of nuclear parameters

For reasons I shall consider later, one element deemed essential to humans and all known forms of life is carbon, more specifically the stable isotope carbon-12. But, as Hoyle pointed out, three parameters must be highly tuned for sufficient carbon-12 to be produced in stars.

On page 98 we saw how the fine-tuning of Epsilon, a measure of how firmly helium nuclei bind together, was needed to start the series of chain reactions leading to the production of elements, and eventually atoms and molecules, essential for life. Hoyle's three parameters show what precise values are needed in the nuclear chain reactions to produce carbon. The reactions are represented by the following equations:

---

*   See page 121 and following pages.

†   It used to be thought that 91 elements occurred naturally, with the remaining 27 synthesized artificially. As of 2014 claims have been made that minute traces of 7 of these have subsequently been found in nature. I will use "about 95".

$$2He^4 + 0.99 \text{ MeV} \rightarrow Be^8$$
$$Be^8 + He^4 + 7.3667 \text{ MeV} \rightarrow C^{12} + 2\gamma$$

These equations mean that 0.99 million electron-volts of energy are needed to fuse two helium-4 nuclei into the unstable isotope beryllium-8. Next, 7.3667 million electron-volts are needed to fuse another helium-4 nucleus with beryllium-8 to produce the stable carbon-12 nucleus plus two high-energy gamma rays. For these reactions to proceed three conditions must be fulfilled.

1. The lifetime of Be^8 ($\sim 10^{-17}$sec) must be sufficiently long compared with the $He^4 + He^4$ collision time ($\sim 10^{-21}$sec) to allow the first reaction to occur, but it must not be so stable that the chain of reactions stops here; it is.
2. Hoyle predicted that the yield of carbon would be negligible unless the reactions were resonant, with the vital resonance level of the carbon-12 nucleus lying near 7.7 MeV. A resonance in nuclear fusion is an energy peak at which the reaction is maximally efficient. Experiments later confirmed that the resonance level of the carbon-12 nucleus is 7.6549 MeV. This lies just above the 7.3667 MeV energy needed to fuse helium-4 and beryllium-8 and so enables this reaction to proceed.
3. The fusion of carbon-12 with another helium-4 nucleus would produce an oxygen nucleus. However, the oxygen-16 nucleus has an energy level at 7.1187 MeV. This lies just below the total energy of carbon-12 + helium-4 at 7.1616 MeV. If it had been higher, then nearly all the carbon would have been rapidly removed from stellar interiors by conversion into oxygen.[6]

Thus this chain of parameters—the longevity of the unstable beryllium-8, the existence of an advantageous resonance level in carbon-12, and the absence of a disadvantageous resonance level in oxygen-16—were necessary, and remarkably fine-tuned, conditions for the production of sufficient carbon to make the molecules on which the existence of humans and all the other known lifeforms in the universe depends.

I shall consider next how these positively charged nuclei of elements evolved into such molecules. The first stage is the formation of atoms.

Formation of atoms[7]

At the high temperatures, and hence high energies, in a star its material consists of plasma: a chaotic gas of positively charged nuclei of elements, negatively charged electrons, plus neutral neutrons and photons of electromagnetic energy. After a star has burned up its nuclear fuel and undergoes gravitational collapse, the supernova flings most of this plasma into the cold of interstellar space. When the temperature of the plasma falls to 3,000K the nuclei of elements capture electrons to form stable neutral atoms and molecules in accordance with the Principle of Conservation of Energy, the Principle of Conservation of Charge, and the Law of Electromagnetic Interaction.*

While the conservation principles and the law are necessary, they are not sufficient to explain why the negatively charged electrons are not attracted into the positively charged nuclei. This is where quantum theory revolutionized our understanding of matter at the small scale. The laws of quantum mechanics and the Pauli Exclusion Principle provide science's current explanation of how cooled stellar plasma formed the building blocks of human beings.

Laws of quantum mechanics

According to quantum theory something as small as an electron behaves as though it were both a particle and a wave. A negatively charged electron can interact with a positively charged nucleus only by surrounding it in a shell-like orbit, called an orbital, that has a discrete energy, say E_2. The electron can lose a quantum of energy, E, by dropping to a lower-energy orbital, say E_1, represented by

$$E = E_2 - E_1 = h\nu$$

where h is Planck's constant and ν is the frequency of the energy lost as electro-magnetic radiation.

Conversely, an electron can gain energy by absorbing a quantum of energy and jumping from a lower- to a higher-energy orbital.

The value of h is approximately 4.136×10^{-15} electron-volts.second. No theory explains why this should be so. Without these discrete energy values of orbiting electrons every atom would be different and none would be stable.

Each orbital is denoted by a principal quantum number n, a function of the distance of the electron from the nucleus. Three other quantum numbers specify the way that an electron interacts with a nucleus: l, the angular quantum number, denotes the shape of the orbital; m_l, the magnetic quantum number, denotes the orientation of the orbital; and m_s, the spin quantum number, denotes the direction of the electron's spin on its axis of orientation.

The solutions of the quantum mechanical equations describing these interactions yield the probability distributions of the electron about the nucleus. But they still leave an unimaginably large number of different types of atom with different energy states for every element. A hypothesis advanced by Wolfgang Pauli in 1925 not only explained how every atom of the same element is the same but also made predictions subsequently confirmed by experiment that laid the foundations of our current understanding of chemistry.

---

*   See page 105.

Pauli Exclusion Principle

The Pauli Exclusion Principle states that no two electrons in an atom or molecule can have the same four quantum numbers.* Again, no theory explains why this should be so other than that Pauli's insight enabled the developing field of quantum theory to accord with observation and experiment.

It is different from other physical laws because it is nondynamic—it has no function of distance or time—and it has nothing to say about the behaviour of an individual electron but only applies to a system of two or more electrons. This universal law selects a small set of energy states of matter from an otherwise inconceivably vast array of possibilities; it makes every atom of the same element identical; and it dictates how an atom can bond with other atoms, either of the same element or of other elements. Consequently it explains the phase of an element—how it can be gas, liquid or solid, and if the latter, metallic or crystalline—and the periodic table in which elements are grouped according to their physico-chemical properties as shown in Figure 9.3.

114 known and 4 predicted elements are ordered along horizontal rows according to atomic number, the number of protons in the nucleus of the atom. The rows are arranged so that elements with nearly the same chemical properties occur in the same column (group), and each row ends with a noble gas element, which has its outer, or valence, orbital filled with the maximum number of electrons permitted by the Pauli Exclusion Principle and is highly stable and generally inert. The position of an element in the periodic table provides chemists with the most powerful guide to the expected properties of molecules made from atoms of the element and explains how simple atoms evolve into the more complex molecules of which we are made.

Fine-tuning of atomic parameters

The formation of all stable atoms and molecules depends not only on the laws of quantum mechanics and the Pauli Exclusion Principle but also on the value of two dimensionless parameters.

The fine-structure constant, α, is the coupling constant, or measure of the strength of the electromagnetic force that governs how an electrically charged elementary particle like an electron interacts with a photon of light. Its value of 0.0072973525376 is independent of whatever units are used to measure electrical charge.

Similarly, the proton to electron mass ratio, β, has a dimensionless value of 1836.15267247. No theory explains why these two pure numbers have the values they do. If the values were much different, no stable atoms or molecules would form.[8]

---

*   The principle has since been extended to state that no two types of fermion (a class of particles that includes electrons, protons, and neutrons) in a given system can be in states characterized by the same quantum numbers at the same time.

Periodic Table of the Elements

Key: atomic number / **Symbol** / name / standard atomic weight

1	2	3	4	5	6	7	8	9	10	11	12	13	14	15	16	17	18
1 **H** hydrogen [1.007, 1.009]																	2 **He** helium 4.003
3 **Li** lithium [6.938, 6.997]	4 **Be** beryllium 9.012											5 **B** boron [10.80, 10.83]	6 **C** carbon [12.00, 12.02]	7 **N** nitrogen [14.00, 14.01]	8 **O** oxygen [15.99, 16.00]	9 **F** fluorine 19.00	10 **Ne** neon 20.18
11 **Na** sodium 22.99	12 **Mg** magnesium [24.30, 24.31]											13 **Al** aluminium 26.98	14 **Si** silicon [28.08, 28.09]	15 **P** phosphorus 30.97	16 **S** sulfur [32.05, 32.08]	17 **Cl** chlorine [35.44, 35.46]	18 **Ar** argon 39.95
19 **K** potassium 39.10	20 **Ca** calcium 40.08	21 **Sc** scandium 44.96	22 **Ti** titanium 47.87	23 **V** vanadium 50.94	24 **Cr** chromium 52.00	25 **Mn** manganese 54.94	26 **Fe** iron 55.85	27 **Co** cobalt 58.93	28 **Ni** nickel 58.69	29 **Cu** copper 63.55	30 **Zn** zinc 65.38(2)	31 **Ga** gallium 69.72	32 **Ge** germanium 72.63	33 **As** arsenic 74.92	34 **Se** selenium 78.96(3)	35 **Br** bromine [79.90, 79.91]	36 **Kr** krypton 83.80
37 **Rb** rubidium 85.47	38 **Sr** strontium 87.62	39 **Y** yttrium 88.91	40 **Zr** zirconium 91.22	41 **Nb** niobium 92.91	42 **Mo** molybdenum 95.96(2)	43 **Tc** technetium	44 **Ru** ruthenium 101.1	45 **Rh** rhodium 102.9	46 **Pd** palladium 106.4	47 **Ag** silver 107.9	48 **Cd** cadmium 112.4	49 **In** indium 114.8	50 **Sn** tin 118.7	51 **Sb** antimony 121.8	52 **Te** tellurium 127.6	53 **I** iodine 126.9	54 **Xe** xenon 131.3
55 **Cs** caesium 132.9	56 **Ba** barium 137.3	57-71 lanthanoids	72 **Hf** hafnium 178.5	73 **Ta** tantalum 180.9	74 **W** tungsten 183.8	75 **Re** rhenium 186.2	76 **Os** osmium 190.2	77 **Ir** iridium 192.2	78 **Pt** platinum 195.1	79 **Au** gold 197.0	80 **Hg** mercury 200.6	81 **Tl** thallium [204.3, 204.4]	82 **Pb** lead 207.2	83 **Bi** bismuth 209.0	84 **Po** polonium	85 **At** astatine	86 **Rn** radon
87 **Fr** francium	88 **Ra** radium	89-103 actinoids	104 **Rf** rutherfordium	105 **Db** dubnium	106 **Sg** seaborgium	107 **Bh** bohrium	108 **Hs** hassium	109 **Mt** meitnerium	110 **Ds** darmstadtium	111 **Rg** roentgenium	112 **Cn** copernicium		114 **Fl** flerovium		116 **Lv** livermorium		

57 **La** lanthanum 138.9	58 **Ce** cerium 140.1	59 **Pr** praseodymium 140.9	60 **Nd** neodymium 144.2	61 **Pm** promethium	62 **Sm** samarium 150.4	63 **Eu** europium 152.0	64 **Gd** gadolinium 157.3	65 **Tb** terbium 158.9	66 **Dy** dysprosium 162.5	67 **Ho** holmium 164.9	68 **Er** erbium 167.3	69 **Tm** thulium 168.9	70 **Yb** ytterbium 173.1	71 **Lu** lutetium 175.0
89 **Ac** actinium	90 **Th** thorium 232.0	91 **Pa** protactinium 231.0	92 **U** uranium 238.0	93 **Np** neptunium	94 **Pu** plutonium	95 **Am** americium	96 **Cm** curium	97 **Bk** berkelium	98 **Cf** californium	99 **Es** einsteinium	100 **Fm** fermium	101 **Md** mendelevium	102 **No** nobelium	103 **Lr** lawrencium

Figure 9.3 Periodic Table of the Elements

As recognized by the International Union of Pure & Applied Chemistry in 2013. Claims for the discovery of the four elements that would complete the last row await confirmation.

Evolution of atoms

The atoms formed by the cooled plasma of nuclei and electrons ejected into interstellar space evolve into more complex forms by bonding. What drives this process is attaining the most stable and lowest energy state possible. In practice this means the nearest electronic configuration to the one in which the valence orbital shell contains the maximum number of electrons permitted by the Pauli Exclusion Principle.

Atoms of the noble gas elements have this configuration naturally and are stable. Other atoms attain stability by bonding with one or more identical or different atoms, in one of four different ways.

Methods of bonding

Ionic bonding (exchange of electrons)
Here an atom gives one or more of its valence electrons to an atom of an element that lacks a complement of electrons in its valence shell. For example, stable common (or table) salt is produced when a highly reactive sodium atom, Na, reacts with a highly reactive chlorine atom, Cl, by donating its single valence electron to produce a stable positively charged sodium ion with a new, lower-orbital, valence shell filled with eight electrons like the noble gas neon, and a stable negatively charged chlorine ion with a full valence shell like the noble gas argon; the product is represented $Na^+ Cl^-$. The donor positive ion and the recipient negative ion attract each other by the electric force.

Covalent bonding (sharing of electrons)
Instead of receiving an electron from an atom of a different element, a chlorine atom can share an electron with another chlorine atom to form a diatomic chlorine molecule,* Cl_2. Filling its valence orbital shell by sharing produces a less stable configuration than filling it exclusively with an electron, and so molecules of chlorine gas are more reactive than salt.

Atoms of one element can share one or more electrons with atoms of another element to form a compound molecule. A hydrogen atom has a single electron in its valence shell when two are permitted by the Pauli Exclusion Principle. An oxygen atom has six electrons in its valence shell when eight are permitted. Two atoms of hydrogen share their electrons with the six in one oxygen atom's valence shell to produce the stable water molecule H_2O.

Metallic bonding
This occurs when atoms of the same element each lose an electron to form a lattice of positively charged ions held together by a pool of free electrons. Unlike

---

* A molecule is the smallest physical unit of a substance that can exist independently, consisting of one atom or several atoms bonded by sharing electrons. See Glossary for the distinction between *molecule*, *atom*, *ion*, and *element*.

in covalent bonding the electrons are free to move, and so a metallically bonded substance conducts electricity.

Van der Waals bonding

This is electrostatic bonding between electrically neutral molecules that arises because, due to the shape of the molecule, the distribution of electrical charge is not symmetrical. The microscopic separation of the positive and negative charge centres leads the positive end of one molecule to attract the negative end of an identical molecule, and so on. At normal atmospheric pressure, between a temperature of 0° and 100° Celsius, van der Waals bonding keeps water molecules bound to each other in a fluid state, water. At a higher temperature, the higher energies break these bonds and water molecules exist as separate entities in a gaseous state, steam. At a lower temperature, with less thermal agitation, these bonds are sufficient to hold the molecules in a solid, crystalline state, ice.

Bonding thus not only gives rise to the chemical properties—how atoms interact with other atoms—but also to the physical structure of substances.

Crystalline structures

With the exception of helium, at an appropriate temperature all atoms, molecules, and ions exist in a solid state, usually crystalline, in which the atoms, molecules, or ions are bonded not just to one other but to very many others in a lattice structure: a regularly ordered, repeating pattern extending in all three spatial dimensions. All four types of bonding form crystalline structures, and the particular structure and bonding method determine the solid's physical properties. Ionically bonded sodium and chlorine ions, for example, form crystals of common (or table) salt, while covalently bonded carbon atoms exist as soft graphite and as hard diamond.

Uniqueness of carbon

Carbon, the fourth most abundant element in the universe, has unique bonding properties. These derive in part because carbon has a high value of electronegativity, the relative ability of an atom to attract valence electrons. A carbon atom has four electrons in its outer, or valence shell, when eight are permitted by the Pauli Exclusion Principle; it is capable of bonding to four other atoms at once. It possesses the greatest tendency to form covalent bonds and, in particular, has a remarkable tendency to bond with itself; it is capable of forming not only a single bond, with one pair of shared valence electrons, but also a double bond (two pairs) or even a triple bond (three pairs.) Another special property of carbon is its ability to bond in sheets, ring structures, and long chains that constitute strings of carbon and other atoms.

These properties generate a uniquely vast range of large and complex molecules—called organic molecules—many of which are found in all forms of life identified thus far. Consequently, carbon is essential for life as we know it; without carbon we would not exist.

Molecules in space

Spectroscopic analysis indicates that interstellar space provides limited conditions for the evolution of complex molecules from the nuclei and electrons ejected by supernovae. Molecules detected include simple diatomic ones like hydrogen, H_2, and carbon monoxide, CO, through organic molecules containing up to 13 atoms, like acetone, $(CH_3)_2CO$,[9] trans-ethyl methyl ether, $CH_3OC_2H_5$,[10] and cyanodecapentayne, $HC_{10}CN$,[11] but so far no more complex molecules have been discovered. Depending on the temperature of the region of space these interstellar atoms and molecules are found in the gaseous or solid ("interstellar dust") state.

Molecules of similar complexity have been found in our solar system, in a class of meteorites known as carbonaceous chondrites. According to radiocarbon dating these are typically about 4.5 billion to 4.6 billion years old and are thought to represent material from the asteroid belt. These rocks, which consist mainly of silicates, contain a variety of organic molecules including simple amino acids, the building blocks of proteins.[12]

The most complex molecules have been found on a planet. Conditions on the surface of the Earth have provided the environment for the evolution of the most complex molecular system yet known, human beings. I shall consider how this process began in Part 2: The Emergence and Evolution of Life.

Conclusions

1. While the way in which helium atoms were formed is disputed, the generally accepted scientific theory, well supported by spectroscopic observation, is that simple nuclei of hydrogen evolved into the nuclei of all other naturally occurring elements by the mechanisms of fusion chain reactions and neutron capture plus decay inside stars and in supernovae. These processes were driven by successive transformations of gravitational potential energy into kinetic energy when stars of different sizes began contracting once the fusion fuel of each stage of the chain had burned up, plus the energy generated by catastrophic gravitational collapse once stars ran out of nuclear fuel and became supernovae.

2. Positively charged nuclei together with negatively charged electrons were flung into the cold of interstellar space by supernovae. Here they evolved into neutral atoms and molecules, driven principally by energy flows in accordance not only with the Principles of Conservation of Energy and of Charge and the Law of Electromagnetic Force, but also with the laws of quantum mechanics and the Pauli Exclusion Principle.

3. While the two conservation principles are generally assumed to be axiomatic, the quantum laws and the Pauli Exclusion Principle are not. Indeed, in many ways quantum theory is counterintuitive. It does,

however, explain how atoms form and evolve into the successively more complex molecules found in interstellar space, on meteorites thought to originate from the asteroid belt in our solar system, and on the surface of the Earth.

4. Though necessary, even the conservation principles and quantum theory are not sufficient to explain how the complex organic molecules, of which we and all known forms of life consist, evolved. If the values of three nuclear parameters were slightly different, insufficient carbon would have been generated in stars to produce organic molecules; if the values of two dimensionless constants—the fine-structure constant and the proton to electron mass ratio—were slightly different, no atoms or molecules at all would have formed. No theory explains why these parameters have the values they do.

5. On the small scale matter has evolved from a simple to successively more complex states.

In the next chapter I shall examine this prevailing pattern of increasing complexification in more detail and, in particular, what caused it and by what mechanisms it was achieved.

CHAPTER TEN

Pattern to the Evolution of Matter

> If your theory is found to be against the Second Law of Thermodynamics I can offer you no hope; there is nothing for it but to collapse in deepest humiliation.
>
> —SIR ARTHUR EDDINGTON, 1929

Chapter 8 shows that, *if* cosmology's current orthodox explanation is correct, then from an initial extremely dense and hot burst of radiation, matter emerged as a seething, disordered plasma of fundamental particles spontaneously forming from, and annihilating into, radiation energy. This evolved on the universe scale into a hierarchy of complex dynamic stellar systems, galaxies, local groups of galaxies, clusters, and superclusters separated by large bubble-like voids. Elements of each level are not identical, and each higher-level order is not simply a larger version of the lower level: the universe is a complex, dynamic whole. Since the late 1990s orthodox cosmology has claimed that the decelerating expansion of the universe changed to an accelerating expansion after about 9 billion years, and this implies that complexification on the large scale ceased at this time, roughly 5 billion years ago.

Chapter 9 outlines the established scientific account that, on the small scale, simple nuclei of hydrogen evolved into the complex nuclei of some 95 elements that in turn evolved into increasingly complex atoms and molecules, and the complexification continued to produce the most complex things in the known universe, human beings.

I shall now consider whether this pattern of complexification over time of all matter—on both small and large scales—accords with known scientific laws in an attempt to see what caused it and how it was achieved.

Consistency with known scientific laws

Principle of Conservation of Energy

The Principle of Conservation of Energy was developed from the First Law of

Thermodynamics, which was derived in the nineteenth century from experiments with heat engines by James Prescott Joule and others. Thermodynamics means heat movements, and the law said that energy is always conserved in such movements. More precisely

> **Thermodynamics, First Law of**   The increase in the internal energy of a system using or producing heat is equal to the amount of heat energy added to the system minus the work done by the system on its surroundings.

It has since been expanded to encompass all three forms of energy so far identified.

1. *Energy of motion*. This includes the kinetic energy of moving bodies, heat (caused by the motion of molecules of a substance), electrical energy (caused by the motion of electrons), and radiation energy (caused by the motion of electromagnetic waves and certain particles).
2. *Stored energy* (potential energy). This includes the following: elastic energy in a stretched spring that can be converted into kinetic energy if the spring is released; gravitational potential energy of a mass that can be converted into kinetic energy if the mass is released and falls towards another mass to which it is attracted by that mass's gravitational field; energy in a chemical bond that can be converted into heat through a chemical reaction.
3. *Rest mass energy* ($E = mc^2$). This follows from Einstein's Special Theory of Relativity; it can be converted into energy of motion, as with nuclear fusion in the core of stars where, for example, in the proton-proton chain reaction the rest mass of four protons is fused into the lesser rest mass of a helium-4 with the difference converted into the energy of heat and light radiation.

Where a law applies to all types of phenomena it is better termed a principle. Although the border is sometimes grey, the following definitions differentiate between the two.

> **law, scientific or natural**   A succinct, general statement capable of being tested by observation or experiment and for which no repeatable contrary results have been reported that a set of natural phenomena invariably behaves in an identical manner within specified limits. Typically it is expressed by a single mathematical equation. The result of applying such a law may be predicted from knowing the values of those variables that specify the particular phenomenon under consideration.

For example, Newton's Second Law of Motion states that the acceleration of an object is directly proportional to the net force exerted on it and is inversely proportional to its mass. If you know the mass of an object and the net force on it, you can apply the law to predict the acceleration of the mass; Einstein's Special Theory of Relativity introduced the limit that a force cannot accelerate an object to a speed equal to, or greater than, the speed of light.

> **principle, scientific or natural**   A law considered to be fundamental and universally true.

While the First Law of Thermodynamics describes heat energy and the work it does (the kinetic energy it is transformed into), the Principle of Conservation of Energy applies to all forms of energy.

> **Conservation of Energy, Principle of**   Energy can neither be created nor destroyed; the total energy of an isolated system remains constant, although it may be transformed from one form into another.

Consequently, regardless of the conditions in any interaction, the total energy in an isolated system is always the same before and after if we do the accounting correctly.

Applying this principle to the transformations of energy that cause increases in the complexity of matter in the universe as a whole raises three questions.

Question 1: What is the initial value of the energy?
Cosmology's orthodox account, the Big Bang, says the universe was created out of nothing and so the initial energy must have been zero; at the same time it says the universe erupted into existence as a massive burst of energy. The attempt to explain this contradiction by the net zero energy conjecture produces a logical contradiction.* Other attempts that invoke a prior collapsing universe or similar cannot be tested, still less provide an estimate of energy. The fact is that we have no means of ascertaining empirically or rationally what the initial value of the universe's energy is when accounting for energy transformations in the universe as a whole.

Question 2: What is the final value of the energy?
The flat geometry assumed in the Big Bang model means that the universe is infinite in extent, and so it is impossible to ascertain its final energy. This applies to other conjectures in which the universe is infinite in extent. The arbitrary addition of unknown dark energy to account for an assumed change to an accelerating rate of universe expansion would render the question even more unanswerable if that were possible.

Models with a closed geometry that lead to a universe ending in a Big Crunch would produce a final value equal to the initial value of energy in the universe, but

---

* See page 47 and page 95.

no such models are currently being developed; for orthodox cosmologists they imply that the universe consists of an even larger amount of unknown dark matter that more than offsets the arbitrarily specified unknown dark energy.

The only reasonable conclusion is that we do not know whether or how the Principle of Conservation of Energy can be applied to the universe.

If we set aside the initial and final values, then observations do indicate that isolated processes within the universe conform to this principle. This leads to the third question.

Question 3: Are these energy transformations reversible?
The Principle of Conservation of Energy is symmetric. It allows molecules of cold air to collide with molecules of hot water in a glass and gain kinetic energy by transfer from the water molecules so that the air surrounding the glass warms while the water cools; it also allows those cold air molecules to transfer part of their kinetic energy to the hot water molecules, thus spontaneously heating up the water still further while the air surrounding the glass becomes even colder. It allows the mass-energy of a nuclear bomb to be transformed by detonation into heat and radiation energy, and also for that heat and radiation energy to transform back into mass-energy to reassemble the bomb. Such reversals are not observed naturally. This led to the Principle of Increasing Entropy.

Principle of Increasing Entropy
This principle was developed from the Second Law of Thermodynamics, which was formulated in many different ways during the nineteenth century to constrain the First Law of Thermodynamics to what is observed. It may be expressed as

> **Thermodynamics, Second Law of** Heat never passes spontaneously from a cold body to a hot body; energy always goes from more usable to less usable forms.

It is a statistical law of probabilities that has nothing to say about the behaviour of individual particles. For example, it doesn't say that one cold air molecule cannot lose kinetic energy by colliding with a molecule of hot water. It does say, however, that the probability of many such interactions happening, thus spontaneously heating up the hot water still further and cooling the surrounding cold air, is negligible.

In 1877 Ludwig Boltzmann expressed it as an equation, using the concept of entropy introduced in 1862 by Rudolf Clausius. This concept has since been developed to encompass more general phenomena than thermodynamic systems; it has also been extended to incorporate the idea that the organization of a system gives us information about it. It may be defined now as

> **entropy** A measure of the disorder or disorganization of the constituent parts of a closed system; a measure of the energy that is not available for

use. The lower the entropy the greater the organization of its constituent parts and, consequently, the more energy that is available for use and the more information that can be gained by observing its configuration. At maximum entropy the configuration is random and uniform, with no structure and no energy available for use; this occurs when the system has reached a state of equilibrium.*

In an idealized cyclical system energy transformations are reversible, and in such cases the mathematical equations show entropy returning to its starting value at the end of each cycle and thus staying the same. However, all observed energy transformations are irreversible and entropy increases.

The Second Law of Thermodynamics, which applies to heat and mechanical energy changes, can be extended to a general principle describing all known energy changes in terms of entropy. This may be defined in non-mathematical terms as follows.

> **Increasing Entropy, Principle of**   During any process in an isolated system the entropy either stays the same or, more usually, increases, i.e. disorder increases, available energy decreases, and information is lost over time as the system moves towards a state of equilibrium.

Put simply, any change in a system isolated from other matter or energy tends to produces a state that is more and more probable with a consequent loss of usable energy and of information.

All the evidence, however, demonstrates that the evolution of matter both on the large and the small scales contradicts this principle.

Contradictions of the Principle of Increasing Entropy

Local systems
Science explains that increases in order and complexity occur in local, open systems, not isolated ones; such local decreases in entropy are more than offset by increases in entropy elsewhere in the universe. Thus the Earth's biosphere is a system in which organic molecules became more complex, evolving into cells and increasingly more complex lifeforms through to the most complex things known in the universe, human beings. The Earth's biosphere is an open system, and this local increase in order—and therefore decrease in entropy—is driven principally by heat and light energy from the Sun, which consequently loses energy and undergoes an increase in entropy as it irreversibly burns up its nuclear fuel.

Attempts to explain the mechanism of such decreases in entropy have produced various complexity theories and systems theories. These owe much to the studies

---

*   See Glossary for a full definition of *entropy* and its equation.

of self-organizing systems conducted between 1955 and 1975 at the University of Brussels by the chemist Ilya Prigogine, for which he was awarded the Nobel Prize, plus the work of Manfred Eigen, also a chemist, during the 1970s at the Max Planck Institute for Physical Chemistry at Göttingen. Put simply, they maintain that a complex structure can emerge from an open, disordered system by a flow of energy and matter through that system, and this flow can maintain the complex structure in a stable—but not fixed—state far from equilibrium despite the continuous change of components of the system. In this state the system is sensitive to small changes: if the flow increases, the structure encounters new instabilities from which new structures of greater complexity may emerge that are even further from equilibrium.

Vortices in fluids provide supporting evidence: they include cyclones in the Earth's atmosphere and the elaborate patterns seen on the surface of Jupiter. Figure 10.1 demonstrates a vortex created by airflow from the wing of an aeroplane.

Figure 10.1 Airflow from the wing of an aeroplane made visible by a technique that uses coloured smoke rising from the ground

This illustration is strikingly similar to the spiralling nebula hypothesized to produce a star and its planetary system and also to a spiral galaxy. A common pattern does not necessarily mean a common cause, but it is suggestive.

The universe
By definition the universe comprises all the matter and energy that exists. Even if we allow the speculated existence of other universes in a multiverse, such universes have no physical contact with our universe, which is therefore an isolated system. Hence the explanation that decreases in entropy for local, open systems are more

than offset by increases in entropy elsewhere in the universe cannot be used for the observed increases in complexity of the universe itself.

Determining whether or not the increasing complexity shown in the evolution of matter on the scale of the universe is consistent with the Principle of Increasing Entropy raises four interrelated questions for cosmology's current orthodox model.

Question 1: What is the initial state, i.e. does the universe begin with maximum or minimum entropy?
According to Paul Davies, then Professor of Theoretical Physics at the University of Newcastle upon Tyne in the UK, "What comes out of a [Big Bang] singularity is either totally chaotic and unstructured, or it is coherent and organized."[1] If anyone should think that Davies is hedging his bets, he does say six pages previously that a universe starting from a singularity at the Big Bang "was not one of maximum organization but one of simplicity and equilibrium". This describes a state of maximum entropy. Perhaps the reason Davies then hedged his bet was that Roger Penrose comes to the opposite conclusion: "The fact that [the Big Bang] must have had an absurdly low entropy [highly ordered state] is already evident from the mere existence of the Second Law of Thermodynamics."[2] Penrose is assuming that the Second Law of Thermodynamics (more precisely, the Principle of Increasing Entropy) applies to the universe, and therefore its initial state *must* be one of minimum entropy. This is a circular argument and is invalid.

Moreover, it contradicts the Third Law of Thermodynamics.

> **Thermodynamics, Third Law of**   The entropy of a perfectly ordered crystal at a temperature of absolute zero is zero.

Cosmology's current orthodox model posits the universe's initial matter as a seething, disordered plasma of fundamental particles spontaneously forming from, and annihilating back into, radiation energy, far too hot for ordered structures to form. This clearly is a state of very high entropy, with an overall balance, or equilibrium, between matter and energy at near infinite temperature immediately following a Big Bang singularity that has maximum entropy, as Davies initially concluded.

Question 2: Is the evolution of matter at the scale of the universe entropic or antientropic, i.e. does entropy increase or decrease?
Figure 10.2 illustrates the dilemma for cosmology's current orthodox model.

Figure 10.2 (a) shows molecules of a gas in a sphere that are constrained in an ordered arrangement, i.e. in a state of low entropy. Released from their constraint and provided that the temperature is above absolute zero, these molecules move at different speeds and collide with each other. The electrical repulsion of the

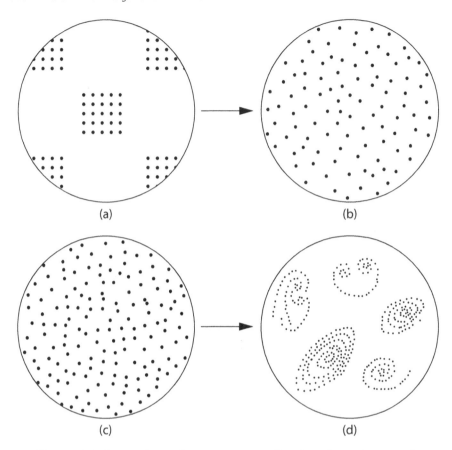

Figure 10.2 Comparison of entropy at a small scale and a universe scale

Sphere (a) contains molecules of a gas in an ordered arrangement (low entropy) that soon
produces an equilibrium state of disorder (high entropy) due to molecular collisions as shown
in sphere (b) in accordance with the Principle of Increasing Entropy. Sphere (c) contains
primordial molecules that do the opposite. From an initial state of disorder (high entropy),
and despite an expanding universe, they form hierarchical patterns of greater order—galaxies,
local groups, etc.—(low entropy) as shown in sphere (d).

negatively charged electrons on the outside of each molecule is so much greater
than the gravitational attraction between the molecules that the collisions rapidly
result in the molecules spreading out to fill the whole of the sphere, reaching
thermodynamic equilibrium at a uniform temperature as shown in sphere (b).
This is a state of maximum entropy.

Sphere (c) shows primordial molecules (about 75 per cent hydrogen and 25
per cent helium by mass) shortly after the hypothesized Big Bang. They constitute
all the matter in the universe and are spread over the whole of space existing at
this time. They are in a state of extremely high entropy (disorder) while the space

between them expands. If they were in thermodynamic equilibrium they would be a state of maximum entropy.

Pages 116 to 123 give the orthodox account, plus a critique, of how this primordial molecular cloud evolved by gravitational attraction into the structures of galaxies, local groups of galaxies, clusters, and superclusters as shown in sphere (d).

The process shown in sphere (c) → sphere (d) is the opposite of that shown in sphere (a) → sphere (b): it is antientropic.

Penrose, however, claims it is entropic.

Both he and Stephen Hawking[3] refer to gravitating bodies "clumping together" to produce disorder. But observations clearly show that galaxies and so on are more ordered and structured than the clouds of material from which they formed.

Penrose concludes that "an initial uniformly spread system of gravitating bodies represents a relatively low entropy, and clumping tends to occur as the entropy increases. Finally, there is a vast increase in entropy as a black hole forms, swallowing most of the material."[4]

Here Penrose is using "entropy" for gravitating bodies to mean precisely the opposite of the meaning for non-gravitating bodies. His argument rests on the assertion that "gravity just seems to have been different".[5] We are in danger here of again falling through Alice's looking glass where a word "means just what I choose it to mean—neither more nor less".

Moreover, if there is a vast increase in entropy as a black hole forms, and if the conditions of matter at the Big Bang, deduced by running the expansion of the universe backwards, are similar to those in a black hole, as he and Hawking argued,[6] then the Big Bang represents a state of very high, if not maximum, entropy. This directly contradicts his assertion quoted above that the Big Bang had "an absurdly low entropy".

As Ellis remarks

> Despite what is often said in discussions of the Second Law of Thermodynamics, claiming that left to itself matter will always tend to a state of increasing disorder, the opposite is in fact shown in the natural history of the universe. By a spontaneous process, first ordered structures (stars and galaxies) grew through the process of gravitational attraction; then second generation stars with planets arose and provided the habitat for life....Thus there is in fact an extraordinary propensity of matter to create order, spontaneously generating higher and higher levels of hier-archical structure and concomitant emergent order.[7]

This is an antientropic process by the generally agreed definition of entropy.

This contradiction between conjecture and evidence for the Big Bang orthodox model does not confront hypotheses in which the universe is eternal and infinite in extent because such a state is one of constant entropy on the scale of the universe and so is consistent with the Principle of Increasing Entropy ("either stays the same or, more usually, increases..." see definition on page 145).

Question 3: What causes this increase in complexity (reduction in entropy) at the scale of the universe?
Davies says the cosmic gravitational field is ultimately responsible for generating order via the cosmic expansion. He concludes that it

> presumably suffers some disordering tendency as a result [of its creating order]...but then we have to explain how the order appeared in the gravitational field in the first place....The issue turns on whether or not the second law of thermodynamics applies to gravity as well as matter. Nobody really understands this. Recent work on black holes [by Jacob Bekenstein and Stephen Hawking] suggests that it does.[8]

On this question of whether the Principle of Increasing Entropy applies to gravity, which science currently explains as a warp in the space-time fabric caused by mass, the only reasonable conclusion is that arrived at by Ellis.

> A further unresolved issue is the nature of gravitational entropy....There is as yet no agreed definition of gravitational entropy that is generally applicable; until there is, cosmological arguments relying on entropy concepts are ill-founded.[9]

The answer to this question is that currently we simply do not know what causes this observed increase in complexity, and hence reduction in entropy, as the universe expands.

Question 4: What is the final state, i.e. does the universe end in disorder?
The problem of evidence contradicting cosmology's current orthodox model posed at the end of the answer to Question 2 may be resolved if it were the case that, although complexity and order have increased so far, the universe ends in disorder. That is, although the evolution of matter in the universe has been an antientropic process up to now, it will ultimately prove to be entropic.

However, as Chapter 8 concluded, whether the universe remains at its current level of dynamic complexity in perpetuity, or is maintained at this level through perpetual cycles as quasi-steady state cosmology proposes, or whether it has begun an entropic process that will lead eventually to maximum disorder in a heat death, or whether it will contract back to maximum disorder in the singularity of a Big Crunch, is an open question.

Until about 85 years ago, on the basis of available evidence science concluded that the universe was eternally unchanging. For the next 35 years, on the basis of new evidence, science was divided between the views that the universe was eternal, infinite, and undergoing a steady state expansion, or else that the universe began from nothing as a point that exploded in a Big Bang and has been expanding at a decelerating rate ever since. 50 years ago evidence that arguably was not new convinced orthodox cosmology to adopt the Big Bang idea. About 20 years later orthodox cosmology concluded on the basis of conflicts of the evidence with the Big Bang model that the primordial universe underwent an indeterminate but incredibly brief and incredibly large inflationary expansion before embarking on a very much smaller decelerating expansion. A further 15 years later orthodox cosmology interpreted evidence to conclude that after two-thirds of its life the universe's decelerating expansion changed for reasons that are not understood to an accelerating expansion, although nowhere near its inflation rate.

It is a bold cosmologist indeed who will confidently predict what orthodox cosmology will conclude about the evolution of matter on a large scale in 10 years time, still less in 100 years time, and even less in 1,000 years time, which periods are negligible compared with orthodox cosmology's currently estimated 13.8 billion year lifetime of the universe.

Projections about the fate of the universe are purely speculative and cosmology cannot provide an answer to this question. Hence the current orthodox model cannot say whether the Principle of Increasing Entropy can be applied to the universe as a whole.

I shall incorporate the conclusions reached in this chapter in the conclusions for the whole of Part 1 given in the next chapter.

CHAPTER ELEVEN

Reflections and Conclusions on the Emergence and Evolution of Matter

Cosmologists are often wrong but never in doubt.

—Attributed to Lev Landau, 1908–1968

Reflections

When I began this quest I was under the impression that the Big Bang was scientific theory firmly established by evidence, and so ascertaining what we know about the emergence and evolution of inanimate matter would be relatively straightforward. But the more questions I asked the more it became clear that my initial assumptions were wrong.

This was no bad thing. I had approached the subject with an open mind, seeking what we know or can reasonably infer from observation and experiment on the one hand as distinct from hypothesis, speculation, or belief on the other hand. If empirical evidence failed to support, or even contradicted, my assumptions, well, that is how science works: scientific understanding is not set in stone but evolves as evidence is revealed and thinking develops.

However, what I found disturbing was that some cosmologists responded very defensively when asked to correct any errors of fact or omission or unwarranted conclusions in draft sections. Having read only half of one chapter, one said "in my view, the general public should not be given such a skewed view of what science has and has not achieved".

Another responded that "I don't really feel that someone who hasn't studied cosmology in its full physical and mathematical detail can really write a credible critique of modern cosmology." When I reminded him that the draft chapter described critiques of the current orthodoxy made by several distinguished cosmologists, he dismissed two as having "no remaining credibility in cosmology". One of those two had been awarded the Royal Astronomical Society's Gold Medal two years previously.

To justify his assertion that the draft chapter was "fundamentally flawed", he pointed to a section which concluded that inflation theory was not confirmed by data from the WMAP space-based telescope. He said "Most people think WMAP is highly consistent with inflation." Had he checked the sources I'd cited in support of my conclusion, he would have seen they included part of a textbook and also an article in *Nature* he had written that reached the very same conclusion. He ended by saying that he didn't want to spend any more time aiding what is an attack on cosmology.

I think part of this rather sad, defensive response to questioning either from within or from outside cosmology derives from the problems facing cosmology as an explanatory means described in Chapter 6. Part also arises because the discipline of cosmology began with a mathematical theory—Einstein's field equations of general relativity applied to the universe—and has been led primarily by theorists whose principal instrument is mathematics. Some conflate mathematical proof with scientific proof, which is altogether different. Introducing into their equations a range of arbitrary parameters whose values they can freely adjust to achieve consistency with observation is a reasonable way for theorists to develop a mathematical model, but it does not constitute empirical proof as generally accepted by the scientific community as a whole. This is strikingly evident when mutually contradictory models claim consistency with the same observational data.

In the absence of observational or experimental tests that can unambiguously validate or disprove a particular hypothesis, its proponents are left with a set of mathematically consistent equations that they *believe* represents reality as perceived by our senses. Adherents of such beliefs tend to act no differently than adherents of any other belief: anyone who does not agree either does not understand or is wrong or has no credibility. (This last term of disapproval is revealing when used by a scientist.) When adherents of the majority belief gain power in academia by training apprentice cosmologists and by deciding which apprentices will be offered research posts, which research will be funded, and which papers will be published, that belief becomes institutionalized; as with any institution, the pressures to conform to the current orthodoxy in order to secure and maintain a career are enormous. The consequences are that other approaches to the same problems are not researched, innovation in thinking becomes stifled in what theoretical physicist Lee Smolin calls "groupthink" (exemplified by the "most people think" counterargument to my conclusion), and scientific progress slows down.

If the pattern in scientific progress is followed, eventually a maverick will make a breakthrough and a Kuhnian paradigm shift will occur.* But in the field of

---

* Philosopher of science Thomas Kuhn argued that social and psychological factors influence how scientists move from adherence to an existing theory to embracing a revolutionary new one. See Kuhn (2012)

cosmology, when empirical testing requires massive expenditure, as with satellite-based detectors or particle colliders, the opportunities for an academic reject like Einstein are few.

> **paradigm**   The prevailing pattern of largely unquestioned thought and assumptions in a scientific discipline within which research is carried out and according to which results are interpreted.

In the meantime orthodox cosmology tends to behave less like a science and more like a religion that stops short of physical violence in its response to dissenters from within and infidels from without. When examining alternatives to current orthodoxy in Chapters 3 and 5, I noted examples of the treatment of cosmologists who disagree with the orthodox beliefs or the often-unstated assumptions underlying them or the orthodox interpretation of raw data. Smolin gives many more in his book *The Trouble with Physics*. I do so not to attack cosmology, but rather to show how cosmology works in practice and to suggest that such actions are contrary to the ethos of science. Moreover, asserting that cosmological speculations are established scientific theory opens the door for more outlandish speculations and beliefs to claim the mantle of science.

Nor is my purpose to criticize individual cosmologists. Some are the brightest minds in science and mathematics. But intellectual brilliance did not prevent Newton and many other great scientists from highly questionable behaviour towards other scientists.

Having said that, I express my profound thanks to those cosmologists who have been kind enough to correct errors, omissions, and unwarranted conclusions in early drafts. The Acknowledgements show them to be among the most outstanding minds in cosmology. Any remaining errors are, of course, my responsibility and, as I make clear, my indebtedness to them does not imply that they necessarily agree with all my conclusions or indeed with each other. It was, however, an enjoyable and stimulating experience to engage in reasoned debate, from which this book benefited greatly.

My attempt at an impartial evaluation of what we know or can reasonably infer from observation and experiment about the emergence and evolution of matter leads to the following.

Conclusions

1. Cosmology, the branch of science that studies the origin, nature, and form of the universe, puts forward the Big Bang model as its orthodox theory. However, the basic model does not meet science's test for a robust theory because, among other things, it conflicts with observational evidence. (Chapter 3)

2. Two major modifications made to the basic model in order to resolve these

conflicts produced the Quantum Fluctuation Inflationary Big Bang model, but the central claim of these modifications is untestable by any known means. (Chapter 3) Moreover, these modifications produce a logically inconsistent model if it is held that the Big Bang is the start of everything or else a model that contradicts this basic tenet if a quantum vacuum and an inflation field existed prior to the conjectured Hot Big Bang. (Chapters 4 and 7)

3. Furthermore, this modified model fails to explain whether or not and, if so, how the Big Bang was a singularity, a point of infinitely small volume and infinite density. It also fails to convincingly explain, still less predict, the observed ratio of matter and radiation in the universe, and the nature of "dark matter" (believed to account for 27 per cent of the universe) and the mysterious anti-gravity "dark energy" (believed to account for 68 per cent of the universe and for a change in the universe's expansion after two thirds of its lifetime from a decelerating rate to an accelerating rate), both of which are now additional add-ons to make the orthodox model consistent with the orthodox interpretation of observational data. (Chapter 4)

4. This further modified model fails to explain why the universe took the form it did when other forms were possible (Chapter 7) and, crucially, where everything came from. (Chapters 4 and 7)

5. The current orthodox interpretation of observational evidence maintains that the universe is expanding, but knowing whether this expansion can be traced back to a creation event at a point of origin or whether it is a phase in a cyclic, eternal universe depends upon extrapolating known physics beyond its limits and also on questionable assumptions underlying the interpretation of raw data. (Chapters 3, 5, and 6)

6. Other hypotheses advanced to modify still further or else supersede the orthodox model are either untested or untestable by any known means. (Chapter 5) Those that seek to explain where the universe came from by positing the existence of either a prior collapsing universe or a multiverse do not answer the origin question because they do not explain where the prior universe or the multiverse came from. (Chapters 3 and 5)

7. An eternal universe in which energy and matter are continually recycled would provide a more reasonable explanation than creation out of nothing, and several such conjectures have been advanced. The various cyclic models posit an unending future number of increasingly longer cycles or increasingly more massive universes but, by their own logic, running back the clock leads to a singular beginning for the cycles, and so they are not eternal. The quasi-steady state cosmology model avoids this problem, but posits the continual creation of matter and energy by a series of non-singular mini-bangs; the conjectured source of energy for these mini-bangs is no more reasonable than that conjectured to create the universe

out of nothing in one Big Bang. Plasma cosmology's static evolutionary universe model proposes an eternal universe that does not require the creation of matter-energy out of nothing, but so far fails to explain what caused the initial disordered plasma of an eternal universe to exist, and what caused known physical forces to exist and interact in such a way as to produce successively more ordered, complex states of matter. (Chapters 5 and 7)

8. Cosmology is different from other branches of science in that we have only one universe, it is incomparably large, and we are part of it. These differences produce practical problems of detection limits and measurement difficulties, problems of data interpretation and validity of underlying assumptions, and inadequacy not only of cosmological theory but also of the underlying theories of relativity, quantum mechanics, and particle physics when applied to the whole universe. Consequently we do not know with confidence such things as the age of the universe or its claimed rate of expansion. (Chapter 6)

9. Despite these formidable problems, cosmologists often make assertions that have little scientific justification. Their language frequently reflects that of a belief system rather than that of a science, and the response of institutional cosmology to reputable scientists who have different interpretations of data or who advance alternative conjectures is too often reminiscent of a Church dealing with dissenters. (Chapters 3, 5, and 6)

10. As detection techniques improve, and interpretation and theories develop in response to new data and new thinking, the practical, interpretative, and theoretical limitations will be pushed back, and cosmology will provide us with a greater understanding of the origin of the matter and energy of which we consist. However, until cosmologists produce a new definition of science and scientific method that is acceptable to both the scientific community and the wider intellectual community, many cosmological "theories" must be classified as untestable conjectures and hence are philosophical rather than scientific. (Chapter 6)

11. A particular theoretical problem is the concept of infinity, both spatially and in time. Whether this mathematical concept corresponds to the reality of the physical world as perceived by the senses of finite beings like ourselves is a metaphysical question. If any theories incorporating infinity cannot be tested by systematic observation or experiment, this places them outside the realm of science as currently understood. (Chapters 6 and 7)

12. Although science is concerned with understanding the material universe, forms of existence other than material—the laws of physics, for example—either describe or else regulate how the material universe comes into being and how it functions. These laws are expressed by mathematical relationships. Accordingly some theoretical physicists deduce that mathematical

forms exist as an objective reality outside of space and time and regulate the formation and functions of the material universe. Such reasoning is outside the realm of science. It is essentially indistinguishable from the ancient Upanishadic insight of Brahman or the ancient Chinese insight of the Dao as the ultimate reality that exists out of space and time, from which everything springs, and according to which the natural world functions. (Chapters 6 and 7)

13. If we acknowledge the unique problems of cosmology and subject the various cosmological speculations to the test of reason rather than the stricter empirical test of science, none is compelling, some are flawed by internal inconsistencies, and several are little more reasonable than some beliefs that science regards as superstition. (Chapter 7)

14. Neither science nor reasoning offers a convincing explanation of the origin of the universe, and hence the origin of the matter and energy of which we consist. Most likely it is beyond their ability to do so. Nonetheless, science continues to supply a wealth of provisional explanations—to be changed by future discoveries and thinking—of how the universe functions at both large and small scales. (Chapter 7)

15. If the primordial universe consisted of an extremely dense, hot, seething, disordered plasma of fundamental particles spontaneously forming from, and annihilating into, radiation energy, then it has evolved on a large scale to form complex dynamic structures in a hierarchy ranging from stellar systems with orbiting planets through rotating galaxies, rotating local groups of galaxies and clusters of galaxies, to sheet-like superclusters separated by bubble-like voids. (Chapter 8)

16. Compelling observational evidence shows that at the same time matter evolved on the small scale in a pattern of increasing complexity, from simple nuclei of hydrogen atoms to organic (carbon-based) molecules of up to 13 atoms found in interstellar space and simple amino acids of around the same number of atoms found in asteroids in our solar system. (Chapter 9)

17. Science has deduced certain physical and chemical laws that either describe or else determine how matter interacts in invariable ways throughout the observable universe. Without conforming to such laws matter could not have evolved into the complex form of humans. (Chapters 6, 7, 8, 9, and 10)

18. In addition to these physical and chemical laws, the values of six cosmological parameters need to be fine-tuned to produce a universe that enables matter to evolve to the complexity of the atoms and molecules of which we consist. Moreover, the values of two dimensionless constants need to be fine-tuned to permit the evolution of *any* atoms or molecules. The values of three parameters in stellar nucleosynthesis need to be fine-tuned to enable the production of sufficient carbon for the evolution of those

organic molecules essential for the existence of humans and all known lifeforms. (Chapters 7 and 9)

19. No theory explains how these laws arise or why these parameters have their critical values. Several types of multiverse conjecture propose that, in a cosmos of an unimaginably large, or an infinite, number of universes, it is overwhelmingly probable that there came into existence a universe with precisely the properties of our own; we just happen to exist in that one (the anthropic universe). Such conjectures cannot be validated or disproved by observation or experiment and are outside the realm of science. Moreover, all current types of multiverse conjecture are based on questionable logic. (Chapter 7)

20. Initial complexification on the small scale to produce some 95 naturally occurring elements was caused by energy transformations that began inside stars, particularly between gravitational potential energy and rest-mass, kinetic, heat, and light energies. Further transformations of these energies constrained by quantum theory produced more complex atoms and then atoms bonded into molecules. Hypotheses based on self-organizing systems explain, with some empirical support, how an open system of molecules can complexify still further by a flow of energy and matter through the system that produces a stable system far from equilibrium. (Chapters 9 and 10)

21. Orthodox cosmology so far fails to provide a satisfactory explanation of what caused complexification on the scale of the universe. Its account of how initial inhomogeneities in matter arose—through the conjectured inflationary expansion of quantum fluctuations in the Big Bang or else in a pre-existing vacuum—is so flexible as to be untestable, and claimed validation by observed density ripples in the cosmic microwave background have encountered serious challenges by many reputable cosmologists. Its explanation of how such inhomogeneities generated gravitational fields of sufficient strength in different clouds of primordial atoms and molecules to produce the rotating stellar systems and rotating galaxies, local groups of galaxies and clusters of galaxies, plus sheet-like superclusters separated by bubble-like voids is the bottom-up-unknown-cold-dark-matter-unknown-dark-energy model. Observations of red giant stars and iron and other metals in galaxies that are very young according to the model, plus observations of structures very much larger than predicted, throw significant doubt on that model. Moreover, cosmologists are not agreed on how first-generation stars formed and which were the first structures in the universe: globular clusters, supermassive black holes, or low-mass stars have been proposed. (Chapters 8 and 10)

22. This pattern of increasing complexity contradicts science's Principle of Increasing Entropy (disorder). Current orthodox cosmology explains that increases in complexity (decreases in entropy) occur in local, open

systems and are more than offset by increases in entropy elsewhere in the universe. But, since the universe is an isolated, and hence a closed, system, this cannot explain the increase in complexity (reduction in entropy) in the universe as a whole. This problem does not confront quasi-steady state cosmology's model of an infinite universe continually undergoing cyclical phases because entropy is the same at the beginning and the end of each cycle; like orthodox cosmology, though, it does have a problem with the Principle of Conservation of Energy because its mini-bangs involve creation of matter-energy from nothing. Plasma cosmology proposes an eternal universe, but one that begins in a state of disordered plasma or some unspecified prior state and evolves to produce more complex states of matter such as galaxies, clusters of galaxies, and superclusters from magnetically confined vortex filaments. Like the orthodox model this is an anti-entropic process that conflicts with the Principle of Increasing Entropy. Moreover, it is unclear how a universe that has a beginning can be eternal. (Chapters 5 and 10)

23. The problem may be resolved for the current orthodox model if
    a. complexification ceases, reverses, and the universe contracts back to end in the maximum disorder of a Big Crunch, or if the universe expands to the absence of any order in the thermodynamic equilibrium of a heat death, but projections about the ultimate fate of the universe are speculations that have changed radically several times in the last hundred years; or
    b. the universe is an open system kept far from equilibrium by a through-flow of energy, but this requires that the universe does not consist of all that there is and doesn't explain where this energy comes from; or
    c. a form or forms of energy not yet identified in cosmology are involved in energy transformations and associated changes in complexity.

PART TWO

The Emergence and Evolution of Life

CHAPTER TWELVE

A Planet Fit for Life

In our century the biosphere has acquired an entirely new meaning; it is being revealed as a planetary phenomenon of cosmic character.

—Vladimir Vernadsky, 1945

The broader question of how life of various, possibly exotic, kinds might emerge in the universe is outside the scope of this quest, which focuses on established scientific theories and supporting evidence of how we humans evolved. Hence Part 2 examines how life emerged and evolved on Earth.

Part 1 showed that the most complex matter currently known to have evolved on a small scale in interstellar space and on asteroids in our solar system consists of organic molecules of up to 13 atoms. The Earth, by contrast, is home to the most complex things known in the universe: human beings. But only in a tiny zone of the Earth—on, just below, and just above the surface—do humans and other lifeforms exist. This is the biosphere, perhaps better called the biolayer.

I begin Part 2 by considering what conditions are necessary for the emergence and evolution of life as we know it and then examine the scientific theories that explain how such conditions developed on Earth to create its biosphere.

Conditions necessary for known lifeforms

The various conditions that scientists deem necessary for the emergence and evolution of known lifeforms may be grouped into six: essential elements and molecules, mass of the planet, biospheric temperature range, sources of energy, protection from hazardous radiation and impacts, and stability.

Essential elements and molecules

All known lifeforms consist of a great variety of highly complex molecules. The only element capable of forming molecules of such complexity is carbon,* and hence life, at least when it begins, requires the element carbon.

---

*⁾ See page 138.

All biologists consider water in its liquid form essential for life. The reasons derive from the size, shape, and particular distribution of electrical charge in a water molecule, as shown in Figure 12.1.

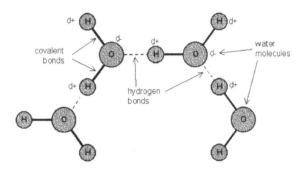

Figure 12.1 Hydrogen bonds in liquid water molecules
where **d⁺** and **d⁻** represent dipolar electrical charges

A water molecule is electrically neutral, but within the molecule electrical charge is distributed unevenly. The oxygen atom attracts a greater share of the electrical charge and is electronegative; the two hydrogen atoms are correspondingly electropositive. These dipolar charges enable the electropositive hydrogen atoms within a water molecule to form particularly strong Van der Waals bonds* of attraction with electronegative atoms of other molecules; these are known as hydrogen bonds. Figure 12.1 shows such hydrogen bonds with other water molecules. Their strength means that water can remain in liquid form at much higher temperatures than can other small molecules like methane, carbon dioxide, or ammonia, which dissociate into separate molecules as gas at corresponding temperatures. Water exists as a liquid under normal atmospheric pressure across the unusually wide temperature range of 0° Celsius to 100° Celsius (273 Kelvin to 373 Kelvin), which is the ideal range for biochemical reactions necessary for reproducing and maintaining lifeforms.

The ability of these small molecules to form hydrogen bonds also makes liquid water a powerful solvent both for ionic compounds, like salts, and for organic molecules that also have an uneven distribution of electrical charge, like amino acids, which are so fundamental to known lifeforms that they are called the building blocks of life.

The hydrogen bonds can prove stronger than the covalent bonds joining the single oxygen atom to two hydrogen atoms, breaking the latter so that two liquid water molecules form a positive hydronium ion, H_3O^+, and a negative hydroxyl ion, OH^-.

$$H_2O + H_2O \rightarrow H_3O^+ + OH^-$$

* See page 138.

Such ions in liquid water make it a good conductor of electricity.

These solvent and conducting properties make water a good medium within which essential biochemical processes take place, for example, dissolving nutrients and transporting them across semi-porous membranes inside living organisms and also dissolving and transporting away waste products.

The elements essential for maintaining life range from these lighter ones of hydrogen, oxygen, and carbon through to heavy elements. For example, molybdenum helps nitrogen fixation, a biochemical reaction in which generally unreactive nitrogen molecules are changed into more reactive nitrogen compounds, a key step in plant metabolism, which is an interrelated series of biochemical reactions whereby plants obtain the energy and nutrients to sustain their existence. Because animals eat plants, and humans eat both plants and animals, nitrogen fixation forms a key step in the animal and human metabolic chains.

Much heavier radioactive elements, like uranium and thorium, generate heat by their radioactive decay. As we shall see later, most geophysicists deem that heat from these elements produces the movement of tectonic plates, and corresponding movement of continents, which plays an important role in the evolution of life.

Mass of planet

Quantifying this factor is difficult because it depends on many other factors such as distance from, and luminosity of, its star, and the planet's size and hence its density.

Broadly speaking, if the mass of a planet is too small its gravitational field will not be strong enough to retain either gases for an atmosphere or volatiles, like liquid water, on its surface. A low-mass planet will have cooled relatively quickly since its formation and lack the tectonic activity that promotes biological evolution. A majority consensus is that a planet's mass should be no less than one-third of the Earth's.[1]

Less of a consensus exists on the upper limit. The conventional view is that a planet much larger than about 10 Earth-masses would capture significant amounts of gas from its nebular disc followed by runaway growth into a gas giant inimical to known lifeforms.[2] On the other hand the Science and Technology Definition Team for NASA's Terrestrial Planet Finder project concluded that planetary habitability is not necessarily much affected by large mass.[3]

Temperature range

The temperature must not be so high that bonds in complex organic molecules would break or so low that essential biochemical reactions would proceed too slowly to produce more complex molecules and sustain essential metabolic reactions. Biophysicist Harold Morowitz estimates a temperature range of 500° Celsius.[4] However, since biologists deem liquid water essential for life, this reduces the range considerably. Under different pressures and certain other circumstances, liquid water can exist somewhat below or above the 0° to 100° Celsius temperature range at Earth's normal atmospheric pressure.

Sources of energy

A planet must have sources of energy to produce the temperature essential for maintaining life. The principal source on the surface is electromagnetic energy radiated from its star. Earlier calculations multiplied a star's luminosity (energy radiated per second) by the distance of a planet from the star in order to estimate its surface temperature. The range of distances from the star that produces a surface temperature of between 0° to 100° Celsius was defined as the circumstellar habitable zone (CHZ).

This, however, is far too simplistic. First, of the radiation reaching the planet, a fraction (called the albedo) is reflected back by such things as clouds and ice. Second, the amount of energy arriving at the surface depends on how long the surface faces the star, which depends on the planet's period of rotation about its axis; part of this incoming solar energy at visible wavelengths that heats the planet's surface is radiated away at longer, thermal wavelengths into the cold night sky when the planet faces away from the star. Third, the amount of energy radiated away depends on what gases are in the planet's atmosphere; gases like carbon dioxide, water vapour, and methane absorb this thermal energy and radiate part of it back to produce a greenhouse effect.

Furthermore, other energy sources may exist, such as heat from the planet's interior, which is either generated from radioactive decay or is residual heat from planetary formation processes. The direct effect on the surface of such geothermal heat is small, but it generates volcanic activity, which in turn supplies carbon dioxide to the planet's atmosphere to increase the greenhouse effect.

Protection from hazardous radiation and impacts

Not all radiation from a star can be used to promote essential biochemical reactions. Above certain intensities the frequencies of some radiation, for example ultraviolet, can irreparably damage organisms, while solar flares—intense bursts of electromagnetic radiation and high-energy electrons, protons and other ions— can strip away a planet's atmosphere. For lifeforms to exist the biosphere needs shielding from such hazardous radiation.

The biosphere needs protection, too, from the impact of comets or asteroids of such size as to destroy complex lifeforms.

Stability

These conditions must remain stable over a sufficiently long period to enable simple organisms to emerge from essential molecules and to evolve into lifeforms as complex as humans.

Formation of the Earth and its biosphere

I shall start by summarizing what science tells us about the Earth and its biosphere now and then examine theories explaining how they developed.

Characteristics of the Earth

The Earth is approximately spherical, with a slight bulge at the equator, and has a mean diameter of 12,700 kilometres. It has a mass of about 6×10^{24} kilograms and a strong magnetic field. It orbits its star, the Sun, in a slightly elliptical plane once every 365 and a quarter days at a mean distance of 149 million kilometres. It rotates once every 24 hours around its axis, which is tilted 23.5 degrees from the perpendicular to the elliptical plane of orbit as shown in Figure 12.2.

Orbiting the Earth at an average distance of 384,400 kilometres is its only natural satellite, the Moon. With a diameter of just over a quarter that of Earth (and almost three-quarters that of the planet Mercury), the Moon is unusually large compared with its planet.

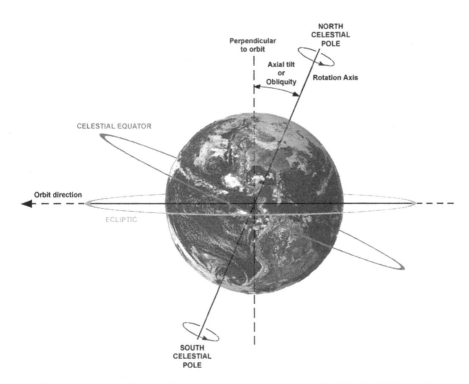

Figure 12.2 The Earth, showing its axis of rotation tilted by 23.5° from the perpendicular to the orbital plane

Internal structure

The impossibility of directly accessing all the Earth's interior, combined with differing interpretations of data such as wave patterns from earthquakes, leaves some areas of geology still in dispute. Figure 12.3 shows the current consensus view of the Earth's internal structure.

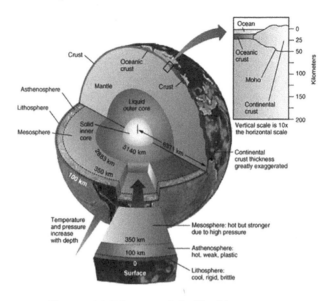

Figure 12.3 Diagram of the Earth's structure

Hydrosphere
Approximately two-thirds of the Earth's surface is covered by water in the form of oceans, seas, and rivers. The depth of this hydrosphere varies from 0 to about 5 kilometres.

Atmosphere
The gaseous layer of the Earth extends up to 10,000 kilometres above the solid and liquid surfaces, but the first 65–80 kilometres contain 99 per cent of the total mass of the Earth's atmosphere. It comprises by density approximately 78 per cent nitrogen, 20 per cent oxygen, 1 per cent argon, and the remaining 1 per cent a mixture of other gases, including water vapour, whose concentration increases with temperature, and 0.003 per cent carbon dioxide; a layer about 19–50 kilometres above the surface contains ozone, a form of oxygen with three, instead of two, atoms per molecule.

Magnetosphere
In principle a magnetic field extends indefinitely. In practice the Earth's magnetic field produces significant effects up to tens of thousands of kilometres from the surface and is called the magnetosphere. The solar wind deforms the usual symmetrical shape of a magnetic field, as shown in Figure 12.4.

Biosphere
The biosphere comprises all environments capable of supporting life as we know it. A thin shell ranging from about 5 kilometres below to 5 kilometres above sea

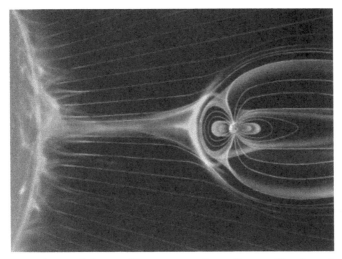

Figure 12.4 Shape of the Earth's magnetosphere distorted by the solar wind

level, it covers part of the lithosphere (the solid outer layer of the Earth), most of the hydrosphere, and part of the atmosphere.

Formation
Evidence

Evidence of how the Earth and its outer spheres developed is hard to find. Nothing on the surface remains from the planet's formation: rock has been eroded by weather, or metamorphized by heat and pressure that produced structural and chemical changes, or melted by greater heat when pushed into the interior by a process known as subduction; the dating of ancient surface rocks is controversial, but the oldest are thought to be around 3.8 billion years, some 700 million years younger than the Earth's formation. The hydrosphere has continually recirculated through evaporation as water vapour followed by precipitation as rain, hailstone, or snow. The atmosphere has been completely transformed by biological processes I shall examine later.

Scientists have developed hypotheses about the Earth's formation by inference, mainly from studies of stellar formation at different stages elsewhere in our galaxy, examination of meteorites thought to be representative of planetesimals that aggregated to form the Earth, the dating of rocks and other compounds from the surface of the Earth and magma ejected from its interior, physico-chemical analyses and dating of rocks from the Moon and Mars, plus other data from probes to most planets in our solar system.

Accurate dating is critical in determining evolutionary processes. Comparative dating, based on observations of stars in various stages of evolution combined with studies of nuclear fusion, is used to account for changes in the size and luminosity of the Sun over time.

Radiometric (otherwise known as radioisotopic) dating provides a powerful and accurate method of absolute dating of samples if certain parameters can be determined. For example, scientists have experimentally established the half-life of carbon-14 as 5,730 years. This means that in 5,730 years half of any quantity of carbon-14 will undergo radioactive decay into stable nitrogen-14. Scientists have also established the ratio of carbon-14 to the more common, stable isotope, carbon-12, in naturally occurring carbon compounds. Ascertaining the different ratio in a rock sample shows how much carbon-14 has been lost by radioactive decay; a simple calculation then determines how long this process has taken and hence how old the sample is. Similar techniques are used with radioactive elements like uranium-238, with a half-life of 4.6 billion years, and thorium-232, with a half-life of 14 billion years.

Hypotheses resulting from these observational and experimental data and computer modelling explain the formation of the Earth and its spheres as follows.

The planet

The general nebular hypothesis outlined in Chapter 8* provides an account of the planet's formation. In this particular case, about 4.6 billion years ago the vast majority of a rotating disc some 24 billion kilometres across had spiralled into its massive central ball of hydrogen, increasing its density and temperature to the point when its core ignited in nuclear fusion to generate the Sun.

The disc formation wasn't perfect. In the turbulent conditions some lumps of material, mainly ice and dust, continued in highly eccentric orbits around the Sun at an angle to the disc; these became known as comets.

Of the material left in the rotating disc, the solar wind—ions and electrons radiating from the burning Sun—pushed the remaining light hydrogen and helium gases to the cold, outer regions of the disc where they clumped into planetary nebulae—not dense or hot enough for fusion—that eventually formed the four outer gas giant planets.

Heavier molecules consisting of hydrogen combined with other elements, such as molecules of methane (CH_4), water vapour (H_2O), and hydrogen sulphide (H_2S), remained in the inner regions of the disc where grains of dust, mainly silicates, collided, sometimes bouncing off each other and sometimes sticking together. When they stuck they formed slightly more massive grains with slightly stronger gravitational fields that attracted lighter grains and gas molecules. Pebble-sized clumps formed and grew into planetesimals† up to several kilometres across. Chaotic collisions between planetesimals split some apart and fused others to produce larger bodies.

---

* See page 121.
† A planetesimal normally refers to a small body formed during the early phase of planetary formation rather than simply to the size of a body in a solar system.

At the border between the inner and outer regions, where the planetesimals were widely spaced and too far from the nearest gas giant, Jupiter, to be drawn into its orbit, they remained orbiting the Sun in what is known as the asteroid belt.

Nearer to the Sun, where planetesimals were much closer to each other, many more collisions occurred, and four rocky protoplanets grew in mass to dominate their orbital belt of the disc. Each grew by a violent process of accretion as its increasingly powerful gravitational field voraciously pulled in the remaining planetesimals in its region while it travelled round the Sun. These planetesimals smashed into, and fused with, a protoplanet, producing great heat; this, combined with the heat generated by the radioactive decay of some of its elements, like uranium and thorium, heat converted from gravitational potential energy as its gaseous matter condensed, plus heat from the Sun, was enough to melt most of each protoplanet. Once it had swept up the vast majority of planetesimals in its belt, the accretion phase was over, probably 400–700 million years after the Sun ignited. A molten rocky planet, with an atmosphere of vaporous silicates, was born.* Earth, the third such planet in distance from the Sun, thus entered its Hadean (hell-like) Eon of accretion some 4.56 billion years ago and emerged with its current mass about 4.0 to 3.9 billion years ago.[5]

Iron core

The heat generated by the bombardment of planetesimals in the Earth's accretion phase fractionated the molten mixture. The heavier elements sank to the centre of the planet. By far the largest proportion was iron. Thus the core of the planet became predominantly iron, with some nickel, the next abundant heavy element. Analyses of iron-rich meteorites, plus densities inferred from seismic data, support this hypothesis.[6]

When the heat-generating accretion eon ended, the core began to cool from its centre, which solidified under intense pressure, while the outer core remained molten.

Magnetic field

Evidence from surface rocks indicates that in the last 330 million years the Earth's magnetic field has flipped more than 400 times, that is, the north pole has changed into the south pole and vice-versa. The time between reversals varies from less than 100,000 years to tens of millions of years; the last reversal occurred 780,000 years ago.

During the last century the magnetic north pole moved 1,100 kilometres. Moreover, tracking annual movements since 1970 has shown that this rate

---

* A protoplanet becomes a planet when it has swept up the planetesimals and other debris in its orbital zone and has achieved approximately constant mass and volume. See Glossary for a more detailed defininition of a *planet*, which takes into account the 2006 redefinition by the International Astronomical Union.

is accelerating, and the pole is now moving at more than 40 kilometres each year.[7]

Scientists don't really understand how the magnetic field formed. A majority of planetary geoscientists support the dynamo hypothesis, by which the movement of an electrically conducting material across an existing magnetic field reinforces the original magnetic field. In this case the rotation of the Earth causes the liquid iron outer core, which is electrically conducting, to circulate around the axis of rotation thus generating a magnetic field with a north pole at the axial (or geographic) north pole and a south pole at the axial south pole. However, this doesn't explain how the magnetic poles have reversed so erratically or why the poles move at accelerating rates; Pennsylvania State University atmospheric scientist James Kasting suggests this is because the dynamo is a chaotic system.[8]

However, in 2009 Gregory Ryskin, associate professor of chemical and biological engineering at Northwestern University, Illinois, advanced the controversial hypothesis that fluctuations in the Earth's magnetic field are due to movements of the oceans. The salt in seawater allows it to conduct electricity, generating electrical and magnetic fields. In the north Atlantic changes in the strength of ocean currents correlate with changes in magnetic fields in Western Europe.[9] Historically, tectonic plate movements caused changes in the route of ocean currents, which could have produced the flipping of the magnetic poles.

Crust, lithosphere, and mantle
At the end of Earth's accretion phase heating from bombardment stopped and the surface cooled as it rotated away from the Sun and radiated heat away into the night sky.

The vaporous and molten silicates condensed, with the lighter silicates at the top and silicates richer in heavier elements below. The outer, lighter silicates cooled and solidified first, forming the rigid crust, followed by the cooling and solidification of the top layer of the mantle, the layer between the crust and the outer core. The majority of the mantle remained hot, but the pressure of material above it forced it to solidify, albeit it in a ductile form.

Uniform cooling from the outside of such a molten ball ought to lead to smooth layers of consistent depth for the crust, the top layer of the mantle, and so on. However, in the crust we see mountains, valleys, volcanoes, and plains, all separated by vast and deep oceans. Various ideas were advanced to explain this topography. By the end of the nineteenth century geologists had adopted an idea suggested by one of the most illustrious of their number, the Austrian Eduard Suess: as the Earth cooled it crinkled, rather like a baked apple. Although vertical motion took place, the view that the continents and oceans were permanent features of the Earth's surface became geology's orthodox theory.

Drawing on previous observations that continents are shaped rather like jigsaw pieces that fit together, a 32-year-old Austrian, Alfred Wegener, put forward in

1912 an alternative hypothesis, called continental drift, in which a supercontinent, Pangaea, broke up and continent-sized pieces drifted apart. Orthodox geology rejected this as preposterous.

Wegener worked on his idea and in 1921 published a revised and expanded version of his book to justify the hypothesis. This included evidence of identical geological structures on the eastern coast of South America and the western coast of Africa, together with identical fossil remains on these and other widely separated continents. Orthodoxy ignored or dismissed his geological evidence; Wegener was, after all, a meteorologist by profession. As for the fossil evidence, they invoked the idea of land bridges between continents that have since disappeared without trace. Wegener's idea was clearly absurd: what force could possibly move such massive things as continents?

In 1944 an English geologist, Arthur Holmes, offered an explanation. Heat generated by the radioactive decay of elements in the interior provides the energy to split up and move continents. Apart from a small number of supporters of continental drift, the vast majority of geologists stuck to the orthodox line and continued to treat the continental drift hypothesis as fanciful nonsense.

Over the years evidence accumulated. Oceanographers found that the oceanic crust was remarkably young, on average 55 million years old, compared with the continental crust that averaged 2.3 billion years, with some rocks as old as 3.8 billion years. They discovered a series of huge mountain ridges more than 50,000 kilometres long that towered up from the ocean floor and snaked round the globe rather like a seam on a baseball. The crests of these mid-oceanic ridges consisted of the youngest rock, which showed the same magnetic polarity that we observe today. On either side stripes of rock alternated in magnetic polarity. These were identical to the sequence of magnetic reversals already known from continental lava flows and were used to date the stripes of rock, showing that the age of the rock increased with distance from the crest (see Figure 12.5). The time sequence of magnetic reversals provided powerful support for the idea that magma from the interior thrust up through the crest of a mid-oceanic ridge, splitting the previously formed crust in two and pushing it laterally away from the ridge; when it cooled it recorded Earth's magnetic polarity at that time.

Lawrence Morley, a Canadian geologist, drew together strands of the evidence to produce a coherent explanation of all the data, but scientific journals adhered to the orthodox view and rejected his paper in 1963. The *Journal of Geophysical Research* dismissed it as "not the sort of thing that ought to be published under serious scientific aegis".[10]

Drummond Matthews of Cambridge University and his graduate student Fred Vine had better luck. Working independently of Morley they had come to the same conclusion, and later in 1963 *Nature* published their findings. Vine and Matthews dropped the phrase "continental drift" because it was clear that more

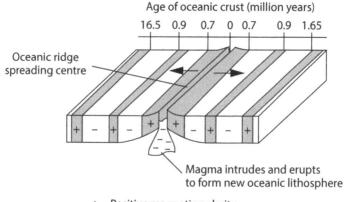

+ = Positive magnetic polarity
− = Negative magnetic polarity

**Figure 12.5 Stripes of rock of alternating magnetic polarity either side
of an oceanic ridge**

The sequence of magnetic reversals is identical to the sequence of reversals seen in continental
lava flows and is used to date the stripes of rock.

than continents had moved. By 1968 their theory was accorded the title "plate tectonics", and in a Kuhnian paradigm shift the vast majority of geologists rapidly adopted this as the new orthodoxy.[11]

It is still developing and some elements remain open questions. However, unlike cosmology's orthodox model, it qualifies as a scientific theory. It has made unique retrodictions, for example where identical fossils of certain ages may be found that have been confirmed by investigation; it has made unique predictions, for example the location of earthquake and volcanic zones and the movement of landmasses, that have been confirmed by observation.

Drawing on evidence from such diverse fields as palaeontology, oceanography, seismology, and, more recently, GEOSAT (Geostationary Satellite) mapping, this unifying theory currently explains geological phenomena in terms of the formation, movement, and interaction of about seven large and several smaller blocks of the rigid lithosphere, called tectonic plates, that float on the asthenosphere (see Figure 12.6).

The Pacific plate, for example, is grinding past the North American plate at an average rate of about 5 centimetres each year; this produces the San Andreas fault zone, about 1,300 kilometres long and in places tens of kilometres wide, that slices through two thirds of the length of California.[12]

Although there is no direct evidence, most geologists consider that the force driving the plates is the slow movement of the hot, ductile mantle that lies below the rigid plates. Heated principally by the radioactive decay of elements like uranium and thorium, the mantle is thought to move in circular convection currents as shown in Figure 12.7.

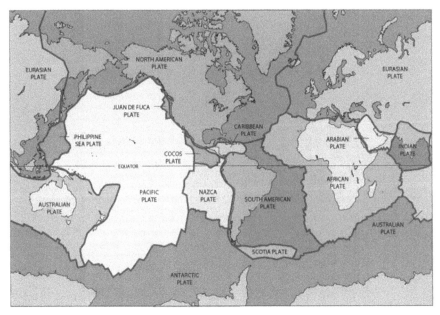

Figure 12.6 The Earth's tectonic plates

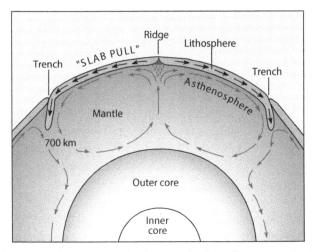

Figure 12.7 Convection currents in the mantle thought to move tectonic plates

Geological phenomena are caused by the collision of plates. Figure 12.8 illustrates the slow collision of an oceanic plate with a continental plate.

For example, hot magma from the interior forces up through a fault line in the oceanic Nazca plate to create a sub-oceanic mountain range and push the plate sideways. The eastern part of this plate slowly collides with the western side of the continental South American plate. As it does so it is pushed beneath the

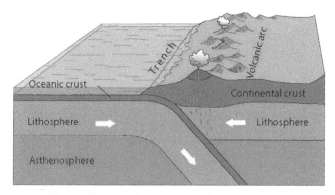

Figure 12.8 Slow collision of an oceanic plate with a continental plate
The oceanic plate is subducted (i.e. drawn down) beneath the continental plate creating a
trench in the ocean floor, uplifting the continental plate into a mountain range and generating
volcanic activity.

South American plate in a process labelled subduction, disappearing down the
Peru-Chile trench at the eastern edge of the ocean and into the mantle, where
the heat melts it. As it is pushed beneath the South American plate it forces up
the latter to create the Andes Mountains, the backbone of the continent, and the
weakened plate along this line forms a zone where strong destructive earthquakes
and the geologically rapid uplift of mountain ranges are common.

Accordingly the oceanic crust is young because it is continually recycled over
geological time as the oldest section is forced down below a continental plate,
where it melts in the interior, while it is replaced by molten rock from the interior
emerging from a fault line as the crest of a mid-oceanic ridge.[13]

Geology's current orthodox theory provides an evidence-based explanation
of the lithosphere and its evolution since the Pangaea super-plate broke up
some 225 to 200 million years ago. As for how Pangaea formed in the first place,
ideas have been advanced that this is only the latest in a cycle of supercontinent
formation, breakup, formation, and breakup that began well over 3 billion years
ago.

This still leaves the question of what was the original form of the crust and the
lithosphere. Measurements of laser pulses to a reflector array erected on the Moon
by America's Apollo 11 astronauts in 1969 show that the Moon is moving away
from the Earth at a rate of about 3.8 centimetres every year.[14] This suggests that
the Moon was very much closer to the Earth around 4.5 billion years ago. The very
much stronger gravitational field of a very much closer Moon may have pulled the
lighter silicates together as they condensed to form the first supercontinent spread
round the equator. However, such ideas are necessarily speculative because no
evidence remains.

I have devoted some space to the theory of plate tectonics because, as we shall
see later, a body of scientists considers that the movement of these plates, and in

particular the associated movement of continental landmasses, plays a vital role in biological evolution.

Hydrosphere and atmosphere

No evidence remains to show what atmosphere replaced vaporous silicates when the planetesimal bombardment phased out and the surface cooled. In 2001 Stephen Mojzsis and colleagues claimed that liquid water was present as early as 4.3 billion years ago, some 200 million years after the planet formed, based on their analyses of tiny zircon crystals within rocks of the Jack Hills of Western Australia.[15] In 2005 two Earth scientists, E B Watson of the Rensselaer Polytechnic Institute, Troy, New York, and T M Harrison of the Australian National University, Canberra, also used analysis of zircon to propose an even earlier date, 4.35 billion years ago.[16] Their interpretation of the data was challenged separately by two other Earth scientists from the Australian National University, Canberra, Allen Nutman[17] and Andrew Glikson.[18] Prior to these claims, metamorphism by heat and pressure of the oldest sedimentary rocks on Earth implied that liquid water existed on the surface at an estimated age of 3.8 billion years.

Precisely *how* the hydrosphere and the atmosphere formed has long been a matter of controversy. In the 1950s and 1960s geologists followed William Rubey's view that volatiles had been trapped in the interior during the Earth's formation and subsequently released through volcanic vents, a process labelled outgassing, to produce the oceans and atmospheric gases.

Members of the emerging discipline of planetary science thought that the proto-Earth's orbital zone was too close to the Sun, and therefore too hot, for the planetesimals from which it formed to contain volatiles. They favoured the view that the large volume of water on the Earth's surface came from outside. The oceans contain about one billion trillion (10^{21}) kilograms of water. A typical comet contains perhaps one million billion (10^{15}) kilograms of ice, and so it would take only a million (10^6) comets to hit the Earth and have their ice melt with the heat of impact to produce the oceans.[19]

This view was supported by some computer models of planetary accretion that showed the early Earth being bombarded not just by debris in its orbital zone but also by planetesimals much further from the Sun and by comets. The cometary ice hypothesis became the orthodox explanation of the Earth's oceans.

However, surveys in 2000 of the comets Halley, Hyakutake, and Hale-Bopp show that these comets have about twice the amount of deuterium (an isotope of hydrogen) compared with hydrogen in their ice as does liquid water in the Earth's oceans. This seems to rule out the cometary hypothesis. Nonetheless, one of its proponents, Michael Mumma, a NASA cometary scientist, is undaunted. He maintains that these comets come from the wrong region of the solar system, and comets from the Jovian region will show the correct ratio. This remains to be tested.

An alternative hypothesis that claims to account for the ratio of deuterium to hydrogen proposes that the bulk of water now on Earth came from a few large planetary embryos originally formed in the outer part of the asteroid belt and accreted by the Earth in its final stage of formation.[20]

As for the early atmosphere, volcanoes today release gases from the mantle. These gases may have been recycled from subducted surface rocks: when limestone and chalk, for example, are pushed into the mantle the heat dissociates their calcium carbonate content to produce carbon dioxide. However, the relative concentrations of the inert noble gases in volcanic gases are about the same as they are in the atmosphere. This suggests that gases like hydrogen sulphide, sulphur dioxide, carbon dioxide, water vapour, nitrogen, and ammonia had either been trapped in the interior during planetary formation or released in the interior by the dissociation of compound molecules like calcium carbonate due to the intense heat. As the surface cooled these gases escaped through vents created by faults in the lithosphere. No free oxygen, and no ozone, would have been present in such an early atmosphere.

This process could also account for the hydrosphere: as the cooling continued, outgassed water vapour condensed as liquid water in violent storms to form the rivers and oceans.

Without direct evidence, however, all these ideas remain hypotheses.

The Moon

For many years three speculations competed to explain how the Earth came to have such an unusually large moon. One said that it was just coincidence that a large planetesimal formed and grew in size due to accretion near the proto-Earth. Another said that the gravitational field of the Earth captured a large, passing planetesimal and trapped it in its orbit. A third said that a large mass spun off from the rotating molten proto-Earth and cooled and condensed to form the Moon.

Physico-chemical analyses and radiometric dating of rock samples brought back from the Moon by Apollo astronauts in the 1970s provided the first hard evidence, which was inconsistent with the three speculations. It revealed, among other things, that lunar rocks were more typical of the Earth's mantle than of primitive meteorites (and hence planetesimals), that they were strongly depleted in elements more volatile than potassium compared with the Earth's mantle, that they had very little iron, and that the Moon's oldest surface rocks—at some 4.4 to 4.5 billion years—were older than the Earth's surface rocks.[21]

The evidence prompted two groups of scientists to formulate in 1975 the giant impact hypothesis, which they then spent 10 years of computer simulations to develop. It says that a huge planetesimal, about the size of Mars, hit and fused with the proto-Earth. In the heat of the collision both molten iron cores joined together like two drops of water and sank to the centre of the proto-Earth, while the impact sent up to 70 per cent of the surface material of the Earth hurtling into space where it was trapped by the new, larger proto-Earth's gravitational field. Very rapidly, perhaps in as little as a year, this molten debris aggregated under its own gravitational fields and condensed to form the Moon.[22]

The hypothesis explains not only the abnormally large size of the Moon compared with its parent planet, together with the chemical composition and low density of the Moon, but also why the Earth is the densest planet in the solar system, why it has an iron core that, from what evidence we can deduce, appears proportionately larger than that in the other rocky planets of comparable size, Venus and Mars, and also why it has a thinner crust than these two other planets.

Although some questions remain unanswered, the giant impact hypothesis is now scientific orthodoxy.

Biosphere
At the beginning of the chapter I gave six conditions considered necessary for the emergence and evolution of life as we know it. Science's current account of how these formed the Earth's biosphere is as follows.

1. *Essential elements and molecules*
   The stellar nebular hypothesis explains the presence of the essential elements while different hypotheses, mainly bombardment in the Hadean Eon by water-rich comets and/or asteroids and/or protoplanets, suggest how the Earth possesses such a large volume of surface water.

   Some planetary scientists suggest that a significant proportion of organic molecules present in comets and meteorites survived intact when these hit the surface, and so the early water collections were seeded with organic compounds.[23] Since we know that organic molecules as complex as amino acids formed on comets and asteroids,* it seems to me feasible that they also formed independently on Earth; the evidence is lacking simply because nothing is left of the original surface of the Earth, whereas comets and asteroids are thought to be largely unchanged since their formation.

2. *Mass of planet*
   While the nebular hypothesis explains the formation of gas giants in the outer disc, too massive for life on their surfaces, and the formation of rocky planets in the inner disc, it does not explain why the Earth grew to a mass that could support life, while another rocky planet, Mercury, ended up far too small at 0.055 that size. Computer simulations of terrestrial planetary formation devised by Sean Raymond of the University of Colorado in 2004 imply that planetary size is a random feature of the accretion process.

3. *Protection from hazardous radiation and impacts*
   Protection from hazardous radiation was achieved for the new Earth by its atmosphere and its powerful magnetic field. The magnetosphere deflected away ionizing radiation from the Sun as illustrated in Figure 12.4. The early Earth did not have protection, currently supplied by the ozone layer

---

*   See Glossary for the distinction between *comet, meteorite,* and *asteroid*.

of the atmosphere, against any high intensity ultraviolet radiation fatal to higher lifeforms that evolved later.

The gravitational field of the gas giant, Jupiter, at more than 300 Earth masses and 5 times the Earth's distance from the Sun, shields the Earth and the other inner planets from most comets. George Wetherill, Director Emeritus of the Department of Terrestrial Magnetism at the Carnegie Institute, estimated that without Jupiter's mass and location the Earth would receive 1,000 to 10,000 times the current number of cometary impacts.[24]

4. *Sources of energy*

Four sources of energy were available to drive essential biochemical processes: conversion of the kinetic energy of planetesimals to heat energy as they hit the surface of the Earth; conversion of gravitational potential energy to heat energy as molten matter gravitated towards the centre of the Earth; energy generated by the radioactive decay of elements like uranium and thorium; and energy radiated by the Sun.

5. *Temperature range*

The temperature at or near the Earth's surface resulted from all four sources of energy. Its range was principally controlled in the long term by solar radiation, which depends on five factors.

First is the amount of energy emitted by the Sun (its luminosity).

Second is the distance of the Earth from the Sun, which varies from 147 to 152 million kilometres because of the elliptical shape of its orbit during its 365.25-day journey round the Sun.

Third, the albedo, the proportion of incident solar radiation reflected away, varies: tenuous stratus clouds have a meagre albedo while thick stratocumulus clouds have an albedo of up to 80 per cent, fresh snow 40–70 per cent, and dry sand 35–40 per cent; the Earth's current overall albedo is about 35 per cent. However, we don't know what it was in the first two billion years or so of the Earth's existence.

Fourth, of the solar energy reaching the surface, some is absorbed and then radiated away as heat at longer wavelengths, and this depends on the length of time the surface faces the cold night sky, which in turn depends on the Earth's 24-hour period of rotation about its axis and, in specific locations, on the latitude and the 23.5° tilt of the Earth's axis.

Fifth, part of this heat energy radiated away is reflected back to heat up the surface, and this depends on the gases that make up the atmosphere. Currently carbon dioxide accounts for 0.003 per cent of the atmosphere and is principally responsible for reflecting back enough of this radiated energy to produce a mild greenhouse effect.

All these five factors currently produce a biospheric temperature that ranges from about –50° Celsius to +50° Celsius depending on time of year, location, and distance from the surface, while the mean ocean surface temperature averaged over a year is 15° Celsius.[25] Thus at most times of the year most places on, just below, or just above the surface of the Earth have a temperature well suited to the biochemical reactions that promote and maintain life as we know it.

6. *Stability*

The final condition, stability, principally requires that the biosphere has a temperature maintained within the range that enables water to exist as a liquid on the planet's surface and is protected from hazardous radiation and impacts over a period long enough for primitive lifeforms to emerge and evolve into more complex ones that evolve eventually into humans. That such stability has been achieved in the Earth's biosphere over some 4 billion years is attributed to several factors.

Computer modelling of planetary formation shows that, once formed, the principal planets maintain a generally stable orbit round the Sun due mainly to conservation of angular momentum.

However, this of itself would not produce a stable biospheric temperature range because the early Sun was considerably dimmer and cooler than it is now. Based on studies of stellar formation and different stages of stellar evolution in other parts of our galaxy and studies of the solar nuclear fusion processes that generate energy, calculations indicate that at the beginning of the Earth's Archaean Eon, 3.8 billion years ago, the Sun radiated 25 per cent less energy than it does now. If everything else had been the same, the surface temperature then would have been −18° Celsius:[26] all water would be ice, with no liquid water necessary for biochemical processes essential for reproducing and maintaining life.

To account for the existence and evolution of life in this eon requires additional sources of energy to warm the biosphere. These could be residual heat from the planetary formation process, greater heat from radioactive decay of elements in the interior, or more radiated heat reflected back from the Earth's atmosphere to increase the greenhouse effect.

We don't know the composition of the Archaean Earth's atmosphere but, since the atmosphere of the Earth at its formation 4.5 billion years ago was clearly very different from the 78 per cent nitrogen and 21 per cent oxygen of today (see *Hydrosphere and atmosphere* above), it is reasonable to infer that 3.8 billion years ago it was nearer the former. A greater proportion of greenhouse gases, like carbon dioxide and methane, could have warmed the biosphere sufficiently for liquid water to exist. However, if the different atmosphere had a different albedo, this factor must also be taken into account in any calculation.

Moreover, if a different atmosphere produced a larger greenhouse effect to account for liquid water on the Earth during an eon when the Sun radiated 25 per cent less energy, this implies a finely tuned reduction in the greenhouse effect as the Sun's radiated energy increased over 4 billion years.

Several ideas have been advanced to account for this phenomenon, including the Gaia hypothesis, first proposed by the British independent scientist James Lovelock in the 1960s, that posits regulation by biotic feedback. For example, photosynthesizing bacteria in the oceans that absorb carbon dioxide from the atmosphere flourish as sunlight and temperature increase, thus reducing the greenhouse effect; as the temperature falls their rate of growth falls, leaving more carbon dioxide in the atmosphere, thereby increasing the greenhouse effect and the surface temperature. It was given empirical support by chemical analyses of long ice cores drilled through ice sheets in Antarctica.[27]

Kasting, however, argues that a biotic feedback mechanism cannot be the primary regulator of temperature, not least because the living biosphere is not a large enough carbon reservoir. He claims the most important long-term regulation of surface temperature by the greenhouse effect is caused by negative feedback between atmospheric carbon dioxide and surface temperature through what is known as the carbonate-silicate cycle. This involves carbon dioxide dissolving in water and then breaking down silicate rocks through weathering. Rivers carry the products to the oceans where they eventually form an oceanic tectonic plate that is pushed beneath a continental plate, and carbon dioxide is subsequently released back into the atmosphere through volcanoes.[28]

Whichever explanation is correct, one or, more likely, several feedback mechanisms have regulated a stable biospheric temperature range.

Within this range the temperature varies from one time of the day to another, one location on the surface to another, one distance from the surface to another, and one part of the solar year to another. Such temperature variations generate weather patterns in different zones that change on a daily and seasonal basis. Moreover, these have varied as the Earth's axial tilt and the ellipticity of its orbit have changed, while the Sun's radiation has increased and the composition of the Earth's atmosphere has changed dramatically, generating climatic changes over periods of tens of thousands of years.

Earth's biosphere is a system that has been in a stable state, but far from thermodynamic equilibrium, for some 4 billion years.

Is the Earth special?

The orthodox view

Ever since Galileo used a telescope to prove Copernicus's hypothesis that the Earth was not the centre of the universe and that the Earth and other planets orbit the Sun, scientists have assumed that the Earth is not special.

More sophisticated instruments have changed our understanding of what the universe comprises, and so today the orthodox view is that the Earth is just an ordinary planet orbiting an ordinary main-sequence star, which is in an ordinary orbit round the centre of an ordinary galaxy of some hundred billion stars, which is part of a local group in an ordinary cluster of galaxies in orbit round the centre of an ordinary supercluster of galaxies, which forms part of the observable universe of an estimated hundred billion or more galaxies, which is part of a universe much of which is beyond the visible horizon. Orthodox cosmologists believe that even this universe is only a microscopic part of an inflated universe.

It is difficult to demote the Earth to a place of less significance in the cosmos.

The strongly held view that the Earth isn't special, combined with the knowledge that it is the home of intelligent life, led radio astronomer Frank Drake in 1961 to estimate the number of other intelligent civilizations in our galaxy alone. He devised an equation that multiplies 7 parameters. His value for the first of these, the rate of formation of stars suitable for the development of intelligent life, was a crude, if not simplistic, estimate; his values of the others were guesses based on extrapolating then contemporary views of the only known example, the Earth. Multiplying these values produced 10 civilizations in the galaxy currently detectible by their emitting radio and television signals as we do. Although the number resulted from a series of guesses rather than evidence-based estimates, the word "equation" seemed to give it an aura of scientific respectability. Drake's Equation, as it is now known, generated great excitement and prompted the SETI (Search for Extraterrestrial Intelligence) programme. This was adopted by NASA and funded for a time by the US Congress.[29]

More than 50 years searching likely locations has produced no evidence of any other intelligent life. Other searches have failed so far to find evidence of any forms of life, however primitive, in our solar system or elsewhere.

Nonetheless, the current orthodox view was summed up by the then President of the Royal Astronomical Society, Michael Rowan-Robinson, in the 2004 edition of his book *Cosmology*:

> It is evident in the post-Copernican era of human history, no well-informed and rational person can imagine that the Earth occupies a unique position in the universe. We shall call this profound philosophical discovery the *Copernican principle*.[30]

Evidence questioning the orthodox view
Evidence began accumulating many years before Rowan-Robinson's book that the Earth, if not unique, may have resulted from a special concurrence of factors that enable the emergence and evolution of complex life as we know it.

Galactic habitable zone
The most recently proposed factor is location in the galactic habitable zone (GHZ). This idea was advanced in 2000 by two scientists at the University of Washington, palaeontologist and Professor of Biology and of Earth and Space

Sciences Peter Ward and Professor of Astronomy Don Brownlee,[31] and developed by them and others, notably astronomer Guillermo Gonzalez, formerly of the University of Washington and now of Ball State University, Indiana.[32]

Based on studies of giant molecular clouds and stellar formation in different parts of the Milky Way, together with radiation activity, this hypothesis applies the circumstellar habitable zone idea to the galaxy. It proposes that there are very few zones where complex life can evolve and that these zones change with time.

The outer regions of the galaxy lack sufficient essential elements and molecules for rocky planets to form with the ingredients for life. These elements are present near the centre of the galaxy, but chaotic gravitational interactions in such a concentration of stars create conditions too unstable for any planetary biosphere. Moreover, during the billions of years it takes for complex carbon-based life to evolve, stars and any planetary systems would be hit by deadly bursts of radiation, either from supernova explosions or from matter just before it is sucked into the giant black hole thought to be at the galactic core. This region of fatal radiation has shrunk with time because these bursts probably were more common in the galaxy's early phase.

The spiral arms of our galaxy are also inimical to life. These are the regions of active star formation where giant molecular clouds condense chaotically to produce several stars. Energetic ultraviolet radiation from massive young stars evaporates neighbouring gas and dust discs before planet formation can commence, while these large stars end their lives relatively quickly in supernova explosions.

The only zones where conditions are right and remain stable long enough for biological evolution are between spiral arms in the plane of the galactic disc, where our own solar system is currently located about half way between the galactic centre and the visible outer edge of the disc. Assessing where these zones are is complicated by the fact that stars in these zones orbit the galactic centre, but not necessarily in the same plane or at the same velocity as do the spiral arms.

Nikos Prantzos of the Institut d'Astrophysique de Paris has criticized the hypothesis as being imprecise and unquantified. We saw how initial quantifications of the circumstellar habitable zone hypothesis were too simplistic and failed to take account of many factors that determine planetary surface temperature.* On the galactic scale many of the determinants are poorly understood, and quantifying them is currently beyond the ability of astrophysicists; in fairness, its proponents have made clear that the hypothesis is in its development stage. Nonetheless the basic proposal is sound: zones providing the six conditions for the emergence and evolution of complex life as we know it constitute only a small part of our galaxy, and the physical location of these zones changes with time as the galaxy evolves.

---

*   See page 166.

The Earth has been in such a zone for 4.5 billion years and so cannot be considered just a typical planet orbiting a typical main-sequence star in a typical orbit round the centre of a typical galaxy.

Suitability of star
Even if a stellar system is in a galactic habitable zone, it doesn't mean that the star is suitable for the emergence and evolution of carbon-based life. Stars are classified by a letter—O, B, A, F, G, K, or M—ranging from large to small. Within each category they are further subdivided by number. Our Sun is a G2 star.

Only about 2 to 5 per cent of all stars are as large as the Sun. A planet orbiting a star smaller than about K5 will probably be tidally locked, that is, it always shows the same side to the star; consequently the facing side is likely to be too hot and the dark side too cold for complex life to evolve. Complex life is also unlikely to evolve on a planet orbiting a star more massive than F0 because the main sequence lifetimes of these larger stars are relatively short and the stars emit large amounts of damaging ultraviolet radiation.

This leaves about 20 per cent of main-sequence stars within a suitable mass range. However, about two-thirds of all stars occur in binary or multiple star systems.[33] A binary stellar system, especially where the two stars are reasonably close to each other, is unfavourable for life: any planets would almost certainly have such a highly elliptical orbit that their surface temperatures would change from being far too cold to far too hot to support life in a single revolution round the two stars.

Circumstellar habitable zone
For those suitable stars that remain long enough in a galactic habitable zone, life as we know it can only emerge and evolve if the star has one or more planets that remain long enough in a zone in which the surface temperature of the planet permits the existence of liquid water, commonly called the circumstellar habitable zone (CHZ). As we saw previously, early attempts to define such a zone were far too simplistic.

A more sophisticated attempt was made in 1993 by the atmospheric scientist regarded as the leading authority on planetary habitability, James Kasting, together with Daniel Whitmire and Ray Reynolds. Their one-dimensional climate model addresses the solar energy reflected away (the albedo effect) and the greenhouse effect, plus other factors, but it doesn't address the effect of the rotational period of the planet on the amount of solar energy the surface receives and subsequently radiates away.[34] Although a considerable improvement on previous models, the assumptions and approximations made highlight the difficulties of quantifying such complex, interactive systems involving climatic feedback processes. The Science and Technology Definition Team for NASA's Terrestrial Planet Finder project concluded that one-dimensional climate models are incapable of accurately simulating the effects of clouds (water vapour or carbon dioxide) on planetary radiation and hence are not reliable; it did, however, use a more sophisticated version of Kasting's model to suggest a CHZ of between 0.75 AU and 1.8 AU* for the Sun, with scaling for other stars.[35]

Exoplanets

1994 saw confirmation of the 1992 announcement by two radio astronomers of the first planets detected outside our solar system. Two planets about three times the mass of the Earth, plus probably a Moon-mass object, orbit an old, rapidly spinning neutron star, which is a small, dense remnant of a large star that has ended its main sequence life by catastrophic gravitational collapse and supernova explosion.[36] It is one of the least likely places astrophysicists thought they would find planets, and their proximity to the powerfully radiating neutron star means that the six conditions for the evolution of known lifeforms cannot be fulfilled.

The first planet orbiting a main-sequence star was detected the following year. Telescopes didn't have sufficient resolution to observe it directly, but astronomers calculated the planet's mass and distance from its star by the wobble its gravitational field caused in the star's distance from the Earth as the planet passed in front of, round, and behind the star. 51 Pegasi b is a gas giant like Jupiter that orbits the star 51 Pegasi every four days, meaning that it is closer to its star than our innermost rocky planet, Mercury, is to the Sun.

By the end of 2008 astronomers had detected 330 extrasolar planets, now labelled exoplanets. These undermined the nebular model of planetary formation, which predicts that planets orbit their star in the same direction with almost circular orbits in or near the star's equator, with comparatively small, dense inner planets comprising mainly rock and iron (like Mercury, Venus, Earth, and Mars) and enormous gas giants made mostly of hydrogen and helium at much greater distances from the star (like Jupiter, Saturn, Uranus, and Neptune). Not only are there "hot Jupiters"—gas giants that speed around their star closer than Mercury to the Sun where the temperature is too hot for solid ices and gases to form a planetary core—but other planets follow wildly elliptical orbits, with one orbiting its star's pole instead of its equator and another orbiting in a direction opposite to its star's spin.

NASA subsequently launched the space-based Kepler telescope that detects the slight dimming of a star's light when a planet passes in front of it, enabling it to identify much smaller planets than previous techniques were capable of. As of August 2014 it had found 978 exoplanets in orbit around 421 stars, with a further 4,234 candidate planets awaiting confirmation.[37]

This prompted a widespread belief that the Earth certainly wasn't special and in a galaxy of some hundred billion stars there must be very many habitable planets, on some of which intelligent life would have evolved.

---

*   AU means Astronomical Unit, the mean distance of the Earth from the Sun.

However, the stellar systems identified by Kepler further undermine the nebular model of planetary formation. In addition to hot Jupiters it found giant planets with idiosyncratic orbits, while the most common planets orbiting some 40 per cent of nearby Sun-like stars are the so-called Super-Earths—a range of planets more massive than Earth but smaller than Neptune (17 Earth masses)—most of which orbit their star too closely for life to exist.

Astrophysicists have advanced various proposals to modify the nebular hypothesis of planetary formation in the light of this evidence. Several conjecture that all types of planet grow to full size in the middle to outer part of the nebular disc and spiral in as viscous gas in the disc slows their orbits; this would explain hot Jupiters. However, simulation models show such migrating planets continue to spiral into their star. No one has yet been able to explain why they stop their death spiral and stabilize in their observed orbits.[38] In fact, because we haven't observed them in sufficient detail over a sufficient period, it may be that they are still spiralling in. Moreover these models do not explain why our solar system is so different, other than by making the arbitrary assumption that the nebular gas in our disc wasn't sufficiently viscous to cause the outer gas giants to spiral in.

Current techniques are not capable of detecting Earth-mass planets around most stars. It may be that more sophisticated instruments will find many planets resembling Earth in mass, size, and stable circumstellar habitable orbit.

On the other hand, our solar system does not contain the most common planets detected so far, the Super-Earths. Perhaps it is rare for a small rocky planet like Earth to remain in a stable orbit—speculatively due to the gravitational influence of an outer gas giant like Jupiter that also shields it from cometary bombardment—at the optimum distance from its star over some 4 billion years to enable complex life to evolve.

Impact of a Mars-sized planetesimal

Science's orthodox explanation is that the impact of a Mars-sized planetesimal produced Earth's exceptionally large Moon.

Recent studies suggest that large impacts may not have been so rare in the final stages of planetary accretion when the four rocky protoplanets were sweeping up the remaining large planetesimals in their orbital zones. Nonetheless, it required the impact of a planetesimal with sufficient mass at just the right relative velocity and angle to create the Earth's abnormally large iron core and an abnormally thin crust, cause the Earth to spin faster, cause the Earth's axial tilt of 22° to 24°, and produce its abnormally large Moon that stabilized this tilt, slowed the Earth's rotation, and caused the tides. These consequences significantly affect the emergence and evolution of life.

a. *abnormally large iron core*

An abnormally large iron core means that only the inner core has cooled enough to solidify, leaving a large outer liquid core for 4.5 billion years to generate a correspondingly powerful magnetosphere that protects the surface of the Earth from fatal ionizing radiation and its atmosphere from the solar wind.

Analyses of rocks collected by Mars landers indicate that Mars once had a magnetic field, but surveys find no magnetic field now. This prompted the hypothesis that Mars once had a liquid iron core but, like the Earth's inner core, it cooled and solidified; without a protective magnetosphere nearly all the Martian atmosphere was blown away by the solar wind. Probes have also failed to detect any significant magnetic field on Venus (the planet is thought to maintain a dense atmosphere because its high surface temperature volatilises compounds like water that would be liquid on Earth).

b. *abnormally thin crust*

The Earth's abnormally thin crust enables the movement of tectonic plates, thought to be unique in the solar system;[39] evidence from probes, for example, suggests that the tectonic plates of Venus are locked.

As we saw, Kasting argues that a favourable biospheric temperature range has been maintained over billions of years despite an increase in the Sun's radiated energy principally by regulation of the greenhouse effect of carbon dioxide gas through the mechanism of plate tectonics.

Moreover, members of a species who are separated by continental drift due to plate tectonics find themselves in different physical and climatic conditions; the two branches of the species will evolve differently in these different environments.

In these ways the thin crust enables the evolution of species.

c. *changed rotation*

Computer modelling suggests that the giant impact caused Earth to spin much more quickly about its axis, perhaps once every 5 hours, and the newly formed Moon was very much closer.

The proximity of the large Moon created an enormous gravitational pull on the surface of Earth, which rose and fell by about 60 metres as the Moon passed overhead. This drag on the Earth's crust slowed down the Earth's rotation and forced the Moon to move further and further away, as evidenced by the NASA laser measurements.* The Earth's rotation has now slowed to once per 24 hours. This longer day means that the surface receives more solar energy during the day and radiates more heat away during the night. Hence there is a greater variation of surface temperature about its mean value compared with a faster spinning planet. This enables a greater range of biochemical reactions, but still within the range for liquid water, leading to greater variety of complex molecules, which increases the possibilities for greater molecular complexification. After life has emerged

---

*   See page 176.

the same effect produces a greater variety of environments, which promote greater evolutionary diversity.

By contrast, Earth's "sister planet" of nearly the same mass, Venus, rotates on its axis once every 243 Earth days. Combined with its orbital period around the Sun, this produces a solar day on Venus of 117 Earth days. Such a long daily exposure to solar radiation helped create a runaway greenhouse effect to produce a surface temperature that now stays around 470° Celsius during both day and night, which is hot enough to melt lead.[40]

d. *axial tilt of 23.5°*
The 23.5° tilt of the Earth's rotational axis (see Figure 12.2) generates the seasonal variations in temperature, which are different at different latitudes, but still within the biospheric range. This variety in seasonal climates at different places on the surface produces a variety of changing environments that promote biological evolutionary diversity.

e. *stabilization of axial tilt*
The proximity of the abnormally large orbiting Moon generates a strong gravitational field that stabilizes the Earth's axial tilt.

The stabilization isn't perfect; the axial tilt varies between 22° and 24° over 41,000 years[41] and contributes to the Earth's periodic ice ages as proposed in what is known as the Milankovitch hypothesis. However, this variation is small. A 1993 study concluded that without the Moon the tilt would vary chaotically from 0° to 85° within a timescale of tens of million of years, and this would wreak havoc on the Earth's climate. A 2011 computer simulation has questioned this, suggesting that the tilt would vary between 10° and 50° over 4 billion years.[42] Nonetheless, such tilt changes would result in wide-ranging temperature changes, making it extremely difficult for complex life to evolve.

f. *tidal flows*
The Moon's powerful gravitational field is also responsible for tidal flows in the Earth's oceans and seas. Such tides not only change physical environments—and thus promote biological evolutionary diversity—by eroding coastlines, they also bring material from the sea to the shore and take material from the shore into the sea, promoting dynamic ecosystems.

The result of all these factors, combined with a significant increase in the Sun's radiated energy and a huge change in the composition of Earth's atmosphere, is a changing flow of energy through a system that has remained stable but far from thermodynamic equilibrium over some 4 billion years. Drawing on the work of Ilya Prigogine, Morowitz and others consider that such a system generates physical and chemical complexification that leads to the emergence of life followed by its evolution.[43]

Conclusions

Part 1 led to the conclusion that a universe in which matter can evolve into known lifeforms requires a set of physical and chemical laws that either describe or else determine how matter interacts, the fine-tuning of six cosmological parameters and two dimensionless constants, and the fine-tuning of three parameters in stellar nucleosynthesis that enable the production of sufficient carbon for essential organic molecules.

This chapter leads to the following conclusions.

1. Six conditions are necessary for the organic molecules of up to 13 atoms found in interstellar space and on asteroids to evolve into things as complex as humans: a planet with essential elements and molecules, sources of energy, a minimum and probably a maximum mass, protection from hazardous radiation and impacts, a narrow temperature range just below, at, and just above its surface, and stability of this biosphere over billions of years.

2. A concurrence of galactic, stellar, and planetary factors provides these six conditions on Earth.

   2.1 Its parent star is single, has a mass within the narrow range required for stability over 4.5 billion years, and is located in the relatively small and changing galactic habitable zone over such a period.

   2.2 It formed as a rocky planet comprising or subsequently acquiring essential elements and molecules.

   2.3 Its mass lies within the range that supports a biosphere.

   2.4 Its location within a narrow circumstellar habitable zone is atypically shielded over 4.5 billion years from life-destroying cometary bombardment by the gravitational effect of an outer gas giant.

   2.5 As the planet was forming it was impacted by a planetesimal with sufficient mass at just the right relative velocity and angle to produce several features favourable for the evolution of complex life: an abnormally large iron core that generates a protective powerful magnetosphere; an abnormally thin crust enabling the movement of tectonic plates; and an abnormally large moon producing an optimal rotation, a stable axial tilt, and tidal flows in its oceans.

   2.6 The planet has one or more feedback mechanisms that maintain a surface temperature range favourable for biochemical reactions and enable liquid water to remain on its surface for some 4 billion years despite a large increase in energy radiated by its evolving parent star.

3. Together these factors produce a changing flow of energy through a physico-chemical system that remains stable but far from thermodynamic equilibrium over some 4 billion years to generate the complexification necessary for the emergence and evolution of varieties of lifeforms.

4. These factors contradict the view of orthodox cosmology that the Earth is just an ordinary planet orbiting an ordinary star in an ordinary galaxy of some hundred billion stars forming part of the observable universe of an estimated hundred billion galaxies.
5. The Earth, if not unique, is a rare location in the galaxy, if not the universe, in possessing the conditions necessary for the emergence and evolution of lifeforms as complex of humans.

The Earth evolved from a planet with a hot surface and a probable atmosphere of hydrogen sulphide, sulphur dioxide, carbon dioxide, water vapour, nitrogen, and ammonia, which is poisonous to humans and provides no protection from ultraviolet radiation, to one with an average surface temperature over a year of around 15° Celsius, blue seas, fluffy clouds, and an atmosphere primarily of nitrogen and oxygen together with an ozone layer to block harmful ultraviolet radiation thus enabling the emergence and evolution of life on its surface. But what is life?

CHAPTER THIRTEEN

Life

Then, what is life? I cried.
—Percy Bysshe Shelley, 1822

The essence of life is statistical improbability on a colossal scale.
—Richard Dawkins, 1986

If we are to find how life emerged on Earth we need to know what life is. To most of us this seems self-evident. The cat that rubs against my leg is alive; the piece of burnt toast on the plate in front of me is not alive. But defining what it is that distinguishes the living from the nonliving is less easy.

As Chapter 9 noted, we consist primarily of atoms of hydrogen, oxygen, carbon, and nitrogen. But these are no different than the atoms of hydrogen, oxygen, carbon, and nitrogen found in water, in the air, and in burnt toast. Indeed, a stream of such atoms enters and leaves our bodies as we breathe, drink, eat, sweat, urinate, and defecate.

If it is not the atoms that are different in living things then is it their arrangement in more complex molecules? But a person who has just died possesses the same level of molecular complexity as he did the moment before he died.

I shall try to arrive at an understanding of what life is by examining what the ancient world thought it was, what explanations developed in science, some contemporary explanations that claim to reconcile ancient insights with modern science, orthodox science's response to these, the characteristics of life currently advanced by orthodox science, and some significant definitions of life.

The ancient world's understanding of life

The earliest attempts at understanding the essence of life were made in India. Seers used meditation—a disciplined introspection that seeks understanding by becoming one with the object of study—and their insights were recorded in the Upanishads. Different Upanishads characterize life as *prana*.

This Sanskrit word probably comes from the prefix *pra-* meaning forth (possibly used here as an intensifier) and the root *na* meaning to breathe.[1] It could be interpreted literally as breath, but the Prashna Upanishad makes its full meaning clear. Six seekers after truth each ask the seer Pippalda a question. In answer to the first, who created the universe, Pippalda replies that the Lord meditated and brought forth *prana* and *rayi* (matter), and from this duality spring the dualities of male and female, Sun and Moon, and light and dark. Like other Upanishads the rest of the Prashna Upanishad makes clear by metaphor, simile, and parable that *prana* is the fundamental and vitalizing energy of each individual's body just as it is the fundamental and vitalizing energy of the universe.[2] Ayurvedic medicine seeks to achieve a balance in *prana*, while yoga aims to enhance its flow through the body.

Michael Nagler, Emeritus Professor of Classics and Comparative Literature at the University of California, Berkeley, argues that *prana* means living energy, and that all the vital signs by which we try to identify the presence of life from without are tokens of the capacity of the body to direct, conserve, and employ energy at a high level of complexity. That is life from the biological standpoint.[3]

Responding to the rejection of vitalism by modern science, he asks: if life is not energy, what is it?

A similar understanding is given in the Daoist texts of China, collections of insights and philosophical reflections that date from around the sixth century BCE onwards. Here the ancient Chinese word 氣, *qi* (or *chi* depending on the Westernising convention), means both breath and life-giving spirit.[4] Manipulating the flow of *qi*, the body's vitalizing energy, plays a central role in traditional Chinese healing.

This understanding was transmitted to the Japanese when they adapted Chinese culture for their own society. Their word for breath and spirit is *ki*.

The Hebrews also used one word with a double meaning: *ruach* as wind or spirit. In the Tanakh, their scriptures, it is used to mean the Spirit of God.

In the Homeric poems of Greece dating from the eighth century BCE, *psyche* is the breath or spirit that leaves the hero's body when he dies.[5] The Stoics, from the third century BCE to the first century CE, spoke of *pneuma* (breath, soul, or vital spirit) and God and the organizing principle of nature as basically the same thing, echoing Heraclitus's insight of a fire-like intelligence or soul that animates the universe.[6]

Similarly the Romans used the Latin word *spiritus* to mean both breath and spirit; the Western Christian Church adopted the word to name the third person of the Trinity as the Holy Spirit.

Thus the explanation that a vital spirit or life energy animates (literally "breathes life into") matter, and distinguishes it from the inanimate, was widespread in the ancient world. Moreover, this life energy—generally called vitalism   is the ground of all energy in the universe.

In different cultures and religions it became interpreted as either the creation of a Supreme Godhead beyond all form, or God itself, or an aspect of God, or, anthropomorphically, that which God or a god breathed into inanimate matter, like clay, to give it life.

It was not just mystical insight that gave rise to vitalism, but early science deduced its existence.

The development of science's explanation of life

Much of Western science and medicine is rooted in its rediscovery from the twelfth century onwards of Aristotle. Unlike many other ancient Greek thinkers, like Pythagoras and Heraclitus, Aristotle was not a mystic. He drew a distinction between inanimate mineral things on the one hand and vegetable and animal things on the other hand, and he held that the life of the latter consists in its *psyche* or soul. For Aristotle, however, *psyche* was not some ethereal spirit. It was the form, or organization, of the physical characteristics that endows it with life and purpose; the soul and the body are two aspects of a single living thing. Several mediaeval commentaries, however, interpreted translations (mainly Arabic) of Aristotle to justify a less materialistic vitalism.[7]

Such a vitalism was advocated in the emergence of modern science during the sixteenth and seventeenth centuries to oppose the extension to biology of Cartesian mechanism, which considered an organism simply as a machine. The vitalists argued that matter could not explain movement, perception, development, or life. Even mechanists of the eighteenth century like John Needham and the Comte de Buffon were compelled by their experiments in developmental biology to invoke a vital force analogous to gravity and magnetic attraction.

Vitalism played a pivotal role in the development of chemistry in the eighteenth and early nineteenth centuries. It gave rise to an Aristotelian distinction between substances that were organic (extracted from animals and vegetables) and inorganic (minerals). Georg Stahl burned wood and argued that the loss in weight between the wood and its ashes was due to the vital force that had been irretrievably lost.

Opposition to vitalism grew from the mid-nineteenth century onwards, led by physicians and physiologists who advocated mechanical materialism and by chemists who synthesized compounds found in nature from their constituent chemicals, thus refuting Stahl's claim.

Vitalism nevertheless continued to be held in the nineteenth century by such notable scientists as Louis Pasteur, who concluded that fermentation was a "vital action". By the early twentieth century, however, orthodox medicine, biology, and chemistry had rejected vitalism on the grounds that all such claimed phenomena could be explained by reducing them to physical and chemical components that obeyed physical and chemical laws. Nonetheless, vitalism was championed throughout the first half of the twentieth century by French philosopher and

Nobel Literature laureate Henri Bergson and the eminent German embryologist Hans Driesch.[8]

Claimed reconciliations between ancient insights and modern science

Alternative medicine

The last fifty years or so have seen the growth in the West of alternative medicine based on ancient Eastern healing techniques. Such treatments as acupuncture and acupressure are designed to unblock restrictions in the flow of *qi*, while therapies like reiki (from the Japanese *rei* meaning unseen or spiritual and *ki* meaning life energy, hence "universal life energy") also aim to transmit or channel the universal life energy to the patient either by touch or by the movement of the practitioner's palms close to the patient's body.

Since 1972, when a Western medical journal first reported Chinese use of acupuncture as an analgesic instead of drug-induced anaesthesia in major surgery, parts of the Western medical profession have successfully trialled and used acupuncture for many conditions, including the management of acute and chronic pain, recovery from post-stroke paralysis, and relief from the effects of respiratory diseases. How it works is an open question.

The Eastern holistic approach has been adopted to a greater degree in the West by the nursing profession. The best-known treatment, taught at over a hundred universities and nursing and medical schools in the United States and Canada, is Therapeutic Touch. The touch is not with the patient's body but with the body's vital energy field, or aura, that practitioners believe extends several inches to several feet from the body. The techniques employed are indistinguishable from those of reiki and are also practised under different names like healing, spiritual healing, or psychic healing.

I agreed to be a recipient on two occasions. On the first I experienced a sensation of warmth and on the second a tingling as the practitioner passed her hands over my body. I do not know the nature of these sensations, which are claimed to be characteristic.

The founder of Therapeutic Touch, New York University professor of nursing Dolores Krieger, identified the energy as *prana*. She maintains that Therapeutic Touch, like hypnotherapy, works most effectively on the autonomic nervous system. "In the final analysis it is the healee (client) who heals himself."[9] The healer or therapist, in this view, acts as a human energy support system until the healee's own immunological system is robust enough to take over.

Claimed successes are largely anecdotal, with very few systematic trials. In 1973 Krieger demonstrated that the mean haemoglobin values of 46 subjects increased after Therapeutic Touch treatment compared with no significant change for 29 in a control group.[10] In 1998 the University of Pittsburgh Medical Center reported significant improvements in function and pain relief for osteoarthritis of the knee in patients treated with Therapeutic Touch compared with placebo and control groups, although only 25 patients completed the study.[11]

On the other hand the *Journal of the American Medical Association* famously published on 1 April 1998 a trial purportedly designed by 11-year-old Emily Rosa. She invited 21 Therapeutic Touch practitioners each to put both their hands through a hole in a screen and sense her living energy field by saying near which one Rosa had chosen on the toss of a coin to place her own hand. The practitioners achieved a success rate of 44 per cent, close to the 50 per cent rate they would be expected to achieve by chance alone.[12]

Field hypotheses
During the same period a small number of scientists and philosophers developed the view that orthodox biology is still stuck in the Newtonian mechanistic approach that characterized physics in the nineteenth century and hasn't taken account of the early twentieth century revolution in physics, especially quantum theory with its concepts of fields and non-localization. They are attracted by the idea that the cosmos is a whole and that all its parts exist in a state of dynamic, coherent interdependence.

Quantum theorist David Bohm, who was influenced by the Indian mystic and philosopher Jiddu Krishnamurti, was one of the first to try to develop a holistic model of the universe based on scientific principles. He called this "Undivided Wholeness in Flowing Movement" in which "the flow is, in some sense, prior to that of the 'things' that can be seen to form and dissolve in this flow". He went on to say that "Life itself has to be regarded as belonging in some sense to a totality" and that it is somehow "enfolded" in the whole system.[13]

Former professor of philosophy, systems theorist, and classical pianist Ervin László claims support from the latest discoveries in the natural sciences for a field that instantaneously connects and correlates all things in the cosmos. This is a rediscovery of the ancient mystical insight of the *akasha* (a Sanskrit word meaning space), the most fundamental of the five elements of the cosmos that holds the others (air, fire, water, and earth) within itself but is at the same time outside of them. He claims that the akashic field is the fundamental medium of the cosmos, and equates it with the quantum vacuum of the universe,* which orthodox quantum theorists consider has a ground state energy from which matter can spontaneously emerge.[14] According to László the akashic field is the originating ground of all things in the universe, and hence the source of all life as well as that which connects all life.

Plant biologist Rupert Sheldrake was for seven years Director of Studies in Biochemistry and Cell Biology at Clare College, University of Cambridge. He went to India where he worked on crop physiology, held discussions with Krishnamurti, and spent 18 months at the ashram of Bede Griffiths. He also drew

*   See page 46.

on the vitalist ideas of Bergson and Driesch when developing his hypothesis of formative causation, according to which memory is inherent in nature: most of the so-called laws of nature are more like habits that depend on non-local similarity reinforcement.[15]

Sheldrake argues that natural systems, or morphic units, at all levels of complexity—atoms, molecules, crystals, cells, tissues, organs, organisms, and societies of organisms—are animated, organized, and coordinated by non-local morphic fields that contain an intrinsic memory. A particular specimen (e.g. an individual cell) of a morphic group (e.g. liver cells), which has already established its collective morphic field by its pattern of past behaviour, will tune into that group's morphic field and read its collective information through a process of morphic resonance that guides the specimen's own development. This development feeds back by resonance to the morphic field of the group, thus strengthening it with its own experience and adding new information so that the field itself evolves.

Sheldrake suggests that there is a continuous spectrum of morphic fields, including morphogenetic fields, behavioural fields, mental fields, and social and cultural fields. Thus morphic fields function as evolving universal databases for both living and mental forms, and lead to a vision of a living, evolving universe with its own inherent memory.

What most of these different ideas—whether in medicine or biology—have in common is a view that orthodox science rooted in reductionism is incapable of explaining what life is, a belief that the cosmos is a dynamic whole of interdependent parts connected by a universal energy field similar to that perceived by mystics in the ancient world, a belief that living organisms are connected by an interdependent relationship with this field, together with a conviction that this concept is supported by evidence and is compatible with advanced scientific principles, particularly quantum field theory.

Orthodox science's response

The response of contemporary orthodox science to such proposals in biology is summed up by a 1981 opinion piece written in *Nature* by its then editor, John Maddox, on Sheldrake's hypothesis.

> Sheldrake's argument is in no sense a scientific argument but is an exercise in pseudo-science. Preposterously, he claims that his hypothesis can be tested—that it is falsifiable in Popper's sense—and indeed the text includes half a dozen proposals for experiments that might be carried out to verify that the forms of aggregation of matter are indeed moulded by the hypothetical morphogenetic fields that are supposed to pervade everything. These experiments have in common the attributes of being time-consuming, inconclusive...and impractical in the sense that no

self-respecting grant-making agency will take the proposals seriously.... The more serious objection to his argument is that it says nothing of any kind about the nature and origin of the crucial morphogenetic fields and contains no proposals for investigating the means by which they are propagated. Many readers will be left with the impression that Sheldrake has succeeded in finding a place for magic within scientific discussion.[16]

Opposition to the use of these holistic field ideas in medicine has been led by, among others, Linda Rosa, mother of the precocious Emily. Two years before Emily's experiment was published (the paper was written by her mother and two others in addition to Emily), she published a *Survey of Therapeutic Touch "Research"*, alleging methodological and other flaws in all the positive studies.

Victor Stenger, Emeritus Professor of Astronomy at the University of Hawaii, is an ardent materialist who also campaigns against what he calls pseudoscience employed in the Western medical profession. He concludes that there is not a shred of evidence uniquely requiring the existence of a vital energy or bioenergetic field. Everything can be reduced to electromagnetic interactions explained by well-tested orthodox physics and chemistry. Until this can be shown not to be the case, with the same degree of experimental significance as that demanded in physics, parsimony requires that other explanations of the evidence be rejected.[17]

I think Stenger is right in that many claims are disprovable and that the evidence reviewed so far in this quest is insufficient to confirm other claims about the existence of a cosmic field that accounts for the phenomenon of life.

However, absence of evidence is not evidence of absence. A John Maddox of the eighteenth century would have dismissed as preposterous the idea that there existed a cosmic field that accounted for electrical and magnetic phenomena (science currently considers that the electromagnetic field is infinite in range). Moreover, while orthodox science accepts the idea of a cosmic quantum vacuum field, it cannot yet explain its nature.

Orthodox science's approach to defining life

This brings us to the question of how current orthodox science defines life. According to Edward O Wilson, one of the few world authorities who has thought deeply about science outside his specialist field, the study of ants, "reductionism is the primary and essential activity of science".[18]

Most of our vast wealth of knowledge about, and our understanding of, nature since the first scientific revolution in the West is due to this analytical technique of breaking a thing down into its component parts and studying them. Without it we wouldn't know that the page of the book you are reading consists principally of cellulose, which is a linear polymer of molecules consisting of atoms of carbon, oxygen, and hydrogen, each of which consists of a positively charged nucleus orbited at relatively huge distances by smears of negatively charged electrons, while it is the motion of free electrons that produces electricity and magnetic fields.

The spectacular successes of reductionism in physics in the nineteenth and early twentieth centuries were repeated in biology in 1953 when James Watson and Francis Crick showed how the double helix structure of the DNA (deoxyribonucleic acid) molecule explains heredity.

Reductionism, however, can no more explain life than breaking down a Shakespeare play into individual words and then into individual letters of the alphabet can explain the characters, the emotions, and the drama of the play.

We can agree that a human being is alive. We may further agree that each of the hundred trillion cells that make up a human is alive. But if we examine the components of each cell, is each chromosome alive, is each protein alive? The answer is clearly no, just as it is for each atom that is a component of each protein.

The failure of reductionism* to explain life led scientists to turn to the concept of emergence articulated by the English philosopher John Stuart Mill in 1843 and subsequently developed in hundreds of versions. At its simplest, it means that the whole is more than the sum of its parts, just as the picture shown by a jigsaw only emerges when the parts of the jigsaw are organized in a precise way.

The following definition, incorporating three broad categories relevant to the questions this investigation seeks to answer, gives the meaning I use (others use the same words to define and categorize in various ways).

> **emergence**   The appearance of one or more new properties of a complex whole that none of its constituent parts possesses.
>
> "Weak emergence" is where novel properties at the higher level are explained solely by the interaction of the constituent parts.
>
> "Strong emergence" is where novel higher-level properties can neither be reduced to, nor predicted from, the interaction of the constituent parts.
>
> "Systems emergence" is where novel higher-level properties causally interact with lower level properties; this top-down as well as bottom-up causality frequently forms part of a systems approach that, in contrast to the reductionist approach, sees each component as an interdependent part of the whole.

The avowedly reductionist scientist Francis Crick conceded that invoking emergence may be necessary, but claimed that weak emergence is sufficient to explain life.

> While the whole may not be the simple sum of the separate parts, its behaviour can, at least in principle, be understood from the nature and

---

* I am using reductionism in its scientific sense rather than any version of metaphysical reductionism.

behaviour of is constituent parts plus the knowledge of how these parts interact.[19]

Paul Davies, theoretical physicist and cosmologist who now heads Arizona State University's Center for Fundamental Concepts in Science, applies this to life by arguing that

> the secret of life will not be found among the atoms themselves, but in their pattern of their association—the way they are put together....Atoms do not need to be 'animated' to yield life, they simply have to be arranged in the appropriate complex way.[20]

British neurobiologist Donald Mackay, however, had previously challenged weak emergence in this context by using the example of an advertising display, like one of those in New York's Times Square consisting of coloured light bulbs that flash on and off in a programmed way to spell out the message "Things go better with Coke." An electrical engineer can reduce this system to its component parts and explain how and why each light is flashing, and how the flashes are co-ordinated. But understanding the interactions of the different electrical parts gives no understanding of, still less predicts, the message that drinking Coca-Cola makes life better. This requires a different level of explanation. It is an example of strong emergence.[21]

Davies's explanation of life is insufficient. The hundred trillion cells in a live human body are not merely arranged in an appropriately complex way; the cells interact with each other and are dependent on each other to form a whole living human. Following this logic through, life is a systems emergent property.

This doesn't tell us, however, what this emergent property is. Specifically, what are the characteristics of this emergent property we call life that differentiate it from what is not life?

Claimed characteristics of life

For something that seems intuitively obvious there is no agreement among scientists or philosophers as to what the characteristics of life are. Most offer a checklist. One or more of the following items appear in the majority of lists, while the number of items included in each author's checklist varies considerably.

  a. reproduction
  b. evolution
  c. sensitivity (response to stimuli)
  d. metabolism
  e. organization
  f. complexity

These words mean different things to different people, and so it is important to be clear what each characteristic means if we are to decide which items are necessary or sufficient to define life.

Reproduction

Reproduction appears in most lists. It is not, however, a sufficient condition. If you drop a crystal of salt into a saturated saline solution, the crystal will reproduce: a much larger crystal grows and as it does so it reproduces precisely the same structure as the original.

Neither is reproduction a necessary condition. Mules do not reproduce; neither do worker ants and many varieties of garden plant, and yet they are alive.

Hence reproduction is neither a sufficient nor a necessary characteristic of life.

Evolution

Evolution, too, appears in many lists, but of itself is too vague to be useful. A coastline evolves over time in response to weathering by sea, wind, and rain, but no one considers that a coastline is alive. For this reason some prefer "adaptation". But a coastline may be said to adapt to a changing environment.

NASA's Exobiology Program is more precise, defining life as "a self-sustained chemical system capable of undergoing Darwinian evolution".[22] Yet many species, like cyanobacteria, coelacanths, and some types of crocodile, have existed for at least hundreds of millions of years without undergoing any change in physical characteristics. If a NASA space probe were observing Earth, how long would it wait before concluding whether or not crocodiles were capable of undergoing Darwinian evolution? Since individual members of species in evolutionary stasis are alive, evolution by natural selection is neither a necessary nor a sufficient characteristic of life.

Response to stimuli

A light meter responds to light shone on it and responds by registering a change in the number of lumens shown in its display. So too do myriad kinds of detectors. Hence sensitivity is not sufficient as a characteristic of life.

A person in a deep coma or an animal in deep hibernation may fail to respond to stimuli, and so sensitivity is not a necessary characteristic.

Metabolism

Most definitions of metabolism say that it consists of biochemical processes within a living organism. If a living organism is necessary to define metabolism, it is a circular argument to say that metabolism is a characteristic of life.

If the processes can be abstracted from living organisms, then it may be that such processes are characteristic of life (and I shall consider some attempts when examining significant definitions later), but metabolism of itself is meaningless as a characteristic of life.

Organization

Although featuring on many checklists, organization is too general a term to be useful. It can mean the static arrangement of parts of a whole, like the ions in a crystal, or the active coordinating of separate elements into a systemic whole, like the operation of the divisions and personnel of the Ford Motor Company. Neither a crystal nor the Ford Motor Company is generally considered to be a living thing, and so organization is not a sufficient characteristic of life.

It may be argued that organization is a necessary condition of life, but then it is a necessary condition of anything that is not chaos. Hence organization of itself is not a necessary condition except in a trivial sense.

Complexity

Complexity, too, appears on many checklists and suffers from the same drawbacks as organization if it is used to characterize life: a dead body and the Golden Gate Bridge are both complex but neither is alive. Complexity is not a sufficient nor, by itself, a necessary condition except in a trivial sense because it is a condition of anything that is not absolutely simple.

If none of these characteristics is either necessary or sufficient of itself, does a combination of two or more of them define life? Davies asserts that "the two distinguishing features of living systems are complexity and organization".[23] But this doesn't take us much further: the Ford Motor Company possesses the characteristics of complexity and organization.

Six pages later Davies says that

> What happens [in metabolism] is that there is a flow of energy through the body. This flow is driven by the orderliness, or negative entropy, of the energy consumed. The crucial ingredient for maintaining life is, then, negative entropy.[24]

What I think he means is that, following Prigogine,* a flow of energy through a living system maintains its complex structure in a dynamic yet stable state far from thermodynamic equilibrium in opposition to the Principle of Increasing Entropy,† and this antientropic feature is a characteristic of life. It is also a characteristic of certain non-living systems, like vortices in fluids.‡ Moreover, entropy is a measure of the degree of disorder in an isolated system; a living body is an open system through which energy flows.

---

*   See page 145.
†   See page 145.
‡   See page 145.

Definitions of life

In 2004 Philip Ball, then a consultant editor at *Nature*, thought the whole exercise of trying to define life is pointless and a waste of philosophers' and scientists' time. Arguing that there are no boundaries between what is living and what is not living, he cites the case of viruses. These reproduce, evolve, and are organized and complex (compared with, say, an amino acid), but they are parasites. Outside a living cell a virus is inactive; only within an appropriate host cell does it becomes active, taking over the cell's metabolic machinery in order to reproduce new virus particles, which can then infect other cells.

Since viruses depend on a host cell for their activity they cannot be considered candidates for the first lifeforms to emerge on Earth, nor are they independent lifeforms. Contrary to Ball's view that "No one knows whether to call viruses living or not",[25] most sources define viruses as particles that are either active or inactive as distinct from alive or dead.[26]

Ball also notes that in August 2002 Eckard Wimmer and colleagues at the State University of New York looked up the chemical structure of the polio virus genome on the internet, ordered segments of the genetic material from companies that synthesize DNA, and then strung them together to make a complete genome. When they mixed this with the appropriate enzymes, this synthetic DNA provided the seed from which polio virus particles grew. He has little doubt that biologists will soon be able to fabricate things like cells that are generally considered to be alive, thus reinforcing his view that trying to define life is pointless.

While the mechanical ticking of boxes in a checklist of often-vague characteristics is not productive, it does not follow that there is no distinction between what is alive and what is not alive. Some promising attempts have been made to define life in terms of a system that interacts in specific ways with its environment, and I examine now what I think are the most significant ones. (Interestingly, one is advanced by a theoretical physicist, one by a former theoretical physicist, and one by a biologist whose conjecture is rooted in theoretical physics. Most biologists who work in university faculties now usually labelled Life Sciences are focused on narrow fields within narrow sub-branches of biology, such as the study of retroviral vectors within the sub-branch of molecular biology, and show little interest in what it is that defines their branch of science.)

Smolin's self-organized system

Theoretical physicist Lee Smolin draws on self-organizing complex systems ideas put forward by Ilya Prigogine, John Holland, Harold Morowitz, Per Bak, and Stuart Kauffman to construct what he calls a developing theory to define life. (I shall consider some of these underlying ideas in the next chapter since they claim to explain the emergence of life.)

It leads him to propose that life on Earth may be defined as:

1. a self-organized non-equilibrium system such that
2. its processes are governed by a program which is stored symbolically in the structures of DNA and RNA (like DNA, a nucleic acid but usually forming a single strand of nucleotides*), and
3. it can reproduce itself, including the program.[27]

However, only single-celled organisms reproduce themselves. An animal, for example, does not reproduce itself and its DNA program; sex with a mate produces an offspring that is different from either parent and has a different DNA program. And, as discussed previously, several species, like mules and worker ants, cannot produce offspring at all although they meet criteria (1) and (2). Hence this attempt falls short of being comprehensive.

Capra's web of life

After rejecting reductionism as a method to explain life, former theoretical physicist Fritjof Capra attempted a synthesis of systems theories developed from the pioneering ideas that the Russian physician, philosopher, economist, and revolutionary, Alexander Bogdanov, put forward before the First World War (although the West did not learn of them until much later), the proposals for self-organization that Smolin drew upon, cyberneticists' patterns of circular causality underlying the feedback concept, and the mathematics of complexity to propose an "emerging theory" (perhaps better labelled a developing hypothesis) of life. The work of Chilean neuroscientists Humberto Maturana and Francisco Varela exerted a major influence on his thinking.[28]

Capra argues that living systems are defined by three totally interdependent criteria: pattern of organization, structure, and life process.

Pattern of organization

The pattern of organization is the configuration of relationships that determines the system's essential characteristics; for a living thing the pattern is autopoiesis. Etymologically this means self-making or self-production, but Capra uses the definition given by Maturana and Varela in 1973 as a "network pattern in which the function of each component is to help produce and transform other components while maintaining the overall circularity of the network".[29] I think this basically means a closed network of processes that continuously maintains itself.

Structure

The structure is the physical embodiment of the system's pattern of organization. For a living thing it is a dissipative structure as defined by Prigogine, namely

---

* See page 217 for definition.

a system maintained in a stable state far from thermodynamic equilibrium by a through-flow of energy. Whereas an autopoietic network is organizationally closed, it is structurally open as matter and energy continuously flow through it.

Life process

The life process is the activity involved in the continual embodiment of an autopoietic pattern of organization in a dissipative structure; it is cognition, as defined initially by the anthropologist, linguist, and cybernetist Gregory Bateson in the 1970s and more fully by Maturana and Varela as the process of knowing.

Autopoiesis and cognition are thus two different aspects of the same phenomenon of life: all living systems are cognitive systems, and cognition always implies the existence of an autopoietic network.

Capra claims that autopoiesis is the defining characteristic of life. A problem arises because Maturana and Varela use the jargon of general systems theories plus mathematical models to produce an abstract and generalized description of what they regard as the simplest living system, a cell. While I think it important to avoid characteristics that are so specific as to produce circular arguments (as with "metabolism"), Maturana and Varela use autopoiesis in so abstract a manner as to define the invariant feature of a living system without referring to things like function or purpose. They assert that the behaviour of a system is something ascribed to it by someone observing it in interaction with its environment and is not characteristic of the system itself.

However, if a particular interaction of a system with its environment invariably occurs, then surely this is a characteristic of the system. Here, according to Maturana and Varela's own descriptions, one invariable interaction of an autopoietic system with its environment has the purpose of repairing and maintaining the system. Without this purpose the lifeform dies.

Avoidance of purpose

This avoidance of purpose as a characteristic of living things is widespread among scientists; purpose rarely, if ever, appears on their checklists. Some scientists appear to confuse theology with teleology—an explanation of events in terms of purpose—and are afraid to use it lest someone accuse them of implying a Divine Designer of the purpose. It implies no such thing. Others shy away from the term because purpose generally implies intent. The actions of animals are indeed intentional, as with you turning the pages of this book in order to read its contents, or a hawk swooping down with the intention of catching a mouse. However, the actions of more primitive lifeforms, like a plant opening its leaves in sunlight to generate energy by photosynthesis, or a bacterium swimming towards a source of food, are generally thought of as internally directed or instinctive responses to stimuli rather than intentional acts. Hence it is probably better to use the term "internally directed action" as a characteristic of all lifeforms; no physical or chemical law says that you, or the hawk, or the bacterium, or the plant should act in this way, and no nonliving thing, like water or a rock, undertakes internally directed action.

McFadden's quantum life

Johnjoe McFadden, Professor of Molecular Genetics at the University of Surrey, England, is one scientist who doesn't eschew such a characteristic. Indeed he defines life as the capability to perform directed action against prevailing exterior forces.

In support he cites, among other examples, a salmon swimming against the flow of inanimate water in a river under the force of gravity because the salmon has the purpose of reaching its spawning ground upstream.

Life, McFadden claims, defies determinism, the principle at the heart of Newtonian mechanics that says the present or future state of any system is determined solely by its past: if you know the precise configuration of any system, apply the laws of physics and chemistry to it, then in principle you can calculate its future behaviour. We cannot account for life by classical science alone, which cannot explain how living creatures are able to direct their actions according to their own internal agenda, as does the salmon.

His solution is not to invoke some Divine Designer who invests living things with purpose but to propose that it is the motion of fundamental particles governed by the non-deterministic laws of quantum theory that explains how living things can act contrary to the classical laws of nature.[30]

I shall consider this idea in more detail in the next chapter because McFadden goes on to propose a quantum theory of the emergence and evolution of life.

Working definition of life

In order to understand how life emerged from the atoms and molecules that made up the newly formed Earth we need to be clear what life is. As we have seen, arriving at a reasonable definition is far from easy and there is no agreement among either scientists or philosophers. The working definition I propose is

> **life**   The ability of an enclosed entity to respond to changes within itself and in its environment, to extract energy and matter from its environment, and to convert that energy and matter into internally directed activity that includes maintaining its own existence.

A lifeform may have the ability to produce descendants, but this is not a necessary characteristic.

Conclusions

This definition rejects the argument that there is no distinction between what is alive and what is not alive: non-living things are not necessarily enclosed and do not possess the characteristic functions and internally directed activity of living things. Just because the boundary is indistinct, as with a virus, does not mean that there is no boundary. The change from non-living to living represents not simply a difference in degree but a difference in kind. It is a qualitative difference, analogous to a phase change, just as gaseous water is qualitatively different from liquid water and is not simply hotter water, even though the bubbling surface of boiling water is not a distinct boundary.

How and when this change took place is something I shall examine in the next chapter.

CHAPTER FOURTEEN

The Emergence of Life 1: Evidence

He who thus considers things in their first growth and origin...will obtain the clearest view of them.

—ARISTOTLE, FOURTH CENTURY BCE

I shall try to establish how life emerged on Earth by examining evidence of the earliest lifeforms on Earth, whether life began once on Earth—and hence all current lifeforms are descendants of a common ancestor—or different forms of life began in different places at different times, and what are the characteristics of the earliest lifeforms. In the following chapter I shall evaluate the many hypotheses advanced to explain how such lifeforms emerged from the inanimate Earth.

Direct evidence

While there is no generally agreed definition of life, most scientists agree that the earliest lifeforms must have been the simplest. This is a self-sustaining prokaryote, a cell that doesn't have its genetic material enclosed within a nucleus. Biologists and geologists seek evidence for the earliest appearance of such lifeforms from two sources: fossils and extremophiles, which are organisms that currently live in the extreme conditions thought to resemble those of the early Earth.

Fossils

Fossils are the mineralized or otherwise preserved remains of lifeforms. Usually they are found in sedimentary rock, but they can also be preserved from decomposition by very low temperatures, or desiccation, or an anoxic (without oxygen) environment.

Two problems confront scientists trying to find the earliest fossils. First, very few organisms are ever fossilized: most are either eaten—dead or alive—by other organisms or else they decompose after death. Second, the very few fossils that do exist are usually formed when the organism is rapidly covered after death by sediment, like sand or mud, that is then compressed and concretized into sedimentary rock. But very little sedimentary rock remains from the first billion years or so of the Earth's existence and nearly all of this is metamorphized by processes likely to destroy any fossilized remains.

Until 1993 the earliest evidence was in layers of chert, a fine-grained sedimentary rock exposed in the Gunflint Range of western Ontario in Canada. From 1953 to 1965 botanist Elso Barghoorn of Harvard University and geologist Stanley Tyler of the University of Wisconsin discovered structurally preserved fossils of well-defined morphology of 12 new species, including complex branched micro-organisms, whose age, determined by radiometric dating of the chert layers, was around 2 billion years.[1] If these complex structures evolved from simpler lifeforms then life must have existed earlier.

Bill Schopf, a palaeobiologist from the University of California, Los Angeles, set out to find these older lifeforms by investigating the much older sedimentary rocks of the Pilbara range in Western Australia, which very accurate uranium-lead radiometric dating had put at 3.465 billion years old. These rocks had been metamorphized, but in 1993 he announced that he had discovered 11 microfossils of differing structures thought to be different species of cyanobacteria (blue-green bacteria, formerly classified as algae). He supported his claim by analysis of the carbon isotope content of the fossils. This technique is similar to the radiocarbon dating method described earlier.* The most common isotope, carbon-12, is more reactive than carbon-13 and takes part in photosynthesis, in which carbon dioxide in the atmosphere is metabolized by organisms into organic carbon compounds. Consequently the carbon-13 to carbon-12 ratio in biogenic carbon is 3 per cent lower than in inorganic carbon, and this ratio is preserved through metamorphic processes that destroy microfossils. Schopf claimed that his specimens showed this characteristic ratio.[2]

The announcement caused a sensation at the time. The clear implication was that less complex lifeforms must have existed even earlier, which would place them late in the Hadean Eon when Earth was being bombarded by asteroids and other debris left over from planetary formation.

The hunt was on to find evidence. PhD student Stephen Mojzsis from the Scripps Institution of Oceanography at the University of California San Diego headed for the Isua rocks of western Greenland that had been dated at 3.8 billion years old. These rocks had been metamorphized even more than the Pilbara. Estimates suggested that within a billion years of deposition these sediments had been subjected to temperatures of 500° Celsius and pressures greater than 5,000 atmospheres, which would have destroyed any fossils. Nonetheless, Mojzsis and his team claimed in 1996 that they had found evidence of life not only in these rocks but also in rocks 50 million years older on the nearby Akilia Island, putting the earliest life at 3.85 billion years ago, within the Hadean Eon.

The traces of carbon consisted of incredibly tiny globules—a trillionth of a gram—but Mojzsis directed an ion-microprobe at them and measured their

---

*   See page 170.

isotopic composition by magnetic-sector mass spectroscopy. They all showed the telltale signature of carbon-13 depletion. Furthermore, the team had found these traces of carbon embedded within grains of apatite; although this mineral is a common minor constituent of rocks, it is also found in organisms, and so the evidence was doubly suggestive of life.[3]

Although questions were raised about the age of the Akilia rock, for six years palaeobiologists accepted the evidence that life had been detected in the Hadean Eon.

In 2002, however, Christopher Fedo of the Department of Earth and Environmental Sciences, George Washington University and Martin Whitehouse of the Laboratory for Isotope Geology at the Swedish Museum of Natural History challenged the Mojzsis claim that the carbon had been found in layers of sedimentary rock known as Banded Iron Formation. Their analysis of the rock concluded that it was igneous, formed by ancient volcanic activity, and so could not possibly contain organic relics. Their challenge also implied that it was invalid to assume that carbon-13 depletion was caused only by biological activity.[4] Mojzsis and colleagues in turn challenged Fedo and Whitehouse's data and their interpretation of it.[5]

In the same year Martin Brasier and colleagues at Oxford University announced that they had examined Schopf's samples in detail and challenged Schopf's claim that the morphology of his specimens was indisputably that of cells, still less cynanobacterial cells. Brasier maintained that the carbon blobs had probably been formed by the action of scalding water on minerals in the surrounding sediment.[6] Schopf defended his claim, but was undermined by his former research student Bonnie Packer, who alleged that Schopf had been selective in the evidence he presented and had ignored her protests.[7]

In 2006 and 2007 the Mojzsis team involved in the Akilia Island claim hit back against their critics, but the general view among the cross-disciplinary scientists working in this field is that the evidence so far does not sustain the claim that the first record of life on Earth is 3.85 billion years ago. Current evidence only supports the hypothesis that organisms called extremophiles probably existed around 3.5 billion years ago, some 1 billion years after the Earth formed.

Extremophiles

Mojzsis's initial claim to have found life within the period of late heavy bombardment reawakened interest in the study of extremophiles. Four kinds offer promising clues about organisms that might have existed in the extreme conditions of this period: surface thermophiles, sub-oceanic thermophiles, cave acidophiles, and subterranean thermophiles.

Surface thermophiles

Thermophiles are organisms that exist at very high temperatures. In 1967 microbiologist Thomas Brock from the University of Wisconsin isolated algae and bacteria from the hot scum on the surface of a volcanic spring at Yellowstone National Park, Wyoming, where rainwater seeping through the surface rocks meets the hot magma below that blows out superheated water and steam to form hot pools.

Sub-oceanic thermophiles
Exploring the Pacific Ocean floor in a specially constructed diving bell, geologist John Corliss of Oregon State University and marine geochemist John Edmond of the Massachusetts Institute of Technology discovered in 1979 sea anemones, mussels, giant clams, miniature lobsters, and snake-like pink fish with bulging eyes that all live in total darkness under immense pressure in the relatively cooler waters surrounding water superheated to 400° Celsius by magma from the Galapágos Rift, some 2.5 kilometres below the surface.[8] On the hot walls of lava-encrusted chimneys strains of bacteria can grow at temperatures as high as 121° Celsius.[9] Denied sunlight, they extract energy for their own maintenance from hydrogen sulphide.

Cave acidophiles
An acidophile is an organism that exists in conditions of high acidity. In the first decade of the twenty-first century Diana Northup of the Biology Department of the University of New Mexico and Penny Boston of the Department of Earth and Environmental Science, New Mexico Institute of Mining and Technology examined the Cueva de Villa Luz (the Cave of the Lighted House) near Tabasco, Mexico. It has a noxious atmosphere high in concentrations of carbon monoxide and the foul-smelling hydrogen sulphide, while its walls drip with sulphuric acid the strength of battery acid. Hanging like stalactites from the cave roof are colonies of bacteria named snottites because they have the consistency of snot or mucous. Genetic evidence suggests these snottites are ancient. They extract their energy from chemosynthesis of volcanic sulphur compounds and the drips of warm sulphuric acid.[10]

Because the conditions in the cave resemble those thought to exist on the early Earth, some microbiologists suggest they represent the earliest lifeforms.

Subterranean thermophiles
Whether sub-oceanic thermophiles, surface thermophiles, or cave acidophiles resemble Earth's earliest lifeforms is debateable. As far as the first is concerned, this depends on when deep oceans existed on Earth. Both Mojzsis's claim of 4.3 billion years ago and Watson and Harrison's claim of even earlier, 4.35 billion years ago, have been challenged.*

Princeton University microbiologist James Hall believes that surface thermophiles and cave acidophiles would not have survived the late heavy

---

*   See page 177.

bombardment and they must represent second-generation life. He and others in the new scientific field of geomicrobiology (drawing on geology, geophysics, hydrology, geochemistry, biochemistry, and microbiology) have turned their search to deep below ground where life would have been sheltered from impacts.

They obtain evidence by piggybacking exploratory oil drills and mine shafts. The latter has provided the deeper and more productive source. In 2006, for example, Princeton University geoscientist Tullis Onstott led a multidisciplinary team that accompanied mining engineers drilling into 2.7-billion-year-old rock 2.825 kilometres below the surface in the Mponeng gold mine of South Africa's West Rand. When the drill hit fissures in the rock, out poured hot foul-smelling saltwater containing thermophilic microbes that died on exposure to oxygen. Rather than deriving energy from sunlight they maintained their existence by extracting energy from sulphur compounds and hydrogen resulting from the decomposition of water by the radioactive decay of uranium, thorium, and potassium. Their metabolism was considerably slower, and presumably less efficient, than that of surface microbes. Analysis of the water showed that it has been isolated from the surface for many millions of years, and the hydrocarbons in the environment do not derive from living organisms as is usual.[11]

Although genetic analyses indicate that some of these extremophiles are ancient, this doesn't constitute empirical proof that such lifeforms did exist during the extreme conditions thought to exist in the Hadean Eon, which spans some 700 million years after the Earth's formation 4.5 billion years ago. It is also possible that they evolved from other lifeforms by adapting to the environment of high temperatures, high acidity, high pressures, and absence of sunlight which had developed near magma vents as oceans deepened, and to which they had been transported by water flows.

Indirect evidence

In the absence of irrefutable empirical evidence, I turn to hypotheses about the earliest life on Earth and whether there was a single ancestor of all current life or whether different forms of life originated.

Several scientists have suggested that life may have begun at different times but these lifeforms were extinguished by asteroid bombardment. Without a shred of evidence in support, such ideas are pure speculation; moreover, they do not address the question of whether any such multiple starts were from identical or different common ancestors.

Genetic analyses

Carl Woese of the University of Illinois found that the genes encoding the RNA of ribosomes (protein-manufacturing units of a cell) are ancient and exist in all kinds of organism. In 1977 he published a genetic analysis of small subunit ribosomal RNA of a wide range of cells and classified them by their molecular similarity. Assuming that differences represent evolutionary change, he claimed this genetic tree of life depicts their evolutionary lines of descent more accurately than the incomplete fossil record or subjective views on the size and shape of organisms.*

Figure 14.1 shows molecular biologist Norman Pace's updated version of Woese's phylogenetic tree, which groups all cells into three domains.

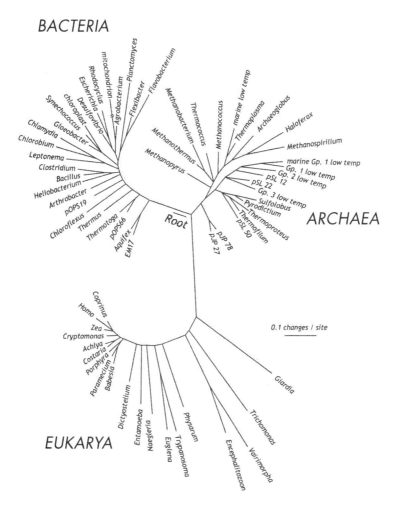

Figure 14.1 Universal phylogenetic tree

The position and length of each branch is determined by comparisons of ribosomal RNA

---

* Biologist Lynn Margulis, then Distinguished Professor at the University of Massachusetts, and others challenged the use of just one characteristic for this purpose. I shall consider this in more detail when examining the evidence for biological evolution.

Different authors use different names for these three major groups of lifeforms. Following Woese's groundbreaking work the names shown in the diagram were adopted in order to clearly distinguish between the two types of prokaryote, the domains Archaea and Bacteria, and I shall use them for consistency's sake; however, more recently some evolutionary biologists have reverted to the original names of archaebacteria and eubacteria following analyses of complete genomes that I will consider later.

Bacteria Extremely small, single-celled organisms whose genetic information, encoded in a folded loop of double-stranded DNA, is not enclosed in a membrane-bound nucleus (and hence they are prokaryotes). In addition to this nucleoid, the cell may include one or more plasmids, separate circular strands of DNA that can replicate independently and are not responsible for the reproduction of the organism. Most reproduce by splitting in two and producing identical copies of themselves. They occur in a variety of shapes, including spheres, rods, spirals, and commas.

Archaea Prokaryotes that differ from Bacteria in their genetic make-up and the composition of their plasma membranes and cell walls. They include most extremophiles. Although structurally similar to Bacteria, their chromosomal DNA and their cell machinery more closely resemble those found in Eukarya.

Eukarya Organisms whose cells incorporate a membrane-bound nucleus, which contains the genetic information of the cell, plus organelles, which are discrete structures that perform specific functions. Larger and structurally and functionally more complex than prokaryotes, they comprise single-celled organisms, like amoeba, and all multicellular organisms, like plants, animals, and humans.

Most eukaryotic cells replicate to produce identical copies of themselves. However, a type of eukaryotic cell in a multicellular organism, called a gamete, can fuse with a gamete of a different organism to produce a daughter organism possessing genetic characteristics of each parent. This sexual reproduction thus mixes different parental genes in the daughter cell compared with the asexual reproduction of prokaryotes.

Universal common ancestor?

Nearly all our knowledge of biology has derived from analysing plants, animals, and humans. On the universal phylogenetic tree shown, however, fungi represented by *Coprinus*, plants represented by *Zea*, and animals and humans represented by *Homo* constitute just three small and peripheral sub-branches of one of the 12 genetically distinct branches of Eukarya (some of which also have sub-branches), separated from the far more numerous distinct branches of the domains of Bacteria and Archaea.

Only around a hundred genes are common to all currently known living organisms, but analyses that allow for lineage-specific gene losses suggest that a last universal common ancestor (LUCA), denoted Root in the diagram, perhaps possessed ten times as many genes as that.[12]

The nature of LUCA is currently a matter of dispute among evolutionary biologists. Woese concluded that

> The ancestor cannot have been a particular organism....It was a communal, loosely knit, diverse conglomeration of primitive cells that evolved as a unit, and it eventually developed to a stage where it broke into several distinct communities, which in turn become the three primary lines of descent [bacteria, archaea, and eukarya].[13]

Since the beginning of the twenty-first century significant horizontal gene transfers (also called lateral gene transfers) have been discovered not only between related prokaryotes but also between prokaryotes that are not closely related on phylogenetic trees.[14]

By 2009 several evolutionary biologists were proposing that there was no universal common ancestor; life emerged as "a population or populations containing diverse organisms. Furthermore, these organisms probably did not live at the same time."[15]

As far as the emergence of life is concerned, the problem with the population idea is that if each member of the population is an organism, namely a lifeform as generally understood and conforming with the definition of life given on page 206, then it doesn't tell us from what or how each member emerged; if each member is not an independent lifeform, then the population idea simply reinforces the conclusion of Chapter 13 that the boundary between what is alive and what is not alive is indistinct; but an indistinct boundary is a boundary nonetheless.

In 2010 bioinformaticist Douglas Theobald of Brandeis University undertook a statistical comparison of various alternative hypotheses and concluded that

> the model selection tests are found to overwhelmingly support UCA [a universal common ancestor] irrespective of the presence of horizontal gene transfer and symbiotic fusion events.[16]

This reinforces the current orthodox opinion in biology that life emerged once on Earth and there is a single common ancestor, although molecular biologists dispute where this single common ancestor lies on the genealogical tree and how the three major groups are related to it. The current consensus, but by no means unanimous view, is that the root lies between Archaea and Bacteria.

The fact that some archaea are extremophiles living in conditions thought to obtain on Earth around 3.5 billion years ago and that their genomes imply they are ancient suggests that the earliest lifeform might have been an archaeon or an ancestor of an archaeon. University of Oxford taxonomist specialist Tom Cavalier-Smith vigorously opposes this view, claiming that the cell machinery of Archaea shows they are distant descendants of Bacteria.[17] The evidence is by no means conclusive and it is unlikely ever to be so.

Size, complexity, structure, and functioning of the simplest cell

In order to evaluate the ideas for how life emerged on Earth we need to appreciate the difference between the size, complexity, structure, and functioning of the molecules that evolved on the surface of the early Earth, or else evolved on asteroids and comets that deposited them on the Earth's surface during bombardment, on the one hand, and the size, complexity, structure, and functioning of the simplest independent form of life on the other hand, and that is a single-celled prokaryote.

Size
Most prokaryotes range from between one thousandth to one hundredth of a millimetre long, and occur in a variety of shapes, including spheres, rods, spirals, and commas.

Components and structure
Figure 14.2 shows the components and structure of a simple bacterium, which are the same as those of an archaeon (their biochemical and stereochemical differences need not concern us here).

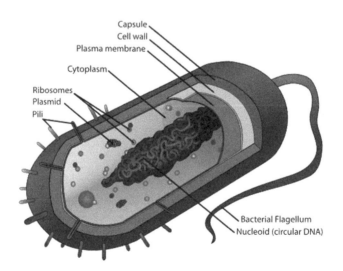

Figure 14.2 **Structure of a bacterium**

If we examine its component parts and how they interact, the key to the functioning of such a simple lifeform is its DNA.

DNA

DNA   Deoxyribonucleic acid, located in cells, contains the genetic instructions used in the maintenance and reproduction of all known independent organisms and some viruses.

Each DNA molecule normally consists of two long chains of four nucleotides in a characteristic sequence; the chains (usually referred to as strands) are twisted into a double helix and joined by hydrogen bonds between the complementary bases adenine (A) and thymine (T) or cytosine (C) and guanine (G) so that its structure resembles a twisted ladder.

When DNA is copied in a cell the strands separate and each serves as a template for assembling a new complementary chain from molecules in the cell.

DNA strands also act as templates for the synthesis of proteins in a cell through a mechanism that makes another nucleic acid, RNA, as an intermediary.

RNA   Ribonucleic acid resembles DNA in that it consists of a chain of four nucleotides in a characteristic sequence, but uracil (U) replaces thymine (T) alongside adenine (A), cytosine (C), and guanine (G) as the bases of the nucleotides, and the strands are single, except in certain viruses.

gene   The fundamental unit of inheritance, which normally comprises segments of DNA (in some viruses they are segments of RNA rather than DNA); the sequence of the bases in each gene determines individual hereditary characteristics, typically by encoding for protein synthesis. The segments are usually split, with some parts located in distant regions of the chromosome and overlapping with other genes.

chromosome   A structure that contains the genetic information of a cell. In a eukaryotic cell it consists of threadlike strands of DNA wrapped in a double helix around a core of proteins within the cell nucleus; in addition to this nuclear chromosome, the cell may contain other small chromosomes within, for example, a mitochondrion. In a prokaryotic cell it consists of a single tightly coiled loop of DNA; the cell may also contain one or more smaller circular DNA molecules called plasmids.

In the simplest archaeon or bacterium, the chromosome usually takes the form of a single loop of double-stranded DNA that is folded in order to fit into the cell, as shown in Figure 14.2.

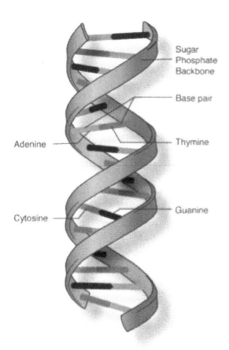

Sugar
Phosphate
Backbone

Base pair

Adenine

Thymine

Cytosine

Guanine

Base pairs
Adenine bonds with Thymine
Guanine bonds with Cytosine

Figure 14.3 Schematic representation of DNA structure

Each DNA molecule consists of two long strands that wind around each other to form a
double helix. Each strand has a backbone made of alternating groups of sugar (deoxyribose)
and phosphate groups. Attached to each sugar is one of four bases: adenine, cytosine, guanine,
or thymine. The two strands are held together by hydrogen bonds between complementary
bases, adenine forming a base pair with thymine, and cytosine forming a base pair with
guanine so that the structure resembles a twisted ladder.

Physical chemist and philosopher Michael Polanyi pointed out that whereas
the base pairing ability of DNA (A-T and C-G) is fully determined by the laws of
chemistry, the base *sequence* in DNA is not. DNA is able to form every conceivable
sequence of bases of any length and any composition. The information that
determines how a cell functions, repairs itself, and replicates itself, is contained
in the particular sequence, and this is irreducible: it cannot be predicted from a
knowledge of its constituent parts, the way they interact, and the laws of physics
and chemistry.[18]

The hypothetical common ancestor probably had between 800 to 1,000
genes. The bacterium *Mycoplasma genitalium* has some 470 genes consisting
of around 580,000 DNA base pairs. However, this is a parasite that depends on
other cells to carry out much of its biosynthetic work. Hence it is reasonable
to suppose that the simplest cell that could have been the common ancestor
would have had a chromosome of at least 600,000 DNA base pairs in order to
function independently. Since each base is part of a nucleotide consisting of

the base, a sugar, and one or more phosphate groups comprising a total of at least 30 atoms, this means that its chromosome consists of at least 36 million atoms arranged in a very specific and complex shape that changes as the cell functions.

Ribosome
The cell makes proteins for its own repair and maintenance by a DNA strand acting as a template for making messenger RNA from molecules in the cell. The messenger RNA then carries the DNA's genetic information, coded in the sequence of its bases, to a ribosome.

> **ribosome**   A round particle composed of RNA and protein in the cytoplasm of cells that serves as an assembly site for proteins by translating the linear genetic code carried by messenger RNA into a linear sequence of amino acids.

> **protein**   A molecule consisting of a chain of between 50 to several thousand amino acids that provides structure, or controls reactions, in all cells. A particular protein is characterized by the sequence of up to 20 different kinds of amino acid that comprise the chain plus the three-dimensional configuration of the chain.

The simplest archaea synthesize proteins of between 50 to 300 amino acids long.

> **amino acid**   A molecule consisting of a carbon atom bonded to an amino group (–NH2), a carboxyl group (–COOH), a hydrogen atom, and a fourth group that differs from one amino acid to another and often is referred to as the -R group or the side chain. The -R group, which can vary widely, is responsible for the differences in chemical properties of the molecule.

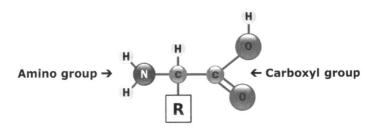

Figure 14.4 Structure of amino acid

Amino acids normally occur in two forms, or optical isomers, in which the positions of the -R group and the carboxyl group are switched. This phenomenon is called chirality, and one form is called the right-handed, or D, form (from the Latin *dexter*, right) while the other is the left-handed, or L, form (from the Latin *laevus*, left). Nearly all amino acids in cells are the L form.

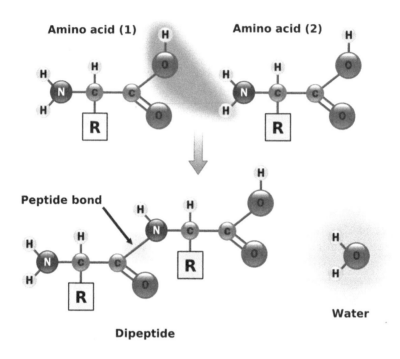

Figure 14.5 The chemical linking of two amino acids by means of a peptide bond

The carboxyl group of one amino acid can combine chemically with the amino group of another amino acid by releasing a molecule of water to form what is called a peptide bond.

> **peptide**   A chain of two or more amino acids formed by chemically linking the carboxyl group of one amino acid to the amino group of another amino acid.

In cells this reaction does not occur directly, but by a sequence, or pathway, of intermediate chemical reactions assisted by enzymes.

> **enzyme**   A biological catalyst, or chemical that speeds up the rate of a chemical reaction without being consumed by the reaction. Such catalysts are essential for the functioning of an organism because they make possible processes that would otherwise occur far too slowly without the input of energy (measured by an increase in temperature) to activate the reaction but which would damage or destroy the organism.

Nearly every enzyme is a protein consisting of a chain of between 62 and some 2,500 amino acids long, and each has a specific and elaborate three-dimensional structure that enables it to catalyze a specific biochemical reaction.

By convention a chain of 50 or so amino acids is termed a protein, although the distinction between a polypeptide and a protein is arbitrary. All proteins are formed from combinations of only 20 different amino acids out of some 500 known amino acids.

Cytoplasm
All these activities take place within the cytoplasm of the cell.

> **cytoplasm** Everything outside the cell nucleus and inside the cell membrane. It consists of a gelatinous, water-based fluid called a cytosol, which contains salts, organic molecules and enzymes, and in which are suspended organelles, the metabolic machinery of the cell.

A prokaryote doesn't have a cell nucleus and its principal organelles are DNA and ribosomes.

Plasmid
A plasmid, shown in Figure 14.2, is a circular molecule of DNA that replicates within a cell independently of the chromosomal DNA. Plasmids occur in many prokaryotes, have different functions, but are not essential for cell growth.

Cell enclosure
The simplest cell is enclosed—its cytoplasm is separated from the external environment—typically by three layers. The **cell capsule** is the outermost protective layer. Inside this is the semi-rigid **cell wall** that stabilizes the **plasma membrane** (also called the cell membrane) that surrounds the cytoplasm.

These layers are semi-permeable to allow the exchange of water and gases while controlling the exchange of particular molecules with the outside environment in order that the cell can repair and maintain itself while protecting the cell against destructive chemicals. They are made of a variety of molecules that invariably include proteins; the latter account for much of the cell's structure and, as we have seen, are made by a complex process starting with DNA.

External parts
Figure 14.2 shows two things on the outside of the cell enclosure. **Pili** are hair-like appendages that connect a cell to another of its species, or to another cell of a different species. A pilus builds a bridge between the cytoplasms of the connected

cells; this enables the transfer between the cells of plasmids, which can add new functions to a cell.

The **flagellum** is a tail that moves with a whip-like action to propel the cell through a fluid, often towards a source of energy or molecules needed to maintain its own existence.

Shifting protein shapes

To add to the complexity of the simplest cell, it is not sufficient that each one of its many proteins and enzymes consists of a combination of up to 20 different amino acids in a chain of between 50 to 300 such amino acids in a characteristic sequence, but in order to function each chain must have the correct shape. Most proteins fold into unique three-dimensional structures, and they shift between several shapes when they take part in biochemical reactions.

Figure 14.6 and Figure 14.7 illustrate the complex shape of one protein.

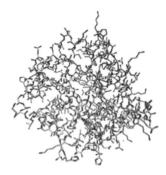

Figure 14.6 Two-dimensional representation of the three-dimensional structure of the protein triosephosphate isomerase showing atoms represented by differently shaded lines

Figure 14.7 Simplified representation of the protein triosephosphate isomerase showing the backbone shape

Conclusions

1. Conclusive fossil evidence of life, complex branched microorganisms, has been identified in rock dated to 2 billion years ago, some 2.5 billion years

after the Earth formed. Assuming that these microorganisms evolved from simpler lifeforms, life must have existed before then. Claims have been made that life dates from 3.5 billion, or even 3.8 billion or 3.85 billion years ago, putting it in the Hadean Eon when Earth was bombarded by asteroids and other debris left over from the formation of the solar system, but the evidence and its interpretation are disputed. The current best estimate is probably 3.5 billion years ago.

2. The discovery of extremeophiles, organisms currently living in extreme conditions similar to those thought to have obtained in the Hadean Eon, suggests that life *could* have existed at that time, during the first 700 million or so years after the Earth formed, but does not prove it.

3. Because of the paucity of the fossil record in general, and because nearly all sedimentary rock from the first two billion years has been subducted or metamorphized, it is almost certain that we will never find firm evidence of the first lifeforms and when they emerged on Earth.

4. The simplest, and presumably the earliest, form of life is an independent prokaryote, a single cell in which its genetic information encoded in DNA is not enclosed in a nucleus within the cell.

5. Genetic analysis of a wide range of cells strongly suggests, but does not prove, that life emerged naturally only once on Earth, and that all current lifeforms—from the numerous types of bacteria to humans—have evolved from a single common ancestor.

6. In order to function independently the simplest prokaryote needs a chromosome in the form of a folded loop of double-helical strands of DNA consisting of at least 36 million atoms configured in a specific structure. These strands unwind in order to act as templates to synthesize identical DNA strands needed for the cell to replicate itself and also to synthesize proteins needed to repair and maintain itself. This latter synthesis is a complex process involving the production of a messenger strand of RNA that carries the DNA's genetic information encoded in the sequence of its nucleotide bases to a ribosome, a manufacturing site comprising RNA and protein within the cell's liquid interior. The ribosomes use this genetic information to produce amino acids from molecules in the cell. The amino acids, which are made in only one of two possible stereo isomers, the L form, link together in chemical reactions catalyzed by enzymes, which are complex proteins. Proteins are formed from combinations of up to 20 different amino acids from some 500 known amino acids. Different amino acid chains constitute the different proteins needed to act as catalysts and to repair and maintain the cell, including its DNA. The simplest prokaryote requires proteins consisting of chains of between 50 to 300 of these limited forms of amino acids in characteristic sequences, but in order to function each chain must adopt shifting structures in order to create the required products. All of this is enclosed by semi-permeable layers of proteins

and other complex molecules that separate the cell from the outside environment, protecting the cell from damaging molecules while allowing specific molecules to enter and leave the cell as it needs and disposes of them in its synthesizing.

7. Any hypothesis of the emergence of life needs to explain how such a size of cell with such a complexity of components, functions, and changing configurations emerged from the interactions of atoms and simple molecules consisting of up to 13 atoms on the surface of the newly formed Earth.

CHAPTER FIFTEEN

The Emergence of Life 2: Hypotheses

Anyone who tells you that he or she knows how life started on Earth some 3.45 billion years ago is a fool or a knave.

—STUART KAUFFMAN, 1995

If attempting to find out how life emerged on Earth makes me a fool or a knave, then I'm in the good company of those who think they know.

Any scientific hypothesis of the emergence of life on Earth needs to explain how the most complex molecules of up to 13 atoms found in interstellar space and on asteroids—and by implication found on, or deposited on, the early Earth—evolved to the size, complexity, structure, and functioning of the simplest lifeform described in the previous chapter. That is, how inanimate matter became life.

Oparin-Haldane primordial soup replicator
The basic orthodox explanation is that advanced independently by Russian biochemist Alexander Oparin in 1924 and the English geneticist J B S Haldane in 1929.

According to Oparin's version the atmosphere of the early Earth consisted of hydrogen, methane, ammonia, and water vapour. Energy from sunlight and lightning caused these molecules to combine to form a mixture of simple organic compounds. Over the course of millennia these compounds accumulated in the oceans to form a warm, dilute primordial soup that eventually yielded a new kind of molecule, one that could replicate itself.

This replication was inefficient, and produced variations. From these variations an UltraDarwinian mechanism selected the ones most efficient at self-replication. (I use the term UltraDarwinian when natural selection is invoked beyond, or outside, the Darwinian evolution of species.) These self-replicators recruited proteins to make their replication more efficient, leading eventually to an enclosing membrane and to the first cell.

In 1953 Stanley Miller, a young research student working in the Chicago University laboratory of chemist Harold Urey, attempted an experimental test of this hypothesis. He boiled water in the bottom of a flask that he had filled with hydrogen, methane, and ammonia. To simulate the energy from lightning and the Sun he subjected the mixture to electrical discharges. At the end of a week his flask contained tarry deposits and traces of at least three amino acids.[1] Biochemists hailed this as experimental support for the Oparin-Haldane hypothesis: if amino acids could be produced in a week in a flask, then over thousands of years in a large ocean these amino acids would have polymerized to form peptides and complex proteins, eventually yielding the first cell.

However, more than 60 years of experiments with different primordial soup recipes and different conditions have failed to produce anything remotely like a self-replicator, still less a cell.

Biochemists usually explain this by the inability of laboratory experiments to replicate the vast timescales thought to be needed, but Johnjoe McFadden gives five reasons why life will never be generated this way, and consequently why the hypothesis is wrong.[2] First, Miller's assumption about the composition of the Earth's early atmosphere is now considered incorrect, and what is currently thought to be the primordial atmosphere is less favourable for such reactions.

Second, the main product of such reactions is gunk, a tarry mass of mainly hydrocarbons, which is the inevitable result when the ingredients can take part in a vast range of possible reactions.

Third, the amino acids produced in these simulated early Earth conditions are a mixture of right- and left-handed forms, or optical isomers. Linking amino acids to produce peptides and proteins does not work when both forms are present. Moreover, amino acids occur in cells only in the left-handed form, and no one has suggested a mechanism by which the primordial soup enriches for the left-handed form of amino acids.

Fourth, the reactions take place in an aqueous solution, which makes it very difficult for amino acids to polymerize, or link together, to produce proteins. The chemical linking of two amino acids by a peptide bond involves the loss of a water molecule, as shown in Figure 14.5. With so many water molecules in the solution the natural tendency is for a water molecule to break a peptide bond in a reaction called hydrolysis, which is the reverse of that shown in Figure 14.5; for this reason biological polymers tend to break apart, albeit slowly, in aqueous solutions. Hydrolysis is prevented in cells because the linking of amino acids does not take place directly but through a series of reactions that are catalyzed by enzymes. But since these enzymes are themselves proteins they cannot be used to create the first proteins.

Fifth, Darwinian natural selection depends upon gradualism: each small step in the evolutionary ladder must have arisen by random mutation, it must be viable in order to produce descendants, and it must represent a tiny improvement on its progenitor in adapting to the environment. The simplest cell examined in the previous chapter, a single-celled prokaryote (see Figure 14.2), could not have arisen by blind chance. Where did the proto-cells come from? If the proto-cells arose by natural selection then each ancestor must have been viable, but there is no fossil record of any proto-cells.

I should add that if current attempts by Craig Venter and colleagues at the J Craig Venter Institute to produce a living cell by stitching together its component parts prove successful, this will not provide experimental support for the Oparin-Haldane hypothesis; it will simply demonstrate that life can be produced by intelligent design where the intelligence is Venter's.

Self-replicating RNA

Since the probability that the first self-replicating independent cell was produced by the random reactions of simple molecules in the primordial soup is virtually nil, biochemists searched for a plausible primordial replicator. The favoured candidate is self-replicating RNA molecules that don't require enzymes to function.

Here a self-replicating molecule of RNA inefficiently produces copies of itself. UltraDarwinian natural selection favours those products that are the most efficient in producing descendants. Eventually these more efficient versions catalyze the linking of amino acids to produce proteins in order to aid their replication and form a protective membrane, which makes them still more efficient. Finally they generate DNA, which proves a more stable repository of the genetic information, thus producing the first cell.

This hypothesis, usually termed the RNA World, is now the orthodox explanation for the origin of life on Earth. Experimental support is provided by the discovery that short segments of RNA called ribozymes can act as enzymes to catalyze many biochemical reactions: they can join up two RNA molecules and polymerize up to six activated RNA bases on an RNA template. Furthermore, many viruses, like the flu virus, have an RNA rather than a DNA genome, which suggests that an enzymatic RNA molecule catalyzed its own replication and that RNA in modern cells is the evolutionary remnant of the original RNA self-replicator.

According to McFadden, however, this hypothesis suffers from the same problems as the primordial soup hypothesis considered above: RNA polymerization excludes water and does not occur naturally in an aqueous solution; ribozyme-catalyzed RNA polymerization does not work with a mixture of left- and right-handed RNA nucleotide bases, and nobody has suggested a prebiotic mechanism that would enrich one chiral form of RNA bases over another.

Experimentally, nobody has designed, still less discovered, a self-replicating RNA molecule. An RNA molecule consists of three parts: the bases A, U, C, and G, together with a ribose sugar and a phosphate group. They comprise about 50 atoms that have to be structured in a highly specific way. One strand of Figure 14.3, with U substituting for T, gives an indication. Although experimenters have achieved some success in synthesizing the bases and phosphate groups from simpler molecules, they have done so only through a series of carefully controlled reactions in order to avoid the junk generated by simulating natural conditions. Moreover, they have been unable to produce a solution enriched in ribose sugar.

Graham Cairns-Smith, an organic chemist and molecular biologist at the University of Glasgow, maintains that there is no reasonable possibility that an RNA replicator could have emerged from a primordial soup. He has estimated that there are about 140 steps that go into the synthesis of an RNA base from simple prebiotic compounds. For each step a minimum of six alternative reactions could occur instead of the desired reaction. The probability that the right outcome occurs by chance is 6^{140} or 10^{109}.[3]

Self-replicating peptide

This led some biochemists to try and find a simpler primordial replicator than RNA. In 1996 David Lee and colleagues at California's Scripps Research Institute designed a short peptide thirty-two amino acids long that could act as an enzyme to stitch bits of itself together and replicate.[4]

McFadden dismisses the self-replicating peptide as a candidate for a primordial replicator because Lee and colleagues used activated peptide fragments to minimize the tar-yielding side reactions and offered no plausible explanation of how such activated amino acids emerged by chance from the primordial soup.

Two-dimensional substrate

The overwhelming improbability of orthodox biochemistry's RNA self-replicator, or even a self-replicating peptide, emerging by chance from the molecules present in a primordial soup prompted ideas about the much greater probability of much simpler self-replicators being formed on a two-dimensional surface rather than in an aqueous solution.

Clay replicator

Cairns-Smith pursued this idea from the mid-1960s and in 1985 published a book setting out his hypothesis.[5]

We saw when considering reproduction as a proposed characteristic of life that a crystal of a simple salt can reproduce itself in a saturated solution of that salt.* Cairns-Smith proposes that the primordial replicator was such a simple crystal. Its information is encoded in its structure: in effect the crystal structure is a precursor of the organic gene.

Clay is a dense sediment comprising minerals and silicates, mainly of aluminium, and was probably common on the early Earth. These silicate crystals reproduced, and natural selection favoured mutant crystals that modified their environment and enhanced their replication. Their charged surfaces attracted dipolar organic molecules like amino acids and nucleotides, effectively catalyzing their polymerization to proteins and the components of RNA and DNA. Eventually there was a genetic takeover in which more stable products of these surface reactions, organic polymers like RNA and DNA, inherit the crystal's information and gradually

---

*   See page 201.

displace it as the primary genetic material, while other organic polymers, like proteins, formed a protective membrane: the primitive crystal self-replicator sheds its clay enclosure and emerges as a cell.

In 1996 chemist James Ferris of the Rensselaer Polytechnic Institute, Troy, New York, together with biologist colleagues appeared to offer experimental support when they incubated separate solutions of amino acids and nucleotides in the presence of mineral surfaces. They obtained polymers of amino acids and of nucleotides up to 55 monomers long; without the mineral surfaces no polymer longer than 10 monomers was produced because hydrolysis prevented further lengthening of the chains.[6] This, however, does not constitute experimental confirmation because the amino acids and the nucleotides were artificially activated rather than naturally occurring.

Moreover, there is no empirical support for the claim that the first replicator in the pathway of reactions to the cell was a clay replicator. There are no clay replicators in contemporary clay deposits resembling those of the early Earth or in any fossil record. Given the problems of identifying cell fossils earlier than 2 billion years ago, I think it impossible that fossils of any clay precursors of cells can ever be identified.

Without evidential support the clay replicator idea remains an intriguing conjecture. However, it lacks a convincing explanation for a key step. The order and symmetry of a crystal is a repetitive periodic arrangement with low information content. By contrast a cell is an aperiodic, interactive, complex entity with high information content, as are each of its components like RNA, DNA, and proteins. The conjecture fails to explain how such high information content is inherited from the low information crystal.

Iron pyrites replicator
Günter Wächtershäuser, a chemist turned patent lawyer, proposed in 1988 a later version of the two-dimensional idea. He suggests that the formation of crystals of iron pyrites from iron and hydrogen sulphide in sulphide-rich waters near sub-oceanic vents generates the electrons that could have chemically reduced carbon dioxide to organic compounds. The charged surface of the crystals could bind the organic compounds and promote various reactions leading to the formation of amino acids and nucleotides, replicators, and eventually life.[7]

Wächtershäuser and colleagues from the Regensburg Institute for Microbiology demonstrated that the formation of iron pyrites can be coupled to the polymerization of amino acids. However, they have yet to provide experimental support for the key first step, the conversion of carbon dioxide into carbon compounds through iron pyrites formation,[8] and so this, too, remains conjecture.

Extraterrestrial origin
The problems of explaining how the simplest independent cell could emerge from a primordial soup on Earth led some scientists to turn to an idea that has recurred

throughout the centuries and was popularized in science fiction. According to the version of panspermia proposed in 1903 by the Swedish chemist and Nobel Prize winner Svante Arrhenius, microbes ejected from planets with life travelled through space and alighted on Earth.

Another Nobel Prize winner, Francis Crick, co-discoverer of DNA's double-helical structure, together with Leslie Orgel, then research professor at the Salk Institute for Biological Studies and a principal NASA investigator, concluded that this was highly unlikely to have happened by chance. Instead in 1973 they proposed directed panspermia, whereby an advanced civilization in the galaxy deliberately targeted planets like Earth with microorganisms.[9]

While satisfying science fiction fans, the lack of evidence failed to satisfy the scientific community. In 1978, however, Fred Hoyle and his former student Chandra Wickramasinghe, then Head of the Department of Applied Mathematics and Astronomy at University College Cardiff, claimed evidence for panspermia. For years astronomers had failed to account for certain spectral lines of interstellar dust, which was thought to consist principally of ice crystals. Hoyle and Wickramasinghe announced that these spectral lines were consistent with bacteria.[10]

It prompted them to propose that sudden outbreaks of diseases to which we had little resistance and which often had localized origins, such as syphilis in the fifteenth century and AIDS in the twentieth century, were caused by bacteria and viruses being deposited on Earth by comets. This is consistent with the opinion of the eighth century English historian and monk, Saint Bede, who declared that comets presaged "a change of sovereignty, or plague". As recently as 2003 Wickramasinghe and colleagues claimed in the medical journal, *The Lancet*, that SARS (severe acute respiratory syndrome) was caused by microbes from a comet.[11] In support of such claims Wickramasinghe cites evidence that air samples gathered in 2001 from the stratosphere by the Indian Space Research Organization contain clumps of living cells, arguing that no air from lower down would normally be transported to a height of 41 kilometres.

More recent, and more sophisticated, spectral analysis has indeed identified organic molecules in interstellar dust. But, as we saw,* the largest of these consist of 13 atoms and are very far from anything bacterial. The stratospheric samples include two bacteria known on Earth. If we apply the test of reasonableness, however, a simpler and far more likely explanation is that these bacteria were swept up 41 kilometres from the Earth's surface by meteorological events rather than travelling billions of kilometres through interstellar space. Medical researchers have dismissed the extraterrestrial origin of diseases idea by offering more compelling evidence that Wickramasinghe fails to take into account.[12]

In any case, all these panspermia ideas simply postpone the origin of life on Earth question just as the various multiverse and progenitor universe speculations postpone the origin of matter question.

---

* See page 139.

In their 2007 version, however, Wickramasinghe, his daughter, and another colleague from Cardiff University's Centre for Astrobiology—the unofficial centre for panspermia research—rose to the challenge. They claim that life originated inside comets.[13]

This claim is not without its problems. It depends on three assumptions.

1. Supernovae debris incorporated in comets contains radioactive elements whose decay heat maintains water in liquid form inside comets. No evidence supports this assumption.
2. Comets also contain clay, from which livings cells evolve as proposed by Cairns-Smith. Although the Deep Impact mission of 2005 showed that Comet 9P/Tempel contains clays and carbonates, which usually require liquid water to form, no evidence supports Cairns-Smith's conjecture, still less that it occurs inside comets.
3. The lifetimes of friendly prebiotic environments within comets exceed those of localized terrestrial regions by four or five powers of ten. No evidence supports this assumption.

Based on these assumptions they argue that, since the combined mass of comets throughout the galaxy overwhelms that of suitable terrestrial environments by another 20 powers of ten, it follows that the totality of comets around G-dwarf Sun-like stars offers an incomparably more probable setting for the origin of life than any that was available on the early Earth.

However, it is illogical to deduce this probability by comparing the mass of all comets with the mass of the Earth because the claim is that life originated not inside something whose mass equals that of all comets in the galaxy but inside one comet, and the mass of an average comet is 6×10^{10}, or 60 billion, times less than that of the Earth. What should be compared is the life-forming suitability of the interior of a single comet—and most cometary scientists deduce that this is principally solid ice—and the surface of the primordial Earth.

The website announcing their claim says that life originated in one comet and the "emergent life then quickly spreads like an infection from comet to comet, star-system to star-system, encompassing ever-increasing volumes of the Universe".[14] They do not explain *how* life in the interior of one comet orbiting one star spreads to other comets, especially those orbiting other stars.

Intelligent design

The problem with using the term intelligent design is that, since the mid-1990s, it has become inextricably associated with Intelligent Design, a claimed scientific theory whose proponents are funded by, or who are staff members of, the Discovery Institute, an American think tank established and financed by Christians with the purpose of proving that God created life.

I fear it is impossible to divorce ideas from beliefs, whether those beliefs are in God or in materialism, especially when the evidence, as in this case, is so scarce and open to different interpretations. Consequently I think it instructive to consider intelligent design ideas that either do not posit a Judaeo-Christian God or else contradict the beliefs of their proponents.

Computer simulation

Oxford University philosopher Nick Bostrom says that, since we are now able to create computer simulations of worlds and people, a far more technologically advanced "posthuman" civilization will be able to simulate people who are fully conscious, and what we perceive to be life could be such a computer simulation.[15] Lest this be thought a philosopher's fanciful whim, Bostrom provides an equation to prove there is a high probability that this is indeed the case.

However, like Drake's Equation calculating the probability of intelligent civilizations in our galaxy,* Bostrom's Equation is a multiple of separate probabilities, each of which is based on questionable assumptions or guesses. Moreover, after extrapolating the growth of computer power over the last 60 years into an unspecified and unknowable future, his next sentence begins "Based on this empirical fact…" A projection, however reasonable—and this one is questionable—is not an empirical fact.

This may not be a whim, but it is no more than a speculation that cannot be falsified by empirical test. And, like Crick and Orgel's directed panspermia, it fails to tell us the origin of these posthuman designers and hence the ultimate origin of life on Earth as we perceive it.

Irreducible complexity

Michael Behe, Professor of Biochemistry at Lehigh University in Pennsylvania, says he believes all lifeforms on Earth have evolved from a common ancestor, although the Darwinian hypothesis does not explain the differences between species. However, individual components of the first cell, and the elaborate and interconnected biochemical pathways of their production, are irreducibly complex: if they are missing just one of their parts they cannot function. They could not have evolved by an [Ultra]Darwinian mechanism because that depends upon natural selection from a variety of mutants at each step in the pathway, and each step must be viable. The evidence he cites includes the systems that target proteins to specific sites within the cell and the bacterial flagellum.† This latter, for example, consists of a dozen or more proteins; no intermediate stage is viable as a functioning unit.

He says he was forced to conclude that the first form of life, the common ancestor cell, could only have resulted from intelligent design. To reconcile this with biological evolution he suggests that this first cell contained all the DNA

---

*   See page 183.
†   See page 222.

necessary for subsequent evolution. He does not identify the designer, but says that orthodox science has rejected this conclusion because of its possible theological implications.[16]

Orthodox evolutionists were quick to condemn Behe's 1996 book, *Darwin's Black Box: The Biochemical Challenge to Evolution*. In his review in *Nature* University of Chicago evolutionary biologist Jerry Coyne finds a clue to Behe's reasoning by identifying him as a Roman Catholic. Most scientists, however, don't dismiss Newton's work on mechanics because he believed in alchemy or Kepler's work on astronomy because he believed in astrology. A more substantial criticism is that Behe fails to take sufficient account of mechanisms other than sequential steps for the production of cell components, like the co-option of components evolved for other purposes, duplicated genes, and early multifunctional enzymes.

In the case of the bacterial flagellum, for example, microbiologist Mark Pallen and evolutionary biologist Nicholas Matzke point out that there is not just one bacterial flagellum but thousands, perhaps millions, of different flagella today; hence "either there were thousands or even millions of individual creation events...or...all the highly diverse contemporary flagellar systems have evolved from a common ancestor". Evidence for the evolution of bacterial flagella includes the existence of vestigial flagella, intermediate forms of flagella, and the pattern of similarities among flagella protein sequences: almost all of the core flagellar proteins have known homologies with non-flagellum proteins, suggesting that flagella evolved from combinations of existing cellular components.[17]

Both Coyne and Brown University biologist Kenneth Miller note that Behe concedes that some components of the first cell could have evolved by an [Ultra] Darwinian mechanism but requires that all biochemical features would have to be explained by natural effects before intelligent design is disproved. Because of the difficulties in obtaining evidence, it is impossible to prove that all are not. Hence, they argue that Behe's claim is not falsifiable and therefore not scientific.

Science's inability to explain
Behe's claim is a particular example arising from a more general problem, the inability of science to explain certain phenomena. This was articulated by Fred Hoyle, a confirmed atheist. It was atheism that motivated Hoyle to search for an alternative to the Big Bang theory, but when he came to consider how life emerged on Earth he famously compared the random emergence of even the simplest cell to the likelihood that "a tornado sweeping through a junk-yard might assemble a Boeing 747 from the materials therein".[18]

By the time he gave the Royal Institution's Omni Lecture in 1982 he had concluded

> If one proceeds directly and straightforwardly in this matter, without being deflected by a fear of incurring the wrath of scientific opinion, one arrives at the conclusion that biomaterials with their amazing measure of order

must be the outcome of intelligent design....[P]roblems of order, such as the sequences of amino acids in the chains [that constitute cell proteins]... are precisely the problems that become easy once a directed intelligence enters the picture.[19]

I'm not aware that Hoyle converted to any religion, but his later writings suggest he considered that some superior intelligence governing the universe is inferred by phenomena science cannot explain. And herein lies the problem. Just because science cannot explain a phenomenon now, it does not follow that science will never be able to explain the phenomenon. Equally, it does not follow that science *will* be able to explain the phenomenon in the future, as some materialists like Richard Dawkins assert.

Since evidence can neither prove nor disprove intelligent design as the origin of life on Earth, the test of reasonableness is best applied by examining consistency with other evidence, which in this case is the pattern of human understanding of natural phenomena. Historically, most humans have invoked a supernatural cause for natural phenomena they do not understand at the time. Thus the Greeks of the warring city states of around the tenth to the fifth centuries BCE did not understand what caused lightning and thunder, and so they attributed these powerful and awe-inspiring occurrences to the most powerful god in a pantheon of superhumans who reflected the hierarchy of their own society.

As science developed in a Western society that was predominantly Christian, its empirical reasoning filled gaps in our understanding of natural phenomena and progressively removed the need for supernatural explanations. Thus the Earth was no longer something God had created as the centre of the universe and the Sun was no longer something God had created to illuminate the Earth between periods of darkness.

The realm of the gaps—and hence the realm of the transcendent creator God—continued to diminish as science's explanatory power increased, and God was continually pushed back towards being the ultimate, rather than the direct, cause of natural phenomena. (I am using here the mainstream Christian concept of God because science from the sixteenth century developed mainly in the Christian West. Other religions and cultures have different views of God or gods, and some hold that a creative Cosmic Spirit is both immanent and transcendent, rather than immanent for only 33 years in one person of a transcendent Trinitarian God.)

While there is no guarantee that this pattern will continue, the most reasonable approach to understanding the emergence of life on Earth is that, while keeping an open mind, we should seek a natural explanation rather than invoke a supernatural cause like God or intelligent design.

Anthropic principle

We met the Anthropic principle idea in Part 1 when discovering that unless precise physical laws of unknown origin operate, and several cosmic parameters and dimensionless constants have very finely-tuned values that no law can explain, then a universe that enabled the evolution of humans would not exist. Here the anthropic principle is invoked to explain the puzzling emergence of life from simple molecules in the primordial soup on Earth.

The anthropic idea is generally credited to theoretical physicist Brandon Carter in 1974 while he was at the University of Cambridge.[20] The most comprehensive work on this subject was undertaken by cosmologists John Barrow and Frank Tipler, whose 1986 book includes all the relevant laws, parameters, and constants and the different approaches to the idea.[21] I shall consider their definitions of three versions of the anthropic principle.

Weak anthropic principle (WAP)

> The observed values of all physical and cosmological quantities are not equally probable but they take on values restricted by the requirement that there exist sites where carbon-based life can evolve and by the requirements that the universe be old enough for it to have already done so.

Barrow and Tipler engage in scientific and philosophical discussion of such things as the Bayesian approach to the age and size of a universe necessary for the evolution of carbon-based lifeforms like us who can observe the universe. What all this reduces to is that the characteristics of the universe we observe are such as to enable us to observe it. It is a tautology and explains nothing.

Strong anthropic principle (SAP)

> The universe must have those properties which allow life to develop within it at some stage in its history.

This is distinguished from the WAP by the word "must". According to Barrow and Tipler there are three interpretations of it.

a. *The intelligently designed universe*

> There exists one possible universe designed with the goal of generating and sustaining observers.

This extends the arguments for intelligent design of the first cell to the whole universe and is countered by similar objections. I won't repeat these, but simply reiterate my conclusion that this is outside the empirical realm of science and, while keeping an open mind, we should seek a natural explanation rather than invoke a supernatural cause like God or some other unknown or unknowable intelligent designer.

b. *Participatory universe*

> Observers are necessary to bring the universe into being.

This is based on John Wheeler's conscious-dependent interpretation of quantum mechanics that I considered on pages 87 to 88. It invokes the philosophical conjecture of backward causation, which holds that an effect can precede in time its cause. This idea has proponents and opponents among philosophers. No one has proposed how it can be falsified by experiment or observation, and so the idea lies outside the realm of science.

The participatory universe idea means that the universe did not exist before our Palaeolithic ancestors observed the universe's wave function and so collapsed it into an observable reality. That is, our Palaeolithic ancestors created the observed universe.

Applying the tests of reasonableness to this conjecture, I think Antony Flew and other philosophers are right to argue that backward causation is a logical contradiction, and so the participatory universe is internally inconsistent. It is also inconsistent with the generally accepted interpretation of observations that show the universe is at least 10 billion years old.

c. *The multiverse*

> An ensemble of other universes is necessary for the existence of our universe.

The application of this interpretation of the SAP means that, however improbable it is for the first cell to emerge from the interactions of simple molecules in the primordial soup, it must have occurred on one planet in one of an unimaginably large number, if not an infinity, of universes; we just happen to be on that planet in that universe.

I examined the four principal categories of multiverse proposals when considering why the universe came into existence with one specific form when many other forms are logically possible, and concluded that none is testable, and so none is scientific.* Correspondingly, those who argue that the intelligently designed cell is not falsifiable, and therefore not scientific, cannot logically argue that the multiverse idea is scientific unless they can show how it can be tested.

Barrow and Tipler propose a third version not considered by Carter.

---

* See pages 99 to 102.

Final anthropic principle

> Intelligent information-processing must come into existence in the universe, and, once it comes into existence, it will never die out.

This is a metaphysical speculation about the future rather than a physical principle.

A further argument against the anthropic principle in general is given by Roger Penrose, who says "it tends to be invoked by theorists whenever they do not have a good enough theory to explain the observed facts".[22]

Quantum emergence

I have cited McFadden's challenges to many of the proposals considered above. He considers that none of the primordial soup hypotheses is valid because, among other reasons, the essential first self-replicating entity arises through thermodynamics. However, the random motion of molecules will inevitably produce such a multiplicity of reactions that the odds against the random construction of a self-replicating entity are impossibly high.

For example, even after making a series of favourable assumptions, he calculates that the odds against the simplest known self-replicating peptide emerging from the molecular soup by random reactions is 10^{41} against, that is, virtually nil.

McFadden believes that the production of such a peptide needs to be directed. This does not require an intelligent designer; in the right conditions it can be achieved by the mechanism of quantum mechanics rather than thermodynamics.[23]

As we saw in Chapter 6 when considering the reality of the quantum world,* quantum theory says that a subatomic particle is both a particle and an information wave that is non-localized: it is infinite in extent and contains all possible existence states for that quantum entity; this is known as a quantum superposition state.

McFadden does not favour Everett's multiverse interpretation† because of its horrendous violation of the parsimony test, and agrees with the view that only when the wave function is measured does it collapse into the probability of a physical particle having position, momentum, and energy, thus entering the classical world that we perceive.

But neither does he favour the Copenhagen interpretation‡ or Wheeler's conscious observer interpretation§ to explain the measurement and collapse of the quantum wave. He also rejects Bohm's pilot wave interpretation (the details of which needn't concern us here).

Instead he favours Zurek's decoherence interpretation, according to which entities remain in the quantum state only as long as their wave functions are

---

* See pages 87 to 88.
† See page 99.
‡ See page 87.
§ See page 87.

coherent; as soon as interference takes place they decohere and collapse into classical reality. The entire world appears as classical reality because any open system is continually bombarded with photons, electrons and other particles, and quantum entanglement with so many entities produces decoherence and collapse of superposition states. Thus entanglement with the environment, not an observer, measures the quantum system and causes its collapse.

In order to make the quantum decoherence idea work for the emergence of a self-replicating peptide from a primordial soup, McFadden makes three key assumptions:

1. the primordial molecular soup is microscopically small and is trapped in a tiny structure, like the pore of a rock or an oil droplet, that serves as a kind of proto-cell protecting the coherence of quantum states inside;
2. new molecules, including a fresh supply of amino acids, diffuse into and out of this proto-cell;
3. the system remains in the quantum state, and so instead of a classical addition of a single amino acid to make a single peptide product, each amino acid addition produces a quantum superposition of all possible resulting peptides.

In the process of adding an amino acid each peptide couples with its environment and so decoheres its quantum state into a classical state. Thereafter

> it would have been free to drift once more into the realm of quantum superposition and await the next measurement....This process of drifting into the quantum realm, measurement, collapse into a classical [particle] state, and drifting back into the quantum realm would have continued... to elongate the quantum superposition of possible peptides until such time when the system irreversibly collapsed into a classical state.

However, this whole mechanism depends on McFadden's assertion that

> most importantly, while the peptide remained a single molecule, *it could always re-enter the quantum realm after measurement* [his italics].

This repeated drifting back into the quantum state is questionable. It is not the same molecule that is doing it. In contradiction to his statement that this hypothesized peptide "would have emerged unscathed from the measurement process", an amino acid is added at each stage, thus altering its molecular composition.

According to McFadden, the process irreversibly collapses into the classical state when a self-replicating peptide is produced in the superposition state. He concedes that the probability that this particular peptide, rather than any of the others in the superposition state, is the one that collapses out into classical reality is the same as that of it being produced in the thermodynamic conditions of the molecular soup, 10^{41} to one against in the example chosen.

He suggests that one explanation is to invoke Everett's quantum multiverse, which he had previously dismissed as "preposterous": every possible quantum superposition collapse occurs in a different universe, and we just happen to be in the universe where the classical peptide produced is the self-replicating one. However, the odds against this happening anywhere else in the same universe are 10^{41} minus one. Hence, if life is found anywhere else in our universe (and many astrobiologists think it may be found as close as Europa, one of Jupiter's moons), this would disprove the hypothesis.

As an alternative explanation McFadden invokes the inverse quantum Zeno effect, by which a dense series of quantum measurements of a system along a particular path can draw the system along that path rather than the vast number of paths possible without the measurements.

In order for this to work here McFadden makes two further assumptions:

4. the sequence of electron and proton movements within and between molecules that constitute the chemical reactions leading to the first self-replicator are essentially no different at the quantum level than electrons and photons moving in empty space;
5. the quantum measurements performed by proto-enzymes on peptides in the superposition state are essentially no different than the quantum measurements performed by polarized lenses on photons.

Accordingly these measurements guide the chemical system along a path that reduces the odds from 10^{41} to one against to produce the first self-replicating peptide. Thereafter the inefficient self-replicating peptide produces mutants that, by UltraDarwinian natural selection, lead to a gradual increase in self-replicating fitness. They become more efficient by recruiting lipid membranes to shelter from the outside environment and evolve into more efficient enzymatic proteins, and so on until the much more efficient cell emerges.

The last two assumptions are reasonable in that they are consistent with other, empirically supported, scientific tenets. The overall hypothesis, however, has no empirical support. On the contrary, whenever researchers have tried to mimic this process in the laboratory by using enzymes to copy DNA or RNA molecules, more efficient replicators sometimes evolve after many hundreds of cycles but these are smaller, simpler molecules. The system never evolves in the opposite direction, towards greater complexity, which is needed to produce a cell and which is the direction that biological evolution takes. The same occurs with computer simulations.

McFadden, however, is undaunted. He believes the problem is isolating the reactants from the environment in order to maintain quantum coherence. He speculates that the first cellular life was a simple self-replicator sheltering within nanometre-scale microspheres, not very different from nanobacteria found in terrestrial subterranean rocks.

Hence laboratory experiments need to mimic these conditions that retain quantum coherence, using such things as a carbon nanotube, a one-atom thick sheet of carbon rolled into a cylinder whose diameter is approximately 50,000 times smaller than that of a human hair.

So, too, computer simulations will only work on quantum computers, which utilize quantum mechanical phenomena, such as superposition and entanglement, to perform operations on data rather than digital switching as at present.

Both technologies are in their infancy.

Self-organizing complexity

In developing their definitions of life, both Smolin and Capra drew on complexity theory as expressed by Stuart Kauffman, a medical doctor turned biochemist who is associated with the Santa Fe Institute, an inter-disciplinary organization dedicated to the study of complex systems.

In 1995 Kauffman proposed that life developed from a primordial soup containing billions of different kinds of molecules by a process of self-organizing complexity.[24] He assumes that in such a soup a molecule A catalyzes the production of another molecule, B, which thereby becomes more plentiful in the soup. B in turn catalyzes the production of C, which catalyzes the production of D, and so on, producing a series $A \rightarrow B \rightarrow C \rightarrow D \rightarrow E \rightarrow F \rightarrow G$ etc. He further assumes that one molecule in this series, say F, also catalyzes the production of A, giving catalytic closure of the cycle $A \rightarrow B \rightarrow C \rightarrow D \rightarrow E \rightarrow F \rightarrow A$, which he calls an autocatalytic set. This continually perpetuates itself by feeding on the raw material in the primordial soup with the aid of energy from sunlight or volcanic vents, thereby increasing the concentration of these molecules in the soup.

He then assumes that one molecule of this set, say D, also catalyzes the production of another molecule, say A, as well as E, and so on. In this way a network of self-sustaining autocatalytic sets arises.

He illustrates what he maintains is the characteristic growth pattern of such networks by analogy with buttons and threads. Randomly choose a pair of buttons and link them with a thread. If you continue to link pairs of buttons randomly by separate threads, then you will inevitably link some buttons to ones that are already linked to other buttons.

The number of buttons in the largest cluster of linked buttons is a measure of how complex the system has become, illustrated by Figure 15.1, where Kauffman generalizes the phenomenon by labelling each button a node and each link an edge.

The size of the largest cluster grows slowly at first because most buttons don't have many links. But when the number of threads approaches and then exceeds half the number of buttons, the size of the largest cluster increases extremely rapidly because, with most buttons now in clusters, there is a high probability that each new cluster will link a smaller cluster to the largest cluster. Very quickly

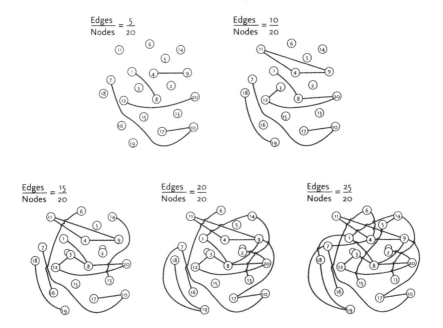

Figure 15.1 Kauffman's "button model" of network formation
where a button is called a node and an edge represents the thread linking buttons
in a network.

a single supercluster forms a network in which the great majority of buttons are linked. After that the size of this largest network grows only slowly because very few buttons are not already part of that network.

Figure 15.2 illustrates what Kauffman calls this network phase transition by analogy to the phase transition between, say, water and ice. A highly complex system, like a network of interlinked parts of a cell, suddenly emerges from the component networks of the cell, which have grown through the linking of self-sustaining autocatalytic networks of molecules. This highly complex system is stable because there is little scope for further change.

According to Kauffman, when this mechanism operates in a primordial soup it removes the need to build a long chain of unlikely chemical events one after the other. Life, as a super-complex self-sustaining autocatalytic set, emerges suddenly as a phase transition. "Life crystallizes at a critical molecular diversity because catalytic closure itself crystallizes."

This hypothesis is consistent with the network, rather than genealogical tree, idea of evolution discussed earlier. However, Kauffman concedes that "Scant experimental evidence supports this view as yet."

Applying the tests of reasonableness, the self-organizing complexity proposal is internally consistent and provides external consistency in that it is consistent with computer models. Unfortunately, that is all.

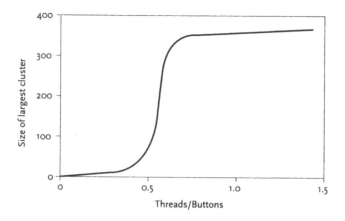

Figure 15.2 Kauffman's network phase transition

As the number of threads (links) increases, a sharp transition occurs between the state in which there are many unconnected buttons and a few links to the state in which almost every button is part of the network.

The model doesn't explain how the catalytic molecules, *A*, *B*, *C*, etc. arose in the first place. As we saw above, even with favourable assumptions the probability that the simplest known self-replicating peptide emerging from the molecular soup by random reactions is 10^{41} against.

Edward O Wilson concludes that, while they may be on the right track, the problem with complexity theories in this context is that they are too divorced from biological details.

> The basic difficulty, to put the matter plainly, is an insufficiency of facts. The complexity theorists do not yet have enough information to carry with them into cyberspace. The postulates they start with clearly need more detail. Their conclusions thus far are too vague and general to be more than rallying metaphors, and their abstract conclusions tell us very little that is really new.[25]

In McFadden's opinion,

> the spontaneous emergence of autocatalytic sets is feasible only in computers, where each set can be isolated from the jumble of reactions around them. In real chemical soups, each component gets caught up in a thousand side reactions that inevitably dilute and dissipate any emerging autocatalytic sets.

He goes on to make a theoretical, and what he considers a more important, objection to complexity theory as a hypothesis to explain the emergence of life. Self-organization shown by cyclones and other examples is generated by the random interaction of billions of molecules. They are phenomena of huge numbers of particles and have a structure only on a macroscopic scale; at a molecular level there is only chaos and random motion. But cells have ordered structures all the way down to fundamental particles. The macroscopic structures of living cells are not generated by random incoherent motion.[26]

I think Wilson is right to say that, as currently developed, self-organizing complexity theory is too abstract and reliant on computerised models to provide an empirically supported hypothesis. The hope is that refining such models in the light of empirical data will lead to deep new laws that account for the emergence of life from inanimate matter.

Emergence theory

Smolin and Capra also draw on emergence theory as articulated by Harold Morowitz, Professor of Biology and Natural Philosophy at George Mason University in the USA, who believes that deeper laws will explain emergent complexity from biochemical processes just as at the quantum level the Pauli Exclusion Principle selects from an otherwise inconceivably vast array of possibilities a small set of energy states of electrons in an atom that explains the emergence of 118 elements.*

For many years Morowitz has searched for unifying features in the major biochemical reactions that living organisms carry out in their cells because he considers that metabolic-like reactions are the core, and therefore most ancient, characteristic of life, although in his seminal book[27] he never defines metabolism or life.

If I understand correctly Morowitz speculates that the process for the emergence of life began when the through-flow of energy in a network of autocatalytic sets of primordial molecules led to the emergence of metabolism (presumably this is what complexity theorists term self-sustaining autocatalytic sets).

Morowitz identifies several problems. First, "starting with the 20 naturally occurring amino acids, a chain of 100 amino acids can be one of 20^{100} possibilities....this is greater than 10^{101}....Many of these sequences will be catalytic for a wide array of reactions." They need to be selected for.

Next, the problems of building structures like prokaryotic cells with specific functions are architectural (three-dimensional structures that determine where chemical reactions take place), chemical (which molecules and how they react with others), and informational (a set of processes whereby "sequences of amino acids can be encoded in sequences of nucleotides", namely DNA and RNA). All these functions must be selected from a range of unimaginably large possibilities.

*   See page 135.

The emergence of life consequently involves "many, many emergences" that take place within a time span that, geologically, is almost instantaneous.

Morowitz concedes that understanding the intermediate emergences "is subject to experimental and theoretical study in the world of chemical networks, physical study of macromolecules, and a better understanding of system properties". That is, as yet there is neither empirical support nor adequate theory behind the proposal.

It is an enormously attractive idea that one or more biochemical equivalents of the Pauli Exclusion Principle explain the pathway from random simple molecules to the emergence of an independent lifeform, but neither Morowitz nor anyone else has suggested what such a deeper law, or laws, might be. Thus far emergence theory offers nothing new except an alternative, and perhaps more sophisticated, description of the problems couched in terms of the need for one or more selection principles to explain the phenomena.

Conclusions

1. Biochemistry's orthodox account of how life emerged on Earth from the most complex molecules of up to 13 atoms found on or deposited on the newly formed Earth is that energy from the Sun and lightning caused such molecules in the atmosphere of the newly formed Earth to produce the 20 naturally occurring amino acids which dissolved in the oceans. Over hundreds of thousands, if not millions of years, random reactions produced the first self-replicator, an RNA molecule that could act as a catalyst to reproduce itself and also produce proteins. UltraDarwinian natural selection favoured mutant descendants that were more efficient at self-replication, leading to the recruitment of proteins to form a protective membrane and, eventually, molecules of DNA that were more stable repositories of genetic information, and thus the first cell. However, this hypothesis, and also one proposing that the primordial replicator was a self-replicating peptide, is unsupported by any empirical evidence. More than 60 years of attempts to generate life from these primordial molecules in conditions thought to have existed on the early Earth have failed to produce anything remotely like a self-replicator, still less a living cell.

2. Moreover these hypotheses are invalid because, among other reasons, there is an overwhelming improbability that random reactions in an aqueous solution could have produced a self-replicating RNA molecule or a self-replicating peptide, still less ones with only the left-handed amino acid isomers found in cells.

3. The proposal that the first cell evolved by a process of self-organizing complexity likewise fails to explain how the catalytic molecules on which the proposal is based emerged from overwhelmingly improbable random reactions of molecules in a primordial soup. Its subsequent account

is currently too abstract and divorced from biological detail to offer a convincing explanation for the emergence of life.

4. The proposal that the pathway to the first cell began with self-replicating crystals of clay present on the early Earth, and that progress along this pathway was effectively catalyzed by reactions taking place on the charged two-dimensional surface of the clay crystals, is an intriguing conjecture that lacks adequate explanations for key steps and also experimental support; there is also the almost certain impossibility of obtaining fossil evidence. That is also the case for the proposal that the pathway to life was enabled by the two-dimensional catalytic effect during iron pyrites formation near sub-oceanic thermal vents; its first step requires the conversion of carbon dioxide to carbon compounds for which currently there is no empirical support.

5. The proposal that life came to Earth from outer space, either in bacterial spores or bacteria inside comets or asteroids, fails to tell us how that life originated. Moreover, various such hypotheses either have been proven false by conflicting evidence or else rest on highly questionable assumptions for which there is no supporting evidence. The same applies to the claim that life originated inside comets.

6. The proposal that the first cell is irreducibly complex and could only have been caused by intelligent design is not supported by evidence; it is not falsifiable and so is not a scientific explanation.

7. Invoking the anthropic principle, too, does not provide a scientific explanation. Its weak version explains nothing. The three interpretations of the strong version are not testable and are either unconvincing or no more reasonable than many other non-scientific beliefs and insights.

8. The proposal that the first organic self-replicator emerged from primordial molecules in a microscopically small natural enclosure by quantum mechanics rather than thermodynamics depends on a particular, albeit reasonable, interpretation of quantum theory. However, as currently developed, it contains internal inconsistencies and is not yet supported by any experimental evidence.

9. The idea that biochemical equivalents of the Pauli Exclusion Principle explain the emergences of intermediate steps in the pathway from random simple molecules to the emergence of a living cell is attractive, but no one has suggested what such a deeper law, or laws, might be or how they arose.

10. As with the emergence of matter, it is very probably beyond the ability of science to explain the emergence of life.

If science cannot now, and very probably never will, explain how life emerged, can it tell us how life evolved?

CHAPTER SIXTEEN

Development of Scientific Ideas about Biological Evolution

> If I were to give an award for the single best idea that anyone ever had, I'd give it to Darwin, ahead of Newton and Einstein and everyone else.
>
> —DANIEL C DENNETT, 1995

> In science the credit goes to the man who convinces the world, not to the man to whom the idea first occurs.
>
> —SIR FRANCIS DARWIN, 1914

Before evaluating science's current account of the evolution of life, I think it necessary to restate what I mean by evolution because many scientists equate that term with biological evolution and, of those, many conflate the phenomenon of biological evolution with just one of its several possible mechanisms, natural selection. The word has a wider meaning, as we have seen when considering the evolution of matter in Part 1.

> **evolution**   A process of change occurring in something, especially from a simple to a more complex state.

I shall begin by summarizing how the main ideas about life developed into the current orthodox explanation of biological evolution. The next two chapters will examine what current evidence shows, while the following chapters will evaluate the current orthodox explanation of the evidence and hypotheses that seek to modify or challenge that explanation.

Pre-evolutionary ideas

Aristotle
In the thirteenth and fourteenth centuries the West rediscovered accounts of Aristotle's thoughts from the fourth century BCE and found that he had

undertaken an extensive classification of living things. From a basic division into plants and animals he divided animals into those with red blood and those without (corresponding to the modern distinction between vertebrates and invertebrates). He further grouped animals with similar characteristics into genera and then divided these into species.

Creationism

As science developed in the West from natural philosophy nearly all its practitioners were Christians. Most eighteenth century scientists* believed in the truth of the Judaeo-Christian bible whereby God created each species and these remained unchanged ever since. From 1701 the Authorized (King James) Version of the bible included in its margins the widely accepted calculation of the scholar Archbishop Ussher that this creation event occurred 6,000 years ago.†

Linnaeus

Limited progress was made on Aristotle's classification of species, or taxonomy, until 1735 when the Swedish medical doctor and botanist known by his Latin name of Carolus Linnaeus (almost all scientific writing was in Latin) published *Systema Naturae* (The Natural System) that grouped specimens by their physical characteristics. In the course of subsequent and expanded editions he categorized the natural world in a hierarchy that started with three kingdoms of animals, plants and minerals, and progressed to the highest ranked species, man. Table 16.1 shows where humans fit into his classification.

This caused controversy because it implied that we were simply one part of nature and close to monkeys.

Linnaeus, however, considered his role was to classify the different species that God had created, although late in life he was troubled by the fact that cross breeding different species of plants produced hybrids that had not existed before. He stopped short of proposing evolution and concluded that such hybrids were the product of species that God had created at the beginning of the world.[1]

The Linnaean taxonomic system is still used today, although biologists, zoologists, palaeontologists, anthropologists, geneticists, and molecular biologists have expanded the hierarchy and developed the basis of classification from just morphology.

---

* The term "scientist" as now understood did not come into usage until the nineteenth century, but I shall use this word to describe those who practised what we now recognize as science, namely an empirically based approach as distinct from the purely conceptual approach of natural philosophy.

† See page 11.

Table 16.1 Linnaeus's classification of man

Classification	Name	Brief description	Examples
Imperium [Empire]		Everything in nature	Animals, plants, and minerals
Regnum [Kingdom]	*Animalia* [Animals]	Living things that are mobile	Mammals, birds, amphibians, fish, insects, worms
Classis [Class]	*Mammalia* [Mammals]	Animals that nourish their young with milk produced by mammary glands	Humans, monkeys, apes, bats, dogs, cats, cows
Ordo [Order]	Primates	The first order of mammals in the natural hierarchy	Humans, monkeys, apes, bats
Genus	*Homo* [Human]	All human species	Modern humans and *Homo troglodytes* (cave-dwelling men)
Species	*Homo sapiens* [wise man]	The highest form of God's creation	Modern humans
[Unnamed]	Race	Varieties of man	European, American, Asian, African, Monstrous (Alpine dwarf, Patagonian giant, etc.)

Development of evolutionary ideas

De Maillet

In the eighteenth century a few intellectuals, influenced by geologists suggesting the Earth was considerably older than inferred from the bible, speculated that species changed. Probably the first modern advocate of biological evolution and a common ancestry for radically different animals was the well-travelled French diplomat and natural historian Benoît de Maillet. His dating of the Earth's age as 2.4 billion years and his idea that all life began in shallow waters are set out in *Telliamed*, published posthumously in 1748.[2]

Buffon

The French natural historian Georges Louis Leclerc, Comte de Buffon, compiled a major work, the 44-volume *Histoire Naturelle*, between 1749 until his death in 1788, with the final eight volumes completed by a colleague in 1804. A product of the French Enlightenment, Buffon drew on the ideas of Isaac Newton and the detailed observations of many scientists to propose that the whole of the natural world, from the formation of the Earth to the production of different species, comprised natural phenomena that could be explained by natural forces obeying natural laws.

He prefigured the science of biogeography by observing differences in species in different locations. His ideas on the transformation of species are exemplified by his noting, on rather questionable evidence, that animals in the New World were smaller and weaker than their counterparts in the Old World; he attributed this *dégénération* to the transforming agents of climate, nurture, and domestication. Such biological transformism, however, did not result in new species.[3]

Erasmus Darwin

Erasmus Darwin, the grandfather of Charles Darwin, was a physician, poet, abolitionist, and radical freethinker who made important contributions in medicine, physics, meteorology, horticulture, and botany. He acknowledged many sources, including "the ingenious Mr Buffon", when he set out his views on biological evolution in the first volume, published in 1794, of his *Zoonomia, or, the Laws of Organic Life*, which had the distinction of being banned by the Pope. This goes beyond Buffon by speculating that all warm-blooded animals arose from one "living filament...endued with animality, with the power of acquiring new parts...possessing the faculty of continuing to improve by its own inherent activity" and that these improvements are passed on from one generation to the next.[4]

In *Temple of Nature; or, The Origin of Society*, a 1,928-line poem supplemented by extensive notes and published posthumously in 1803, he portrays the glory of humanity, including its mental powers and its sense of morality, as having its origins in the first microscopic life beneath the sea. This evolves into different species in response to a striving for perfection in different environments. The last note even speculates that the universe will end in a big crunch only to emerge and evolve again as a cycle in an eternal universe operating under "immutable laws impressed on matter by the Great Cause of Causes".[5]

Hutton

James Hutton is best known for his revolutionary explanation of how geological processes occur slowly over long timescales, contrary to the prevailing biblically derived account of the age of the Earth. He was probably the first to propose survival of the fittest as the cause of evolutionary change. In 1794, 65 years before Charles Darwin's *The Origin of Species*, he published in Edinburgh *An Investigation of the Principles of Knowledge*, a philosophical treatise of three volumes running to 2,138 pages. Based on his experiments in plant and animal breeding, Chapter 3 of Section 13 in Volume 2 describes what Darwin later called natural selection:

in conceiving an indefinite variety among the individuals of that species, we must be assured, that, on the one hand, those which depart most from the best adapted constitution, will be most liable to perish, while, on the other hand, those organized bodies, which most approach to the best constitution for the present circumstances, will be best adapted to continue, in preserving themselves and multiplying the individuals of their race.

As one example, Hutton says that in dogs that relied on nothing but swiftness of foot and quickness of sight for survival, those most defective in those qualities will most likely perish while those best equipped with such qualities will be best adapted to survive and multiply similar individuals of their race. But if an acute sense of smell were more necessary for survival then, applying the same "principle of seminal variation", the natural tendency "would be to change the qualities of the animal, and to produce a race of well scented hounds, instead of those who catch their prey by swiftness".[6]

Lamarck

The French invertebrate zoologist and palaeontologist Jean-Baptiste Lamarck had many ideas similar to those of Erasmus Darwin, although no evidence suggests the two were aware of each other's work.

As a result of his researches he wrote his most important book, *Philosophie zoologique*, which he presented to the Institut National des Sciences et Arts in 1809. Its repetitive and sometimes obscure 903 pages contain concepts that are central to modern evolutionary thought. From his study of animals he concluded that:

1. many forms of life must have been replaced over time because they appear as fossils but are no longer seen today, while extant forms of life do not appear as fossils;
2. animals can be classified according to increasing complexity;
3. living animals exhibit a wide diversity of form;
4. living animals are particularly well fitted to their particular environment.

He explained these phenomena by proposing that small granules of living matter are generated by the action of heat, light, and moisture on inorganic matter, which today is called abiogenesis, the origin of life from inanimate matter. Such living matter possesses an inherent power of acquiring successively more complex organization. Abiogenesis was not a unique event in the Earth's history, as is generally considered today, but continually occurs, and the different species we see result from different lineages starting their process of complexification and perfection at different times. *Homo sapiens* is the oldest lineage because it has reached the highest stage of complexification.

He hypothesized that adaptation to different environmental conditions disrupts this smooth progression in complexity to produce the great diversity of species within each genus. Hence this *transformisme*, or evolution, progresses from simple to more complex in branching family trees of which humans are at the apex.

These materialist proposals challenged the central Christian beliefs that God created each lifeform and that man held a unique place in God's creation.

In support of his evolutionary theory Lamarck gave four categories of evidence that Charles Darwin was to employ in *The Origin of Species* fifty years later: the fossil record; the great variety of animal and plant forms produced under human cultivation (Lamarck even anticipated Darwin by mentioning fantail pigeons); the presence of vestigial, non-functional structures in many animals; and the presence of embryonic structures that have no counterpart in the adult.

Lamarckism, however, later became associated not with the phenomenon of biological evolution but with its proposed cause, set out in two laws in *Philosophie zoologique*. The first states that organisms change in response to a changing environment: environmental change causes changes in the physiological needs of organisms, which causes changes in their behaviour, which in turn causes increased use or disuse of their body parts, which causes enlargement or shrinkage of those parts. The second law states that such changes are inherited. As a consequence, very small changes in morphology build up over generations to produce major transformations. As an example, Lamarck said that giraffes began stretching their necks in order to reach the only available sustenance, leaves at the top of trees; this gradual lengthening of the neck was inherited and increased each generation to produce the species we see today.

The idea that the environment causes heritable changes in organisms has only recently found empirical support in epigenetics, which I shall consider in a later chapter.

Lamarck's ideas were largely ignored or ridiculed during his life, and he died a pauper in 1829.[7]

Geoffroy

The French naturalist Étienne Geoffroy Saint-Hilaire was a deist who believed that a God had created the universe but then left it to operate under natural laws with no intervention from him. A friend and colleague of Lamarck, he defended and expanded Lamarck's theories. Collecting evidence from comparative anatomy, palaeontology, and embryology, he argued for the underlying unity of organismal design. By analogy with the development of simple embryo to complex adult, he used the term "evolution" to apply to the transmutation of species in geological time.[8]

Wells

In 1813 William Wells, a Scottish-American physician and printer, read a paper before the Royal Society. According to his biography written by J H S Green and published in *Nature*, Wells "not only gave a theory of variation, selection, descent with modification, and of the origins of races of man; he also realized the importance of disease in selection—a factor not mentioned by Darwin in *On the Origin of Species*.[9] Wells's paper was published posthumously in 1818.

Grant

Robert Edmond Grant, a well-travelled radical biologist and friend of Geoffroy, was a leading member of Edinburgh's Plinian Society. Charles Darwin joined in 1826 when starting his second year of medical studies at Edinburgh University and became Grant's keenest student. Grant was an enthusiastic advocate of the transformist ideas of Lamarck and Geoffroy, and speculated that "transformation" might affect all organisms that evolved from a primitive model, suggesting a common origin for plants and animals.[10]

Matthew

In 1831 Patrick Matthew, a Scottish landowner and fruit grower, published a book *On Naval Timber and Arboriculture*. In an appendix Matthew extended his idea for how artificial selection might improve cultivated trees to a universal law of natural selection in which the "progeny of the same parents, under great differences of circumstances, might, in several generations, even become distinct species, incapable of co-reproduction".[11]

Wallace[12]

Alfred Russel Wallace's education at the local grammar school in Hertford in southeast England was cut short at the age of 14 when his parents withdrew him because of financial difficulties. While working as a surveyor and engineer he became fascinated by the natural world. Inspired by the chronicles of naturalists, including Charles Darwin's 1839 journal of his *Beagle* voyage, he set off in 1848 on his own voyage to South America to collect plant, insect, and animal specimens.

He had also been impressed by Sir Charles Lyell's *Principles of Geology*, which showed how slow, ongoing processes could bring about major changes, and by *Vestiges of the Natural History of Creation*, a highly controversial work of popular science published anonymously in 1844* that advocated an evolutionary origin for the solar system, the Earth, and all living things. These appear to have convinced him of the transmutation of species, and he wanted to investigate its causes. He planned to finance the enterprise by selling specimens to collectors back in the United Kingdom.

In 1854 he began an eight-year expedition to the Malay archipelago (present-day Malaysia and Indonesia), during which he began a correspondence with Darwin. In 1858 he sent Darwin an essay entitled "On the Tendency of Varieties to

---

* The 12th edition published in 1884 showed the author to be Robert Chambers, LL.D.

Depart Indefinitely From the Original Type" with the request that Darwin review it and pass it on to Lyell if he thought it worthwhile.

Darwin was devastated. Two years before he had written to his friend Lyell admitting that "I rather hate the idea of writing for priority, yet I should certainly be vexed if any one were to publish my doctrines before me."[13] On 18 June 1858 he sent the essay to Lyell with a letter saying "I never saw a more striking coincidence; if Wallace had my MS. sketch written out in 1842 he could not have made a better short abstract!....So, all my originality...will be smashed."[14]

A week later Darwin wrote another letter to Lyell pointing out that

> There is nothing in Wallace's sketch which is not written out much fuller in my sketch, copied out in 1844, and read by Hooker some dozen years ago. About a year ago I sent a short sketch, of which I have a copy, of my views... to Asa Gray, so that I could most truly say and prove that I take nothing from Wallace....It seems hard on me that I should be thus compelled to lose my priority of many years' standing.

Amid protestations of his own unworthiness, he beseeched Lyell for his opinion on whether it would be dishonourable of him to publish now, and asked him to forward the material to another of his confidantes, the botanist Joseph Hooker. He wrote in similar self-deprecatory tones to Hooker, nonetheless offering to write a more accurate version of his 1844 sketch for the Linnean Journal.[15]

Lyell and Hooker responded by offering the solution implicit in Darwin's requests to them. On 1 July 1858 the two made a joint presentation to the Linnean Society entitled "On the Tendency of Species to form Varieties; and on the Perpetuation of Varieties and Species by Natural Means of Selection". It consisted of extracts from the 1844 sketch Darwin had sent to Hooker plus part of the letter Darwin had written to Asa Gray in 1857, followed by Wallace's 20-page essay. Prefatory remarks by Lyell and Hooker declared that they were not solely considering the relative claims to priority of Darwin and Wallace but presenting the material in the interests of science generally.[16] Without Darwin being seen personally to claim priority, it had the desired result of publicly recording him to have been the first with the idea of natural selection while acknowledging that Wallace had independently, but later, arrived at the same view.

Wallace wasn't consulted about this, but seemed content when informed of the event. It gave a professional lacking a university education, who earned money by collecting and selling specimens, recognition by, and access to, members of the scientific establishment; in Victorian England this consisted of an upper class of gentlemen with private means, like Lyell and Darwin.

Darwin offered the younger man friendship and Wallace responded by becoming one of Darwin's staunchest supporters, defending *On the Origin of Species by Means of Natural Selection, or The Preservation of Favoured Races in the Struggle for Life*, the book Darwin rushed to publish the year following the Linnean presentation.

This friendship did not prevent Wallace disagreeing with Darwin on a number of issues, including the extent to which natural selection could explain the development of humankind's higher moral and intellectual faculties, the extent to which sexual selection could explain sexual dimorphism, Darwin's belief in the inheritance of acquired characteristics, and Darwin's idea of pangenesis (see below).

Wallace went on to publish in 1864 a paper "The Origin of Human Races and the Antiquity of Man Deduced from the Theory of 'Natural Selection'", which Darwin had not then publicly addressed, and in 1889 a book, *Darwinism*, which explained and defended natural selection. In it he proposed that natural selection could drive the reproductive isolation of two varieties by encouraging the development of barriers against hybridization, thus helping generate new species.

Charles Darwin

An original thinker?

As the 2009 celebrations of the 200th anniversary of his birth and the 150th anniversary of the publication of *The Origin of Species* demonstrated, Charles Darwin is generally portrayed as the first to develop the theory of biological evolution and certainly the first to propose natural selection as its cause.

After his five-year voyage round the world on *HMS Beagle* ended in 1836, Darwin spent the next 23 years trying to make sense of the specimens he had collected and what he had seen, seeking many more specimens and accounts from other naturalists, conducting experiments, and reading avidly. His notebooks during this period, which refer repeatedly to "my theory", give no indication that he drew on the ideas of others. Yet, as Cardiff University's Paul Pearson points out, perhaps it was no coincidence that Wells, Matthew, and Darwin had studied in Hutton's city of Edinburgh, renowned for its learned societies and radical thinkers, and "it seems possible that a half-forgotten concept from his student days resurfaced afresh in [Darwin's] mind as he struggled to explain the observations of species and varieties compiled on the voyage of the *Beagle*".[17] Moreover, his great friend and confidant, the geologist Sir Charles Lyell, was the chief proponent of Hutton's anti-biblical geological theory of uniformitarianism.* Furthermore, Grant, Darwin's mentor in Edinburgh, was "an enthusiastic Lamarckian transmutationist and introduced the young Darwin to Lamarck's transformism as well as to his invertebrate zoology".[18] Four years after leaving Edinburgh, Darwin consulted Grant, then Professor of Comparative Anatomy

---

* The theory that the same physical and chemical processes have occurred in the Earth's surface since the origin of the Earth and account for all geological phenomena.

at the recently founded London University,* prior to setting out on his *Beagle* voyage.

Apart from acknowledging that he had been induced to publish by learning that Wallace had arrived at the same conclusions independently, Darwin makes no mention of anyone else's evolutionary ideas in the first and second editions of *The Origin of Species*. The third edition of 1861 contains a brief prefatory sketch outlining "Recent Progress of Opinion on The Origin of Species". In this he declares that he is not familiar with the writings of Buffon, gives a brief account of Lamarck's (but only after Lyell had urged him to give Lamarck due credit[19]), and offers a nuanced summary of Geoffroy's views as given by his son, saying that Geoffroy was cautious in drawing conclusions and did not believe that existing species are now undergoing modification.

He does not refer to Hutton. He mentions his grandfather only once, and that in a dismissive footnote saying he "anticipated the erroneous grounds of opinion and the views of Lamarck".

Among other references in the historical sketch he mentions one paragraph of Grant's 1826 paper in the *Edinburgh Philosophical Journal* which "clearly declares his belief that species are descended from other species, and that they become improved in the course of modification".[20]

Darwin credits Matthew for having in 1831

> precisely the same view on the origin of species as that...propounded by Mr. Wallace and myself in the 'Linnean Journal', and as that enlarged in the present volume. Unfortunately the view was given by Mr. Matthew very briefly in scattered passages in an Appendix to a work on a different subject.

He goes on to quote a letter from Matthew saying

> To me the conception of this law of Nature came intuitively as a self-evident fact, almost without an effort of concentrated thought. Mr. Darwin here seems to have more merit in the discovery than I have had....He seems to have worked it out by inductive reason, slowly and with due caution to have made his way synthetically from fact to fact onwards.[21]

In the fourth edition of 1866 he concedes that Wells was the first to recognize in 1813 the principle of natural selection "but he applies it only to the races of man, and to certain characters alone".

Comparing the texts of what these predecessors had written with Darwin's account of them, it is difficult to avoid the conclusion that Darwin was engaging in

---

*   Founded in 1826 as the first university in England to admit students who weren't members of the Church of England, it was granted a charter in 1836 as University College London.

nineteenth century spin: minimising their contributions* in order to claim credit for the ideas of biological evolution, or descent with modification as he called it, and of natural selection as its cause.

Darwin's contribution
Darwin's contribution to biological evolutionary theory is four-fold. First, he collected considerable evidence undermining the orthodox view that God had created each species separately and argued persuasively, but by no means conclusively, that each species in a genus had evolved gradually from a common ancestor. Second, he hypothesized that the principal, but not sole, cause of such evolution was natural selection. Third, he proposed that sexual selection was another cause. Fourth, his promotion of the phenomenon of biological evolution and its hypothesized cause of natural selection was instrumental in the acceptance of these ideas by the scientific community.

For more than 40 years following the *Beagle*'s return Darwin studied and experimented with plants and animals, publishing 19 books and hundreds of scientific papers. Of this prolific output, which established his reputation as a naturalist, the most significant were *On the Origin of Species by Means of Natural Selection, or the Preservation of Favoured Races in the Struggle for Life*, usually abbreviated to *The Origin of Species*, first published in 1859 and followed by five further editions, and *The Descent of Man, and Selection in Relation to Sex* first published in 1871 with a revised edition in 1874 followed by a final edition incorporating his 1876 *Nature* article "Sexual Selection in Relation to Monkeys".

Evidence
The evidence for descent with modification (it is only in the 1872 and final edition of *The Origin of Species* that Darwin uses the term "evolution") set out in *The Origin of Species* may be grouped into nine categories.

1. *Variation of domestic plants and animals, especially by selective breeding*
   The rapid alteration in the characteristics of domesticated animals and plants, particularly when they are selectively bred for specific characteristics, demonstrates evolutionary change; inherited variations include behaviour, exemplified by the instincts of untrained young retrievers and sheepdogs.

   Darwin selectively bred pigeons and maintained that all the different varieties are descended from the rock pigeon (*Columba livia*). However, he concludes that the great variety in breeds of domestic dog means that they are descended from different wild ancestors.

   All his examples demonstrate variations within species rather than the origination of different species.

---

*   For example, although Matthew's views are contained in an appendix, they are neither scattered nor brief compared with his own contribution to the Linnean Society journal.

2. *Variation in nature*

   After noting the different definitions and classifications of species given by botanists, zoologists, and naturalists, Darwin reflects "how entirely vague and arbitrary is the distinction between species and varieties".* In large genera closely related species tend to form little clusters round other species and have restricted ranges. This, Darwin argues, is consistent with those species originating from earlier varieties, but the pattern cannot be explained if each species is an independent creation.

3. *Fossil record*

   Darwin argues that the sudden appearance of fossils and the lack of intermediate forms is not an argument against evolution. He gives several reasons, including the imperfection of the geological record because fossils normally form only during subsidence, and the extinction of intermediate species by successors better modified to win the struggle for existence.

   He cites examples of fossils whose forms show relationships with living species, like fossil mammals in Australia and the living marsupials on that continent, and argues that the greater variety of living species from ancestral fossil types supports descent with modification.

4. *Classification of species*

   The classification of species in related groups "all follow if we admit the common parentage of allied forms, together with their modification through variation and natural selection, with the contingencies of extinction and divergence of character".

5. *Similarities of organs performing different functions in species of the same class*

   Such organs as a man's hand, a mole's hand, a horse's leg, a porpoise's flipper, and a bat's wing are constructed on the same pattern and include similar bones in similar relative positions. This suggests they are descended from the organ of a common ancestor and were subsequently modified to adapt to different environments.

6. *Similarity of embryos*

   The early embryos of mammals, birds, lizards, and snakes are strikingly similar, both as a whole and in the mode of development of their parts; they lack the variations shown in their later life. Such embryos, Darwin argues, are similar in structure to the less modified, ancient, and usually extinct adult progenitors of the group, although this is speculation because the evidence is lacking.

---

* Unless otherwise stated, in quoting from *The Origin of Species* I use the sixth edition of 1872 because it is the final edition and incorporates Darwin's most considered views. All editions, and all Darwin's other works, are available online, thanks to the outstanding work of John van Wyhe, Director of The Complete Works of Charles Darwin online http://darwin-online.org.uk/

7. *Developmental changes*

    Darwin extends this argument to the early development of the young, particularly their transition stages. He notes, for example, that while the young rely on their parents for nourishment they tend not to show the adaptive variations they acquire in later life. For example, the plumage of immature birds of related genera resemble each other. His line of reasoning incorporates metamorphoses, like the transition from caterpillar to butterfly and from larva to fly.

8. *Rudimentary, atrophied, and aborted organs*

    Things like teeth in foetal whales, mammals whose male members possess rudimentary mammary glands, wings of ostriches, and other examples all point to characteristics of the ancestors from which they have evolved. Creation theory, by contrast, has difficulty in explaining such features.

9. *Geographical distribution of species*

    Darwin thought powerful evidence for descent with modification and its principal cause was provided by variations of species in locations separated by migration barriers and also in different environments.

    The iconic example of what is now called biogeography is Darwin's finches. The Encarta World English Dictionary 2008 defines them as

    > finches of Galapagos Islands: the birds of the Galapagos Islands on which Charles Darwin based his theory of natural selection through observation of their feeding habits and corresponding differences in beak structure. Subfamily Geospizinae.

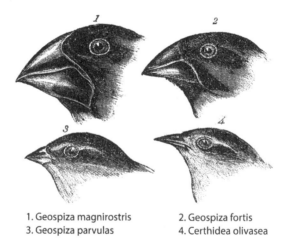

1. Geospiza magnirostris
3. Geospiza parvulas

2. Geospiza fortis
4. Certhidea olivasea

Figure 16.1 Finches from the Galápagos archipelago

This is amplified and taught in many schools and colleges. For example a tutorial at Palomar College in California says

Darwin identified 13 species of finches in the Galápagos Islands. This was puzzling since he knew of only one species of this bird on the mainland of South America, nearly 600 miles to the east, where they had all presumably originated...

Each of the various islands had its own species of finch, differing in various ways, and in particular in beak shape. Darwin's idea was that these finches were all descendants of a single kind of ancestral finch, but that the different environments of the different islands had given advantages to different characteristics in its finch population. If one island had an abundance of insect food, for instance, but relatively few seeds and nuts, finches that happened to be born with fine beaks that could pick out insects from small crevices would do better than birds with large, hard beaks. The fine-beaked birds would succeed in rearing more young than their coarse-beaked relatives, and as the offspring would tend to inherit their parents' characteristics, that beak shape would gradually spread through the finch population of the island. The reverse would happen on islands where there were more seeds than insects, and where the birds whose beaks were better adapted to seed-crushing would flourish. Eventually the separated populations would become too different to interbreed, and would be separate species.[22]

This, unfortunately, is another Darwin myth.

At the time of the *Beagle* voyage Darwin thought that most of the bird specimens he had collected were not finches at all; some he labelled blackbirds, others "gross beaks", and another a wren; moreover he failed to label which birds came from which island. He gave them to the Zoological Society of London, which gave them to John Gould, an ornithologist, for identification. It was Gould who identified the specimens as distinct species of finches (his drawings are reproduced in Figure 16.1). Darwin later examined the collections of three other *Beagle* shipmates, including those of Captain FitzRoy.[23]

The myth was promulgated by British ornithologist David Lack in his 1947 book *Darwin's Finches*.[24] However, Harvard University's Frank Sulloway examined the primary sources. In 1982 he demonstrated, among other things, that Lack wrongly assumed that FitzRoy's accurately labelled specimens were really Darwin's and confused the locations of several specimens.

When he visited the Galápagos islands not only did Darwin fail to notice that different islands had different finches, he did not match different beak shapes to different diets. According to his biographers Adrian Desmond and James Moore, after his return to England, Darwin

remained confused by the Galapagos finches...unaware of the importance of their different beaks....He had no sense of a single, closely related group becoming specialized and adapted to different environmental niches.[25]

Although images of these finches were used as a symbol in many of the 150th anniversary celebrations of the first publication of *The Origin of Species*, Darwin never mentioned them in any edition of that book.

Two chapters of *The Origin of Species*, nonetheless, do offer examples from biogeography, which was Wallace's main source of evidence for biological evolution. In them Darwin points out similarities in species across a whole range of climatic conditions on the American continent and how different these species are from those existing in comparable conditions in Africa and Australia. He attributes this difference to natural barriers to migration.

As for the Galápagos islands, Darwin observes that nearly every water and land species shows a close affinity to species on the South American mainland, some 500–600 miles away, despite major geological, topographical, and climatic differences. Where conditions are similar, in the volcanic Cape Verde archipelago off the distant West African coast, the species are very different.

A creationist explanation requires God to have separately created innumerable variations of species in different locations. A more reasonable explanation is that each such species from the South American mainland migrated to the Galápagos archipelago and subsequently developed their relatively slight variations compared with the much greater variations from species in far distant locations.

Natural selection
Darwin argued that the chief, but not sole, cause of such variations was natural selection, a principle by which the best competitors in any given population of organisms have the best chance of surviving and breeding, thus transmitting their characteristics to subsequent generations.

In forming this view he was greatly influenced by the 1826 revised 6th edition of *An Essay on the Principles of Population* by Thomas Malthus, an English clergyman and economist. Essentially Malthus maintains that while food supplies increase arithmetically populations naturally increase geometrically; hence, if unchecked, a population will outgrow its food supply and famine will result.

All the editions of *The Origin of Species* give an imaginary illustration of how natural selection works. In a pack of wolves the swiftest and slimmest would be the most effective predators of deer and hence would have the best chance of surviving.[26] This is strikingly similar to Hutton's illustration of the survival of the fittest mentioned earlier, namely that in dogs that relied on nothing but swiftness of foot and quickness of sight for survival those best equipped with such qualities will be best adapted to survive and multiply while those lacking such characteristics will perish.

By the final edition of *The Origin of Species* Darwin defined natural selection as follows.

> Owing to this struggle [for survival], variations, however slight, and from whatever cause proceeding, if they be in any degree profitable to the individuals of a species…will tend to the preservation of such individuals, and will generally be inherited by the offspring. The offspring, also, will thus have a better chance of surviving, for, of the many individuals of any species which are periodically born, but a small number can survive. I have called this principle, by which each slight variation, if useful, is preserved, by the term Natural Selection, in order to mark its relation to man's power of selection.[27]

Darwin did not know the causes of these variations but, unlike the current orthodoxy, considered they were not random.

> I have hitherto sometimes spoken as if the variations—so common and multiform with organic beings under domestication, and in a lesser degree with those under nature—were due to chance. This, of course, is a wholly incorrect expression.[28]

Natural selection acts on these variations.

> We have seen that man by selection can certainly produce great results, and can adapt organic beings to his own uses, through the accumulation of slight but useful variations, given to him by the hand of Nature. But Natural Selection, as we shall hereafter see, is a power incessantly ready for action, and is as immeasurably superior to man's feeble efforts, as the works of Nature are to those of Art.[29]

Darwin claims that not only is natural selection an immeasurably superior power to artificial selection but (unlike the view of current biological orthodoxy) it also causes evolutionary improvement.

> It leads to the improvement of each creature in relation to its organic and inorganic conditions of life; and consequently, in most cases, to what must be regarded as an advance in organization.[30]

Speciation

Despite its title it may be argued that Darwin's most famous book does not address the origin of species. This is due principally to Darwin's view that natural selection is slow and gradual and there is little to distinguish between varieties and species.[31] He doesn't define what it is that distinguishes variations from species of the same

genus except to observe that species are "tolerably well-defined" when intermediate varieties had been "supplanted and exterminated" during natural selection.[32]

Survival of the fittest
Darwin is ambiguous when discussing what he means by survival of the fittest. On
the one hand he says that it is metaphorical.

> I use this term in a large and metaphorical sense including dependence of
> one being on another, and including (which is more important) not only
> the life of the individual, but success in leaving progeny.[33]

This is echoed in a section of *The Descent of Man* where, after giving examples of
sociability and cooperation in animal species and among humans, he observes
that

> those communities, which included the greatest number of sympathetic
> members, would flourish best, and rear the greatest number of offspring.[34]

Yet by far the greater part of these books compiles "facts" showing that the
struggle is real, not metaphorical. Indeed, only nine pages after the above quote
from *The Origin of Species*, he says

> As the species of the same genus usually have, though by no means invar-
> iably, much similarity in habits and constitution, and always in structure,
> the struggle will generally be more severe between them, if they come into
> competition with each other, than between the species of distinct genera.[35]

The reality of this competitive struggle for survival is made clear in *The Variation
of Animals and Plants under Domestication*, which he intended as a chapter in a
major book on evolution for which *The Origin of Species* was simply a sketch.

> It has truly been said that all nature is at war; the strongest ultimately
> prevail, the weakest fail; and we well know that myriads of forms have
> disappeared from the face of the earth. If then organic beings in a state of
> nature vary even in a slight degree...then the severe and often-recurrent
> struggle for existence will determine that those variations, however slight,
> which are favourable shall be preserved or selected, and those which are
> unfavourable shall be destroyed.[36]

And

> This preservation, during the battle for life, of varieties which possess
> any advantage in structure, constitution, or instinct, I have called Natural

Selection; and Mr. Herbert Spencer has well expressed the same idea by the Survival of the Fittest.[37]

This literal interpretation was widely accepted.

Sexual selection
In response to criticisms that such things as a peacock's tail feathers conferred a disadvantage rather than an advantage in a struggle for limited resources, Darwin proposed that sexual selection was also a cause of biological evolution, particularly for males of the species.

According to this hypothesis, increasingly large and colourful tail feathers attract peahens and enable the cocks thus endowed to produce more offspring than less flamboyant cocks.

For humans,

> I conclude that of all the causes which have led to the differences in external appearance between the races of man, and to a certain extent between man and the lower animals, sexual selection has been the most efficient.[38]

Lamarckian use and disuse
When Lyell urged him to give due credit to Lamarck, Darwin did so reluctantly and replied to Lyell saying

> You often allude to Lamarck's work; I do not know what you think about it, but it appeared to me extremely poor; I got not a fact or idea from it.[39]

This did not prevent him employing four categories of evidence for biological evolution that Lamarck had used. Moreover, in the first edition of *The Origin of Species* he discusses without attribution "Effects of Use and Disuse", writing

> I think there can be little doubt that use in our domestic animals strengthens and enlarges certain parts, and disuse diminishes them; and that such modifications are inherited. Under free nature...many animals have structures which can be explained by the effects of disuse.[40]

This is a lucid summary of the law by which Lamarck is best known. In the preface to the 1882 edition of *The Descent of Man, and Selection in Relation to Sex* Darwin says

> I may take this opportunity of remarking that my critics frequently assume that I attribute all changes of corporeal structure and mental power exclusively to the natural selection...whereas, even in the first edition of the 'Origin of Species,' I distinctly stated that great weight must be attributed to the inherited effects of use and disuse.[41]

Promotion

By contrast with Lamarck's *Philosophie zoologique*, *The Origin of Species* is brief and more readable. Furthermore, while he undermines the creationist model, as did Lamarck, Darwin avoided offending the Christian establishment by such means as including a quote from Bacon below the title:

> let no man…think or maintain, that a man can search too far or be too well studied in the book of God's word, or in the book of God's works; divinity or philosophy

and concluding with an allusion to one of the Genesis creation accounts

> There is grandeur in this view of life, with its several powers, having been originally breathed by the Creator into a few forms or into one.[42]

Yet in his posthumously published autobiography, with his wife's deletions of religious references restored, he says

> I had gradually come, by this time [January 1839, 33 years before the final edition of *The Origin of Species*], to see that the Old Testament from its manifestly false history of the world…was no more to be trusted than the sacred books of the Hindoos, or the beliefs of any barbarian.[43]

Darwin used his status as a Fellow of the Royal, Linnean, and Geological Societies to cultivate friendships with influential members of the scientific establishment. His many letters soliciting support for *The Origin of Species* are a model for lobbyists: flattery to the point of sycophancy, affection, self-deprecation, anxiety, and criticism of his own work frequently excused by ill health.[44] They proved highly effective, as one opponent, keeper of the British Museum's Zoological collection, John Grey, bemoans.

> You have just reproduced Lamarck's doctrine, and nothing else, and here Lyell and others have been attacking him for twenty years, and because *you* (with a sneer and laugh) say the very same thing, they are all coming round; it is the most ridiculous inconsistency.[45]

Darwin's aversion to controversy together with a stubborn determination served him well. By inducing the likes of Lyell, Hooker, and the combative young anatomist Thomas Henry Huxley to advocate his cause among the scientific establishment while he remained distant from the fray in his rural retreat of Down House in Kent, Darwin promoted his ideas without incurring the fate of his former mentor. Robert Grant was ousted from his position in the Zoological Society for his materialist denial of the truths of Christianity and was sidelined by the Christian-dominated scientific community during most of his poorly paid tenure at University College London.[46]

As the quotation by his son shown below the chapter title suggests, Darwin undoubtedly deserves credit for beginning the process that led to biological evolution and its cause of natural selection being adopted by the scientific establishment as its new orthodoxy.

Problems with Darwin's hypothesis

The Origin of Species proved popular, and the international scientific community accepted the phenomenon of biological evolution—descent with modification—within 20 years. However, their acceptance of Darwin's proposed principal cause—natural selection—took some 60 years. This was for two reasons. First, although Darwin had suggested why the fossil record did not support the gradualism that was an essential part of his hypothesis, leading palaeontologists, like his former mentor Adam Sedgwick and Louis Agassiz, founder and director of Harvard's Museum of Comparative Zoology, claimed fossil evidence for gradualist evolution was entirely lacking. Even T H Huxley was sceptical about Darwin's gradualism.[47] Most palaeontologists of the late nineteenth and early twentieth century advocated saltation: the geological record shows new fossil forms appearing, remaining substantially unchanged, and then disappearing to be replaced by new fossil forms; therefore biological evolution proceeds by leaps rather than by gradualism.

Second, the then orthodox view was that inheritance was a blending process: offspring showed a mixture of their parents' characteristics. Thus the offspring of a tall father and a short mother would be intermediate in height. This means that, unless each parent possessed the advantageous variation, only half of that variation would be passed on to each offspring, who would then pass on only half of its half of the variation. Thus the variation would be diluted for successive generations and could never come to dominate the species.

To overcome this objection Darwin proposed pangenesis, an idea originated by ancient Greek philosophers and developed by Buffon. In Darwin's version gemmules, or invisible germs, containing hereditary information from every part of the body, coalesce in the reproductive organs and thereby transmit hereditary attributes. However, Galton failed to find these gemmules in rabbit blood, and the scientific community rightly dismissed the idea. Just as Lamarck was wrong about the continual generation of life from inanimate matter and that adaptation disrupts the process of complexification, Darwin was wrong about how advantageous traits are inherited and thereby come to change a population in a particular environment.

Darwinism

Darwin has become a secular saint and, like his religious counterparts, he has been mythologized. Many books published and TV programmes broadcast in

2009 to celebrate the 150th anniversary of the first publication of *The Origin of Species* gave the impression that Darwinism or Darwin's theory of evolution is an unambiguous theory like, say, Newton's theory of gravitation. In reality, however, Darwinism and Darwinian mean different things to different people. I think it useful to define Darwinism as what Charles Darwin actually proposed in order to differentiate it from subsequent modifications and extensions such as NeoDarwinism and UltraDarwinism.

> **Darwinism**   The hypothesis that all species of the same genus have evolved from a common ancestor. The principal cause of this biological evolution is natural selection, or the survival of the fittest, whereby offspring whose variations make them better fitted to compete with others of their species for survival in a particular environment will live longer and produce more offspring than those less fitted. These advantageous variations are somehow inherited and, over successive generations, will gradually come to dominate the increasing population in that environment while less well-adapted variants will be killed, starved, or driven to extinction. Sexual selection of traits favourable to mating, and use and disuse of organs, are also heritable and cause biological evolution.

Orthogenesis

Orthogenesis is another term that means different things to different people. Essentially it may be defined as follows.

> **orthogenesis**   The hypothesis that biological evolution has a direction caused by intrinsic forces; versions range from those that hold that adaptation also plays a significant role in the evolution of species, through the view that adaptation only influences variations within species, to the view that direction demonstrates a purpose to biological evolution.

As we saw, Erasmus Darwin and Jean-Baptiste Lamarck each proposed that biological evolution occurs because of an intrinsic tendency to increase in complexity.

The German zoologist and anatomist Theodor Eimer used the term "orthogenesis" in 1895 to explain why similar evolutionary sequences occur in different lineages. Eimer thought that nonadaptive evolution of similar forms was widespread and that biological evolution proceeds in predictable directions according to unspecified natural laws similar to those governing the development of individual organisms from simple embryo to more complex adult.

This downgrading of Darwinian natural selection was adopted by many palaeontologists, like the Americans Edward Cope and Alpheus Hyatt, on the fossil evidence that successive members of an evolutionary series become progressively modified in a single undeviating direction with characteristics that have little if any adaptive value and indeed often lead to extinction. Several accepted the Lamarckian explanation of an intrinsic tendency, and also the inheritance of acquired characteristics, as causal.[48]

Kropotkin and mutual aid

Peter Kropotkin is best known in the West as a Russian prince's son who became an anarchist and revolutionary. He was also a scientist who, in 1871, was offered the prestigious post of secretary of the Imperial Geographical Society in St Petersburg. He declined, renounced his privileges, and devoted himself to applying the lessons he had drawn from his scientific observations in order to relieve the exploitation and hardship suffered by the majority of Russian people under Tsarist rule. His weapon was not the bomb but the pen.

Pivotal observations

Nine years previously, at the age of 20, he had travelled to eastern Siberia and northern Manchuria eager to witness the struggle for existence set out in *The Origin of Species*, which had been published three years before and had impressed him deeply. What he found, however, led him to conclude that Darwin's evolutionary hypothesis had been distorted, principally by Darwin's followers.

Kropotkin's careful observations of animal and human life revealed few instances of ruthless competition between members of the same species. Among populations teeming with life in isolated areas subject to harsh conditions, where competition for scarce resources should have been at its most savage, he discovered instead mutual aid "carried on to an extent which made me suspect in it a feature of the greatest importance for the maintenance of life, the preservation of each species, and its further evolution".[49]

He also observed mutual aid among Siberian peasants in autonomous communities, and this informed his belief that such a system of government should replace the centralized State, which he had found repressive and brutal.

Mutual aid

In 1883, during one of his spells in prison, he read a lecture by Professor Karl Kessler, a respected zoologist and dean of St Petersburg University. This argued that mutual aid is a greater factor in the evolution of species than is competition. It corroborated his own findings and, on release from prison, Kropotkin moved to England to pursue this work, drawing on his own observations and those of other field naturalists and anthropologists, much as Darwin had done.

In 1888 Thomas Huxley, Darwin's leading advocate, published an influential essay, "The Struggle for Existence", that argued life was a "continuous free fight" and that competition between individuals of the same species was not merely a law of nature but the driving force of progress.

> the animal world is on about the same level as a gladiators' show. The creatures are fairly well treated and set to fight; whereby the strongest, the swiftest, and the cunningest live to fight another day. The spectator has no need to turn his thumbs down, as no quarter is given.[50]

Kropotkin considered that the opinions of Huxley, whose own expertise lay in comparative anatomy and palaeontology, were not supported by evidence from field naturalists and zoologists. He replied to Huxley in a series of articles that formed the basis of a book. *Mutual Aid: A Factor of Evolution* was first published in 1902, with a revised edition in 1904 and a final edition in 1914, which I use here.

In it Kropotkin says

> If we refer to the paragraph [in *The Origin of Species*] entitled 'Struggle for Life most severe between Individuals and Varieties of the same Species', we find in it none of that wealth of proofs and illustrations which we are accustomed to find in whatever Darwin wrote. The struggle between individuals is not illustrated under that heading by even one single instance.[51]

Kropotkin has no quarrel with natural selection, nor does he deny that the struggle for existence plays an important role in the evolution of species. He declares unequivocally that "life is struggle; and in that struggle the fittest survive".[52] But this is not a competitive struggle between members of the same species. *Mutual Aid* amasses a wealth of evidence from field studies of insects, birds, and mammals rearing and training progeny, protecting individuals from harm, and obtaining food to support the conclusion that "those animals which acquire habits of mutual aid are undoubtedly the fittest [to survive and evolve]".[53]*

Supporting evidence

Among the many insect examples, Kropotkin describes ants and termites that collaborate in divisions of labour to forage and build elaborate communal nests with vaulted granaries and nurseries for rearing young. Although ants of different species engage in terrible wars with each other, within a community mutual aid, and very often self-sacrifice for the common welfare, is the rule. If an ant that is full with partly digested food refuses to regurgitate and share with a hungry ant, it is treated as an enemy. Collaboration allows a colony to overcome other, more powerful insects, like beetles and even nests of wasps.

---

* I shall consider Kropotkin's observations on human societies in Part 3.

Bees collaborate in a similar way. Kropotkin observes that in conditions of great scarcity or great plenty some bees prefer to rob rather than to work for the common good, but over the long term collaboration proves more advantageous to the species.

Such collaboration usually extends only to the hive or nest or colony of insects rather than the whole species, but he notes that ant colonies consisting of more than 200 nests containing two different species have been observed, while some savannahs in South America contain hillocks of termite nests of two to three different species, most connected by vaulted galleries or arcades.

Many species of birds exhibit social behaviour not only for hunting and migrating, where flocks can include other species, but also for pleasure. Kites collaborate in hunting and can dispossess a more powerful bird, like a martial eagle, of its prey. When winter approaches, birds that have lived for months in small groups scattered over a wide territory gather in thousands at a particular place for several days to await late arrivals. These flocks migrate in a "well-chosen direction" towards milder conditions and more plentiful food supplies, with the strongest taking it in turn to lead. In spring they return to the same place and then disperse, usually to the same nests they had left.

As for pleasure, several species of bird sing in concert. He notes one observer reporting that the first of many distinct flocks of about 500 chakars sited around a lake began singing for three to four minutes and then stopped, when the next flock began, and so on with each flock taking its turn until the song circled the lake and the first flock began singing again.

Kropotkin finds parrots the most sociable and the most intelligent of birds. For example, white cockatoos of Australia send out scouts to look for cornfields. When they return the flock flies to the best field and posts sentries to warn of farmers while the rest gorge themselves on the corn. He attributes parrots' longevity to their social life.

Compassion is not a trait expected in a ruthless struggle for existence between members of the same species, but Kropotkin gives several examples, including one of a flock of pelicans bringing food over a distance of thirty miles to a blind pelican.

Kropotkin notes similar but more advanced cooperation among mammals, most of which live sociably, from nests of rodents through herds of elephants and schools of whales to troops of monkeys and of apes.

Squirrels are primarily individualists, finding and storing their own food in their own nests, but they maintain contact with other squirrels and, when food supplies run out, they migrate as a group. Muskrats of Canada live peaceably and play in communities, inhabiting villages of dome-shaped houses built of beaten clay interwoven with reeds, with separate corners for organic refuse. Viscachas, a species of rodent resembling a rabbit, live peacefully in warrens of between 10 and 100 individuals. Whole colonies visit each other at night. If a farmer destroys a viscacha burrow, other viscachas come from a distance to dig out those that are buried alive.

Mammals collaborate in obtaining food and in protecting themselves and weaker members from predators. Wolves hunt in packs while species of wild horse, like mustangs and zebras, form a ring of studs around their herd to ward off attacks from wolves, bears, and lions. When food supplies run out, dispersed herds of ruminants gather to form a large herd that migrates to another region in search of sustenance.

This pattern is typical of all species: collaboration is strongest within the family, and then the group, and then the association of scattered groups that unite to meet a common need.

Kropotkin notes that Darwinists consider intelligence the most powerful characteristic for an individual to possess in the struggle for existence. He points out that intelligence is fostered by sociability, embracing communication, imitation, and accumulated experience, all of which is denied the unsociable animal.

He finds that evolution is progressive. Within each class the most highly evolved combine the greatest sociability with the greatest intelligence: ants among insects, parrots among birds, and monkeys and apes among mammals.

For Kropotkin the principal struggle for existence is not competition for the same resources between individuals of the same species due to a Malthusian population increase. Rather it is the struggle against their circumstances: a changing environment, limited food supplies, harsh weather conditions, and predators. The species that natural selection favours—those whose members survive longest and breed most—adopt strategies to *avoid* competition: they make their own food, as do ants; or they store food, as do squirrels; or they hibernate, as do many rodents; or they enlarge their habitat; or they migrate temporarily or permanently to a new habitat; or they change their diet and habits and, over the course of generations, evolve into a new species better adapted to the changed environment. Those that do not adopt such strategies fail in the struggle to survive and naturally disappear over time without having been killed or starved out by Malthusian competitors as Darwin proposed.

Kropotkin thus challenges the idea that competition between members of a species is the only, or indeed principal, cause of evolution and argues that mutual aid plays a greater role.

Symbiogenesis

While the Russian naturalist Kropotkin was studying animal behaviour and then developing his ideas in England, the Russian botanist Konstantin Mereschkovsky was studying lichens and developing a hypothesis in Russia that he called symbiogenesis, an evolutionary process that begins with symbiosis.

According to biology historian Jan Sapp,[54] symbiosis as a means of evolutionary innovation had been discussed since the late nineteenth century. The dual nature of lichens as fungi and algae, nitrogen-fixing bacteria in the root nodules of legumes, fungi in the roots of forest trees and orchids, and algae living inside the bodies of protists showed how intimate physiological relationships could be established between distantly related organisms, sometimes leading, as in the case of lichen, to the evolution of whole new organisms.

Symbiosis was defined in 1878 by the German Anton de Bary, who studied fungi and algae, as "the living together of unlike-named organisms", which could lead to saltational evolutionary change. French botanist Andreas Schimper coined the term "chloroplast" in 1883 and suggested that green plants originated from symbiosis. In 1893 the Japanese cell biologist Shôsaburô Watase working in the USA used it to explain the origin of all nucleated cells. They had been formed by a group of small living organisms of dissimilar origin reciprocally interchanging their metabolic products in their struggle to survive; it was evidenced by the deep physiological interdependence of cell nucleus and cytoplasm.

In 1909 Mereschkovsky proposed a detailed theory for the origin of the cell nucleus and cytoplasm from two kinds of organism and two kinds of protoplasm that were the earliest forms of life on Earth. He also argued that chloroplasts in cells had originated as blue-green algae, and coined the term "symbiogenesis" to describe the process by which two different kinds of symbiotic organisms merge to form a new, more complex organism. Mereschkovsky maintained that symbiogenesis offered a better explanation for biological evolution than did Darwin's theory.

The idea that mitochondria in cells had a similar symbiogenetic origin can be traced back to the work of the German histologist Richard Altmann in 1890. In 1918 Paul Portier developed the concept of mitochondria as ancient symbionts in his book *Les Symbionts*. In 1927 the American biologist Ivan Wallin advanced a similar view in *Symbiontism and the Origin of Species*, proposing that acquired mitochondria were the source of new genes. For Wallin, three principles governed biological evolution: species originated through symbiosis; natural selection governed their survival and extinction; an unknown principle was responsible for the pattern of increasing complexity.

Such ideas were ignored or dismissed by most biologists who favoured Darwinian competition rather than collaboration and believed that symbiogenesis lacked any empirical support.

Mendel and heritability

Heritability, one problem that blighted Darwin's hypothesis, was solved by an Augustinian monk, Gregor Mendel, in 1865. His seminal paper, "Experiments in Plant Hybridization", was published the following year in *Proceedings of the Natural History Society of Brünn* in Bohemia (now Brno in the Czech Republic) and laid the foundations for the modern science of genetics.

Experiments
Mendel was not simply a monk as he is often portrayed but, in keeping with the tradition of the Abbey of Brünn, he was also a teacher and scientist. Between

1856 and 1863 he set out to test the orthodox account of heritability that said the characteristics of each parent were blended in the offspring. He chose the pea plant, which has several simple characteristics such as the height of the plant, the colour of the pea seed, and whether the pea seed is wrinkled or smooth.

He began by producing pure-breeding specimens, such as plants that always produce yellow pea seeds and plants that always produce green pea seeds. He then crossed such pure-breeding plants and found that, in this example, the first generation plants produced only yellow peas. When he bred these first generation plants among themselves, the second generation showed yellow-pea plants and green-pea plants in the ratio 3 to 1.

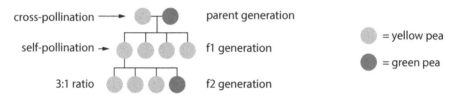

Figure 16.2 **Result of Mendel's experiments with yellow- and green-pea plants**

Experiments with pea plants having different characteristics produced the same kind of result. Mendel concluded that characteristics were not blended, but each was represented by a factor (later called a gene) that comes in alternate forms (later called alleles); in this example, one allele produces the yellow colour in a pea and another allele produces the green colour. Alleles occur in pairs; where paired alleles are different, one is "dominant" (in this case the yellow-producing allele) and masks the effect of the other, "recessive" allele (in this case the green-producing allele). Only when two recessive alleles are passed on does the corresponding characteristic show in the plant.

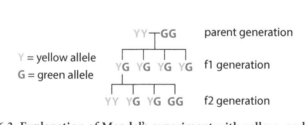

Figure 16.3 **Explanation of Mendel's experiment with yellow- and green-pea plants, with yellow allele dominant**

Mendel's laws

From his experiments Mendel deduced laws of inheritance that may be summarized as follows.

1. The characteristics of an individual are determined by hereditary factors (now called genes) and these occur in pairs.
2. Each gene pair separates during reproduction and only one gene of the pair is passed on by each parent to each offspring.
3. Some genes are dominant and can mask the effects of other, recessive genes.
4. For some characteristics neither gene is dominant.

Which gene is passed on is a matter of chance. As a consequence of Mendel's laws new combinations of genes, giving rise to new combinations of characteristics present in neither parent, are possible. Moreover, a characteristic may not show up in an individual but can show in a later generation because it is passed on unchanged.

This solved the problem with Darwin's hypothesis because it allows genes responsible for advantageous characteristics in an individual to be inherited unchanged and be selected for in a population that is competing to survive.

Darwin and most naturalists didn't read Mendel's paper, although it was given a favourable entry in the *Encyclopædia Britannica* and was cited several times in an 1881 publication on plant breeding by the botanist Wilhelm Olbers Focke, a copy of which Darwin apparently possessed.

NeoDarwinism

Mendel's paper was rediscovered in 1900 by three botanists whose research had led them independently to similar conclusions.

In the 1920s and 1930s statistician and geneticist Ronald Fisher and geneticist J B S Haldane from England plus geneticist Sewall Wright from the USA independently developed a mathematical basis for biological theory by applying statistical techniques; this new sub-branch of biology became known as population genetics. Their theoretical framework integrated Mendelian genetics into Darwin's hypothesis by showing mathematically how natural selection acting cumulatively on small variations could gradually yield major changes in form and function.

Wright also introduced the concept of genetic drift.

> **genetic drift**   The variation in the frequencies of alleles (gene pairs) in a small population that takes place by chance events rather than through natural selection. It can result in genetic traits being lost from, or becoming widespread in, a population irrespective of the survival or reproductive value of those genetic traits.

However, their work made little impact on biologists because it was theoretical and expressed mainly in mathematics; moreover, it showed only how changes could occur within a species rather than how new species could evolve.

What became known as the Modern Synthesis, or NeoDarwinism,* was developed in the late 1930s and early 1940s by the Russian-born American geneticist and experimental zoologist Theodosius Dobzhansky in his 1937 book *Genetics and the Origin of Species*, the German-born American zoologist and population geneticist Ernst Mayr in *Systematics and the Origin of Species* of 1942, the English zoologist and advocate of the mathematical approach to genetics Julian Huxley in *Evolution: The Modern Synthesis* also 1942, and the American who employed the statistical techniques of population genetics to palaeontology George Gaylord Simpson in *Tempo and Mode in Evolution*, 1944.

The aim was to integrate Darwinian natural selection arising from competition with the findings of theoretical and experimental genetics, zoology, and palaeontology in an overarching theory of biological evolution. It changed the agency of evolution from individual organisms to populations. As Dobzhansky put it, "Evolution is a change in the genetic composition of populations. The study of mechanisms of evolution falls within the province of population genetics."

> **NeoDarwinism** The synthesis of Darwinian natural selection with Mendelian and population genetics in which randomly produced genetic variations responsible for characteristics that make individuals in a species population better adapted to compete for resources in their environment survive longer and produce more offspring. These favourable genes are thus inherited in increasingly greater numbers causing the gene pool—the total of all genes in the population—gradually to change over very many generations until a new species emerges. Members of the population lacking genetic variations responsible for such adaptive characteristics are killed, starved, or gradually become extinct in that environment.

Molecular biology

In parallel with these developments in macrobiology following the rediscovery of Mendel's work, microbiologists attempted to identify just what genes are and how they function as agents of hereditary characteristics.

Research on the fast-breeding fruit fly *Drosophila melanogaster* by Columbia University experimental zoologist Thomas H Morgan and colleagues showed in 1914 that genes are arranged in a linear order on chromosomes; moreover genes sometimes undergo spontaneous and permanent changes, or mutations, that result in a change of hereditary characteristic, such as white eyes to red eyes.

---

* Mayr wanted to restrict the term NeoDarwinism to its initial use, which was essentially Darwinism without the inheritance of acquired characteristics, and thereby distinguish it from the Modern Synthesis. However, this distinction was not generally adopted, and "Modern Synthesis" is not a self-evident term for non-biologists. "NeoDarwinism" makes a useful distinction from "UltraDarwinism", which I shall define later.

This work earned Morgan the Nobel Prize in 1933.[55] Biologists adopted the idea that gene mutation was the prime cause of variation in the characteristics of individuals.

Since the chromosomes of eukaryotes were shown to consist of proteins and nucleic acids, usually DNA, three possibilities arose: genes consisted of DNA, or proteins, or a combination of the two. To find out which of the three was correct scientists needed to show how genes pass on the information that determine the characteristics of the offspring and also how they build the cells that grow into the offspring.

For this first function physicist Erwin Schrödinger speculated in 1943 that genes carry their hereditary information in a code consisting of a small number of repeating entities in the way that the simple dots and dashes of the Morse code carry a vast amount of information.[56]

The following year Oswald Avery of New York's Rockefeller Institute Hospital together with colleagues Colin MacLeod and Maclyn McCarty demonstrated that the agent that caused change in bacteria to be passed on to succeeding generations was DNA; this implied that genes consisted of DNA.

Most biologists remained sceptical because they considered that the sequence of nucleotides in a molecule of DNA lacked sufficient variety to convey all the genetic information and so genes must consist of proteins. However, in 1952 Alfred D Hershey and Martha Chase of the Genetics Research Unit of Washington's Carnegie Institute proved that the hereditary material responsible for the fundamental characteristics of viruses that attack and kill bacterial cells is DNA. This prompted the idea that somehow the sequence of the bases along the polynucleotide chain of a DNA molecule could form Schrödinger's code.

By now it had become clear that, in order to show how DNA both replicates itself to pass genes onto progeny and also determines the biochemical reactions that synthesize that progeny, it was necessary to know not only the chemical composition of DNA but also its three-dimensional structure. Molecular biology was the name given to this emerging science that employed X-ray crystallography to discover molecular architecture in the race—and it was a ruthless one—to show how DNA functions as genes.

The race was won in 1953 by a young American geneticist James Watson, who had obtained a PhD, and a somewhat older English physicist Francis Crick, who hadn't yet obtained his. Working as a team at the Cavendish Laboratory in Cambridge, England, they used X-ray diffraction results produced by two physicists at King's College, London, Maurice Wilkins and Rosalind Franklin, who were meant to work as a team but didn't. From these results they constructed a three-dimensional model showing DNA as two complementary helical strands coiled around each other and speculated that the coils could unwind to act as templates to produce copies of themselves.[57]*

---

*   See diagram on page 218.

Their brief paper in the April 1953 edition of *Nature*, together with separate papers from Wilkins and from Franklin, followed by more detailed papers expanding their ideas, opened a new era in molecular biology. Working out the detailed mechanisms, and providing experimental proof, of Watson and Crick's insight into the physico-biochemical nature of genes dominated biology for the next 25 years.

Morgan's observed spontaneous gene mutations were explained by rare copying mistakes from the uncoiled DNA template. Precisely how the differing sequences of DNA's four nucleotides function as Schrödinger's code for conveying genetic information took longer. And it took longer still to determine the details of DNA's second function, how its sequence of nucleotides leads, via intermediary molecules of messenger RNA, to making the proteins and other biochemicals that constitute the cells of which all organisms consist.*

But Watson and Crick's conjectures were eventually vindicated. It was a spectacular triumph for the reductionist method and its focus on genes, which now dominated biology.

In 1958 Crick articulated a key assumption, which he re-stated in 1970 as "the central dogma of molecular biology", namely that each gene in the DNA molecule carries the information needed to construct one protein, which, acting as an enzyme, controls one chemical reaction in the cell, and this is a one-way process: information cannot be transferred back from protein to either protein or nucleic acid.[58] This is now understood to mean generally that information cannot flow from the environment or the organism back to cellular DNA or RNA.

Principles of biological orthodoxy

These key conclusions from molecular biology were incorporated into NeoDarwinism to produce the current orthodoxy.

In 2006 Jerry Coyne, who comes from the school of population geneticists and experimental zoologists who framed the NeoDarwinian synthesis, summarized the modern theory of biological evolution. Principles 1, 2, 3, and 5 below are derived from Coyne's summary,[59] principle 4 is explicit in his other publications and those of other NeoDarwinists, while principle 6 is an updated summary of Crick's central dogma of molecular biology. Although not all evolutionary biologists agree with all of them, these six principles define the paradigm† to which the large majority subscribe.

   1. Living species are descendants of other species that lived in the past.
   2. New forms of life arise from the splitting of a single lineage into two, a

---

* The results, including the intermediary role of messenger RNA, are described on pp. 216–222.
† See Glossary for definition of *paradigm*.

process known as speciation in which members of one lineage are unable to breed successfully with members of the other lineage. This continual splitting leads to a nested genealogy of species, a "tree of life" whose root is the first species to arise and whose live twigs are the vast number of current species. Tracing back any pair of twigs from modern species through the branches leads to a common ancestor, represented by the node at which the branches meet.

3. Such evolution of new species occurs through the gradual genetic transformation of populations of individual species members over thousands or more generations.

4. This transformation results from random genetic mutations in individuals, the mixing of genes from each parent during sexual reproduction that leads to each offspring having a different combination of genes from those of either parent, and the consequent spreading of such genetic mutations over successive generations throughout the population's gene pool.

5. Those randomly produced genetic mutations, or variations, responsible for characteristics that enable individuals in the population to successfully compete for resources in a particular environment, and so survive longer and produce more offspring, are naturally selected by being inherited in greater numbers, while members of the population lacking genetic mutations responsible for such adaptive characteristics in that environment are either killed, starved, or become extinct over generations.

6. Information is a one-way flow from a gene to a protein in a cell.

Consequences of the current paradigm

This gene-focused paradigm that assumes natural selection arises from competition had four principal consequences for science's account of the evolution of life.

First, it rejects or ignores other possible causes of evolution, including

a. sexual selection, which Darwin had advocated;

b. a Lamarckian inheritance of acquired characteristics, which Darwin had accepted;

c. genetic drift, which Sewall Wright had argued could be important in small populations;

d. symbiogenesis, the hypothesis developed by Mereschkovsky, Wallin and others that claimed the merger of different kinds of organisms living together for mutual metabolic benefit provides a better explanation for biological innovation and evolution than does Darwinian competition and gradualist natural selection;

e. cooperation, which Kropotkin had deduced from field observations of animals and argued was a more important cause of evolution than competition;

f. orthogenesis, in the sense of biological evolution progressing along lines of increasing complexity caused by a natural law or intrinsic tendency, which had been advocated by the French palaeontologist, visionary, and Jesuit Teilhard de Chardin in his posthumously published *Le Phénomène Humain* (1955). It is noteworthy that, of the four architects of NeoDarwinism in the late 1930s and early 1940s, Julian Huxley wrote an Introduction to the 1959 English edition endorsing most of Teilhard's ideas save the attempted reconciliation with Christianity, while Theodosius Dobzhansky was a founder-member and President 1968–1971 of the American Teilhard de Chardin Association.

Second, its reductionist focus on genes led to the view, pioneered by Carl Woese, that genetic analysis is the only accurate method of mapping the evolution of species: molecular differences in assumed conserved genes of different species represent their evolutionary relationships in the branching tree of descent from a common ancestor.* As Woese put it, "Molecular sequences can reveal evolutionary relationships in a way and to an extent that classical phenotypic criteria [observable characteristics], and even molecular functions, cannot."[60]

Third, it produced a gene-centred hypothesis of evolution that says the individual gene, not the organism, is the unit of natural selection.

Fourth, when it was discovered in the 1970s that some 98 per cent of the human genome did not consist of genes, defined then as protein-coding sequences of DNA, this was referred to as "junk DNA", which seemed a bold, if not hubristic, view to take of the vast majority of our DNA that didn't fit the model.

In the following chapters I shall consider to what extent this biological orthodoxy and its four consequences remain valid.

---

*   See page 212.

CHAPTER SEVENTEEN

Evidence of Biological Evolution 1: Fossils

> I will give up my belief in evolution if someone finds a fossil rabbit in the Precambrian.
>
> —ATTRIBUTED TO J B S HALDANE

Conflating the phenomenon of biological evolution with just one of its several possible causes can lead to subconscious bias in selecting and interpreting evidence. I shall attempt to avoid this by examining only the phenomenon and seeing whether or not there is a pattern in the evidence. In subsequent chapters I shall consider to what extent the NeoDarwinian orthodox model, together with hypotheses that modify or challenge that model, explains any such pattern.

I shall examine the evidence of fossils in this chapter and the evidence of living species in the next chapter. Because all the evidence refers to species, it is necessary first to clarify just what this term means.

Species

Estimates of the number of living eukaryotic species range from 5 million to 30 million,[1] of which around 2 million have been described.[2]

Only about 4,500 prokaryotic species (bacteria and archaea) have been described,[3] but estimates of their total number in the top 1km of the Earth's crust alone range from 10^8 to 10^{17} (a hundred million to a hundred thousand trillion).[4] No estimates are currently available of prokaryotic species in other locations but, to give an idea of their significance in the biosphere, the total number of prokaryotic cells in the open ocean is thought to be some 10^{29} (a hundred thousand trillion trillion), at least double that in the soil, and 50 times that deep in the oceanic and terrestrial subsurfaces.[5]

One reason for the wide discrepancy in estimates is the lack of agreement on what a species is. As Nick Isaac of London's Zoological Society remarked, "There are almost as many concepts of species as there are biologists prepared to discuss them."[6]

Like Darwin, several biologists have maintained that species are arbitrary constructs. Butterfly specialist Jim Mallet of University College London, whose research focuses on speciation, maintains that "an essential species 'reality' strongly conflicts with our understanding of gradual speciation, and is no longer accepted at all generally".[7] He goes on to argue that "recent genetic studies...have supported the existence of the Darwinian continuum between varieties and species".[8]

Evolutionary biologists Jerry Coyne and H Allen Orr dispute this in their 2004 book *Speciation*, a comprehensive survey of research in this field.

As Ernst Mayr, one of the architects of the NeoDarwinian model, put it:

> The so-called species problem can be reduced to a simple choice between two alternatives: are species realities in nature or are they simply constructs of the human mind?[9]

Since a human being is clearly something different from a mouse, a goldfish, an *E coli* bacterium, or an oak tree, there is a practical need, quite apart from any ontological one, to define what it is scientists are speaking of and studying. This was a major reason for the development of taxonomy: the hierarchical classification of organisms into named groups according to common characteristics, from the most general characteristics to particular ones.

Whatever their differences over what the word means, most scientists use species, rather than a subspecies taxon like variety or superspecies taxa like genus, order, class, or phylum in order to describe organisms and their relationships to other organisms. But the way in which taxonomists, systematicists, naturalists, bacteriologists, botanists, entomologists, zoologists, evolutionary biologists, molecular biologists, geneticists, genomicists, ecologists, and other specialists each define species is very different because they disagree about which characteristics are the defining ones.

When stating what they meant by species, the architects of NeoDarwinism considered that the traditional taxonomic characteristic of morphology—structure and form—was inadequate because many species differ very little morphologically while other species exhibit a wide variety of morphologies. Moreover, male and female of all species usually differ in size and shape and often in colour, as do young and mature, while adults often differ in shape depending on the availability of food.

For these NeoDarwinists the ability of members of a population to exchange genes was the key characteristic: gene flow within a population defines a species and distinguishes that species from another population with which there is no gene flow, and gene flow is achieved by members of the species mating with each other to produce fertile offspring. Ernst Mayr crystallized this view in 1940 as the "biological species concept", which he defined essentially unchanged more than 50 years later as "groups of actually or potentially interbreeding natural populations which are reproductively isolated from other such groups".[10]

Because NeoDarwinists hold that the natural selection of characteristics of a species population in a particular environment is the cause of biological evolution, they maintain that geographical isolation is the principal route to speciation. A population is split in two by either the migration of one part or the formation of a geographical barrier between different parts of a population, as when the Isthmus of Panama closed about 3 million years ago separating marine organisms inhabiting the Atlantic and Pacific coasts. The gene pool of each of the geographically isolated populations changes over generations when natural selection favours members better adapted for competing to survive and reproduce in their particular environment. Eventually members of the different populations are no longer capable of exchanging genes even if contact were to occur. This is termed allopatric speciation, contrasted with sympatric, where a population splits into two or more species sharing the same territory.

The NeoDarwinian biological species concept was widely accepted as orthodoxy until the 1980s, when botanists began to challenge it on the grounds that it did not apply to many plants.

Coyne and Orr, zoologists who research speciation using the fast-breeding fruit fly *Drosophila melanogaster*, follow Mayr in considering that allopatry is the main route to speciation. They slightly modified his definition by recognizing that Darwinian speciation is a process taking a long time and there are intermediate stages when it is not possible to make a clear species distinction. Their updated biological species concept, which is the current orthodox, but by no means unanimous, definition is "groups of interbreeding populations that show substantial, but not necessarily complete, reproductive isolation from other such populations".[11]

Their review of the various isolating barriers divides into premating and postmating. Premating includes

a. ecological, where populations occupy or breed in different habitats;
b. temporal, where populations mate at different times, whether of the day or the year;
c. behavioural, where members of a population choose whether to mate or whom to mate with, also called assortative mating;
d. mechanical, where morphological or physiological differences prevent mating.

Postmating isolation means that the progeny of mating is inviable (aborted or stillborn), or sterile (like a mule resulting from the mating of a horse and a donkey), or is less well adapted than either parent to survive and reproduce, and so its successors become extinct.

There are, however, four principal problems with this current orthodox species definition.

1. *Testability*
   If the principal route to speciation is geographical separation of parts of

a population, then because the two parts are isolated from each other we cannot observe whether or not members of one part are able to breed with members of the other part. The rare exception is when members of one part of the population subsequently migrate back to the territory of the other part.

2. *Asexual reproduction*

The definition hinges on sexual reproduction, but most species reproduce asexually. One such method is parthenogenesis, sometimes called virgin birth.

> **parthenogenesis**   The development of an egg into an offspring without fertilization by a male.

Many invertebrates, like insects, produce progeny this way. For example, bdilloidea rotifers, microscopic water-dwelling invertebrates, consist entirely of females, and yet they reproduce and have developed into some 300 recognized species. Most animals that reproduce parthenogenetically also reproduce sexually. They include bees and ants, plus some vertebrates like species of snakes, fishes, amphibians, reptiles, and birds, but not mammals. The phenomenon is rarer among plants, where natural parthenogenetic development occurs in roses and orange trees.

Another asexual method is binary fission.

> **binary fission**   The splitting of a cell into two, whereby each cell produced is identical to, and usually grows to the size of, the original cell.

By far the overwhelming majority of species on the planet—bacteria and archaea—reproduce in this manner.

Coyne and Orr argue that bacterial reproduction is consistent with their definition because genes are exchanged between bacteria by a process known as horizontal, or lateral, gene transfer, which "can cause rare gene transfer and recombination"; this bacterial equivalent of sexual gene recombination even leads to "a form of reproductive isolation". On the other hand, the Millennium Ecosystem Assessment Report says that

> Species concepts based on gene flow and its limits, such as the biological species concept, are not applicable to asexual taxa. They are also inadequate for 'pansexual' taxa, such as some bacteria, where gene flow can be common between even very dissimilar types.[12]

3. *Successful hybridization*

It follows from the current NeoDarwinian definition that mating between members of two different species produces offspring that are inviable, or infertile, or so less adapted to survive and reproduce that their descendants become extinct.

However, 25 per cent of vascular plants (which include all flowering plants) in Great Britain, 16 per cent of butterflies in Europe, 9 per cent of birds worldwide, and 6 per cent of mammals—including deer—in Europe successfully hybridize to produce fertile offspring.[13]

4. *Polyploid hybridization*

Not only do many plants successfully hybridize, but also hybridization can generate a new species almost immediately instead of gradually over thousands of generations. In the next chapter I shall consider this particular type of hybridization, known as polyploidy, in which the hybrid and its offspring cannot breed with members of either parent species.

Coyne and Orr count at least 25 alternative definitions of species in use today. However, each has at least as many problems as the current orthodox definition.

I'm forced to conclude that in reviewing the evidence for biological evolution there is no alternative but to accept whatever definition of species, and whatever characteristics are used to classify something as a particular species, that the relevant specialists use. This does, however, produce considerable inconsistencies. For example, the characteristics used to classify species of ants are very different from those used to classify species of butterfly. Irish wolfhounds that stand 90cm high at the shoulder are classified as the same species (*Canis lupus familiaris*) as 15cm-high chihuahuas, whereas a Mediterranean gull (*Larus melanocephalus*), average adult length 39cm, is classified as a different species from the Common gull (*Larus canus*), which has an average adult length of 41cm, slightly longer, redder legs, and a black head in summer only. If fossils were the only specimens, the two gulls would be indistinguishable, whereas the two dogs would be classified as different species.

Fossils

The fossil record has expanded enormously since Darwin's time, and especially so in the last 30 years. Nonetheless there are two major problems when evaluating fossils for what they tell us about biological evolution.

Paucity of the fossil record

The same geological and physiological reasons given in Chapter 14 for the lack of fossils of the earliest lifeforms* apply to all lifeforms. In addition, tectonic movements and erosion have destroyed many fossils of younger species in younger strata of rocks.

Richard Leakey and Roger Lewin estimate that only some 250,000 eukaryotic species have been preserved in the fossil record out of a likely 30 billion that have

---

* See page 208.

lived during the last 600 million years.[14] This crude estimate means that only one in every 120,000 eukaryotic species has been fossilized; it takes no account of the vastly greater number of prokaryotic species.

Nor are these fossils representative. Hard parts of organisms, like teeth and bones, are more likely to be fossilized than parts of soft-bodied organisms. About 95 per cent of the fossil record comprises remains of creatures that lived under water, mainly in shallow seas.[15]

Interpretation

Even in those rare cases where the fossil of a whole organism has been found, interpretation is a key problem. For example, after studying Burgess Shale fossils, which date from the Cambrian period, Simon Conway Morris interpreted in 1977 a 25mm-long specimen as the remains of an animal that walked along the bottom of the seafloor on spiny stilts, waving seven dorsal tentacles from its back, as shown in Figure 17.1. It was unique, and Conway Morris classified it as a new species, *Hallucigenia sparsa*, one of several that suddenly appeared in the Cambrian and was never seen again.

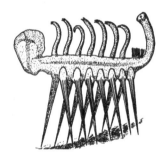

Figure 17.1 First reconstruction of *Hallucigenia sparsa*

An alternative interpretation considered *Hallucigenia* to be an appendage of a larger, unknown animal. Then in 1991 Lars Ramsköld and Hou Xianguang examined specimens found at the Chengjiang site in China and concluded that the Conway Morris reconstruction was upside down: when alive *Hallucigenia* walked on pairs of tentacle-like legs and protected itself with a ferocious palisade of spines, as shown in Figure 17.2.

Hallucigenia was reclassified to the phylum of *Onychophora*, making it a distant ancestor of caterpillar-looking animals now living in tropical rainforests rather than a one-off species.[16] This interpretation was accepted by Conway Morris and most palaeontologists, although there was no agreement as to which end is the head and which the tail.

Normally a fossil consists not of a complete animal but of fragments of bone or teeth, and the problems of interpretation, reconstruction, and classification are correspondingly more difficult.

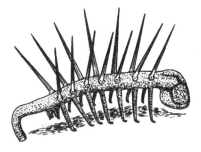

Figure 17.2 Second reconstruction of *Hallucigenia sparsa*

All this provides fertile ground for the Law of Data Interpretation suggested in Chapter 6 by the claims of some cosmologists. For instance, late in 2006 palaeontologist Jorn Hurum persuaded Oslo's Natural History Museum to pay $1million for a small fossil he had not studied in detail. Barely two years later the doyen of TV natural history presenters, Sir David Attenborough, announced to news media assembled in the presence of New York mayor Michael Bloomberg and banner adverts for a book and a TV tie-in series called *The Link*, that the missing link was missing no more. Hurum, who had classified the exceptionally well-preserved 47-million-year-old fossil as *Darwinius masillae*, explained that it was "the first link to all humans". It would, he added modestly, "probably be pictured in all the textbooks for the next 100 years".

Less than 5 months later Erik Seiffert of Stony Brook University in New York, who led the study published in *Nature* of a similar fossil 10 million years younger, said "Our analysis provides no support for the claim that *Darwinius* is a link in the origin of higher primates."[17]

At least *Darwinius* was not a fake, but frauds often go undetected because of interpretation problems. From Piltdown Man, unearthed in 1912 and only discovered to have been a fake 40 years later, a succession of frauds has been uncovered. One of the most recent and most extensive was the systematic falsification of fossil dates perpetrated over 30 years by the then distinguished anthropologist Professor Reiner Protsch before he was forced to retire from the University of Frankfurt in 2005. Archaeologist Thomas Terberger, who discovered the fraud, commented that

> Anthropology is going to have to completely revise its picture of modern man between 40,000 and 10,000 years ago. Professor Protsch's work appeared to prove that anatomically modern humans and Neanderthals had co-existed, and perhaps even had children together. This now appears to be rubbish.[18]

More recent genetic evidence, considered in Part 3, suggests that some Neanderthals did breed with some early humans, but this does not validate Protsch's fake fossil evidence.

The fossil record

Bearing these problems in mind, Figure 17.3 draws on Chapters 12 and 14 plus studies cited later in this section to illustrate the current best estimates of the fossil record's timescale represented as a 24-hour clock.

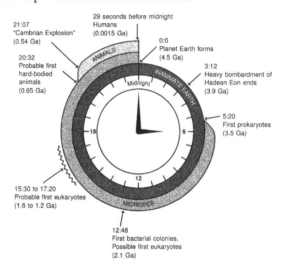

Figure 17.3 Evolution of life from the fossil record represented as a 24-hour clock

This overall picture shows that if the Earth's formation some 4.5 billion years ago starts the clock, the postulated heavy bombardment by asteroids or comets in the late part of the Hadean Eon ends at 3:12 (3.9 billion years ago). Although we will never know precisely when life emerged from the inanimate chemicals on the Earth's surface, the least disputed estimates show the earliest microbial fossils at 5:20 (3.5 billion years ago). Such microbes, mainly bacteria and archaea, were the only form of life for nearly 3 billion years; animals emerge only at around 20:32 (0.65 billion years ago), while humans emerge at 29 seconds before midnight.

Chemical and radiometric analyses of rocks indicate that for most of the Earth's history there was very little oxygen in the oceans, shallows, and atmosphere (see Table 17.1).

Correlations with the fossil record suggest the first microbes were extremophiles that maintained themselves by ingesting chemicals, probably sulphur compounds and hydrogen from the decomposition of water. Eventually ancestors of cyano-bacteria evolved to utilize sunlight as their energy source. Western Australia has one of the most continuous and best-studied records of stromatolites, laminated structures built mainly by cyanobacteria, that dominate the fossil record between about 2 billion and 1 billion years ago.[19]

Some cyanobacteria excrete oxygen as a metabolic waste product. While oxygen was poisonous to early bacteria (called anaerobic), some species adapted to use this oxygen for their metabolism (aerobic bacteria).

Table 17.1 Oxygenation of oceans and atmosphere

Period (billion years ago)	Oxygen in deep oceans	Oxygen in shallow oceans	Oxygen in atmosphere (atmospheric pressures)
3.85 to 2.45	None	Possibly small pockets	None
2.45 to 1.85	None	Mildly oxygenated	0.02 – 0.04
1.85 to 0.85	Mildly oxygenated	Mildly oxygenated	0.02 – 0.04
0.85 to 0.54	Mainly none	Rise in oxygenation similar to atmosphere	Rises to 0.2
0.54 to present	Oxygenation fluctuates considerably	Oxygenated	Rises to 0.3 followed by fall to current level of 0.2

Source: Holland, Heinrich D (2006) "The Oxygenation of the Atmosphere and Oceans." *Philosophical Transactions of the Royal Society B: Biological Sciences* 361: 1470, 903–915

The earliest claimed fossil of a eukaryote—a nucleated cell that contains organelles including mitochondria, which use oxygen to generate energy for the cell—was discovered as thin films of carbon in rocks 2.1 billion years old in the Empire Mine, near Marquette in Michigan, USA. It may have been a bacterial colony but its size of more than one centimetre and its tube-shape suggests it may have been *Grypania spiralis*, a eukaryotic alga.[20]

In 2010 an interdisciplinary team led by Abderrazak El Albani of Université de Poitiers announced the discovery of more than 250 well-preserved macroscopic fossils from 2.1-billion-year-old black shales in southeastern Gabon that they interpret as representing multicellular life. Carbon and sulphur isotopic data indicate that the structures of up to 12 centimetres in size were biogenic, and growth patterns deduced from fossil morphologies suggest cell-to-cell signalling and coordinated responses commonly associated with multicellular organization. Moreover, iron speciation analyses suggest that the organisms probably used oxygen respiration. Like the claimed *Grypania spiralis*, the dating of the fossils coincides with the beginnings of oxygenation of the ocean shallows and atmosphere (see Table 17.1). The researchers don't rule out the possibility that these fossils represent the earliest multicellular eukaryotes.[21] However, as Philip Donoghue and Jonathan Antcliffe of the University of Bristol Department of Earth Sciences point out, without further evidence the assumption must be that they represent bacterial colonies.[22]

More widely accepted as the earliest eukaryotic fossils are a large population of spherical microfossils preserved in coastal marine shales of the Ruyang Group of northern China and labelled as *Shuiyousphaeridium macroreticulatum*, a unicellular organism dating from between 1.6 billion and 1.26 billion years ago.[23]

While oxygen levels in the oceans and in the atmosphere remained low for nearly 3 billion years, from around 0.85 to 0.54 billion years ago oxygenation of the ocean shallows increased significantly and the atmosphere was transformed to its current level of 20 per cent oxygen, presumably due to the rapid spread of oxygen-excreting cyanobacteria. This is consistent with the appearance around 600 million years ago of simple marine animals that extract oxygen from water for their metabolism, followed by fish, and then by land-based animals that metabolize by breathing in oxygen from the atmosphere.

In August 2010 Princeton geoscientist Adam Maloof and colleagues claimed to have discovered the earliest evidence of invertebrate hard-bodied animals. They found shelly fossils beneath a glacial deposit in South Australia that suggest primitive sponge-like creatures lived in ocean reefs about 650 million years ago. If confirmed, it means that animal life existed before, and very probably survived, the severe "snowball Earth" event known as the Marinoan glaciation that left much of the planet covered in ice.[24]

Fossilized imprints in sandstone of a wide range of primitive soft-bodied animals dating from around 600 million years ago have been found at sites around the world. They mark the appearance of Ediacaran fauna, after the Ediacara Hills of South Australia in which they were discovered in 1946.

Most disappear from the fossil record after 542 million years ago, although more recent discoveries indicate that some Ediacaran organisms continued into the Cambrian period. In addition to organisms unlike any known today, the imprints suggest some cnidarians similar to current jellyfish, lichen, soft corals, sea anemones, sea pens, and annelid worms, but whether they were ancestral to extant species is controversial.[25]

Palaeontologists conventionally label the early Cambrian geological period, dating from 545 million years ago, as the Cambrian explosion because it marked the sudden appearance of a wide variety of soft-bodied and hard-bodied members of the animal kingdom—the multicellular eukaryotes that eat other organisms, and require oxygen, to maintain themselves—followed by the equally sudden disappearance of most of them at the end of the period, 485 million years ago. However, since strata in the last 10 per cent of Earth's history were dated primarily by their fossil content, this is a circular argument. More recent discoveries, plus subsequent radiometric dating of rock strata, suggest that many species and lineages appeared before, or disappeared after, this time frame.

Transitional fossils

Creationists claim that the absence of fossils transitional between one species and another disproves biological evolution.

Richard Dawkins counters by asserting that "just about every fossil found can fairly be described as intermediate between something and something else". Echoing Haldane's remark given below this chapter's title, he claims that nothing in the fossil record falsifies biological evolution.[26]

Bacteria, apart from those that evolved into eukaryotes, retain essentially the same, relatively simple, body plans over 3 billion years. For animals, however, the first appearance in the fossil record of a more complex species has never been dated before the first appearance of a less complex species. I venture to suggest, though, that asserting that every fossil is an intermediate between something and something else is perhaps a bit of an oversimplification.

In reply to the creationist claim, evolutionary biologists usually point to the inherent paucity of the fossil record and employ an argument advanced by Darwin: transitional organisms that are less well adapted to an environment than their successors will lose the struggle to survive and reproduce, and so become extinct quickly in geological time; hence there is an even smaller probability of finding fossils of transitional organisms than those of successfully adapted ones.

A few series of fossils, nonetheless, present a case for biological evolution. A relatively rich fossil record exists for the horse family, from *Hyracotherium*, the dawn horse, a mammal about the size of a fox with several spread-out toes and teeth appropriate for omnivorous browsing, which dates from around 50 million years ago, to the only surviving species, the modern single-toed, long-legged horse, *Equus*, with teeth appropriate for grazing. Figure 17.4 illustrates the anatomical changes over time in this lineage.

The diagram does not represent a single linear succession. Fossils of many other extinct equids have been found, suggesting a branching evolutionary tree in which only the lineage leading to the modern horse survived to the present day.

The anatomical transitions are consistent with the hypothesis that a small mammal adapted for walking on the soft, moist ground of primeval forests and browsing on soft foliage and fruit evolved as forests gave way to savannah; here, only the swiftest varieties escaped predators, leading to longer legs and single-toed feet for greater speed, together with teeth better adapted for grazing.

Whales provide another case for biological evolution. Like all mammals they breathe air and produce milk to feed their young. However, they never leave the water, their ears are closed, they have fins instead of legs, and they possess a metabolism that retains oxygen not as a gas but as a chemical compound, enabling some of them to dive more than one and a half kilometres deep and stay submerged for two hours; some, like the blue whale, are massive, up to 150 tons.

According to palaeontologist and anatomist Hans Thewissen of Northeastern Ohio Universities College of Medicine, one of the leading authorities on the emergence of aquatic adaptations in whales, their earliest known ancestors are the family of *Pakicetidae* (Pakistani whales), whose genera comprise *Pakicetus*, *Ichthyolestes*, and *Nalacetus*. He and his team followed up an earlier discovery and found many fossilized bones in Pakistan and northwestern India, thought to have been near the ancient Tethys Sea before the Indian plate collided with the Cimmerian coast to produce the highest peaks in the world.

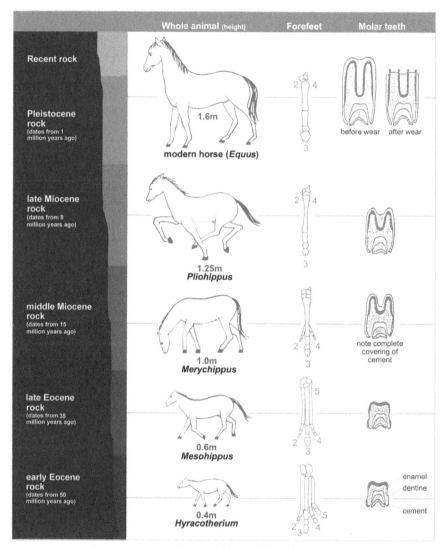

Figure 17.4 Evolution of horse from fossil record

The *Pakicetidae* were found together and so are of the same age. Because the dating of the rocks is problematical, however, Thewissen is only prepared to estimate an age of 50 ± 2 million years ago, but "with low confidence".[27]

Figure 17.5 depicts the bones of *Pakicetus* and *Ichthyolestes*, while Figure 17.6 shows a reconstruction of the former, about the size of a wolf.

They lived on land, but Thewissen claims that they have characteristics not shared by other mammals but only with archaic and recent cetaceans (whales, dolphins, and porpoises): a reduction of crushing basins on their teeth, an increased closing speed of their jaws, and a shape of the postorbital and temporal region of the skull that affects their hearing and vision. He hypothesizes that

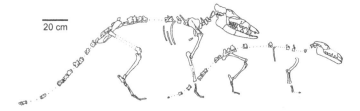

Figure 17.5 Bones of *Pakicetus* (large animal) and *Ichthyolestes* (small animal)
The scale indicates the size of the skeletons, and shows that *Pakicetus* was approximately as large as a wolf

Figure 17.6 Reconstruction of *Pakicetus*

these hoofed mammals changed to a diet of aquatic prey when they began wading in shallow streams. They evolved rapidly. The entire transition from pakicetids to marine mammals took less than 8 million years via ambulocetids (walking whales) that resemble 3-metre-long mammalian crocodiles the fossils of which were found in what was thought to have been marine swampland, protocetids with much reduced limbs, and then basilosaurids with enormous snake-like bodies and whale-like tail flukes, and dorudontids that resemble dolphins.[28]

The interpretation of, and reconstruction from, fossilized bones, together with inferences about how the animals functioned, and deductions from DNA analyses calibrated using the molecular clock technique, plus dating of rock strata by their fossil content, can never provide indisputable proof. However, the evidence is consistent with the hypothesized evolution of small, hoofed, land-based mammals to huge marine-based whales. It is also consistent with current mammals adapted to different degrees for both terrestrial and marine life, like otters, sea otters, and seals.

Other fossils claimed as transitional include finds in the 1990s, mainly from the Yixian formation in China dating from about 150 to 120 million years ago, of more than twenty genera of dinosaurs with fossil feathers.[29] Since then very many other, and much larger, feathered dinosaurs have been discovered, including in

2012 a species named *Yutyrannus huali* about 7 to 8 metres in length and weighing around 1,400 kilograms. Moreover, amino acid sequences from collagen extracted from the remains of a more recent *Tyrannosaurus rex* resemble those of a living chicken. All this evidence is consistent with the hypothesized evolution of birds from reptilian dinosaurs.

Finally, not only does the creationist alternative explanation of the fossil record lack any falsifiable evidence, it is entirely inconsistent with what evidence we do have; moreover it is internally inconsistent.*

Species extinctions

Fossils of species abundant in earlier rock layers but not present in later rock layers imply extinctions. Estimates of the proportion of all species that have ever existed and which became extinct commonly range from 99 per cent[30] to 99.9 per cent.[31]

As we saw at the beginning of this chapter, however, there is no agreement on what a species is. Furthermore, no one knows how many species exist today, not even to the nearest order of magnitude. It defies logic to quantify so precisely from the impoverished fossil record the number of species that existed 3 billion, 1 billion, 100 million, 10 million, or even 10,000 years ago, however sophisticated the mathematical models employed. All we can reasonably infer is that a large majority of species became extinct.

Moreover, evolutionary biologists differentiate between

> **terminal extinction**   A species ceases to exist without leaving any evolved descendants.

and

> **phyletic, or pseudo, extinction**   A species evolves into one or more new species; the first species has become extinct, but the evolutionary lineage continues.

The scientific consensus is that the majority of species extinctions in the fossil record, and all those observed today, are terminal.

Individual species extinctions

According to the NeoDarwinian model individual species extinctions can be phyletic or terminal. The former occurs when the population of a species in one ecological niche evolves very gradually by the accumulation of randomly generated genetic mutations coding for traits enabling its members to compete more effectively for survival and reproduction in that niche, or in another niche, to the point where the population becomes a new species.

---

*   See page 11.

Terminal extinction occurs when members of one species invade the whole of another's territory and either kill off that species or else are better adapted to exploit its food resources, forcing members of the original species either to starve to death or to be so weakened as to reproduce less and consequently die out over generations, or else a major change in environmental conditions occurs too rapidly for a species to adaptively evolve.

Mass species extinctions
Palaeontology's orthodox account says the Earth has experienced at least ten mass extinctions in which a large number species disappear in a short period of geological time. Table 17.2 collates the broadly agreed timetable of the five major mass extinctions considered to have taken place during the last 500 million years.

These events occur at the boundaries between different geological periods because those periods are mainly defined by their fossil content.

While palaeontologists agree that these mass extinctions took place, they do not agree on what, apart from drastic environmental change, was responsible. The numerous proposed causes include periodic ice ages with concomitant drops in global sea levels produced by changes in the Earth's axial tilt*, a deadly burst of high energy particles from a massive solar flare that overwhelmed the Earth's protective magnetosphere, ionizing radiation during temporary losses of the protective magnetosphere due to the flipping of the Earth's magnetic field,† lethal radioactivity from a supernova, tectonic activity, global cooling produced by massive volcanic eruptions generating clouds that cut off sunlight, global warming produced by massive volcanic eruptions generating clouds that create a greenhouse effect, global cooling produced by the impact of a massive meteorite generating clouds that cut off sunlight and/or triggering massive volcanic eruptions, global poisoning of animals and plants that depend on oxygen for their metabolism produced by anaerobic bacteria and archaea that dominated Earth for nearly the first 3 billion years of life, and so on. Some of these conjectured explanations are mutually contradictory, and there is insufficient evidence to confirm them, still less favour one over another.

The Cretaceous-Tertiary (K-T) extinction is the most recent and the one for which most evidence exists. Museums and popular science books usually present as established scientific fact the impact of a massive asteroid that wiped out the dinosaurs and many other species by generating global wildfires, earthquakes, and a global dust cloud rich in sulphuric acid that blocked out the Sun for many months causing global cooling, acid rain, and destruction of food chains. According to planetary scientist Walter Alvarez, co-proposer of the hypothesis, this mass extinction occurred in as brief a period as one to 10 years.[32]

* See page 189.
† See page 171.

Table 17.2 Palaeontology's major mass extinctions

Extinction event	When (million years ago)	Species lost
Ordovician-Silurian	440	All animals and plants then lived in the ocean, and more than 85 per cent of species became extinct, including many families of invertebrate marine animals belonging to groups such as brachiopods, echinoderms, and trilobites.
Late Devonian	360	An estimated 82 per cent of all species were lost, including animals and plants that now lived on land as well as in the sea. The greatest extinctions affected marine animals, including cephalopods and armoured fish.
Permian-Triassic	250	The largest mass extinction event in Earth's history, with 95 per cent of marine species and 70 per cent of all land species, including 8 of 27 insect orders, lost.
Triassic-Jurassic	200	75 per cent of all species perished: most marine reptiles, amphibians, and land-based reptiles, including several groups of archosaurs, advanced reptiles that included dinosaurs; dinosaurs, however, survived.
Cretaceous-Tertiary (abbreviated to K-T because "C" is used as shorthand for Cambrian)	65	Up to 75 per cent of marine genera and up to 50 per cent of all plants and animals, including all non-avian dinosaurs, disappeared.

Sources: American Museum of Natural History; "Mass extinction" *The Columbia Electronic Encyclopedia,* Sixth Edition Accessed 29 October 2008; "Extinction (biology)" *Microsoft Encarta Online Encyclopedia* 2008 Accessed 29 October 2008.

Many scientists questioned this account, but the popular science media reported in 2010 that the matter had been settled:[33] an interdisciplinary panel of scientists had reviewed 20 years' research and concluded that a massive asteroid impacting Chicxulub in Mexico's Yucatán Peninsula did indeed trigger the mass extinction.[34]

However, geophysicists Vincent Courtillot and Frédéric Fluteau accused the review panel of committing "a substantial error and a fundamental misrepresentation of our paper".[35] Princeton geoscientist Gerta Keller and others charged the panel of using

> a selective review of data and interpretations by proponents of this viewpoint. They ignored the vast body of evidence inconsistent with their conclusion—evidence accumulated by scientists across disciplines (paleontology, stratigraphy, sedimentology, geochemistry, geophysics, and volcanology) that documents a complex long-term scenario involving a combination of impacts, volcanism, and climate change.[36]

Evolutionary biologist J David Archibald and 22 other scientists pointed out that the review panel "conspicuously lacked the names of researchers in the fields of terrestrial vertebrates", and that "the simplistic extinction scenario presented in the Review has not stood up to the countless studies of how vertebrates and other terrestrial and marine organisms fared at the end of the Cretaceous".[37]

If the cause of the K-T mass extinction is debateable, what of the phenomenon itself? There are very few fossil specimens of the thousand or so species of dinosaur believed to have existed, and there is only one area where a dinosaur-bearing sedimentary transition across the K–T boundary has been examined, and this extends from Alberta in Canada to the northwestern USA. Records of dinosaurs in this area during the later part of the Cretaceous show a gradual decline in diversity with a drop from thirty to seven genera over the last eight million years of Cretaceous time, which suggests a more gradual extinction. Moreover, if a devastating mass extinction event suddenly wiped out all the dinosaurs, why were other reptiles, like crocodiles, lizards, snakes, and turtles, unaffected?

The fossil record is richest for marine organisms, but a detailed picture across the K-T boundary is known only for planktonic foraminifera and calcareous nanoplankton, and their extinctions occur over an extended period of time, starting well before and finishing well after the boundary. Brachiopods suffered badly across the boundary but, although claims are similarly made for ammonites, there are too few ammonite-bearing sections to show if their extinction was gradual or abrupt.[38]

If questions remain about the nature and cause of the K-T mass extinction, how confident can we be in the details currently presented of the earlier mass extinction events?

Stasis and sudden speciation

As we saw in the last chapter, one of the reasons that delayed the scientific community's acceptance of Darwin's hypothesis of natural selection was its essential gradualism, which wasn't supported by a fossil record that showed the appearance of fully formed species which remained unchanged until they disappeared. The evidence was consistent with saltationism, the hypothesis that biological evolution proceeded by leaps.

Biologists' adoption of NeoDarwinism in the 1940s, however, cemented the population geneticists' theoretical arguments for evolution proceeding by Darwinian gradualism and influenced interpretations of the fossil record.

However, in a 1972 paper entitled "Punctuated Equilibria: An Alternative to Phyletic Gradualism",[39] palaeontologists Niles Eldredge and Stephen Jay Gould challenged the evidence base for the NeoDarwinian explanation. This provoked a sometimes acrimonious debate that continues to the present day. Gould subsequently summarized their conclusions as follows.

> The history of most fossil species includes two features particularly inconsistent with gradualism: (1) *Stasis*. Most species exhibit no directional change during their tenure on earth. They appear in the fossil record looking much the same as when they disappear; morphological change is usually limited and directionless. (2) *Sudden appearance*. In any local area, a species does not arise gradually by the steady transformation of its ancestors; it appears all at once and "fully formed".[40]

I shall examine their punctuated equilibrium hypothesis in a later chapter. Here I want to see what the fossil evidence shows.

NeoDarwinists claim support for their model from the study conducted in 1987 by geologist Peter Sheldon of 3,458 specimens of the extinct marine arthropod group of trilobites (distant relatives of horseshoe crabs and insects) taken from seven strata of sedimentary rock in central Wales that represent about 3 million years. One characteristic differentiating species is the number of pygidial "ribs" (fused segments of the tail). Sheldon found that in eight genera of trilobites the average number of pygidial ribs increased through time. He concludes that, because of the gradual change, which sometimes was temporarily reversed, it was impossible to assign most specimens to a particular Linnaean species. Since taxonomists include a range of morphological forms within a single species, previous assignments to Linnaean taxa could easily have been misinterpreted as evidence of punctuation and stasis.[41]

Eldredge, however, has a different interpretation of Sheldon's data. First, he maintains that lack of ribs does not mean that a trilobite's tail divisions are lost; rather they are just not expressed on the tail's outer surface, rather like extant junior horseshoe crabs. Second, in two lineages of trilobites the number of pygidial ribs oscillates rather than increases gradually, while in three other lineages the number of ribs remains constant for several strata, increases significantly, and then stabilizes, which is consistent with the immigration of related but different stocks in that particular stratum. Third, such "minor anatomical tinkering with rib number—much of it leading in no particular cumulative direction—can hardly account for the much more substantial anatomical differences between closely related stocks".[42]

This view is supported by palaeobiologist Alan Cheetham of the Smithsonian's National Museum of Natural History. In 1986 he undertook a study of a genus, *Metrarabdotus*, species of sessile aquatic invertebrates whose fossils have been found in strata of rocks deposited in a period spanning approximately 11 to 4 million years ago. Some species exist today and provide morphological comparisons. Cheetham measured up to 46 morphological characteristics per specimen in a total of about 1,000 specimens from about 100 populations. He concludes that most of the species did not change in form over long periods of several million years, and most of the new species appeared suddenly without intermediate transitional populations; if there were intermediate forms they lasted on average less than 160,000 years. In at least seven of the cases, ancestor species persisted after giving rise to descendants. All this, he argues, supports the punctuated equilibrium mode of evolution.[43] He later cautioned against using only a change of one characteristic (as Sheldon had done with the number of pygidial ribs).[44]

Eldredge agrees that species exhibit variation, but comments that the fossil record rarely shows progressive transformation in any one direction lasting very long. Bruce Lieberman's studies on the evolutionary histories of two species of archaic shellfish demonstrate that both species changed a little, but after 6 million years both ended up looking very much like they did when they first appeared in the fossil record. This, according to Eldredge, is typical. "What we see…is oscillation. Variable traits usually seem to dance around an average value."[45]

He maintains, moreover, that the small, progressive within-species change observed in the fossil record is simply too slow to account for major evolutionary adaptive changes. The earliest bats and whales took roughly 55 million years to reach their current morphology. But if one extrapolates this rate of change back, bats and whales would have to have diverged from primitive terrestrial mammals long before any placental mammals had evolved.

As for current views, in the *Encyclopædia Britannica Online* 2014 former President of the American Association for the Advancement of Science, Francisco José Ayala, presents the argument of NeoDarwinian orthodoxy.

> The fossil record indicates that morphological evolution is by and large a gradual process. Major evolutionary changes are usually due to a building-up over the ages of relatively small changes….The apparent morphological discontinuities of the fossil record are often attributed by palaeontologists to…the substantial time gaps encompassed in the boundaries between

strata. The *assumption* is that, if the fossil deposits were more continuous, they would show a more gradual transition of form [my italics].[46]

Ayala, who specialized in molecular genetics, rests the case on an assumption. Eldredge, who is a palaeontologist, flatly disagrees.

> Darwin himself...prophesied that future generations of paleontologists would fill in these gaps by diligent search...[but] it has become abundantly clear that the fossil record will not confirm this part of Darwin's predictions. Nor is the problem a miserably poor record. The fossil record simply shows that this prediction is wrong....The observation that species are amazingly conservative and static entities throughout long periods of time has all the qualities of the emperor's new clothes: everyone knew it but preferred to ignore it. Paleontologists, faced with a recalcitrant record obstinately refusing to yield Darwin's predicted pattern, simply looked the other way.[47]

The fossil record indisputably contains many examples of animal fossils that have remained essentially morphologically unchanged for many millions of years. Bacteria have undergone no significant morphological change for more than 3 billion years.

Ayala concedes that

> fossil forms often persist virtually unchanged through several geologic strata, each representing millions of years....Examples are the lineages known as "living fossils"—for instance, the lamp shell *Lingula*, a genus of brachiopod (a phylum of shelled invertebrates) that appears to have remained essentially unchanged since the Ordovician Period, some 450 million years ago; or the tuatara (*Sphenodon punctatus*), a reptile that has shown little morphological evolution for nearly 200 million years, since the early Mesozoic.[48]

The same is true of many of the crocodilians (alligators, crocodiles, caimans, gharials), which have persisted without significant change for some 200 million years, while the ideally preserved remains of 220-million-year-old triopsid crustaceans are indistinguishable from those of the existing horseshoe shrimp, *Triops cancriformis*. These, and more, are living examples of species that are basically unchanged over hundreds of millions of years, despite claimed catastrophic changes in the environment causing mass extinction events. Many more species persist unchanged in the fossil record over tens of millions of years.

Absence of evidence is not evidence of absence. Because of the paucity of the fossil record, the different morphological characteristics used by different taxonomists to classify species, and the possibility that the apparent sudden (in geological time) appearance of fully formed new species was caused by the immigration of a new species, a pattern of stasis would obtain if the underlying mechanism were gradual or punctuated equilibrium. However, there appear to be no undisputed fossil examples of cumulative, gradual change producing distinctly new species. On the other hand there are very many indisputable examples of evolutionary stasis.

We can reasonably conclude that the normal pattern of fossil evidence for animals is one of morphological stasis with minor and often oscillating changes punctuated by the geologically sudden—tens of thousands of years—appearance of new species that then remain substantially unchanged until they disappear from the fossil record or continue to the present day as so-called "living fossils".

Fossil record of animals and plants

Figure 17.7 depicts the currently accepted pattern of the fossil record for animals and plants, which dates from about 650 million years ago.

Evolution of mammals

Many evolutionary biologists maintain that species explosions follow mass species extinctions because they create the opportunity for the survivors to exploit habitats previously dominated by the now extinct species. They evolve new characteristics adapted to these habitats and consequently evolve into a wide variety of new species.

Whether or not the extinction of non-avian dinosaurs occurred suddenly at the K-T boundary enabling an explosion of mammalian species is debateable. However, living mammals certainly show a great morphological diversity, from the bumblebee bat at 30 to 40mm in length and 1.5 to 2 grams in weight through humans to the blue whale weighing over 100 tons. They are characterized as one taxonomic class chiefly by females nourishing their newly born with milk from mammary glands. They are also differentiated from other vertebrates by warm bloodedness: they maintain a relatively constant body temperature independent of environmental temperature, primarily through internal metabolic processes, and this enables them to survive in a wider range of environments.

Neither of these characteristics is normally fossilized, and so palaeontologists have to use other identifiers, which they take from living mammals. These are principally a chain of 3 tiny bones that transmits sound waves across the middle ear for hearing plus a lower jaw that hinges directly to the skull instead of through a separate bone as in all other vertebrates. Because these are so rarely found as fossils, palaeontologists also use many other characteristics, some of which are shared with mammal-like reptiles, making a distinction between the two problematic. Consequently the classification and dating of early mammalian fossils are contentious.

All we can reasonably conclude from currently available evidence is that the first mammalian species probably emerged from therapsid reptiles around 250

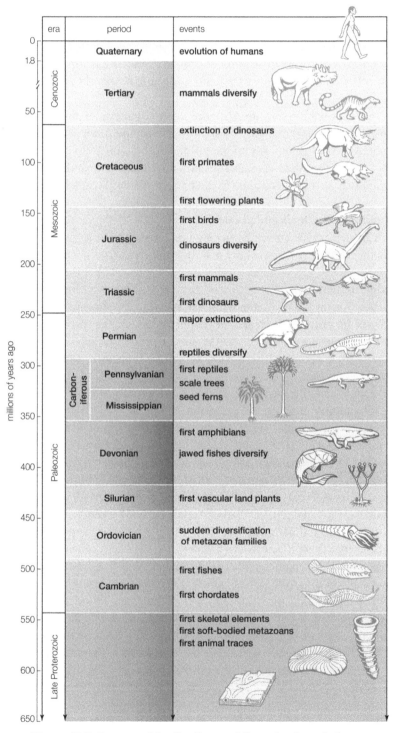

Figure 17.7 Pattern of the fossil record for animals and plants

to 200 million years ago,[49] the vast majority are extinct, one lineage of primitive mammals with some reptilian characteristics—monotremes, like the duck-billed platypus—survives in small numbers to the present day, the common ancestor of marsupial mammals—like the kangaroo—and placental mammals probably emerged around 165 million years ago,[50] while the direct ancestor of placental mammals—the vast majority of living mammals that includes humans—probably emerged around 65 million years ago.[51]

Tracing human evolution from the fossil record
The focus of this quest is to see what we know empirically about the evolution of humans. To give a timetable for the major emergences in biological evolution that resulted in humans is difficult for three reasons. First, the problems of evaluating the fossil record increase the further back in time we go. Second, an emergence is the result of an evolutionary process; it is not possible to determine precisely when each process begins and ends. But this doesn't mean that the result isn't something radically different from its initial state. An analogy is that it is difficult to identify precisely when a rose bud becomes a flower, but that doesn't mean that a flower is not something different from a bud. Third, there is no way of knowing from the fossil record to what extent hybridization resulted in transfers of parts of genomes that gave rise to relatively sudden new morphological characteristics rather than being acquired through the slow, gradual accumulation of genetic mutations in a population gene pool assumed by the NeoDarwinian model.

Table 17.3 (see next page) is an attempt to give the best current estimates of the first appearance of a few of the many significant taxa in, or related to, the branching lineage that ends in modern humans. Consequently I've labelled it indicative: an endeavour to see if there is a pattern in the available evidence, with no claim for comprehensiveness or certainty.

Conclusions

1. The fossil record is extremely small compared with estimates of the number of species that have existed; it is fragmented and unrepresentative of all species, and exceedingly difficult to interpret with a high degree of confidence. Against this background the following pattern in the evidence may be discerned.

2. Fossilized prokaryotes show little morphological change over some three and a half billion years; from about 2 billion years ago they are usually found in colonies.

3. The broad sweep of the fossil record over time is from simple to more complex: prokaryotes appear before eukaryotes, unicellular eukaryotes before multicellular, radially symmetrical before bilateral and cephalized, invertebrates before vertebrates, fishes before amphibians, reptiles before birds, mammals before primates, and apes before humans.

4. This is not a linear progression: grouping animal fossils by fewer common morphological characteristics over time suggests a branching tree of very many lineages that, for the vast majority, end in extinction.

5. A few series of transitional species support such evolutionary lineages.

6. Although the boundaries between progressively more complex organisms are indistinct (as we have found with all emergences), these emergences, or evolutionary transitions, are irreversible: there is no convincing evidence in the fossil record of complex organisms changing into simpler ones.

7. The rate of biological complexification generally increases over time, although the rate is different for each lineage.

8. The normal pattern of fossil evidence for animals is one of morphological stasis with minor and often oscillating changes punctuated by the geologically sudden—tens of thousands of years—appearance of new species that then remain substantially unchanged for some tens or even hundreds of millions of years until they disappear from the fossil record or, less commonly, continue to the present day.

9. While not providing indisputable proof, the fossil record provides strong evidence for the phenomenon of biological evolution of which humans are a product.

Table 17.3 Indicative timetable of the first appearance in the fossil record of significant taxa, with human lineage in bold

Taxon	First appearance (million years ago)
Prokaryotes[a]	**3,500**
Bacterial colonies. Possible first single-celled eukaryotes[b]	**2,100**
Probable eukaryotes[c]	**1,400**
Probable invertebrate animals[d]	**650**
Vertebrates (bilateral and cephalized)[e]	**525**
Bony fishes[f]	420
Four-limbed vertebrates (tetrapods)[g]	**400**
Amphibians[h]	360
Mammal-like reptiles (synapsids)[h]	**310**
Dinosaurs[j]	245
Mammals[k]	**220**
Eutherian mammals[i]	**160**
Birds[f]	160
Placental mammals[i]	**65**
Primates[l]	**55**
Great apes (hominids)[m]	**19**
Protohumans (hominins)[n]	**7**
Humans (genus *Homo*)[o]	**2**
Modern humans (*Homo sapiens*)[p]	**0.15**

Sources: [a]See p. 210, [b]See p. 287, [c]See p. 287, [d]See p. 288, [e]Shu, D G, et al. (2003) "Head and Backbone of the Early Cambrian Vertebrate Haikouichthys" *Nature* 421: 6922, 526–529, [f]Morowitz, Harold J (2004) 109, [g]"Tetrapod" *Encyclopædia Britannica Online* 2014, [h]"Amphibian" *Encyclopædia Britannica Online* 2014, [i]Shedlock, Andrew M and S V Edwards (2009) "Amniotes (Amniota)" 375–379 in *The Timetree of Life* edited by S B Hedges and S Kumar: Oxford University Press, [j]"Dinosaur" *Encyclopædia Britannica Online* 2014, [k]See p. 299, [l]"Primate" *Encyclopædia Britannica Online* 2014, [m]Steiper, Michael E and Nathan M Young (2009) "Primates (Primates)" 482–486 in *The Timetree of Life*, [n]Brunet, Michel, et al. (2005) "New Material of the Earliest Hominid from the Upper Miocene of Chad" *Nature* 434: 7034, 752–755 (which uses the 1825 classification of Hominid), [o]"Homo habilis" *Encyclopædia Britannica Online* 2014, [p]"Homo sapiens" *Encyclopædia Britannica Online* 2014

CHAPTER EIGHTEEN

Evidence of Biological Evolution 2: Analyses of Living Species

> Evolution is an inference from thousands of independent sources, the only conceptual structure that can make unified sense of all this disparate information.
>
> —STEPHEN JAY GOULD, 1998

Very few studies of living species in the last forty years or so have been designed to investigate whether or not biological evolution takes place. Biologists have assumed that it does and for the most part assumed that it is explained by the current NeoDarwinian model. Hence they have focused their investigations on the detailed mechanisms and on advancing hypotheses to explain phenomena not in accord with that model.

Consequently, evidence for the *phenomenon* of evolution in living species is generally only found as a by-product of these investigations. In this chapter I shall consider what may be termed broadly as analyses, which I shall group into eight categories: (a) homologous structures, (b) vestigiality, (c) biogeography, (d) embryology and development, (d) changes in species, (e) biochemistry, (f) genetics, and (g) genomics. In the following chapter I shall examine the behaviour of living species.

Homologous structures

Current evidence reinforces Darwin's findings of similarly structured body parts used for very different purposes by different species. Figure 18.1 illustrates the structural similarity in the forelimbs that humans use for manipulating, cats for walking, lizards for running, climbing and swimming, frogs for swimming, whales as flippers for swimming, and bats for flying. Even birds have a similar structure for their wings, albeit with three rather than five digits.

No engineer would design such similar structures for such different functions. Other homologous structures are evident in many animal body parts. Their degree of similarity reflects the proximity of the species' first appearance in the

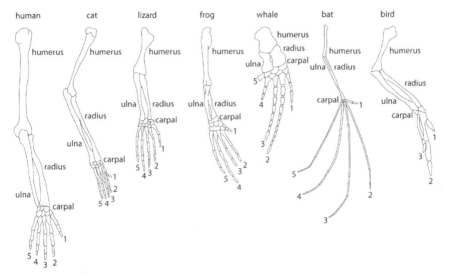

Figure 18.1 Homologous forelimbs in vertebrates

fossil record. Thus they are most similar between humans and chimpanzees, less so between humans and other mammals, still less between humans and birds, and less still between humans and fishes.

The most reasonable explanation is that they are evolutionary adaptations to different environments of common ancestors' body parts.

Vestigiality

Here also current evidence supplies more examples of what Darwin listed as rudimentary, atrophied, and aborted structures. The wings of birds like ostriches, emus, rheas, and penguins cannot be used for flying; their design suggests they are degenerate wings of ancestors that did fly. Ostriches use their wings for balance and for displays in courtship rituals, while penguins use their wings for swimming underwater, consistent with an evolutionary adaptation to a new environment from the wings of a flying ancestor. The most parsimonious explanation of the small internal hind legs of whales is that they are remnants of the hind legs of the land-based mammal whose transition to marine life was examined in the last chapter. So too the human coccyx serves no function, but is best explained as the remnant of an ancestral tail that degenerated and diminished through lack of use as the human lineage adapted to bipedalism.

Vestigiality is not limited to anatomical structures. Whiptail lizards from the all-female species *Cnemidophorus uniparens* exhibit complex mating behaviours even though they reproduce without fertilization by males. Comparisons of human and chimpanzee genomes show dozens of homologous genes in humans that no longer function.

Biogeography

What is now termed biogeography adds more evidence to the geographical distribution of species used by Darwin and Wallace in support of biological evolution and also employs current geological findings on plate tectonics to explain some apparent anomalies.

The general pattern is that continental landmasses have their own distinctive range of animal and plant species, while distant small islands lack the majority of most continental animal species but have large numbers of closely related native species irrespective of environmental conditions.

Thus in Africa we find hippopotamuses, zebras, giraffes, lions, hyenas, lemurs, monkeys with narrow noses and nonprehensile tails, gorillas, and chimpanzees. South America has none of these animals; instead it has tapirs, armadillos, llamas, pumas, jaguars, opossums, and monkeys with broad noses and large prehensile tails. The islands of Hawaii, 3,200 kilometres from the nearest continent, have no native species of reptiles, amphibians, or conifers and only two native species of mammal, a bat and a seal. They are home, however, to about a thousand species of fruit fly and formerly 750 species of snails, although a majority of the latter have since become extinct.

South America separated from Africa 140 million years ago, well before the diversification of early mammal species in the fossil record. The evidence is consistent with early common ancestors of mammals evolving differently in different continental landmasses separated by oceans that prevent migration and interbreeding.

The Hawaiian chain of islands was formed by a series of volcanic eruptions beginning around 66 million years ago and ending less than half a million years ago. The only species that could migrate there are insects, birds and mammals able to fly long distances, together with their parasites and seeds lodged in their plumage or feet, or insects borne on the wind, or seeds and swimming mammals swept there by ocean currents. This explains the absence of large terrestrial mammals that neither swim nor fly, and supports the hypothesis of adaptive radiation: species that did reach the islands found unoccupied ecological niches, with few competitors and predators to prevent them from multiplying and adapting to these new niches, resulting in large numbers of similar species.

The Indonesian islands appear to be an exception, with markedly different fauna on the western, central, and eastern islands. However, they are thought to result from the relatively recent convergence of three lithospheric plates: the northward-moving Australian plate, the westward-moving Pacific plate, and the south-southeastward-moving Eurasia plate. Once joined to the Asian mainland, the western islands have Asian animals, like rhinoceroses, elephants, leopards, and orangutans, while the eastern islands have animals and plants similar to those in Australia from which they separated. The central islands have long been independent and have their own distinctive fauna and flora.

A correlation of the geographical distribution of species with what is known of the geological formation and the environment of the area, plus the fossil record, provides strong evidence of different species evolving from common ancestors.

Embryology and development

After noting similarities in the embryos of very different species, in their development, and in the dependent young that lack the differences of morphology and colour shown in adulthood, Darwin considered this "one of the most important subjects in the whole round of natural history".[1]

However, little progress was made in this field until technological developments at the end of the twentieth century enabled rapid genetic sequencing. This spawned the sub-discipline of evolutionary developmental biology—known as evo-devo to its practitioners—that investigates the genetic mechanisms of embryonic development in different species. It has produced significant and unexpected findings: the same genes control the development of the same organs in species of very different shape, structure, and size. For instance, a gene called Pax-6 in the fruit fly *Drosophila melanogaster* controls and coordinates some 2,000 other genes in making its eyes, which have multiple lenses. An almost identical Pax-6 gene in a mouse has the same role in regulating very different genes that produce its very different, single-lensed eyes. More striking still, if a single mouse Pax-6 gene is inserted into the genome of a fruit fly it will lead to the development of a fruit fly eye, while inserting a fruit fly Pax-6 gene into the genome of a frog produces a frog eye.[2]

Paradoxically, while contradicting the predictions of the NeoDarwinian model (Mayr argued that the eye must have evolved as many as 40 times independently in the animal kingdom), the existence of essentially the same master genes controlling the development of very different kinds of the same organ across a wide range of species supports the hypothesis that these genes are conserved in evolutionary history from a distant common ancestor.

Changes in species

Artificial selection

Bacteria

Biological change can be demonstrated over a 24-hour period for a population of bacteria living in a Petri dish containing a nutrient medium. If a lethal antibiotic is added this causes a mass die-off. However, one or more varieties of the bacterium will be immune to the particular antibiotic and so survive. They replicate rapidly to replace the original population. Whether this constitutes a new species or a variety of the original species is debateable. The current orthodox definition doesn't help because this depends on sexual reproduction and bacteria normally replicate asexually by binary fission, while genetic analyses of bacteria are open to different interpretations.

In 1988 Richard Lenski, then at the University of California, Irvine, began an experiment on 12 populations of a strain of *Escherichia coli* derived from a single bacterium. He followed their genetic mutations and relative fitnesses—selected by how fast they grew on the limited sugar available in the nutrient medium—over 50,000 generations spanning more than 20 years. Most mutations made no difference or were deleterious, but a few correlated with up to a 10 per cent growth advantage over their predecessors. Generally, mutations increased linearly while fitness increased in jumps.

When the experiment reached generation 33,127 his researchers noticed a large increase in turbidity in one flask. The sugar-starved bacteria had begun to use citrate, a pH buffer in the nutrient media of all the flasks, in order to metabolize, and this resulted in a huge population increase. Whether this new strain constitutes a new species is disputed. It co-existed with a minority that continued to use sugar for their metabolism.[3]

Reversible changes

According to evolutionary biologist Francisco José Ayala of the University of California, Irvine, artificial selection generally produces variations that are reversible.

> Breeders choose chickens that produce larger eggs, cows that yield more milk, and corn with higher protein content. But the selection must be continued or reinstated from time to time, even after the desired goals have been achieved. If it is stopped altogether, natural selection gradually takes effect and turns the traits back toward their original intermediate value.[4]

Hybridization

Horticulturalists have long used hybridization to produce new, viable species of plants with specific characteristics. Hybrids produced by cross breeding between two different species are known as F_1 (first generation) hybrids; hybrids produced by breeding among F_1 hybrids are labelled F_2 hybrids, and so on; backcross hybrids are produced when hybrids are cross bred with members of either parental species; backcross hybrids crossed again with members of the same parental species are called second backcross generation.

On average, mammal hybrids are less able to survive and reproduce. Many artificially selected hybrids, like a male donkey with a female horse to produce a mule that is usually stronger than either parent, are sterile. But, contrary to widespread assumptions, this is by no means a general rule; fertility often differs between sexes. For example, crossing domestic cattle with bison to increase beef yield produces beefaloes. Beefalo males produced by the first backcross are usually sterile, but when partially fertile backcross females are backcrossed again the resulting males are frequently fertile. Crossing a dog with a jackal to create a hybrid with a superior sense of smell produces offspring that apparently are as fertile and stable as the wide variety of domestic dogs produced by breeders. Such hybrids show traits falling outside the range of parental variation, known as heterosis, although most hybrids are not heterotic.

In open zoos hybrids have been reported between different species. A liger, the offspring of a lion and a tigress, is usually larger and stronger than either of its parents (positive heterosis), whereas a tigon from a tiger and a lioness tends to be smaller than either parent (negative heterosis). No studies seem to have been made on the fertility and stability over generations of these hybrids. However, the offspring of polar bears and grizzly bears interbred in captivity are fertile.

Polyploidy
In 1912 botanists in Kew Gardens demonstrated the phenomenon of an offspring possessing more than two sets of chromosomes in a cell when they crossed one species of primrose, *Primula floribunda*, with another species, *Primula verticillata*. The resulting hybrid was sterile, but was propagated via cuttings. Later, in three separate years, this otherwise sterile clone produced shoots that gave rise to a fertile plant named *Primula kewensis* that could not breed successfully with members of either parent species. The reason was that its cells had double the number of chromosomes. This chromosomal (as distinct from genetic) mutation can now be induced in plants by a mutagen like colchicine.

An increase in the number of chromosomes compared with those of parents from a single species is known as autopolyploidy, while a chromosome increase that results from hybridization between members of two different species is called allopolyploidy.

Species in the wild
Bacteria
Previous chapters showed that the oldest clearly recognizable bacteria date from around 3.5 billion years ago. Despite this vast time over which random genetic mutations have accumulated, present-day bacteria remain single-celled and morphologically are virtually identical to those ancient fossils.

A growing body of research has overturned the long-held view that prokaryotes—bacteria and archaea—and single-celled eukaryotes, like amoeba, live relatively independent lives and evolve gradually by NeoDarwinian natural selection acting on random genetic mutations passed down through generations by replication.

Many prokaryotic species can incorporate, or lose, DNA, while unicellular eukaryotes can incorporate prokaryotic DNA, by horizontal gene transfer in three principal ways:

1. *Natural transformation*
   Direct uptake of free DNA released into the environment from decomposing cells, disrupted cells or viral particles, or excretion from living cells.

2. *Transduction*

Transfer of DNA from one bacterium to another by an agent like a virus.

3. *Conjugation*

Transfer of mobile genetic elements, like plasmids, by direct contact through pili.[5]

Such same-generation transfers can take place between very different species, like *Escherichia coli*, bacteria that normally inhabit the intestinal tracts of animals, and *Synechocystis* sp. PCC6803, a freshwater cyanobacterium. They can be harmful, neutral, or advantageous to survival and replication. When advantageous they can produce new capabilities or functions for the recipient, like immunity to a toxin. A 2008 analysis of 181 sequenced prokaryotic genomes estimates that cumulative horizontal gene transfers accounted for 81 ± 15 per cent of the genes, which highlights the significance of this mechanism in the evolution of bacteria, archaea, and unicellular eukaryotes.[6]

Hybridization

Hybridization is now recognized not only to play a significant role in natural plant speciation, but also to occur among fishes, birds, and mammals.* In recent years barred owls from midwestern USA have moved westward to the Pacific coast where they have settled in the forest habitat of spotted owls with which they have cross bred to produce fertile sparred owls. Grizzly bears and polar bears not only successfully cross breed in captivity but also in the wild. In recent years grizzlies, normally found in mainland northwestern USA, Alaska, and Canada, have been moving north, probably due to climate warming, into the habitat of polar bears. In 2006 a hunter shot what he thought was a polar bear, but while it had thick creamy white fur, its humped back, shallow face, and brown patches plus subsequent genetic tests showed it was a hybrid with a grizzly father and a polar bear mother.

Polyploidy

Cultured polyploidization has produced fertile new species, particularly in plants, but this means of speciation is probably more extensive in the wild than previously thought. In a 2005 survey of the literature, molecular systematist Pamela Soltis concludes that all flowering plants are probably polyploid or descended from natural polyploids,[7] while in another 2005 review evolutionary geneticists T Ryan Gregory and Barbara Mable note that

> polyploidy is not as common in animals as in plants but neither is it anywhere as near as rare as often assumed. In part the relatively low rate of discovery of polyploidy in animals reflects the low level of effort put

* See page 282.

into finding it….One by one the traditional assumptions about animal polyploidy have faltered in the face of new evidence.[8]

That evidence includes polyploidy in fishes, amphibians, reptiles, and mammals. Milton Gallardo and colleagues of the Universidad Austral de Chile discovered in 1999 that the Plains Viscacha Rat, *Tympanoctomys barrerae*, is tetraploid (having four times the haploid number of chromosomes in the cell nucleus) and that evidence strongly suggests this was caused by ancestral polyploidal hybridization between two different species. They later reported that the Golden Viscacha Rat, *Pipanacoctomys aureus*, is also a tetraploid resulting from ancestral polyploidization.[9]

Peppered moth
Most teachers and textbooks present the peppered moth as the prime example of biological evolution occurring in the wild within a human lifetime.

Before the middle of the nineteenth century in England all *Biston betularia* were white moths peppered with black spots, a form called *typica*. In 1848 a black variety, *carbonaria*, was recorded in Manchester, the heart of Britain's industrial revolution, and by 1895 98 per cent of *B. betularia* in the area were black. The *carbonaria* occurred in many other parts of Britain, reaching high frequencies in industrial centres. In 1896 the lepidopterist J W Tutt hypothesized that the increase in *carbonaria* resulted from differential bird predation in polluted regions.

Nobody, however, had tested this hypothesis, and so Bernard Kettlewell, a medical doctor who was appointed to a research fellowship in Oxford University's Department of Genetics in 1952, began a series of studies to find supporting evidence. According to Kettlewell this change in moth colour was caused by birds eating the moths most conspicuous on their normal resting site, tree trunks. The industrialization of northern England had produced soot and acid rain that darkened trees by first killing the lichens that festooned them and then blackening the naked trunks. The *typica*, previously camouflaged on lichens, thus became conspicuous and heavily predated by birds, while black mutants were camouflaged; over generations they survived longer and reproduced more, thus replacing the *typica*. This phenomenon was termed industrial melanism (darkening). After the passage of Clean Air Acts in the 1950s trees regained their former appearance, and the *typica* regained their predominance in northern England.

Kettlewell supported his conclusions by studies showing a correlation between pollution levels and frequencies of the black variety. Most strikingly, his experiments showed that, after releasing *typica* and *carbonaria* in both polluted and unpolluted woods, researchers recaptured many more of the cryptic, or camouflaged, variety than the conspicuous variety; this differential predation was supported by direct observation of birds eating moths placed on trees. Finally, Kettlewell demonstrated in the laboratory that each form had a behavioural preference to settle on backgrounds that matched its colour. Thus Tutt's 1896 hypothesis was proven, the NeoDarwinian model validated, and textbooks reproduced Kettlewell's photographs of moths on tree trunks.

Cambridge University geneticist Michael Majerus, who worked on ladybirds and moths, was commissioned by Oxford University Press to write a book, *Melanism: Evolution in Action*, to be published 25 years after Kettlewell's *The Evolution of Melanism*. Reviewing this book for *Nature*,[10] Jerry Coyne discovered serious flaws in Kettlewell's work. These include the fact that "*B. betularia* probably does not rest on tree trunks—exactly two moths have been seen in such a position in more than 40 years of intense search....This alone invalidates Kettlewell's release-recapture experiments, as moths were released by placing them directly on tree trunks where they are highly visible to bird predators." The photographs were of dead moths, glued or pinned to tree trunks, and used to illustrate camouflage and conspicuity. Moreover, Kettlewell released his moths during the day, while they normally choose resting places at night; the resurgence of *typica* occurred well before lichens recolonized the polluted trees; a parallel increase and decrease of the melanic form also occurred in industrial areas of the United States where there was no change in the abundance of lichens; and the results of Kettlewell's behavioural experiments were not replicated in later studies: moths have no tendency to choose matching backgrounds.

Coyne refers to numerous other flaws in the work found by Majerus plus additional problems he discovered when he read Kettlewell's papers for the first time. "My own reaction resembles the dismay attending my discovery, at the age of six, that it was my father and not Santa who brought the presents on Christmas Eve."

He concludes that "for the time being we must discard *Biston* as a well-understood example of natural selection in action". He also comments that "It is also worth pondering why there has been general and unquestioned acceptance of Kettlewell's work. Perhaps such powerful stories discourage close scrutiny."

Majerus disagreed with Coyne's conclusions and defended his own views on industrial melanism by referring to other studies, adding

> That said, my own conviction that bird predation is largely responsible is not based purely on empirical data from experiments published in the literature. I "know" that Tutt's differential bird predation hypothesis is correct because I "know" about peppered moths. For....scientists, well-trained in rigour, stringency and experimental controls, for differing reasons this statement must seem insufficient if not heretical. However, I stick by it.[11]

There is little to distinguish this "knowing" from that of creationists "knowing" that God created each species. Perhaps it is one answer to Coyne's question about the uncritical acceptance of Kettlewell's work, but it is the antithesis of science.

In the same year as Majerus's book, University of Massachusetts evolutionary biologist Theodore Sargent and colleagues published a critique of the classical explanation of industrial melanism, concluding that "there is little persuasive evidence, in the form of rigorous and replicated observations and experiments, to support this explanation [that bird predation causes a genetic mutation to become dominant in the population over generations] at the present time".[12] Sargent suggested other causes, like some form of induction triggered by environmental change, might have produced the phenomenon of industrial melanism in a population as a whole, and that this better explains the speed with which the melanic variety replaced the typical form in several studies.

Some creationists seized upon the revelations about Kettlewell's work and accused evolutionary biologists of complicity in offering fraudulent proof of Darwinism. The revelations don't, of course, disprove the phenomenon of biological evolution. What they do is expose either flawed experimental design or else, following the Law of Data Interpretation, a determination to prove a hypothesis in which the experimenter believes rather than to test the hypothesis. The verifiable data raise questions, as Sargent tried to do, about the cause and the mechanism of industrial melanism. Sargent, however, made no headway because evolutionary biologists circled the wagons around the current model as a defence against creationists.

The controversy obscured an important aspect of the phenomenon of industrial melanism: it occurred as the environment changed and it reversed when the environment changed again. Thus the phenotypical change was reversible, and reversible change does not constitute the evolution of a species. What occurred over 150 years—an instant in geological time—was an oscillation in colour, which is less significant than the oscillations around a basic morphology that Eldredge describes as evolutionary stasis in the fossil record.

Ecotypes
Ecotype is a term applied to a variety of a species that adapts to local conditions and exhibits consequent morphological or physiological change, but can nevertheless successfully breed with other varieties. An example is the Scots pine, whose 20 different ecotypes range from Scotland to Siberia and are capable of interbreeding.

The taxonomic status of the tucuxi dolphins (genus *Sotalia*) has been a matter of controversy for more than a century. The genus once comprised five species, but in the twentieth century they were grouped into two, *Sotalia fluviatilis* that lived in rivers and *Sotalia guianensis* that lived in the open sea. Later studies concluded that their differences were due to size only, and from the early 1990s most researchers classified them as a single species, *S. fluviatilis*, that had marine and riverine ecotypes. Because the different ecotypes are geographically separated it is not possible to say whether they can interbreed in the wild. Using phylogenetic species definitions rather than the updated biological species definition, conservationists now argue that each is a different species; they cite molecular analyses which show, for example, that their cytochrome *b* gene differs by 28 out of 1,140 nucleotides.[13] Whether or not a 2.5 per cent difference in the nucleotides of this gene is sufficient to define a species is, of course, a matter of debate, and I will consider in more detail the problems with such molecular analyses in a later chapter. Such ecotypes may be in the process of becoming new species, but that process may be reversible, as in the case of the peppered moth.

"Darwin's finches"

As we saw in Chapter 16, Darwin never used the finches of the Galápagos islands as an example of biological evolution,* but the husband and wife team of Peter and Rosemary Grant, evolutionary biologists from Harvard University, spent more than 25 summers studying these birds, mainly on the island of Daphne Major.

In his Pulitzer-winning 1994 book, *The Beak of the Finch: A Story of Evolution in Our Time*, science writer Jonathan Weiner repeats the discredited myth of Darwin's finches and portrays the Grants' work as "the best and most detailed demonstration to date of the power of Darwin's process". However, Darwin "did not know the strength of his own theory. He vastly underestimated the power of natural selection. Its action is neither slow nor rare. It leads to evolution daily and hourly"[14] rather than gradually over the long timespans predicted by Darwin and the NeoDarwinian model.

In fact the Grants' meticulous measurements of the beaks of individual finches show that the numbers of big-billed and small-billed finches oscillated over 25 years according to whether drought left only large, tough seeds or heavy rainfalls in other years resulted in more small, soft seeds than large, tough ones.

This is not biological evolution. No significant change occurs over the 25 years. The Grants simply found periodic redistributions of gene frequencies—genetic variations already present in the gene pool—in response to changing environmental conditions. It is another example of reversible adaptive change within a species population.

Members of each of six species of finch on the islands of Daphne Major and Genovesa mostly did not mate with members of the other species because of one of the NeoDarwinian pre-mating barriers, behavioural (in this case birdsong). However, when they did mate, their offspring were as fertile as their parental populations, as were the first two generations of the hybrids' offsprings. Other field studies in North America have shown that hybridization of bird species occupying the same territory occurs and that hybrids are viable and fertile to a high degree. This raises the question: are these finches, classified mainly by the size and shape of their beaks, different species or varieties of the same species? The question is particularly pertinent when the Grants' research has demonstrated that changes in beak size and shape are reversible under differing environmental conditions.

---

*   See page 258.

The Grants note that the biological species concept of current orthodoxy was derived mainly from studies of the fruit fly *Drosophila*. While almost 10,000 bird species are recognized according to this definition,

> [i]nterpretations of speciation have been applied to perhaps 500 of them. The genetic basis of variation in premating isolating traits believed to be involved in speciation is known (incompletely) for less than 100 species, and the genetic basis of postmating isolation is virtually unknown for all of them. The knowledge base from which to generalize about the genetics of bird speciation is precariously thin.[15]

Species definition

Any discussion of biological evolution is clouded by the different definitions of species used not only by bacteriologist, botanists, and zoologists but also by specialists within each of these branches of biology, quite apart from very different views on what are species-defining characteristics. If species, rather than the subspecies variety or superspecies genus, class, or order, is to be the basic taxon that marks a clearly defined stage of biological evolution, and which can be used to describe a population and its relationship to other populations, then it seems to me that irreversibility is the key characteristic: nobody would argue that modern humans could evolve back into *Australopithecus afarensis* (or whatever was a common ancestor of humans) with one-third our current cranial capacity.

I suggest the following generic definitions may reduce the confusion.

subspecies, variety, or race   A population of organisms whose defining adult heritable characteristics have undergone reversible change from those of its ancestral population or populations.

species   A population of organisms whose defining adult heritable characteristics have undergone irreversible change from those of the population, or populations, from which it evolved.

speciation   The process by which a population of organisms irreversibly changes its defining adult heritable characteristics from those of the population, or populations, from which it evolved.

The word "process" in the definition of speciation recognizes intermediate stages in which successful breeding can take place between subspecies that may be on the path to becoming a new species.

These generic definitions leave it open for specialists to list the adult heritable characteristics that define a particular species; there may be good reasons for different criteria to apply to different kinds of species. They do not specify whether the cause of the irreversible transformation is artificial selection, or randomly generated genetic mutations naturally selected, or genetic drift, or polyploidy, or hybridization, or any other possible cause that I shall examine in a later chapter. I hope, however, that they achieve the goals of being sufficiently comprehensive to encompass all types of organism, including bacteria and plants, and of distinguishing between species and variations within a species.

Biochemistry

All bacteria, plants, and animals consist of the same or similar chemicals, structured in the same or similar ways, undertaking the same or similar reactions.

Molecules of DNA structured as a double helix are used in the cells of nearly all known lifeforms (the exceptions being molecules of RNA) for their maintenance and reproduction. They comprise a different sequence of the same four nucleotides when many other nucleotides and structures are chemically possible.

The same triplets of these nucleotides act as a pattern for the production of the same amino acids in all organisms. All the various proteins used to build and maintain all organisms are synthesized from different combinations and sequences of the same twenty amino acids, almost invariably as their L-isomer,* although several hundred other amino acids exist.

The series of chemical reactions, known as metabolic pathways, by which the most diverse lifeforms sustain themselves are very similar.†

The most reasonable explanation for the same biochemistry operating in all lifeforms is that it derives from the first form of life on Earth from which all existing lifeforms evolved.

Genetics

In Chapter 14 we saw that all existing lifeforms share some hundred genes, but analyses that allow for lineage-specific gene losses suggest that a last universal common ancestor (LUCA) probably had ten times as many genes as that.‡ The possession of even a hundred genes in common strongly suggests evolution from the first form of life on Earth.

This is reinforced by the more recent discoveries that master control genes, like Pax-6 and the Hox family, regulate the development of very different body plans in a wide range of species and that these genes are interchangeable between species.

---

*   See page 219.
†   See pages 216 to 222 for a more detailed description of the chemicals, structures, and reactions.
‡   See page 214.

In discussing genes, however, it is important to recognize that a gene is not simply a linear sequence of DNA that codes for the production of a particular protein or RNA molecule, as was thought to be the case when the NeoDarwinian model was developed. In 1977 molecular biologists Richard J Roberts and Phillip A Sharp independently discovered that eukaryotic genes are functionally separated into coding segments called exons, which are interrupted by noncoding sequences of DNA called introns. This enables the creation of multiple proteins from one gene by the use or exclusion of different exons.

Although her published conclusions were derided at the time, in 1951 cytogeneticist Barbara McClintock working at the Cold Spring Harbor Laboratory identified "jumping genes", now called transposons. These are segments of DNA able to move, or transpose, themselves either by cutting and inserting, or by replicating and inserting copies of themselves, into different positions on the chromosome and even into other chromosomes. A transposon can disrupt the function of the DNA sequence into which it splices and cause a mutation.

In 1965 McClintock suggested that these mobile elements of the genome might play a regulatory role, determining which genes are turned on and when. In 1969 molecular biologist Roy Britten and cell biologist Eric Davidson speculated that transposons not only play a role in regulating gene expression but also in generating different cell types and different biological structures, based on where in the genome they insert themselves. They hypothesized that this might partially explain why a multicellular organism has many different kinds of cell, tissue, and organ, even though all its cells share the same genome.

Biological orthodoxy dismissed these proposals, but studies from the first decade of the twenty-first century have shown that many transposons are highly conserved among distantly related taxonomic groups and are found in almost all organisms—both prokaryotes and eukaryotes—and typically in large numbers. For example, they make up approximately 50 per cent of the human genome and up to 90 per cent of the maize genome. Moreover, transposons can influence gene transcription.[16]

In 2012 the ENCODE (Encyclopedia of DNA Elements) project, a nine-year concerted study of the human genome by more than 440 researchers in 32 laboratories worldwide, reported that genes could be spread among a genome, with far-flung protein-coding and regulatory regions overlapping with other genes, and that regulatory regions need not be close to the coding sequence on the linear molecule or even be on the same chromosome. Furthermore, while protein-coding DNA constitutes barely 2 per cent of the genome, some 80 per cent of the bases studied—"junk DNA"—showed signs of functional activity. Much of this appears to involve complex collaborative networks that regulate gene expression.[17]

Such research has vindicated McClintock, Britten and Davidson, prompting a re-evaluation of just what a gene is and what is the function of the so-called "junk DNA"—98 per cent of the human genome—which notion persisted for some 50 years.

Genomics

These paradigm-questioning studies were made possible by technological developments in the twenty-first century that enabled the sequencing not merely of individual genes but of an organism's whole genome,* its entire genetic content.

By 2009 the genomes of some 2,000 organisms and many more viruses had been sequenced, allowing much more accurate analyses and comparisons between different species than do individual genes. They showed, for example, that while approximately 98 per cent of a prokaryote's genome code for structural proteins, only about 2 per cent do so in eukaryotes.

A human genome was sequenced in 2003 and this revealed, among many other things, that we do not possess 100,000 genes as previously estimated, but probably only 30,000, since revised down to around 25,000, with multiple copies of some genes. The Human Genome Project commented

> Though the project is complete, many questions still remain unanswered, including the function of most of the estimated 30,000 human genes. Researchers also don't know the role of single nucleotide polymorphisms (SNPs) [single DNA base changes within the genome] or the role of noncoding regions and repeats in the genome.[18]

Table 18.1 compares the genomes of several species.

Not only does a human have roughly the same number of genes as a mouse, but other data show that this is less than half the number of genes possessed by the *japonica* and *indica* strains of rice. These findings provoked a general response typified by science writer Matt Ridley: "Dethronement on this scale has not happened since Copernicus took us out of the centre of the solar system."[19]

Such a response was reinforced by commonly quoted statistics indicating that humans share 98.5 per cent of their genes with chimpanzees, 90 per cent with mice, 85 per cent with 4–6cm-long zebrafish *Danio rerio*, 36 per cent with fruit flies *Drosophila melanogaster*, and about 21 per cent with 1mm-long roundworms *Caenorhabditis elegans*.

This response, however, is based on three fallacies. First, the figure of humans sharing 98.5 per cent of their genes with chimpanzees is taken from an estimated difference of 1.5 per cent between nucleotide sequences of genes considered to have a similar function. Studies that take into account gene losses and gene insertions and duplications imply that humans and chimpanzees differ by at least 6 per cent in their complement of genes.[20]

Second, it fails to compare like with like. For example, plants rely on gene duplication for protein diversity, whereas protein diversity in humans is achieved by a process of alternative splicing: a single gene does several things, and genes are

---

*   See Glossary for more detailed definition.

Table 18.1 Comparison of the genomes of several species

Species	Estimated number of DNA base pairs per genome	Estimated number of genes	Average number of DNA base pairs per gene	Number of chromosomes
Homo sapiens (human)	3.2 billion	~25,000	130,000	46
Mus musculus (mouse)	2.6 billion	~25,000	100,000	40
Drosophila melanogaster (fruit fly)	137 million	13,000	11,000	8
Arabidopsis thaliana (plant)	100 million	25,000	4,000	10
Caenorhabditis elegans (roundworm)	97 million	19,000	5,000	12
Saccharomyces cerevisiae (yeast)	12.1 million	6,000	2,000	32
Escherichia coli (bacterium)	4.6 million	3,200	1,400	1

(Source: Human Genome Project)

constantly broken up and spliced together with a different sequence and function. Consequently, the functioning of human genes produces a far more complex organism than does a rice plant.

Third, it assumes that the degree of relatedness of species is best measured by the number of genes they have in common and ignores around 98 per cent of eukaryotic chromosomal DNA, in particular the regulatory sequences that determine when, to what degree, and how long genes are switched on, and consequently the organism's observable characteristics.

Genome sequencing is now providing evidence that large-scale gene duplication, and even complete genome duplication, events contributed significantly to both gene family expansion and genome evolution.[21]

It was genomics that revealed another inconsistency with biological orthodoxy mentioned previously: horizontal gene transfers occur between prokaryotes, including species that are not closely related.[22] According to Dalhousie University biochemist Ford Doolittle, genomic analyses show that, for prokaryotes at least, horizontal gene transfer played a major role in evolutionary development compared with the vertical transfer of genes from parent cell to daughter cell.[23]

Horizontal gene transfer also takes place with eukaryotes, although much less extensively than with prokaryotes. However, it contributed significantly to the emergence of eukaryotes. Moreover, animal and plant hybridizations are, in effect, massive horizontal gene transfers.

These recent findings in genetics and genomics do not invalidate the phenomenon of biological evolution; to the contrary, they strengthen the evidence for it. They do, however, question the adequacy of the NeoDarwinian model to explain it.

Conclusions

1. Living species form a pattern of increasing complexity from bacteria to humans.
2. Evidence from homologous structures, vestigiality, embryology, biogeography, biochemistry, genetics, and genomics all points to the evolution of living species from a universal common ancestor on Earth.
3. Many species undergo reversible variations in response to environmental changes, but reversible change does not constitute the evolution of species (some varieties are classified by some biologists as new species; these are disputable and blur an understanding of what characterizes a new species).

CHAPTER NINETEEN

Evidence of Biological Evolution 3: Behaviour of Living Species

It has been truly said that all nature is at war; the strongest ultimately prevail, the weakest fail...the severe and often-recurrent struggle for existence will determine that those variations, however slight, which are favourable shall be preserved or selected, and those which are unfavourable shall be destroyed.

—CHARLES DARWIN, 1868

Those species which willingly or unwillingly abandon [sociability] are doomed to decay; while those animals which know how best to combine, have the greatest chances of survival and evolution....The fittest are thus the most sociable animals, and sociability appears as the chief factor of evolution.

—PETER KROPOTKIN, 1914

Zoology spawned the sub-discipline of animal behaviour, called ethology, when naturalists' observations of animals in the wild extended to experiments designed to ascertain the characteristics, causes, mechanisms, development, control, and evolutionary history of their behaviour.

The Austrian Konrad Lorenz and the Dutch-born Niko Tinbergen, who became a British citizen, are generally credited with laying its foundations during the 1920s and 1930s. Lorenz's experiments on ducks and geese showed that a repertoire of behaviours in the newly born are induced by specific stimuli from a parent or parent substitute. Such irreversible behaviour patterns, Lorenz claimed, were as characteristic of a species as its plumage. Lorenz and Tinbergen argued that if a species had a long history of successfully responding to specific stimuli, particularly for surviving and reproducing, then natural selection leads to adaptations enhancing responsiveness to those stimuli. Thus a male stickleback is stimulated to attack another male by its red colour, but to court a female by its swollen, silvery belly. Other types of behaviour, however, could be learned through experience.

Since then ethology itself has branched into a variety of specialized sub-disciplines, like species-specific behaviour, life history theory, evolutionary ecology, behavioural ecology, and sociobiology—which prompted an expansion of research into social evolution—while other trends in ethology interacted with different scientific disciplines to generate new hybrid disciplines like social learning, comparative psychology, cognitive ethology, and neuroethology.

Such specialized, analytical, and theoretical approaches can provide valuable insight into animal behaviour, although there is limited general agreement within these various fields, still less agreement between them. By contrast I want to step back and examine what general evolutionary patterns, if any, arise from species behaviour revealed by these specialized studies.

Unicellular species

Members of many unicellular species exhibit primitive forms of the social behaviours described at the beginning of the twentieth century by Kropotkin for a range of animal species,* such as communicating and working together to construct a communal shelter, produce and rear offspring, forage, defend themselves, attack prey, and migrate to a better environment for survival and reproduction.[1]

Most bacterial species are capable of forming communities such as microbial mats and biofilms, which are often protected by a matrix made from bacterial excretions. Biofilms very rarely contain just one species of bacteria, let alone a single clonal lineage. Dental plaque, for example, can comprise up to 500 species of bacteria. Like the nests of social insects, biofilms can also serve as sites for reproduction.[2]

Myxobacteria, such as *Myxococcus xanthus*, engage in collective attacks on microbial prey, which they overwhelm by force of numbers, break down by bacterial enzymes, and consume.

Communication is exemplified by what is known as quorum sensing. Bacteria release signalling molecules into the immediate environment and also have receptor sites for such molecules. When their receptor sites register a sufficient number—a measure of local population density—this triggers a coordinated response, such as switching on genes that direct the production and excretion of polysaccharides for a biofilm or enzymes that digest prey for the benefit all the population members, or that generate luminescence.[3]

The slime mould *Dictyostelium discoideum*, an amoeba that normally does live a solitary unicellular life in damp soil and feeds on bacteria, illustrates collaborative migration and reproduction. When starved of nutrients, some hundred or so cells aggregate to form a slug that migrates to the surface where it takes the shape of a stalk holding aloft a sorus, a ball of reproductive cells which disperse unicellular spores. The 20 per cent or so of cells that form the stalk do not reproduce, but die in what appears to be an altruistic act.

---

*   See pages 268 to 270.

Bacterial geneticist James Shapiro's comprehensive review of bacterial social behaviour concludes that the collaboration of bacteria, not only within species but also between species, plays a key role in their survival.[4]

Multicellular species

A multicellular organism consists of eukaryotic cells. Each of these cells consists of several membrane-bound parts that perform specific functions, like the nucleus that controls and coordinates the other parts, or organelles, such as the mitochondrion where energy is generated (see Figure 19.1). A eukaryotic cell is, in essence, a collection of organelles descended from different prokaryotes that collaborate to maintain and reproduce the cell.

A group of these cells, called a tissue, collaborate to perform a specific function, like muscle tissue that contracts and expands. Several types of tissue similarly collaborate to form an organ that has a specific purpose, like a heart that pumps blood, and different organs collaborate at higher levels still. Indeed, an organism consists at many levels of parts collaborating to keep the whole alive and reproducing.

Evolutionary biologists David Queller and Joan Strassmann suggest that "the essence of organismality lies in this shared purpose; the parts work together for the integrated whole, with high cooperation and very low conflict".[5] This purposeful activity accords with the definition of life suggested on page 206.

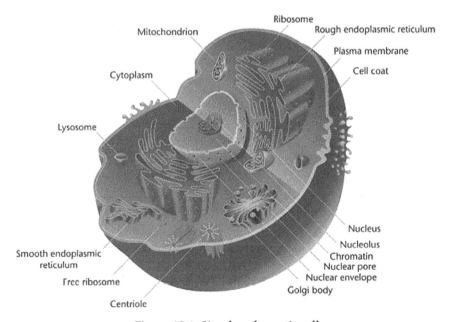

Figure 19.1 Simple eukaryotic cell

Genes

Collaboration extends down to the level of the genes. We saw in the last chapter how the Pax-6 gene controls and coordinates the functioning of up to 2,000 other genes to develop an eye.* This is just one example of a widespread phenomenon of genes working together under the control of other, regulatory genes to develop a specific organ or perform a specific function. One of the most studied regulatory genes is the family of Hox genes, which themselves collaborate to coordinate the development of body parts in nearly every bilateral animal studied so far.[6]

Other recent research includes a 2010 study that concludes intelligence in humans is controlled by a network of thousands of genes, with each making just a small contribution to overall intelligence, rather than a few powerful genes as previously thought.[7]

Collaboration is not the only behaviour found within an organism. When a series of regulatory genes fail, usually through mutation of those genes, a cell can replicate out of control, and such cancer cells compete with and destroy other cells for their own sustenance. However, competition is the exception rather than the rule within an organism.

Plants

Plants aren't normally considered in behaviour studies,† but without collaboration most plants wouldn't be able to propagate because a majority cannot self-pollinate. Although pollen is sometimes borne by the wind to the pistil of a plant, more commonly it is carried by insects, birds, and even mammals (principally bats). This collaboration is usually for mutual benefit: the pollen carrier gains by drinking nectar, or being perfumed, or, in the case of bees, keeping some of the pollen. Some flowers, however, emit the scent of food, like coconut, to attract insects that pollinate them, while species of orchids have flowers that mimic the sight and smell of a female wasp to attract a male wasp.

Insects

Insects are probably more studied than any other animals. A majority of ant, wasp, and bee species (order Hymenoptera) and termites (order Isoptera, although some researchers now classify them as a suborder of Dictyoptera, or cockroaches)

---

* See page 307.
† However, a 2009 review of recent research in plant neurobiology claims that plants are sensory and communicative organisms characterized by active, problem-solving behaviour. It argues that these conclusions are controversial because of a tendency to jealously retain longstanding dogmas in science past their expiry date. (Baluška, František, et al. (2009) "The 'Root-Brain' Hypothesis of Charles and Francis Darwin" *Plant Signaling & Behavior* 4: 12, 1121-1127)

collaborate to such a coordinated degree that they are commonly called the social insects.

While some species of bee are solitary, members of most species live in colonies that, for tropical stingless honeybees, can be as large as 180,000 bees. A colony of bees work together using wax they secrete and mix with plant resin to construct and maintain an elaborate hive consisting of stacks of hexagonal, thin-walled cells, called honeycombs, with separate areas for rearing larvae and for storing honey and pollen, plus a recycling dump, all surrounded by sinuous branching pillars.

The fungus-growing ants of Central America dig nests in the soil that extend hundreds of metres long and up to six metres deep, with some thousand entrances and around a thousand chambers.

Termites typically build protective mounds from a cement-like substance made by worker termites from soil and their saliva. Chambers and tunnels are cooled by an air-conditioning system comprising workers constantly wetting the walls with their saliva together with hollow cavities that allow hot air to rise and escape through tiny holes in the mound's surface. At its heart is the royal cell in which the king—the only fertile male in the colony—and the queen live and breed. Around this are the brood chambers where worker termites take the eggs after hatching. Tunnels lead to food-storage chambers, and above these are fungus gardens where food is cultivated.

Although the particular behaviours of the many species of social insects differ in detail, a general pattern is evident.

a. *Hierarchical colony*
   Social insects live and work together as a hierarchical colony of inter-dependent members in a protective nest, hive, or mound they build for themselves.

b. *Division of labour*
   Only one or more queens (plus a king in the case of termites) usually produce offspring, while the workers typically specialize in tasks such as building and maintaining the nest, foraging, feeding offspring, and policing and defending the colony. A colony of honeybees, for example, may contain 50,000 females, all of whom have developed ovaries and can lay eggs. But 100 per cent of the females and 99.9 per cent of the males are the offspring of just one female, the queen. The other females are the workers, while the males, or drones, usually are stingless, produce no honey, perform no work, and whose only function is to mate with a queen in flight.

c. *Morphological differentiation*
   In many cases colony members develop morphologies suited for their roles, typically large queens capable of increased reproductivity. In Hymenoptera, for instance, a queen may need to be large enough to store and keep alive millions of sperms collected on a single mating flight and then eke these out over a twenty-year period to fertilize the eggs she lays.

d. *Change of reproductive ability*

While queens develop increased reproductive abilities, workers typically lose the ability to mate and so can only lay unfertilized, haploid male eggs. In a few genera of ants and stingless bees the workers are completely sterile.

e. *Control and coordination*

The division of labour is coordinated and enforced. Honeybee workers build a larger cell to rear a queen, which they feed with royal jelly, while they provide other larvae with insufficient nutrients to develop into queens. In the case of Melipona stingless bees, workers seal all larvae into identical cells with identical food; however, when these larvae emerge as adults from their cells, the workers kill off excess queens. In many species worker-laid eggs are eaten by other workers—98 per cent in the case of honeybees—or by the queen. Egg-laying workers can also be attacked.[8]

f. *Altruism*

Some insects exhibit apparently altruistic behaviour. A honeybee worker uses her sting to defend her colony when use and detachment of the sting results in her death. Most evolutionary biologists and ethologists describe the workers' reduction or loss of reproductive ability in favour of a queen's enhanced reproductivity as altruism. However, when such behaviour is enforced, this is not altruism in the generally understood meaning of that word.

> **altruism**   Behaviour characterized by unselfish concern for the welfare of others; selflessness.

Working together voluntarily for mutual benefit is as different from enforced collaboration as the voluntary cooperative societies of peasants in Siberia observed by Kropotkin are from the later collective farms under central control in the Soviet era. I think it useful to distinguish the two, and henceforth I shall use the following terms:

> **cooperation**   Working together voluntarily to achieve commonly agreed aims or for mutual benefit.

> **collectivisation**   Working together involuntarily, whether by instinct or conditioned learning or coercion.

The behaviour of social insects is instinctive and so very probably is inherited and genetically selected for as Lorenz and Tinbergen proposed. As species become progressively more complex, however, we see the emergence and development of learned behaviour.

Fishes

Laboratory experiments from the beginning of the twenty-first century show that fishes introduced into a different shoal copy their new shoal in following routes, preferring particular foods, and using specific places for feeding and mating. Such adoptions of learned behaviour patterns occur far too rapidly to result from the natural selection of genetically determined behaviours best adapted to the environment.[9] According to ethologist Kevin Laland of the University of St Andrews, such socially learned behaviours violate one of the fundamental assumptions of NeoDarwinism. Moreover, they "are maintained as 'traditions' for multiple generations".[10]

Meerkats

Small non-primate mammals found in arid regions of southern Africa, meerkats live in social colonies of 2 to 40 individuals comprising a dominant male and female, who are the parents of over 80 per cent of the pups in the group, plus a variable number of male and females over three months old who assist in rearing the pups and in performing other social functions such as standing sentry to warn of predators while the others are grooming each other, playing, or foraging.

Food acquisition in that environment requires considerable skill, and observations of, and experiments with, meerkats in the wild show not only individual learning through trial and error and the socially learned behaviour of copying[11] but also adult helpers teaching pups how to hunt safely by bringing them scorpions to practise on.[12]

Primates

Like less complex species, some primates kill members of other species for food, but generally their diet is mainly fruit and vegetation.

Within their species primates show competitive and aggressive behaviour frequently resulting in death, especially males competing for females with whom to copulate or for territory and resources. With gorillas and chimpanzees, the most studied primates, such behaviour commonly extends to infanticide, where a male taking over dominance in a group kills young offspring fathered by another male. But primates are also social animals that live in groups, and within the group aggression is often a display rather than an infliction of harm. Collaboration is at least as significant as competition for their survival and reproduction, shown in such behaviours as protection from predators, hunting, rearing offspring, and migration. It is reinforced by mutual grooming and play.

Primatologist Carel van Schaik's studies of orangutan behaviour conclude that intelligence depends more on opportunities for social transmission than on their environment or their genes, and that species with more opportunities for social learning are more intelligent.[13] Laland finds that for nonhuman primates social learning and intelligence, as measured by the invention of more novel solutions to challenging problems, increases with brain size.[14] Cognitive ethologist Simon Reader's comparison of 62 primate species similarly concludes that social learning co-evolves with primate brain enlargement and general intelligence.[15]

These, and many more, studies from the middle of the twenty-first century's first decade onwards extend and reinforce Kropotkin's findings of more than a hundred years ago:

> Mutual aid is as much a law of animal life as mutual struggle, but that, as a factor of evolution, it most probably has a far greater importance, inasmuch as it favours the development of such habits and characters as insure the maintenance and further development of the species, together with the greatest amount of welfare and enjoyment of life for the individual, with the least waste of energy.[16]

and

> Therefore we find, at the top of each class of animals, the ants, the parrots and the monkeys, all combining the greatest sociability with the highest development of intelligence. The fittest are thus the most sociable animals, and sociability appears as the chief factor of evolution.[17]

Such behaviours do not fit the NeoDarwinian model, which is rooted in competition. In a later chapter I shall examine various hypotheses to explain cooperative behaviour by extending that model.

Inter-species associations

Competition for resources plus predation often marks inter-species behaviour. However, associations between members of different species are widespread. They take three forms. Parasitic behaviour occurs when one member benefits while the other is harmed, as with flatworms that infect the eyes of dace in order to find food and shelter while the dace suffer impaired vision. Commensal behaviour, when one gains but the other is unaffected, is rare. Much more common among very many species are cooperative associations, where the benefits are mutual. Pollination was described earlier, while other examples of inter-species cooperation include cleaning associations, such as cleaner wrasse feeding freely on the parasites of much larger predatory fish that would normally eat such small fish.[18]

Conclusions

1. Collaboration is a more significant cause than competition in the development and survival of organisms and is extensive at every level of life.
   1.1. Genes work together for the development of an organism, often regulated by other genes that themselves work together.
   1.2. A unicellular organism consists of parts performing specific functions that work together to maintain and replicate the organism.
   1.3. A eukaryotic cell consists of a nucleus controlling the collaboration of cellular parts, or organelles, that perform specific functions for the maintenance and replication of the cell, while a multicellular organism consists of a hierarchy of specialized cell groups working together with low conflict to maintain the organism's existence and produce offspring.
   1.4. Members of many animal species collaborate in social groups for their collective survival.
   1.5. While members of a species associate parasitically with members of a different species, cooperative association between members of different species for their mutual benefit is also widespread.
2. For unicellular and animal species collaboration principally takes the form of a group within a species communicating and working together to construct a communal shelter, produce and rear offspring, forage, defend themselves, attack prey, and migrate to a better environment for their survival and reproduction. In some cases, several groups within a species, and even within related species, also collaborate for mutual benefit, particularly for migration.
3. For animals a group usually, but by no means exclusively, is based on kinship and has a hierarchical division of labour often dominated by one or more females and a male who reproduce while other members perform specialized tasks.
4. Characteristic behaviour patterns, particularly for surviving and reproducing, are usually instinctive; they are consistent with an evolutionary history of successfully responding to specific stimuli and the inheritance of such responses by descendants.
5. In colonies of simpler species, like insects, collaboration between members specializing in different tasks is usually enforced, with one or a limited number of members gaining reproductive abilities while workers suffer reduced reproductive abilities or sterility. Although a few examples of altruism in the generally understood sense of the word appear to occur, instinctive or coercive collaboration is best described as collectivism to distinguish it from cooperativism in which collaboration is voluntary or for mutual benefit.
6. As the complexity of species increases from fishes through to primates,

social learning, which is more efficient than individual learning by trial and error, increases to supplement instinctive behaviour; such socially learned skills can be inherited.

7. Within each class, increased social learning correlates with increased brain complexity together with increased intelligence measured by the invention of more novel solutions to challenging problems.

CHAPTER TWENTY

The Human Lineage

Science has proof without any certainty. Creationists have certainty without any proof.

—Ashley Montagu, 1905-1999

Phylogenetic trees

In tracing biological evolution, including the lineage leading to humans, evolutionary biologists concluded that the best way is to classify lines, or clusters of lines, of evolutionary descent. This spawned the sub-discipline of cladistics, which presents such relationships as cladograms, or phylogenetic trees. Figure 20.1 shows a simple one.

If each capital letter represents a species, then at branching point, or node, w species **A** diverged into species **B** and **C**. At node x species **B** diverged into species **D** and **E**, while at node y species **C** diverged into species **F** and species **G/H**, and at node z species **G/H** subsequently diverged into species **G** and **H**.

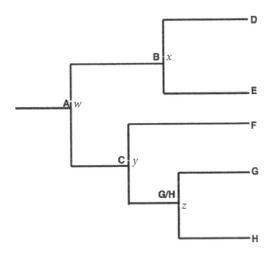

Figure 20.1 Simple phylogenetic tree

Thus species **G** and **H** share a more recent common ancestor, **G/H**, than they do with **F**, while **G, H,** and **F** share a more recent common ancestor, **C**, than they do with **E** or **D**.

D, E, F, G, and **H** all share some general characteristics, indicating that, as a group of species, they all evolved from a more distant common ancestor, **A**. Unless otherwise stated, the lengths of the branches aren't significant: phylogenetic trees depict only branching from common ancestry.

Ideally, traditional Linnaean taxonomic classifications and phylogenetic trees should produce identical groups. However, differences arise for several reasons, including interpretations of fossils and disagreements about what characteristics should be used to determine degrees of relatedness.

Biological orthodoxy adopted Woese's view, outlined at the end of Chapter 16, that a far more accurate way than the fossil record, morphology, or any other characteristic, was genetic analysis.* In practice molecular geneticists analyse the protein or RNA product of one or more homologous genes—genes inherited from a common ancestor—in different species. For example, in the 2014 edition of the *Encyclopædia Britannica Online* former President of the American Association for the Advancement of Science, Francisco José Ayala, explains that cytochrome *c*, a protein involved in cellular respiration, consists of the same 104 amino acids in exactly the same order in humans and chimpanzees, but differs in rhesus monkeys by 1 amino acid, in horses by 11 additional amino acids, and in tuna by 21 additional amino acids. "The degree of similarity reflects the recency of common ancestry."[1]

Evolutionary biologists went further by developing the molecular clock technique. This assumes that random mutation by the substitution of one nucleotide for another occurs at a linear rate that can be calibrated from the fossil record, thus enabling the calculation of divergence times. In Figure 20.1, for example, if **D, E, F, G,** and **H** represent the present time, then the length of the horizontal branch connected to **B** represents how long ago species **D** and **E** diverged from **B**.

This NeoDarwinian approach is based on four assumptions:

1. homologous genes can be unambiguously identified;
2. the branches of the phylogenetic tree only diverge with time and never reunite;
3. the only cause of biological evolution is the gradual accumulation of gene mutations;
4. the rate of genetic change is linear and this reflects the rate of biological evolution.

These assumptions are questionable.

---

*   See page 278.

As we saw in Chapter 18, a gene is not simply a sequence of DNA coding for the production of a protein that determines an organismal trait. How can we be sure that the homologous gene selected is representative of the organism? How does this approach take into account factors like rates of DNA repair or the effect of recessive alleles and gene duplication? Or gene splicing and transposons? Or, significantly, gene regulation, which controls gene expression and thereby determines what the organism consists of, its morphology, and, to some degree, its behaviour?

For the purpose of reconstructing an organism's evolutionary history, how do you choose between the genome found in the nucleus and the genomes found in mitochondria of animal cells or in the chloroplasts of plant cells? How representative of the species is the chosen gene or genes of one organism? How does this approach take into account population sizes and the impact of genetic drift?

More than 25 years before Ayala's 2014 *Encyclopædia Britannica* cytochrome *c* example, biochemist Christian Schwabe had pointed out the consequences of comparing the molecular differences in relaxin, a peptide hormone responsible for widening the birth canal in mammals thereby allowing more highly developed and bigger-brained offspring to be born (depending on the extent to which its gene is switched on). The amino acid sequence of relaxin in humans differs from that in pigs, rats, sandtiger sharks, and spiny dogfish sharks by an average of 55 per cent. Thus if the evolutionary tree were constructed on the basis of changes in the molecular composition of relaxin, then mammals branched from cartilaginous fish (the class that includes sharks) either 450 million years ago (the earliest fossils of cartilaginous fish) or 100 million years ago (then thought to be the earliest fossils of mammals). It implies that pigs and rats have changed very little from this branching point, whereas the evolutionary line that led to humans mutated by 55 per cent. The fossil record, on the other hand, shows a huge divergence in mammalian species in the last 65 million years.

Schwabe concluded

> It seems disconcerting that many exceptions exist to the orderly progression of species as determined by molecular homologies; so many in fact that I think the exception, the quirks, may carry the more important message.[2]

In 2006 Jeffrey H Schwartz, a physical anthropologist from the University of Pittsburgh, offered a comprehensive and withering critique of the underlying assumptions, methodologies, evidential bases, and conclusions of molecular systematics. In his view the "molecular assumption" is biologically untenable. It takes no account of the regulatory sequences that determine morphology in eukaryotes, and "aside from UV-induced point mutation there is no other constant source of mutation in the physical world, and spontaneous mutation rates are low (approximately 1 in 10^{-8} to 10^{-9})", while random mutation is much less likely to occur in germ cells, the only ones to have evolutionary consequences.[3]

Thus interpreting the variation in molecular sequence is far from straight-forward. In practice, dramatic evolutionary divergence can occur with only small accompanying changes in sequence, while minor evolutionary divergence can correlate with large changes in sequence.

Moreover, how does this approach take into account the evolution of the most numerous species on the planet—bacteria and archaea—and horizontal gene transfers between distant species? According to University of Queensland bioinformaticist Mark Ragan in 2009, "A surprising number of gene trees are, in part, topologically discordant with each other and/or with accepted organismal relationships."[4] This echoes Dalhousie University biochemist Ford Doolittle's 2009 observation that "It is now a tedious truism to say that there is much more incongruence between prokaryotic gene trees than we could have imagined there could be two decades ago."[5]

The purpose of this quest is to see if and how the human species evolved, and it would be satisfying to find a phylogenetic tree that mapped how humans evolved from the first lifeform on Earth. Many such trees are produced, such as the one shown in Figure 20.2, described as "the consensus phylogenetic tree of all life".

However, each is different because each is based on different data, different assumptions, different methodologies, different interpretations, and an irretrievably impoverished fossil record.

Investigators using different molecular analyses and molecular clocks give significant differences in the timescales of the same evolutionary divergences. For example, estimates of the branching of the chimpanzee genus *Pan* and the human genus *Homo* range from 2.7 million years ago to 13 million years ago.[6]

Such trees cannot account for speciation that occurs through hybridization when phylogenetic branches merge, or for instant speciation through polyploid-ization, or for the evolution of the most numerous species, bacteria and archaea, for which reticulation, or networking, and horizontal gene transfers between different species played a key role.

Taxonomy of the human lineage

In the absence of a generally agreed phylogenetic tree, Table 20.1 is my attempt to synthesize recent findings in cladistics, cytology, molecular biology, genomics, palaeobiology, palaeontology, bacteriology, evolutionary biology, ethology, and primatology to show what is known of the lineage from the first lifeforms on Earth to humans. Where possible this taxonomy uses several characteristics rather than one, and necessarily involves value judgements on the disagree-ments between different branches of biology and between specialists within each branch. Such disagreements include whether there are 3, 4, 5, or 6 kingdoms (or equivalent ancestral groups, called clades) and whether these should be divided

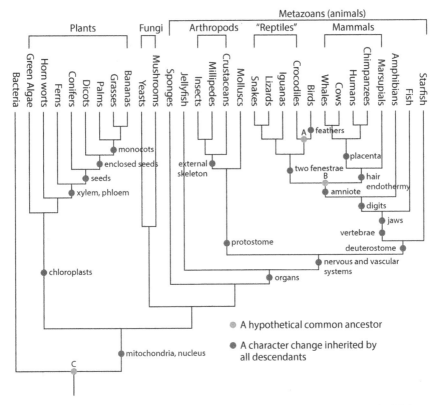

Figure 20.2 A phylogenetic tree claiming to show the evolution of all life from a common ancestor

into a smaller number of superkingdoms labelled domains. For the sake of consistency with preceding chapters of this book, Table 20.1 uses the nomenclature of the three-domain system.

It is worth emphasizing two things: the further back in time, the greater the uncertainty; and, although the boundaries between these taxonomic groups mark the emergence of evolutionary innovations, as with all emergences the boundaries are indistinct.

In the following chapters I shall consider the cause, or causes, of this pattern in biological evolution.

Table 20.1 Taxonomy of the human lineage
The seven principal rankings are shown in bold; additional ranks, such as subclass, are shown in bold italic.

Classification	Name	Cytological, genomic, morphological, and behavioural characteristics	Examples
	Universal Common Ancestor	The first independent enclosed entity, or cell (or possibly network of protocells), comprising several parts that work together in responding to, and extracting energy and matter from, its environment for the purpose of maintaining its own existence and replicating itself.	None.
Domains	Bacteria and Archaea	Extremely small, simple organisms comprising a single cell whose genetic information is normally a folded loop of double-stranded DNA not enclosed in a membrane-bound nucleus; the cell may include one or more plasmids, separate circular strands of DNA that can replicate independently and are not responsible for the reproduction of the organism. Most reproduce by splitting in two and producing identical copies of themselves. The most ancient manufacture their own food by chemosynthesis or photosynthesis, but most living bacteria and archaea sustain themselves by ingesting parts of other organisms, dead or alive. They generally have rigid cell walls and many are capable of incorporating foreign DNA by horizontal gene transfer even from unrelated species. Many exhibit primitive forms of social behaviour to maintain their existence and to propagate.	Existing bacteria and archaea and their ancestors, from which it is thought eukaryotes evolved.

Domain	*Eukarya* [Eukaryotes]	Organisms that consist either of one cell or, more commonly, collaborating groups of cells, each one of which contains a membrane-bound nucleus that controls the action of organelles, the small, self-contained, cellular parts that perform specific functions in maintaining and propagating the organism. Most eukaryotic cells replicate to produce identical copies of themselves, but most multicellular eukaryotes propagate sexually by the fusing of a sex cell from one organism with a complementary sex cell from another organism so that the offspring contains genes from both parents.	Protoctists, fungi, plants, animals.
Kingdom	*Animalia* [Animals]	Multicellular eukaryotes that lack cell walls, or have highly permeable cell walls, and that possess a nervous system; they are mainly motile (capable of independent movement) and eat other organisms in order to maintain their existence.	Sponges (adults are not motile), worms, insects, fishes, amphibians, reptiles, birds, and mammals.
Phylum	*Chordata* [Chordates]	Animals containing at some stage in their development a flexible rod-like structure that supports the body, with a nerve chord above; they usually have an elongated, bilaterally symmetrical body with a mouth and sense organs at the front and an excretory orifice at the rear.	Vertebrates plus closely related marine invertebrates such as lancelets.
Subphylum	Vertebrates	The principal subphylum of chordates whose members have a segmented spinal column, or backbone, and a head. The vast majority of the oldest species produce multiple progeny by external fertilization of eggs. Many	Fishes, amphibians, reptiles, birds, and mammals.

Classification	Name	Cytological, genomic, morphological, and behavioural characteristics	Examples
		relative recent vertebrates are social animals, either living in family or extended family groups or else collaborating at certain times for mutual benefit, like migrating to a habitat with better resources. They exhibit social learning, the degree of which correlates with their complexity.	
Class	*Mammalia* [Mammals]	Air-breathing four-limbed vertebrates that nourish their young with milk produced by mammary glands. They reproduce by internal fertilization of their eggs and, except for the most ancient extant mammals, monotremes that lay usually two reptilian-like eggs, they give birth to live young. They are warm-blooded and able to maintain a relatively constant body temperature regardless of external environmental conditions. Their variety of specialized teeth allows them to chew their food into small pieces. Most species live in social groups.	Monotremes like the platypus, marsupials like kangaroos, and placentals like most mammals.
Subclass	*Eutheria* [Placental mammals]	Comprising about 94 per cent of extant mammals, they conceive their young within the reproductive tract of the mother and nourish their unborn young via a placenta that is attached to the foetus by an umbilical chord. This enables them to give birth to relatively well-developed progeny and greatly increases the survival rate of a fertilized egg compared with that of non-mammals.	From shrews to humans.

Rank	Description	Examples
Order Primates	They are differentiated from other placental mammals by possessing a larger and more complex brain, especially those centres involving vision, memory, and learning, together with complex sensory pathways to the brain, plus stereoscopic vision from forward-facing eyes. Prehensile 5-digit hands and feet, with fingernails rather than claws, enable many to live in trees. Most species normally give brith to a single young and are social animals that live in extended families or troops and exhibit a higher degree of social learning than other mammals.	Lemurs, lorises, tarsiers, monkeys, apes, humans.
Superfamily *Hominoidea* [Hominoids]	Primates with larger and more complex brains, a flatter face with a smaller nose and greater reliance on vision rather than smell; they lack a tail and adopt an upright posture when stationary.	Gibbons, great apes, and humans.
Family *Hominidae* [Hominids]	Large hominoids with larger and more complex brains, they show intelligent behaviour, like the use of very simple tools.	The great apes—orangutans, gorillas, chimpanzees, bonobos—and humans, and their ancestors plus extinct lineages assumed to share a common ancestor.*
Tribe *Hominini* [Hominins]	Hominids that walk upright on two feet.	Modern humans and their ancestors plus extinct genera assumed to share a common ancestor, like *Australopithecus* and *Ardipithecus.**

Classification	Name	Cytological, genomic, morphological, and behavioural characteristics	Examples
Genus	*Homo* [Human]	Hominins that lack a thick covering of fur over their bodies; they possess a significantly larger and more complex brain than other hominins and make tools.	Modern humans and their ancestors plus extinct species assumed to share a common ancestor, like *Homo habilis, Homo erectus* and *Homo neanderthalensis.**
Species	*Homo sapiens* [Modern human]	Humans that possess highly complex neuronal networks, and a brain characterized by a large and dominant neocortex. They use sophisticated technologies and medicine, and communicate by speech, writing and electronic means. They learn by being taught extensively and engage in many non-survival creative activities like visual arts, science, music, literature, and philosophy. They live in a nested range of increasingly larger social groups beyond kinship and practice cooperation and altruism as well as inter-group competition and aggression. They comprise the only species that has extended its habitat to the whole biosphere.	Modern humans.
Subspecies	Race	Interbreeding members of the species distinguished by minor differences of phenotype, or visible characteristics like skin colour.	African, Caucasian, and Mongoloid used to be classified as the principal subspecies, but such differences are increasingly difficult to distinguish.

*Ancestral relationships are unclear and classifications, particularly of chimpanzees and archaic humans, are controversial. Opinions change with discoveries of fossils and their interpretations plus advances in genomic analyses and their interpretations. I shall consider these questions in more detail in Part 3, and especially whether there is a difference in degree or in kind between modern humans and others.

CHAPTER TWENTY-ONE

Causes of Biological Evolution: the Current Orthodox Account

I think "nature red in tooth and claw" sums up our modern understanding of natural selection admirably.

—RICHARD DAWKINS, 1976

Before examining the extent to which current NeoDarwinian orthodoxy explains what causes the pattern of biological evolution, we must have a clear understanding of what biological evolution is. To distinguish this irreversible process from reversible adaptive change (see pages 315 to 316), I use the following definition:

biological evolution   A process of change in organisms that results in new species.

Most evolutionary biologists today conflate, or even equate, evolution with natural selection, but of itself the term natural selection is a statement of the obvious and tells us nothing: organisms that survive and reproduce are the ones selected to survive and reproduce. Similarly for survival of the fittest: the ones that survive are the fittest to survive. These terms only have meaning if they tell us why biological evolution takes place (the cause) or how it takes place (the mechanism).

Many NeoDarwinists say that natural selection causes biological evolution. For example, Luis Barreiro and colleagues of Paris's Institut Pasteur titled a 2008 paper in *Nature Genetics* as "Natural Selection Has Driven Population Differentiations in Modern Humans",[1] while the *Encyclopædia Britannica Online* 2014 states that "Natural selection moderates the disorganizing effects of these processes because it multiplies the incidence of beneficial mutations over generations and eliminates harmful ones."[2] The *Gale Encyclopedia*'s entry on Ernst Mayr is explicit: "Neo-Darwinism stresses natural selection of genetic differences within populations as the fundamental cause of evolution."[3]

Verbs like drive, moderate, multiply, eliminate, and select clearly imply causality, while select in particular also implies purpose. This description of natural selection follows Darwin's meaning of the term, which he used "to mark its relation to man's power of selection". Moreover, it "is a power incessantly ready for action, and is immeasurably superior to man's feeble efforts".*

In artificial selection a pigeon breeder, say, produces a white pigeon from his stock of pigeons that have variations of colour. The human breeder is the cause of the effect, which is a white pigeon. His mechanism—how he causes the effect—is intentionally selecting his palest-coloured pigeons to breed only with each other over successive generations.

How does natural selection cause a similar, if not "immeasurably superior" effect? What is its mechanism?

Most contemporary NeoDarwinists deny that nature has any purpose or intention, and so to be logically coherent they cannot also say that nature selects or that nature causes.

A common defence is that this statement is metaphorical. But metaphorical for what?

If it is claimed it is metaphorical in the way physicists say gravity causes a steel ball dropped from a balcony to fall to Earth, then this begs a question. To an excellent approximation the steel ball obeys Newton's law of gravity. This is a natural law, which is a succinct, general statement capable of being tested by observation or experiment and for which no repeatable contrary results have been reported that a set of natural phenomena invariably behaves in an identical manner within specified limits.† The question begged here is: what natural law is being invoked for biological phenomena? It cannot be the so-called "adaptive principle"—certainly as far as NeoDarwinism's genetic model is concerned—because, as we shall see later, significant genetic changes can occur over a long period with little or no phenotypic change despite changes in the environment.

The current paradigm

Chapter 16 showed how scientific ideas of biological evolution developed into the six principles of the NeoDarwinian/molecular biology model that define the current paradigm:‡

1. Living species are descendants of other species that lived in the past.
2. New forms of life arise from the splitting of a single lineage into two, a process known as speciation in which members of one lineage are unable to breed successfully with members of the other lineage. This continual splitting leads to a nested genealogy of species, a "tree of life" whose root is the first species to arise and whose live twigs are the vast number of current species. Tracing back any pair of twigs from modern species through the

---

* See pages 260 to 261.
† See page 142.
‡ See page 276.

branches leads to a common ancestor, represented by the node at which the branches meet.

3. Such evolution of new species occurs through the gradual genetic transformation of populations of individual species members over tens or hundreds of thousands of generations.

4. This transformation results from random genetic mutations in individuals, the mixing of genes from each parent during sexual reproduction that leads to each offspring having a different combination of genes from those of either parent, and the consequent spreading of such genetic mutations over successive generations throughout the population's gene pool.

5. Those randomly produced genetic mutations, or variations, responsible for characteristics that enable individuals in the population to successfully compete for resources in a particular environment, and so survive longer and produce more offspring, are naturally selected by being inherited in greater numbers, while members of the population lacking genetic mutations responsible for such adaptive characteristics in that environment are either killed, starved, or become extinct over generations.

6. Information is a one-way flow from from a gene to a protein in a cell.

According to these principles the *causes* of biological evolution are random genetic mutations usually ascribed to copying errors in DNA replication to produce gametes (eggs or sperms), competition between members of a growing population for limited resources in their environment that enables successful competitors to live longer and produce more offspring possessing their mutated genes, and sexual reproduction that spreads these mutations throughout the population's gene pool over tens or hundreds of thousands of generations until members of that population are no longer able to breed successfully with members of the original population.

In fact genetic mutations are more likely to be caused by radiation and chemicals plus cellular misrepair of DNA damage than by copying errors, which cellular repair processes usually correct. Spontaneous mutation rates generally are low, approximately 1 in 10^{-8} to 10^{-9}, while random mutation is much less likely to occur in gametes.[4]

If natural selection is more than a statement of the obvious it is, at most, a passive record of effects caused by other things.

natural selection The cumulative effect of small, randomly produced variants inherited over very many generations that enables organisms to survive longer and reproduce more in a particular environment than organisms lacking those variants; it results in an increase in the number of such favourable, or best adapted, variants for that environment and the elimination of unfavourable variants.

According to Darwinism the favourable variants are the observable traits (the phenotype) of individual organisms that enable them to compete more effectively for limited resources in, and hence be better adapted to, a particular environmental niche. According to NeoDarwinism the favourable variants are genetic mutations in the gene pool of a species population.

In 2006 Jerry Coyne, from whom the first five principles of modern NeoDarwinism are drawn, declared that

> A good scientific theory makes sense of wide-ranging data that were previously unexplained. In addition, a scientific theory must make testable predictions and be vulnerable to falsification....Every bit of information we have gathered about nature is consonant with the theory of evolution, and there is not one whit of evidence contradicting it. *Neo-Darwinism, like the theory of chemical bonds, has graduated from theory to fact* [my italics].[5]

Note that Coyne equates the theory of evolution with NeoDarwinism. If he is correct, then NeoDarwinism should explain all the evidence, make testable predictions, and be vulnerable to falsification. But does it?

What NeoDarwinian orthodoxy fails to explain

Evidence reviewed in the preceding four chapters shows several patterns not explained by NeoDarwinism and many that falsify it.

Stasis and rapid speciation
In contrast to the essential gradualism of NeoDarwinism, Chapter 17 concluded

> The normal pattern of fossil evidence for animals is one of morphological stasis with minor and often oscillating changes punctuated by the geologically sudden—tens of thousands of years—appearance of new species that then remain substantially unchanged for some tens or even hundreds of millions of years until they disappear from the fossil record or else continue to the present day.*

Ernst Mayr, one of the architects of the NeoDarwinian model, admitted in 2001 that "The complete standstill or stasis of an evolutionary lineage for scores, if not hundreds, of millions of years is very puzzling."[6]

Harvard University geologist Peter G Williamson put it more bluntly:

---

*   See page 303.

A theory is only as good as its predictions, and conventional neo-Darwinism, which claims to be a comprehensive explanation of the evolutionary process, has failed to predict the widespread long-term morphological stasis now recognized as one of the most striking aspects of the fossil record.[7]

Equally inconsistent with the NeoDarwinian gradualist model is geologically very rapid speciation.

Speciation
From all the evidence surveyed in Chapter 18, no studies of living species show the evolution of new species according to the NeoDarwinian mechanism.

Immediate speciation: polyploidy
The NeoDarwinian model defines a species as a branch of the evolutionary tree that has split from another branch and whose members are unable to breed successfully with members of that other branch. The only unequivocal example of such speciation, whether achieved by artificial selection or observed in the wild, is polyploidization in plants and in some fishes, amphibians, reptiles, and even mammals. This increase in the normal number of chromosomes per cell produces a new species immediately rather than gradually over thousands of generations, thus contradicting the NeoDarwinian mechanism. The phenomenon is probably more widespread than generally assumed.*

Asexual reproduction
The overwhelming majority of species on the planet do not reproduce sexually and pass down genetic mutations vertically from parent to offspring. Most of these are single-celled organisms—prokaryotes (bacteria and archaea) and unicellular eukaryotes—that replicate by producing clones of themselves.†

Horizontal gene transfer
Many prokaryotes undergo horizontal gene transfer in which they gain DNA either from their environment, or via agents like viruses, or through direct contact with other organisms.‡

Coyne and his collaborator H Allen Orr argue that such gene exchange is essentially the same as gene exchange and transmission by sexual reproduction. However, in the latter case the transmission is vertical, from parents to offspring. Not only does horizontal gene transfer take place between members of the same generation but, in some cases, between very different species.

---

*   See pages 309 and 310.
†   See page 282.
‡   See page 309.

Horizontal gene transfer can give the recipient immediate additional functions having a negative, neutral, or positive survival value and is increasingly recognized as having played a major role in the evolution of bacteria and archaea.

Moreover, if some organelles in eukaryotic cells result from the incorporation of previously independent ancestral bacteria followed by horizontal gene transfer to the nuclear genome, as most evolutionary biologists now accept, then horizontal gene transfer also played a major role in the evolution of eukaryotes.

While an estimated 81 ± 15 per cent of prokaryotic genomes have resulted from horizontal gene transfer,* the lower number of sequenced eukaryotic genomes makes it difficult to estimate accurately the contribution of horizontal gene transfers in living eukaryotes. However, a 2007 study of a bacterium living within insects and nematode worms found significant heritable gene transfers to the host genomes.[8] Other studies have shown that the acquisition of foreign genes from bacteria and eukaryotes takes place in fungi and plants. Moreover, massive horizontal gene transfers take place in animal and plant hybridization.[9]

All these examples demonstrate that not only prokaryotes but also multicellular eukaryotes acquire genes that can affect their survival and evolution in ways other than the NeoDarwinian orthodox model of the vertical transmission of random mutatations down very many generations.

Organismal embryology and development

The orthodox model does not explain what causes a single cell formed by the fusion of a sperm and an egg to replicate into a cluster of cells with identical genomes, which then differentiate into cells with specialized functions that form an embryo, which subsequently develops, in the case of humans, into an independent body that is either male or female and weighs on average between 75 and 85kg for an adult male and 55 and 65kg for an adult female, with two legs for upright walking, two arms, and a head containing the most complex thing in the known universe, a human brain.

The discovery of developmental regulatory genes, like the Hox family, which determine the development of body parts in nearly every bilateral animal studied so far, does not explain the phenomenon. These genes, which set in train gene expression cascades, are very similar in fruit flies, worms, fish, frogs, and mammals. The orthodox model does not explain what causes these genes to be switched on or why they activate different cascades to produce the very different body plans of fruit flies, worms, fish, frogs, and mammals.

Genotypes and phenotypes

Neither does the orthodox model explain a related phenomenon: some species with very similar genes have very different phenotypes. For example, humans

---

*   See page 310.

share some 94 per cent of their genes with chimpanzees, 90 per cent with mice, and even 85 per cent with 4-6cm-long zebrafish,* while some species of fruit fly are very different genetically but very similar morphologically.

"Junk" DNA

As noted at the end of Chapter 16, for some 50 years gene-focused NeoDarwinists dismissed as junk the 98 per cent of the human genome that did not consist of genes because it had no place in their model. This was indeed hubristic because we now know that some 80 per cent of this DNA show signs of functional activity; much of this appears to be implicated in collaborative regulatory networks.†

Many adherents to the current orthodox model argue that non-protein-coding regions of DNA are still DNA and so subject to mutation and natural selection in the same way as coding regions. However, in 2007 Michael Lynch, Distinguished Professor of Biology at Indiana University, reviewed the evolution of genetic networks and concluded that, contrary to widespread belief, there is no compelling empirical or theoretical evidence that features of genetic pathways are promoted by natural selection.[10]

Inheritance of acquired characteristics

Chapter 16 also noted that one consequence of biologists' adoption of the NeoDarwinian model was a rejection of other causes of evolution, including the Lamarckian inheritance of acquired characteristics, which Darwin had accepted.

However, particularly since whole genomes began to be sequenced towards the end of the last century, evidence began mounting of precisely this phenomenon. For example, in 2005 Matthew Anway of Washington State University and colleagues showed that if pregnant rats were exposed to the common crop fungicide vinclozolin then their male offspring showed increased infertility. This trait was transmitted through the male germ line to nearly all males of all subsequent generations examined, but no genetic change was involved.[11]

For all types of animals, including humans, it has become increasingly clear that environmental factors, such as diet or stress, can produce traits that are transmitted to offspring without a single change in gene sequences taking place.

Theoretical geneticists Eva Jablonka and Gal Raz of Tel Aviv University compiled examples of non-genetic inheritance of acquired characteristics: 12 in bacteria, 8 in protists, 19 in fungi, 38 in plants, and 27 in animals. These are often less stable than characteristics associated with genetic variations but, Jablonka and Raz argue, the examples represent only a very small fraction of non-genetic inheritance that probably exists.[12]

---

* See page 318.

† See page 317.

I think molecular embryologist Marilyn Monk of UCL's Institute for Child Health is right to conclude that such heritable changes shown to have been induced by environmental factors not only challenge the sixth principle of the current paradigm, the "central dogma" that information is a one-way flow from from a gene to a protein in a cell, but also the fourth and fifth principles that the sole mechanism of biological evolution is the natural selection of randomly generated genetic mutations.

Collaboration

Another consequence of adopting the NeoDarwinian model, noted in Chapter 16, was ignoring collaboration as a cause of evolution. However, the examination of behaviour in Chapter 19 led to the conclusion that

> Collaboration plays a more significant role than competition in the survival and propagation of life, from the level of genes through unicellular organisms, organelles within eukaryotic cells, eukaryotic cells within multicellular organisms, and the behaviour of plants and animals from insects to primates.

According to University of Oxford evolutionary biologist Stuart West in 2011, "Explaining the apparent paradox of cooperation is one of the central problems of biology because almost all the major evolutionary transitions from replicating molecules to complex animal societies have relied upon solving this problem."[13]

It has been a central problem of biology ever since most biologists adopted the NeoDarwinian model because the only behaviour consistent with that model is competition. Attempts to explain collaborative and altruistic behaviour according to the axioms of NeoDarwinism gave rise to a new sub-discipline of sociobiology. In a later chapter I shall consider such attempts as hypotheses complementary to the NeoDarwinian model.

Progressive complexification

Another significant conclusion of Chapter 17 is that the overall pattern of the fossil record over time is from simple to more complex; the available evidence for animals indicates that complexity increases along each lineage, and this is clearly the case for the lineage leading to humans. The evidence from Chapter 18 shows that living species form a pattern of increasing complexity from bacteria to humans. Studies of behaviour reviewed in Chapter 19 demonstrate that, as the complexity of species increases from fishes through to primates, social learning increases to supplement instinctive behaviour; this also correlates with increasing brain complexity together with increasing intelligence attested by more innovative solutions to problems.

Although Darwin had concluded that "As natural selection works solely by and for the good of each being, all corporeal and mental endowments will tend to progress towards perfection,"[14] most evolutionary biologists today deny any pattern of evidence that implies progress.

According to cyberneticist and interdisciplinary complexity theorist Francis Heylighen,

> Like Maynard Smith and Szthmáry, in their study of the major transitions in evolution, [evolutionary biologists] feel obliged to pay lip service to the current [relativistic] ideology by noting the "fallacy" of believing that there is something like progress or advance towards increasing complexity, and then continue by describing in detail the instances of such increase they have studied.[15]

Before examining objections to the pattern of increasing complexity, I shall try to minimise misunderstandings that arise from different meanings of the same word. Complexity is defined differently by systems theorists, automata theorists, information theorists, complexity theorists, cyberneticists, evolutionary biologists, genomicists, ethologists, and so on; they focus on either qualitative, or quantitative, or structural (subsystems embedded in supersystems), or functional (levels of information processing or control) aspects of complexity. Unsurprisingly their results are different.

I shall attempt to step back and focus on the forest rather than a tree by using an abstract definition, which I hope most specialists agree with.

complex   A whole composed of distinct, interrelated parts.

complexity   The state of being complex.

complexification   The process of becoming more complex.

It follows from this definition that each part is necessarily simpler than the whole consisting of several such parts. Nearly all physical things in the universe are complex and form nested hierarchies of increasing complexity, from subatomic particles through atoms, molecules, unicellular organisms, multicellular organisms, to societies of organisms. Moreover it is generally accepted that, in the history of the universe, simpler components appeared before more complex, composite systems. This is only logical: normally you cannot build a higher order system, like a molecule, from its constituent parts, in this case atoms, until the constituent parts have themselves emerged. Thus evolution leads to more complex systems, gradually adding more levels to the hierarchy. What *causes* this evolutionary phenomenon is another matter. Causes may differ for different evolutionary transitions, and that is why I think it sensible to distinguish between the phenomenon and its cause as we trace the trajectory of evolution.

If we are examining the phenomenon of increasing complexity using the methods of science, it is useful to measure it. I think the best measure is the number of parts multiplied by the number of their connections. In the case of living things, however, all the data are not available.

Evolutionary biologists James Valentine and colleagues at the University of California, Berkeley used the number of cell types not "to measure complexity itself, but...simply as an index of [morphological] complexity", conceding that "we are indisputably lumping cells that have biochemical and no doubt functional differences". This index also omits connections between cells of the same and of different types, and such things as complexity of behaviour. Nonetheless, using just this one measure of morphological complexity and mapping it against the first appearance in the fossil record of a range of animal taxa, Valentine found a progression in complexity of the most primitive families in each taxon from Porifera (sponges) to the most recent, Hominids, even when grouping all the different neurons as one cell type.[16]

However, as Princeton evolutionary biologist John Tyler Bonner observed, "There is an interesting blind spot among biologists. While we readily admit that the first organisms were bacteria-like and the most complex organism of all is our own kind, it is considered bad form to take this as any kind of progression."[17]

The objections are both to the phenomenon of *progression*—the pattern of increasing complexity in biological evolution—and also to the implication of *progress*—such progression results in improvement.

Although they overlap, the arguments against complexification may be grouped into nine, given below followed by a response to each.

a. *Biological evolution is caused by natural selection acting on randomly generated genetic mutations, and so there can be no pattern of increasing complexity.*

The deduction does not follow logically from the premises. There is no reason that natural selection acting on random mutations should not increase complexity (whether natural selection acts to cause biological evolution is another matter considered earlier in this chapter).

b. *Biological evolution results from contingent outcomes of fortuitous events; there is no pattern of evolutionary progression.*

Palaeontologist Stephen Jay Gould advanced this argument after asserting that natural selection is "a principle of local adaptation, not of general advance or progress".[18] This followed his oft-repeated claim that if the tape of life were re-run it would always produce different outcomes. Thus "Humans arose...as a fortuitous and contingent outcome of thousands of linked events, any one of which could have occurred differently and sent history on an alternative pathway that would not have led to consciousness."

This objection to a pattern of complexification is similar to the previous one, essentially substituting fortuitous events for random mutations. It suffers from the same fallacy.

Gould attempts to support his argument by citing as evidence four among a multitude of fortuitous events and contingent outcomes. For example, "If a small lineage of primates had not evolved upright posture on the drying African savannas just two to four million years ago, then our ancestry might have ended in a line of apes that, like the chimpanzee and gorilla today, would have become ecologically marginal and probably doomed to extinction."

But the environment *did* change and bipedalism *did* evolve. To say that bipedalism might not have evolved because the environment might not have changed is to ignore the evidence and indulge in speculation: you may as well speculate that the Earth might never have formed from the material circling the Sun after its formation. In that sense everything is contingent on a sequence of fortuitous events that have occurred since the origin of the cosmos. To assert that any one event could have occurred differently takes us into the multiverse idea, which is untestable: it is philosophical conjecture rather than science.

c. *The pattern of evidence is not one of complexification but of punctuated equilibrium.*
This scales up Eldredge and Gould's punctuated equilibrium hypothesis to the mega level. According to Gould, the pattern is one of 3 billion years of unicellularity followed by the 5-million-year Cambrian explosion of phyletic flowering, while "the subsequent history of animal life amounts to little more than variations on anatomical themes established during the Cambrian explosion".[19]

This, I suggest, somewhat oversimplifies the pattern. Portraying the last 540 million years as little more than variations on anatomical themes established during the 5-minute-year Cambrian explosion conflicts with what he says in the same article: "No one can doubt that more complex creatures arose sequentially after the prokaryotic beginning—first eukaryotic cells, perhaps about two billion years ago, then multicellular animals about 600 million years ago, with a relay of highest complexity among animals passing from invertebrates, to marine vertebrates and, finally…to reptiles, mammals and humans."

Moreover, it equates complexity with anatomical complexity and ignores genomic, developmental, behavioural, neural, communications, and cognitive complexities that have evolved over life's history. Table 17.3 shows not only these increases in complexity but also in the lineage leading to humans that the rate of complexification increases.

d. *Biological evolution shows no overall increase in complexity.*
This objection accepts that biological evolution has resulted in more complex species, but argues that, in Gould's words, there is a "constancy of modal complexity throughout life's history".[20]

Put simply, because the most numerous organisms, prokaryotes, are the simplest, if we calculate the average complexity of every living species on Earth at any point in time, the astronomical number of prokaryotes would dwarf the number of more complex organisms and therefore reduce to near zero their contribution to overall complexity.

While this argument is statistically correct, it is specious. Just as there are astronomically more prokaryotes than animals, so too there are astronomically more molecules than prokaryotes, astronomically more atoms than molecules, and astronomically more fundamental particles than atoms. At any point throughout the universe's history there is a statistical "constancy of modal complexity" represented by fundamental particles. But this doesn't mean that the universe has remained in a constancy of fundamental particles throughout its history and hasn't progressed towards more complex structures.

e. *Biological evolution shows no pattern of increasing complexity; some complex organisms evolve into simpler ones.*
This is termed regressive evolution, and the iconic example is cavefish. Species of fish exist with populations living in open water and also in caves with no daylight. The latter, which are assumed to have evolved from the former, have no sight and no skin colour: they have become simpler.

When evolutionary biologists Richard Borowsky and Horst Wilkens examined this phenomenon in 2002 they found the changes that occur in regressive evolution are no different than those that occur in constructive evolution: allelic frequencies and character states alter over time. However, the cave-dwelling variants also exhibit constructive evolution, like increases in sensory modalities other than vision or increased metabolic efficiency: adaptations to their new environment.[21] Thus, while they lost some functions compared with their open-water progenitors, they gained others. Overall, they didn't become simpler.

Parasites are also cited as examples of regressive evolution, but here again closer examination shows the picture is by no means as clear-cut as usually presented. The most studied examples are dicyemids, tiny parasites that live in the kidneys of cephalopods, like octopuses, squid, and cuttlefish. The dicyemid bodies consist of only 10 to 40 cells, which are fewer in number than in any other metazoans (animals), and are organized very simply: they have neither body cavities nor differentiated organs. According to Osaka University's Hidetaka Furuya, probably the world authority on dicyemids, "It is still not clear whether dicyemids are primitive multicellular organisms or degenerated metazoans."[22]

Even if some instances of permanent net regressive evolution were unequivocally identified, they would not negate the *overall* pattern of biological complexification.

f. *The pattern implies progress; this introduces a value judgement, which has no place in science, which is concerned with objectively collecting facts and drawing logical conclusions and hypotheses from the patterns found therein.*
Progress, as distinct from progression, does indeed involve a value judgement. But it is untrue to say that value judgements have no place in science. To the contrary, science could not operate without value judgements: which data to examine, how to interpret the data, what conclusions to draw, what assumptions should be made in drawing conclusions, how tentative or how strong are the conclusions, whether one hypothesis is better than another at explaining the data, to what extent the data conflict with, or support, the current theory, and so on.

g. *Any pattern, of increasing complexity, or intelligence, or any other human-designed criterion, that shows humans are the most complex or most intelligent, etc. is anthropocentric.*
One of the founders of NeoDarwinism, George Gaylord Simpson, answered this objection succinctly: "To discount such a conclusion in advance, simply because we are ourselves involved, is certainly as anthropocentric and as unobjective as it would be to accept it simply because it is ego-satisfying."[23] The task of the scientist when studying the whole of biological evolution is to apply insofar as is possible the same criteria, like number of cell types, to all species, including humans.
Needless to say, identifying humans as the most complex species known does not imply that the human species is the pinnacle, or endpoint, of evolution, but simply that it is the most complex so far.

h. *The pattern of complexification is not progressive: a simpler system is preferable to a complex one, which is an aberration.*
Biochemist William Bains articulated this argument after conceding that the passage of evolutionary time is accompanied by the emergence of structures having greater morphological and functional complexity, but he argues that "In no other theoretical framework is complexity seen necessarily more advanced than simplicity" and concludes that "man is an 'unevolved' aberration of a much simpler and more efficient biosphere".[24]
The main fallacy here lies in invoking the general argument that simpler systems are more efficient than complex ones but failing to compare like with like. Overwhelmingly the bacteria and archaea we observe today are not those that existed 2–3 billion years ago when the Earth was very different and lacked oxygen. Those prokaryotes evolved by forming new species to cope with each new environmental stress as a whole range of environmental niches developed on the evolving Earth, resulting in the current astronomical number of such species. Animals, on the other hand,

evolved as more complex species, each able to cope with several environmental stresses.

One branching lineage produced the most complex species we know. *Homo sapiens* is capable of surviving more and different environmental stresses than any less complex species. Uniquely, it has made the whole planet its habitat. By using its highly complex brain it has been able to devise ways of surviving anywhere on the planet's surface in every kind of environmental condition, several kilometres below the surface of both the ocean and the ground, and in outer space. It has extended the lifespans of its members by conceiving of, and applying, medical science and has begun to alter its own genetic structure by gene therapy.

No other species of animal, still less any species of bacteria or archaea, has such survival abilities. By NeoDarwinism's own criterion that members of successful species are better adapted for survival, this constitutes progress.

i.   *There is no progress in biological evolution.*
In 1988 Matthew Nitecki edited a book entitled, perhaps ironically, *Evolutionary Progress*, in which he concluded "The concept of progress has been all but banned from evolutionary biology as being anthropomorphic or at best of limited and ambiguous usefulness."[25]

According to one of his contributors, Stephen Jay Gould, "Progress is a noxious, culturally embedded, untestable, nonoperational, intractable idea that must be replaced if we wish to understand the patterns of history."[26]

Less polemically, Bains articulated this view in another publication by asking rhetorically "is there an *a priori* reason for qualitatively distinguishing man's recently acquired features, such as intelligence and thalassemia, from those of *E. coli*, such as the *Lac operon*?"[27]

The answer is yes. Human intelligence is an ability normally possessed in varying degrees by all members of the species. It is different qualitatively from thalassemia, which is a relatively rare disease in humans resulting from an inherited genetic defect that is usually recessive in both parents. Each is qualitatively different from the *Lac operon*, which is a genetic regulatory system for the metabolism of lactose in the *E. coli* bacterium. To imply that a general ability is the same category of thing as a rare disease and also as three genes is a logical fallacy.

Bains goes on to assert that "the survival value of intelligence would be hard to defend even if we could decide what intelligence is".

I shall consider intelligence is more detail later. Suffice it to say here that intelligence enabled early humans to make more complex and efficient tools with which to defend themselves against bigger and stronger predators and to increase their success in hunting for food, to construct shelters against hostile elements, to make clothes from the skins of prey to keep themselves warm, to make fire to keep themselves warm, ward off predators, and cook food, and to plant and raise crops in order to feed themselves more effectively than foraging, among many other survival skills.

Although most contemporary NeoDarwinists reject the concept of evolutionary progress, the founders of NeoDarwinism did not. To the contrary, they recognized that the greatest progress was seen in the human species. Mayr outlined the stages of this progress:

> [W]ho can deny that overall there is an advance from the prokaryotes that dominated the living world more than three billion years ago to the eukaryotes with their well organized nucleus and chromosomes as well as cytoplasmic organelles; from the single-celled eukaryotes to metaphytes and metazoans to...[warm-blooded animals] with a very large central nervous system, highly developed parental care, and the capacity to transmit information from generation to generation?[28]

Even the more sceptical Simpson concluded

> A majority of [criteria for evolutionary progress] do, however, show that man is among the highest products of evolution and a balance of them warrants the conclusion that man is, on the whole but not in every respect, the pinnacle so far of evolutionary progress.[29]

Julian Huxley was a most enthusiastic and longstanding advocate of evolutionary progress, arguing

> there was progress before man ever appeared on earth....His rise only continued, modified, and accelerated a process that had been in operation since the dawn of life.[30]

Dobzhansky was equally unequivocal:

> Judged by any reasonable criteria, man represents the highest, most progressive, and most successful product of organic evolution. The really strange thing is that so obvious an appraisal has been over and over again challenged by some biologists.[31]

Objections to the overwhelming evidence of progressive complexity reduce either to ignoring the evidence, or to fallacious attempts to explain it away, because it conflicts either with an assumed cause of biological evolution or with an ideology that treats all species as equal.

Denial of the evidence because it conflicts with a belief is more puzzling in a scientist than it is in a creationist denying the evidence for the phenomenon of biological evolution.

In the next two chapters I shall examine hypotheses that attempt to complement or challenge the current paradigm in order to explain the patterns in the evidence.

CHAPTER TWENTY-TWO

Complementary and Competing Hypotheses 1: Complexification

If it could be demonstrated that any complex organ existed, which could not possibly have been formed by numerous, successive, slight modifications, my theory would absolutely break down.

—Charles Darwin, 1872

During the first half of the twentieth century biologists were keen to make theirs a respectable science, like physics, the theories of which were expressed in mathematical equations that enabled precise testable predictions. This desire was born in the 1920s and 1930s with Ronald Fisher's mathematical formulations of evolutionary fitness, J B S Haldane's ten papers published under the title of *A Mathematical Theory of Natural and Artificial Selection*, and Sewall Wright's computations of fitness landscapes and genetic drift. Their approach, which provided the theoretical basis of population genetics, was rooted in Newtonian determinism.

The NeoDarwinian model combined this approach with the analytical methods of molecular genetics. The previously qualitative science of observational biology hardened into determinism and genetic reductionism, with hypotheses expressed as mathematical equations or statistical models.

Paradoxically, at the same time physics underwent a paradigm change, moving away from determinism into relativity theory and the indeterminacy of quantum theory with its non-localization, uncertainty principle, quantum entanglements, and holistic interpretations. Nonetheless most of the hypotheses advanced to account for patterns in the evidence that the NeoDarwinian paradigm does not explain adopt that paradigm's overall approach and use of mathematical models. A minority, however, take a holistic view, arguing that lifeforms consist of components at different hierarchical levels that interact with each other and also with their environment.

The various hypotheses range from those that seek to complement or extend the NeoDarwinian model to those that challenge it. In this chapter I shall examine those advanced primarily to explain rapid complexification and in the following chapter those advanced primarily to explain collaboration.

Intelligent Design

We saw in Chapter 15 that LeHigh University biochemist Michael Behe claimed the first cell must have been intelligently designed because of its irreducible complexity, but he accepted that biological evolution subsequently occurred.* Following Behe's book, which helped launch the Intelligent Design movement, Jonathan Wells, former postdoctoral research biologist at the University of California, Berkeley published a book, *Icons of Evolution: Science or Myth?* that considers ten iconic case studies used to teach Darwinian evolution and concludes they have been exaggerated, distorted, or even faked.[1]

The response of evolutionary biologists has been uniformly hostile, with accusations that Wells dishonestly and fraudulently misrepresents evidence in order to promote his religious beliefs.

Six of these case studies have been examined in previous chapters of this book. Wells is correct to say, for example, that (a) the Miller-Urey experiment fails to demonstrate the emergence of life from inanimate matter;† (b) there is no Darwinian phylogenetic tree that proves descent with modification (not least because there is no agreement among biologists about how, and on what basis, a tree of life should be constructed);‡ (c) the evidence of species' stasis in the fossil record contradicts Darwinian gradualism;§ neither (d) "Darwin's finches" on the Galápagos islands¶ nor (e) the darkening of peppered moths in industrial areas shows the evolution of species (as distinct from reversible adaptive change) and, in the latter case, Kettlewell's photographs of conspicuous moths on tree trunks are actually of dead moths glued onto those trees;** and (f) there is no proof of a new species having evolved according to the Darwinian (actually, NeoDarwinian) mechanism.††

However, highlighting inadequate hypotheses and experiments or observations that are inconclusive, flawed, or even faked (though I think Kettlewell imprudently intended his photographs to be illustrative rather than deceptive), does not disprove the *phenomenon* of biological evolution. While each of the nine categories of evidence examined in Chapters 17 and 18 does not by itself offer indisputable evidence for biological evolution, taken together they provide overwhelming evidence that humans evolved from the earliest lifeforms on Earth.

What these cases highlight is the inadequacy of the NeoDarwinian hypothesis as the only explanation of the phenomenon. If these examples are still used in textbooks to prove biological evolution, it illustrates one or both of two things:

* See page 232.
† See page 225.
‡ See page 331.
§ See page 344.
¶ See page 314.
** See page 311.
†† See page 345.

the textbooks are out of date or are based on belief rather than science. Adherents to the current orthodoxy respond with an institutional defensiveness that fails to admit defects in the NeoDarwinian model shown by conflicting data or to give adequate consideration to other hypotheses consistent with those data. Paradoxically, such defensiveness gives ammunition to believers in creationism.

In a more recent book, *The Edge of Evolution: the Search for the Limits of Darwinism*,[2] Behe accepts the phenomenon of biological evolution; indeed he goes on to say that natural selection is the obvious mechanism by which adaptive gene variants spread through a population. He maintains that random mutation, coupled with natural selection, is not a sufficiently powerful engine to drive the evolution of biological innovation and increasing complexity. His argument rests on the low probability of an organism having two or more simultaneous mutations to yield some advantage for the organism, which is correct. However, he concludes that most of the really important mutations must therefore have been directed by an intelligent agent. This is a logical fallacy. If the NeoDarwinian hypothesis fails to explain major biological innovations and increasing complexity—which is true—it does not follow that other scientific, testable hypotheses now, or in the future, cannot provide an explanation; this chapter summarizes the current ones.

In summary, advocates of Intelligent Design fail to offer any testable explanations of their beliefs, which puts Intelligent Design outside the realm of science.

Punctuated equilibrium

In 1995 palaeontologist Niles Eldredge claimed that the three responses of members of a species population to a change in their environment are, in order of probability

1. *Habitat tracking*
   The members migrate to a habitat to which they are well adapted.
2. *Extinction*
   When the members cannot find a suitable habitat.
3. *Very slow and gradual transformation of the population to adapt to the changed environment.*
   This third, and least probable, response is the only mechanism of speciation recognized by the NeoDarwinian model.

NeoDarwinian theoretical evolutionary biologist George C Williams, on the other hand, dismissed habit tracking as a fable, although Eldredge offered considerable evidence in support.[3]

In their landmark paper of 1972 challenging the gradualism of the NeoDarwinian model that conflicts with the fossil record,[4] Eldredge and fellow palaeontologist Stephen Jay Gould drew on Ernst Mayr's idea of speciation being most rapid in a small, isolated subpopulation to propose their hypothesis of punctuated equilibrium. They claim that, for a large species population occupying a large geographical region, new and even beneficial genetic mutations are diluted by the sheer size of the population and factors such as constantly changing environments. This produces minor adaptive variations but an overall stabilization of morphology, which is consistent with minor morphological changes fluctuating about a mean found in the fossil record over tens or hundreds of millions of years.

For a small group on the periphery of the geographical region, however, genetic mutations coding for beneficially adaptive traits will spread rapidly and transform that group's gene pool within a geologically brief period of tens of thousands of years. This results in a new species unable to breed successfully with the main group.

Williams's champion, Richard Dawkins, dismissed this hypothesis as an "interesting but minor wrinkle on the surface of neo-Darwinian theory" that "does not deserve a particularly large measure of publicity".[5] This sparked acrimonious exchanges between Dawkins and the Eldredge/Gould camp that continue to the present day.

Sudden origins

Jeffrey H Schwartz, physical anthropologist and science philosopher at the University of Pittsburgh, is an evolutionary biologist who retains the increasingly outdated view that a scientist's role is to test current orthodox hypotheses and their underlying assumptions against the evidence and to seek alternative explanations if necessary rather than to seek or interpret evidence in order to defend those hypotheses.

He, too, finds NeoDarwinian gradualism incompatible with the evidence. Instead of adopting Eldredge and Gould's hypothesis, however, he worked with University of Salerno biochemist Bruno Maresca to develop an earlier hypothesis of his by examining the expression of stress proteins in cells in response to physical changes in the environment. Published in 2006, their sudden origins hypothesis claims that cells prevent or correct genetic change, principally by DNA repair mechanisms; this produces DNA homeostasis, or self-regulating stability, and explains morphological stasis.

Significant genomic change only occurs when severe stress—such as rapid temperature change, severe dietary change, or physical overcrowding—overwhelms this DNA homeostasis during the formation of sex cells (eggs and sperms) and hence affects organismal development. The collapse of homeostasis most frequently has lethal consequences, but it

> generates in a few individuals, and over a relatively short period of time (a few generations), major, potentially nonlethal (therefore, "useful"?), rearrangements.[6]

Thus morphological innovations do not result from a NeoDarwinian gradual accumulation of mutations over very many generations in a population, or from the rapid spread of adaptively beneficial mutations in an isolated subpopulation, but from the sudden and severe disruption of the normal state of DNA homeostasis.

This hypothesis offers a more persuasive explanation of the cause of morphological stasis followed by significant genomic change, but has been ignored by the competing NeoDarwinian and Eldredge/Gould camps.

However, Maresca and Schwartz are vague about the nature of these "major potentially nonlethal (therefore, 'useful'?) rearrangements" and how they lead to speciation or produce evolutionary transitions or emergences that show a pattern of complexification.

Stabilizing selection

Defending the NeoDarwinian model against the calls of Eldredge, Gould, and other palaeontologists for a fundamental revision of the NeoDarwinian model, botanist and geneticist G Ledyard Stebbins and molecular geneticist Francisco José Ayala argued in 1981 that the transformation of a population's gene pool over very many generations is compatible with both the gradualist and punctualist modes of speciation.

They queried whether the phenomenon of palaeontological stasis is as common as claimed. Where it does occur, they argue that molecular evolution can take place while morphologies (the only changes identified in the fossil record) remain stable, and this is explained by Dobzhansky's idea of stabilizing selection. Here, genes coding for an adaptively successful morphology, like that of sharks or ferns, are continually naturally selected so that the morphology is stabilized, while genes coding for other traits change, even leading to speciation events.[7]

Their arguments are principally theoretical, and I think Harvard University palaeontologist Peter G Williamson is right to conclude

> the wide range of environments presently exploited by extensively distributed but morphologically uniform modern species, and the long-term morphological stasis (up to 17 Myr) exhibited by many fossil lineages in fluctuating environments, strongly argues against the idea that simple stabilising selection is an adequate explanation for the phenomenon of biological stasis.[8]

Neutral theory

Since the publication in 1968 of his paper introducing the idea, Motoo Kimura led a group of population geneticists who argued mainly on mathematical grounds that nearly all the genetic mutations that persist or reach high frequencies in populations over generations are selectively neutral: they have no appreciable effect on fitness, or morphological adaptation.

Supporting evidence includes a species known as the Port Jackson shark. It has remained morphologically much the same for some 300 million years despite drastic environmental changes claimed to have caused three mass species extinctions; it persists today as a "living fossil" despite a rate of genetic change, measured by the number of amino acid differences between α- and ß-globin (150) which is close to that in the human lineage (147). Over the same period, by contrast, humans evolved from mammal-like reptiles through many transitions. In general, the rate of genetic change is fairly constant for all lineages, but the rate of morphological change varies significantly in different lineages.[9]*

Neutral theory is essentially Sewall Wright's hypothesis of genetic drift† that was ignored by the architects of the NeoDarwinian model because it is inconsistent with the transformation of a population's gene pool arising from the natural selection of adaptively beneficial genetic mutations.

H Allen Orr, a defender of the NeoDarwinian orthodoxy, argues that new technology for genome sequencing and new statistical tests for distinguishing neutral changes in the genome from adaptive ones suggest that advocates of neutral theory underestimate the importance of natural selection. He cites an analysis by David J Begun and Charles H Langley at the University of California, Davis of 6,000 genes in two species of fruit fly that diverged from a common ancestor and concludes that positive natural selection—in which the environment increases the frequency of a beneficial mutation that is initially rare—accounted for at least 19 per cent of the genetic differences between the two species.[10] If so, then it is hardly a ringing endorsement of a model whose *only* mechanism for genetic transformation is natural selection (notwithstanding the fallacy that the environment actively selects).

Whole genome duplication

Mounting discoveries of polyploidy not only in plants but also in a range of animal species‡ have reawakened interest in the hypothesis initially advanced in 1970 by molecular geneticist Susumu Ohno that was largely ignored or dismissed at the time.

Based on relative genome sizes and the observation that tetraploid species (possessing four sets of chromosomes in each cell) occur naturally in fishes and amphibians, Ohno speculated that such doubling of the normal number of chromosomes accounts for major evolutionary innovations. It is also called the 2R hypothesis because its later refinement proposed there were two rounds of

---

* See page 333 for similar examples given by Christian Schwabe in 1986.
† See page 273
‡ See page 310.

whole genome duplication (WGD) near the ancestral root of the vertebrate tree. It claimed that such WGD enabled biological innovation to occur much more efficiently and that all vertebrates, including humans, are degenerate polyploids.

These claims are controversial, and proof is difficult because little is known about diploidization, the process by which a tetraploid species degenerates to become a diploid. However, the hypothesis is supported by evidence that a large part of the human genome, and also the mouse genome, contains sets of genes on one chromosome that are duplicated on three other chromosomes. In 2007 pathologist Masanori Kasahara claimed genome-wide analyses of key chordate species now provides "incontrovertible evidence supporting the 2R hypothesis".[11]

Epigenetics

Evidence of the inheritance of acquired characteristics without any genetic change widened interest in epigenetics, which had been employed to investigate embryology and development.

Like Darwinism, the term epigenetics is used to mean different things. Current researchers employ technical terms in their various definitions, which may be summarized as

> **epigenetics** The study of mechanisms of gene regulation that cause changes in the phenotype of an organism but involve no change in the DNA sequences of the genes themselves.

It includes mechanisms by which stem cells differentiate into cells with specialized functions to form the body of an embryo and subsequently develop into an independent organism. Many such epigenetic patterns are reset when the organisms reproduce, but evidence in the last ten years demonstrates epigenetic inheritance.

> **epigenetic inheritance** The transmission from parent cell to offspring cell, in either asexual replication or sexual reproduction, of variations that give rise to variations in the characteristics of the organism but do not involve any variations in DNA base sequences.

Three principal mechanisms have been identified:

a. Chemical change, particularly methylation in which a methyl group (CH_3-) replaces hydrogen (H–) or another group in a DNA molecule or in the proteins called histones that pack DNA into the condensed form known as chromatin in a cell nucleus.
b. Changes in the folding of chromatin in a cell.
c. Specific binding of small nuclear RNAs to DNA sequences.

The differing mechanisms of gene activation or silencing can act independently, or may be interdependent in that once one form of modification occurs to activate or silence a gene it will result in further modifications.

Epigenetic inheritance thus does not involve changes in the genes themselves, as in the NeoDarwinian model, but rather changes in the way genes are expressed. It provides a molecular explanation for the Lamarckian inheritance of acquired characteristics. For example, Matthew Anway and colleagues concluded that the increase in infertility inherited down the male line of rats without any underlying genetic change* is associated with methylation.

Deep homology and parallel evolution

Prompted by the discovery that developmental regulatory genes are very similar across a wide range of species, palaeontologist and evolutionary biologist Neil Shubin of the University of Chicago and colleagues proposed in 1997 the hypothesis of deep homology. This claims essentially that such genes, or ancient versions of them, were possessed by a very distant common ancestor, and so what may appear to be independent evolutionary convergences are actually parallel evolutions. Furthermore, he says, the discovery of many more deep homologies in the following 12 years has strengthened the hypothesis.[12]

In the case of eyes, for instance, Shubin claims that all modern variations of light sensing in bilateral animals can be traced to the existence of photosensitive cells in a common ancestor, with Pax-6 and other transcription factors at the top of a genetic regulatory pathway leading to the production of opsin proteins.

Although Shubin claims that ancient regulatory circuits provide a substrate from which novel structures can develop, he does not explain why they develop or why regulation by the same Hox family of genes in bilateral animals produces very different body plans.

Evolutionary convergence

University of Cambridge palaeontologist Simon Conway Morris claims the adaptive principle and inherency result in biological evolution being constrained to limited outcomes. He supports his hypothesis of evolutionary convergence with examples spanning the entire biological hierarchy from molecules through organs to social systems and cognitive processes.[13]

The number of amino acids is vast, and yet only 20 are found in organisms. If you consider a small protein comprising a chain of 100 amino acids, the potential number of ways in which the protein could be assembled is 20^{100}, or 10^{130}, an astronomical number, and yet only an infinitesimally small fraction of these are

---

*   See page 347.

found in organisms. Similarly, in their metabolic functions they are folded in only a minute number of ways out of all possible ways.*

Among organs, Conway Morris claims that the camera-like single-lens eye has evolved independently at least six times in species ranging from octopuses to mammals, whilst among a wide range of species "rampant convergence" occurs in other organs, like those responsible for smell, hearing, echolocation, and electroreception. He argues that in the constraints that accompany the development of sensory organs "we can approach the wider problem of the evolution of nervous systems, brains, and perhaps ultimately sentience".

Behaviour, too, shows evolutionary convergence. Eusociality, in which only one female is reproductive while other members of the colony carry out specialized functions, has evolved independently not only in many species of ants, bees, wasps, and termites but also in shrimps and naked mole rats. While not all species of insects are eusocial, Conway Morris claims that the ones that are have taken over their environments: they are adaptively more successful and have led to the extinction of non-eusocial equivalents. Agriculture has evolved independently, from fungus-growing ants in Central and South America to humans.

Conway Morris concedes that, because phylogenetic trees are in a constant state of flux, it may be difficult to distinguish between independent convergent evolution and parallel evolution. But, he maintains, the cases he cites come from such very different species as to minimise the probability of parallel evolution. He considers the role of conserved genes that regulate development across a wide range of species, like Pax-6 for eyes, is over-emphasized because Pax-6 is implicated in the development of other organs. It is also expressed in eyeless nematodes and, while the development of eyes is controlled by Pax-6, the eye structures themselves have evolved by convergence.

He concludes that evolutionary convergence enables first-order predictions of the emergence of important biological properties on Earth. One of these is progress.

> What we see through geological time is the emergence of more complex worlds...within the animals we see the emergence of larger and more complex brains, sophisticated vocalizations, echolocation, electrical perception, advanced social systems including eusociality, viviparity [giving birth to live offspring], warm-bloodedness, and agriculture—all of which are convergent—then to me that sounds like progress.

Moreover,

> The constraints of evolution and the ubiquity of convergence make the emergence of something like ourselves a near-inevitability.

---

* See pages 216 to 222.

Animal ecologist Rob Hengeveld of Amsterdam University attacked this hypothesis, arguing that Conway Morris is selective in his examples and ignores the vastly greater number of divergences in biological evolution. Furthermore, many of the claimed convergences are questionable. For example, only a handful of fishes show a torpedo form against the many thousands of species with different shapes.

I think Hengeveld is right to criticize some claimed convergences. Moreover, the claimed independent evolution of agriculture in fungus-growing ants and in humans, for instance, conflates instinctive behaviour in ants with intelligent purposeful behaviour in humans.

Hengeveld is also right to emphasize the pattern of evolutionary divergence, which Conway Morris largely ignores. On the other hand, Conway Morris is correct to identify the pattern of progressive complexification. The two patterns are compatible: there has been a vast number of diverging animal lineages; complexification has occurred in each, ranging from comparatively little through much greater, until stasis (and usually extinction) is reached at the end of each lineage, to the greatest complexity in humans.

Conway Morris attributes the cause of evolutionary convergence to the "adaptive principle" and the "constraints of evolution", which is "the recurrent tendency of biological organization to arrive at the same 'solution' to a particular 'need'", largely due to "inherency". However, this describes a pattern in phenomena, not a cause of the pattern.

If they are to go beyond tautology or description, the adaptive principle and inherency must each be framed as a specific principle, or law applicable to all biological phenomena, which currently they are not.

Emergence theory

We encountered Harold Morowitz's proposal in Chapter 15 as another explanation of the emergence of life.* Morowitz goes on to identify 19 emergences of increasing complexity in the evolution of life, from prokaryotes to philosophy. To explain why relatively few entities emerge at each of these major evolutionary transition stages from the innumerable possibilities, he seeks selection rules, biological equivalents of the Pauli Exclusion Principle that selects from the myriad possible energy levels of electrons in an atom those that produce the small number of elements found in nature.†

Morowitz asserts that in some cases "in the competitive Darwinian domain" the "dominant selection principle is fitness as defined by replication and survival", which is a tautology, and in other cases the selection principle "is not at all obvious".

---

*   See page 243.
†   See page 135.

The idea of one or more underlying selection rules to explain the emergences of increasing complexity that Morowitz identifies is attractive. However, like his explanation of the emergence of life, it has not yet progressed beyond a description of the phenomena.

Self-organizing complexity

Stuart Kauffman advanced this hypothesis as an explanation of the emergence of life.* He applies it to explain not only organismal embryology and development but also the evolution of species.[14]

As for the former, Kauffman says the reason stem cells with identical genomes differentiate into specialized cells, like liver cells, muscle cells, and so on, is that only relatively few of the genes in each stem cell are switched on, and these are different in each cell. He proposes that a genome operates like a network in which genes regulate each another's activity with no one gene acting as coordinator, and this self-organizing behaviour constitutes a dynamic non-linear system.

His computerised model assumes each gene is a node (N), which is either on or off, and each is linked by a number of connections (K) to other genes that regulate whether the node gene is on or off.

Assuming that this NK network system is governed by a particular class of Boolean logic switching rules called canalizing functions (typically, on or off), then when $K = 1$ the system tends to freeze. If K is greater than 2, the system becomes chaotic. But when $K = 2$ feedback results in the system spontaneously achieving stability on the edge of chaos. Specifically, it settles into a pattern that repeats a cycle of states in which the length of each state cycle equals the square root of N. Thus if there were 100 genes in a genome and each were connected to two other genes then it would settle into repeating a state cycle of 10 steps (since 10 is the square root of 100); during each step different chemical machinery is switched on to produce a cell with a different function, like a muscle cell or a liver cell.

Kauffman claims that the number of steps in a state cycle is approximately equal to the number of specialized cell types. Thus in a human genome consisting of 100,000 genes the square root of 100,000 is 316 and so the model predicts the number of cell types should be 316, which is close to the actual number of 254.† (It is now estimated that humans have at most 25,000 genes, and the square root of 25,000 is 158.)

Kauffman compares the number of cell types against the square root of the number of genes of different organisms, from bacteria to humans, and finds a close correlation. He therefore claims empirical confirmation of his hypothesis.

---

*   See page 240.
†   Kauffman actually says 370 rather than 316. The number of cell types depends on criteria for classification; most classifications list between 210 and 260 but group all the neurons of the central nervous system as one type.

It is an attractive idea that the genes of a stem cell spontaneously organize the production of different specialized cells according to an underlying rule that says the number of connections a gene has to other genes is two. It appears to identify a biological equivalent of the Pauli Exclusion Principle.

However, Kauffman's model doesn't explain what causes the genes to start interacting. Neither does it explain what causes the number of connections to be two (nor did the Pauli Exclusion Principle explain why no two electrons in an atom or molecule can have the same four quantum numbers).

Whereas the Pauli Exclusion Principle accurately predicted the existence of new elements that were subsequently discovered, Kauffman's claim of predictive accuracy is questioned by what we now know of genes and genomes. His correlations assume the number of genes in an organism is proportional to the amount of its DNA, but Table 18.1* shows this is not the case.

While genes are indeed regulated by networks, these networks appear to involve not other genes comprising 2 per cent of the genome but a majority of the other 98 per cent of DNA in the genome.

Using similar reasoning, assumptions, and computer models he argues that a complex network of species interacting with each other and with their environment, and subject to the force of natural selection (he appears to accept the misconception that natural selection is a cause), will naturally evolve towards a phase transition at the edge of chaos. Here, in homeostasis, biological evolution can occur rapidly because the network possesses stability but with its parts connected loosely enough for changes in one or more parts to cause a change in the whole interactive network.

Kauffman advanced his idea in 1991, and since then mathematicians, computer scientists, and solid-state physicists have refined the computer models. However, insufficient progress has been made in grounding the various models in biological data. Consequently, like its explanation of the emergence of life, this idea currently offers no more than hope that its models will lead to deeper new laws of biological evolution or else some physico-chemical law, or law of form, that imposes constraints on otherwise limitless biological possibilities.

Genome evolution laws

More recently, in 2011 Eugene Koonin, computational evolutionary biologist at the US National Institutes of Health, argued that research in quantitative evolutionary genomics and systems biology has led to the discovery of four main universal patterns in genome and phenome evolution in all evolutionary lineages for which genomic data are available, including diverse groups of bacteria, archaea, and eukaryotes.

Koonin claims that these highly technical statistical patterns are not only surprising but can be accounted for by simple mathematical models, similar to

---

* See page 319.

those used in statistical physics. One such model is his birth-death-innovation model, which comprises only three elementary processes: (1) gene birth (duplication), (2) gene death (elimination), and (3) innovation (the acquisition of a new family, for example via horizontal gene transfer). Such models do not incorporate natural selection. Hence, Koonin concludes, these patterns are not shaped by NeoDarwinian natural selection but are emergent properties of gene ensembles.

He acknowledges that the process and course of biological evolution critically depend on historical contingency and involve extensive adaptive "tinkering", and so a complete physical theory of evolutionary biology is inconceivable. Nevertheless, Koonin claims, the universality of several patterns of genome and molecular phenome evolution, and the ability of simple mathematical models to explain them, suggest that laws of evolutionary biology comparable in status to laws of physics might be attainable.[15]

If Koonin is right, this approach offers a potentially fruitful avenue of research to try and determine how outcomes in biological evolution are constrained to a very few out of near limitless possibilities. One test would be to use the models to predict the future evolution of rapidly changing species like fruit flies.

Natural genetic engineering

Also in 2011 James Shapiro, professor in the Department of Biochemistry and Molecular Biology at the University of Chicago, challenged the NeoDarwinian orthodoxy that biological evolution is caused by randomly generated genetic changes in otherwise static genomes. He argues that cells have an innate ability to re-organize their genomes in response to hundreds of kinds of inputs.

In support of what he calls "natural genetic engineering", he draws on evidence examined in Chapters 17 to 19, together with hypotheses discussed in this chapter plus his own ideas. Specifically, he claims that successful hybridization between members of different species,* transposons (jumping genes),† epigenetics, horizontal gene transfer,‡ whole genome duplication, and symbiogenesis (the merger of two separate organisms to form a single new organism, examined in the next chapter), together with studies of how cells regulate the expression, reproduction, transmission, and restructuring of their DNA molecules, show that genomes are not read-only memory systems subject to accidental change; they are read-write information storage organelles at all time scales, from the single cell cycle to evolutionary eons. As evolution proceeds, so does evolvability: living organisms are self-modifying beings and are intrinsically teleological.

The current biological paradigm, developed by population geneticists and laboratory zoologists, fails to explain the rapidity and variety of biological

---

*   See page 282.
†   See page 317.
‡   See page 309.

complexification; moreover, gradual random mutation is more likely to degrade rather than create. Integrating biological science with an information- and systems-based approach, Shapiro argues, will produce a new paradigm for the twenty-first century.[16]

Shapiro's criticisms of the current NeoDarwinist orthodox paradigm are well founded. However, he doesn't offer a new paradigm, but rather an agenda for one based on the concept of organismal self-modification that is more consistent with the evidence than is the current paradigm.

Systems biology

Systems biology emerged in the mid 1960s as a multidisciplinary approach that, in contrast to the reductionism of NeoDarwinism, takes a holistic view of natural phenomena and aims to discover emergent properties.

Biologists recruited computer scientists, mathematicians, physicists, and engineers in order to study living things as an integrated and interacting network of genes, proteins, and biochemical reactions that give rise to life and are responsible for an organism's form and functions. These emergent properties, they argued, cannot be attributed to any single part of the system, which is therefore an irreducible entity. As one of its founders, Denis Noble, Professor Emeritus and Co-Director of Computational Physiology at Oxford University, put it

> Systems biology...is about putting together rather than taking apart, integration rather than reduction. It requires that we develop ways of thinking about integration that are as rigorous as our reductionist programmes, but different....It means changing our philosophy, in the full sense of the term.[17]

Initially its research programmes were mainly in medical science, like investigating a body's immune system, which doesn't result from a single mechanism or gene; rather the interactions of numerous genes, proteins, mechanisms, and the organism's external environment produce immune responses to fight infections and diseases.

Systems biology expanded and developed into a subdiscipline of its own at the beginning of the twenty-first century when technological developments enabled the rapid sequencing of whole genomes. Trying to understand the exponentially increasing volume of molecular sequencing data resulted in reliance on bioinformatics and specializations like genomics, epigenomics, transciptomics, proteomics, and so on.

In 2010 molecular biologist and Nobel laureate Sydney Brenner attacked systems biology, saying that it was doomed to failure because deducing models of function from the behaviour of a complex system is an inverse problem that is impossible to solve. What I think he means is that for a complex system like a cell there is an incomprehensibly vast number of possible models and we have no way of deducing which is the correct one.

Systems biologists, however, attempt to find rules that select the functional models from all possibilities. Frequently they do this by using algorithms in computer models that lead to the emergence of one or more or novel properties which map most closely the emergence of observed novel properties at a higher systems level in biology.

Brenner counters by asserting that "most of the observations made by systems biologists are static snapshots and their measurements are inaccurate....Any nonlinearity in the system will guarantee that many models will become unstable and will not match the observations."[18]

As a subdiscipline in its own right, systems biology is only in its infancy and it is far too early to dismiss it. To the contrary, its philosophical approach is right to treat things like cells and higher levels like organisms, and then even higher levels like colonies of ants, as interacting parts of irreducible wholes that interact with their environments. The principal danger in the future is that, in a desire to be as rigorous as the reductionists, systems biologists will become so focused on the specialist investigation of the parts as to lose focus on the whole.

Gaia hypothesis

Broadly similar in concept but wider in its application to the whole Earth is the Gaia hypothesis originated during the 1960s by the independent British scientist James Lovelock. It proposes that the Earth's biosphere, atmosphere, oceans, and soil constitute an interactive, self-regulating system that produces an optimal physical and chemical environment for life on Earth.*

Since he developed the hypothesis in collaboration with Lynn Margulis it has diverged into a variety of forms advocated by various proponents in response to new data and to criticisms of its initial arguments.

As far as biological evolution is concerned, the Gaia hypothesis rejects the NeoDarwinian model of a population's very gradual adaptation to its physical environment. That physical environment is shaped by a network of living systems capable of innovation, resulting in co-evolution with such inanimate networks at a planetary level. As Lovelock puts it, "So closely coupled is the evolution of living organisms with the evolution of their environment that together they constitute a single evolutionary process."[19]

This makes intuitive sense, and many of the Gaia hypothesis's simple computer models have a grounding in atmospheric, oceanic, and biological data at a global level. However, as currently developed, the hypothesis does not provide a sufficiently tangible explanation of the evolution of species.

---

*   See page 182.

Formative causation

Epigenetic hypotheses examined earlier suggest *how* non-genetic changes occur but do not explain *why* they occur: why a methyl group should substitute for hydrogen, for example, or why any epigenetic mechanism causes stem cells to differentiate into liver cells, nerve cells etc. and then develop into an independent adult, a process known as morphogenesis.

Plant cell biologist Rupert Sheldrake's hypothesis of formative causation, encountered in Chapter 13 when I examined ideas about what life is, proposes not only how morphogenesis and biological evolution take place but also why they do: nature is habitual.*

In Sheldrake's view there are no immutable universal laws, which are Newtonian deterministic concepts. (What causes such universal laws? According to Newton it is God.) He proposes that memory is inherent in nature in the form of universal morphic fields that cause the development of forms at all levels of complexity, structures, and organization from atoms through to behaviour and mind, and these morphic fields evolve as the universe evolves.

Thus morphogenesis is governed by a morphogenetic field that imposes a pattern on otherwise random or indeterminate activity. Such a field is generated by the pattern of, say, all previous stem cells differentiating into liver cells, nerve cells, and so on, which occurs when different sets of genes in identical genomes are switched on for specific periods. The morphogenetic field imposes this characteristic developmental pattern through what Sheldrake calls morphic resonance, by analogy with energetic resonance. For example, only light waves of particular frequencies are absorbed by atoms and molecules to produce their characteristic absorption spectra. Only radio waves of specific frequencies are absorbed by atomic nuclei placed in a strong magnetic field, a phenomenon known as nuclear magnetic resonance. In each case the system responds, or resonates, only to select frequencies.

So too each stem cell resonates to the morphogenetic field to produce its characteristic development. The *mechanism* by which the genes are regulated may well be methylation or any other epigenetic mechanism. The differentiating behaviour of the new stem cells feeds back by resonance to the universal morphogenetic field, thus adding new information so that the field itself evolves.

This hypothesis offers an answer to the search for a selection principle in all the cases considered so far where theoretically limitless possibilities are constrained to a small number of observed outcomes. For example, the initial, probably random, differentiation of a stem cell into what becomes a liver cell generates a morphogenetic field that reinforces this tendency. Millions of years later, because the number of previous stem cells differentiating into liver cells is so vast, the pattern imposed on a new stem cell's behaviour is so strong that it seems to be

---

*   See page 196.

a law. But it is not a principle in the sense of a universal law applicable for all time. Small changes in the behaviour of new stem cells are incorporated into the morphogenetic field; these accumulate over time, causing the evolution of that field and thus of the observed law and the way consequent stem cells behave.

Sheldrake applies this concept of formative causation to all aspects of biological evolution, offering an explanation for the phenomena that conflict with NeoDarwinian orthodoxy, and he claims empirical support. In the case of inheritance of acquired characteristics, for example, William Agar and colleagues from Melbourne University reported in 1954 an experiment designed to eliminate flaws in previous experiments to train rats. His team introduced a species of rat into a tank of water with two possible exits for escape. The "right" exit was dark and the "wrong" exit was illuminated. Rats that chose the wrong exit received an electric shock. The researchers alternated right and wrong exits during repeats of the experiment, and most rats slowly learned to use the right exit.

Over a period of 20 years Agar measured the rates of learning of trained and untrained rats for 50 successive generations. He found a marked tendency for rats of the trained line to learn more quickly in subsequent generations, in conformity with the epigenetic explanation of Lamarckian inheritance of acquired character-istics. Yet he *also* found the same tendency in the line of untrained rats, which is not explained by Lamarckian inheritance, still less by the NeoDarwinian model.[20]

Sheldrake is open-minded enough to say that, while consistent with the hypothesis of formative causation, it doesn't prove it because there may be some unknown reason for this improved rate of learning over generations of untrained rats.

The lack of experimental testing of the hypothesis since Sheldrake introduced it in 1981 may be attributed to a lack of open-mindedness in the scientific establishment.*

Having considered hypotheses that either extend or challenge the current principles of NeoDarwinism primarily in order to explain the empirical pattern of rapid complexification and biological innovation, I will do likewise for hypotheses that primarily seek to explain the ubiquitous pattern of collaboration in nature.

*   See page 197.

CHAPTER TWENTY-THREE

Complementary and Competing Hypotheses 2: Collaboration

One of the greatest problems for the biological and social sciences is to explain social behaviours such as cooperation.

—Stuart West, 2011

Much of what Darwin said is, in detail, wrong. Darwin if he read [The Selfish Gene] would scarcely recognize his own original theory in it.

—Richard Dawkins, 1976

Life did not take over the globe by combat, but by networking.

—Lynn Margulis, 1987

Sociobiology

Attempts to explain collaboration and altruism in terms of the NeoDarwinian model have occupied entomologists, zoologists, ethologists, population geneticists, and other biologists ever since the formulation of the model because the only behaviour recognized by that model is competition. The problem was brought into sharp focus by the social behaviours of many insect species, and it was one of the world's leading entomologists, Edward O Wilson, who formally launched the new subdiscipline of sociobiology in 1975 by publishing a book, *Sociobiology: the New Synthesis*. In it Wilson reviewed hypotheses advanced by specialists in different fields and called for their integration to produce a "systematic study of the biological basis of all social behavior", including human. A central axiom is that an individual's behaviour is shaped by its genes and thus is heritable and subject to natural selection.[1]

Group selection
Long before the NeoDarwinian synthesis, Darwin had recognized the conflict

between his hypothesis, which is rooted in competitive self-interest, and human morality. He concluded that, while men of a high standard of morality have little or no advantage over others within the same tribe, a tribe containing such men ready to aid one another and to sacrifice themselves for the common good would be victorious over most other tribes "and this would be natural selection".[2]

Although Darwin had been considering human behaviour ruled by reason, in his 1962 book *Animal Dispersion in Relation to Social Behaviour* zoologist Vero Wynne-Edwards applied this idea of group selection to animal behaviour, arguing that many behaviours are adaptations of the group rather than adaptations of the individual.

Theoretical evolutionary biologist George C Williams attacked this hypothesis in his influential 1966 book *Adaptation and Natural Selection* in which he asserted that "group-related adaptations do not, in fact, exist".[3]

Williams did not criticize the conflation of instinctive animal behaviour and intentional human behaviour. To the contrary, he applied the reductionist approach to animals and humans, declaring "the goal of an individual's reproduction is...to maximise the representation of its own germ plasma, relative to that of others in the same population".[4] He helped lay the foundations of a radical gene-centred view of behaviour adopted by most sociobiologists.

Kin-related altruism or inclusive fitness

I don't know whether it is an urban myth, but this hypothesis is said to have its origins in a British pub in the mid 1950s. Geneticist J B S Haldane was asked if he would sacrifice his life for his brother. After a few scribbled calculations he announced he would only sacrifice his life for at least two brothers or eight cousins. His reasoning was that, because a brother shares half his genes, and first cousins share one eighth of his genes, he could only ensure the survival of copies of his genes if at least two brothers or eight cousins survived as a result of his death.

He published these conclusions in 1955,[5] but it was another theoretical geneticist, William D Hamilton, who formalized the hypothesis in 1964[6] as a complex mathematical model based on assumptions about genes that would lead individuals to behave according to a formula that became known as Hamilton's Rule:

$$r \times B > C$$

where r is the genetic relatedness (degree of kinship) of the altruist to the beneficiary, B is the benefit (in number of offspring) gained by the recipient of the altruistic behaviour, and C is the cost (in number of offspring) suffered by the individual behaving altruistically.

In siblings r is 1/2 because they have half their genes in common, while for first cousins r is 1/8 (one eighth of their genes in common). Kin-related altruism claims to explain altruistic behaviour in humans and animals, like worker ants giving up their own reproductivity in order to assist the enhanced reproductivity of the queen, which is their mother.

Hamilton called it "inclusive fitness", which theoretical geneticists defined as the ability of an organism to pass on its genes to the next generation. According to Hamilton it could do so directly by sexual reproduction, passing on half its genes to each offspring, or it could do so indirectly by aiding the reproduction of identical genes in other organisms. Thus its evolutionary fitness included both direct and indirect methods.

George Price, an American physical chemist by training, produced a mathematical theorem extending and proving Hamilton's idea. But Price recognized a mathematical proof that altruistic behaviour resulted from genes preserving copies of themselves through close kin wasn't empirical proof that this is how altruism operates in the real world. To demonstrate this he began giving away his possessions to people with whom he had no genetic relatedness: the homeless, alcoholics, and drug addicts. Eventually he was evicted from his rented accommodation and, after living in various squats in north London, he committed suicide.[7]

The kin-related, or inclusive fitness, hypothesis has six principal problems.

First, it does not address the behaviour of the most numerous species on the planet, single-celled organisms, like prokaryotes, that replicate by producing clones of themselves and can use horizontal gene transfer to gain genes from, or pass genes on to, different species with whom they have no kinship.

Second, the mathematical models assume that all the members of a group will behave the same way if they share the same genes. In reality the relationships between genes and behaviour is far more complex. Behavioural variation between groups can be large even when genetic variation between them is small.[8]

Third, most animal behaviours cited as examples are not altruistic. In insect societies, for example, workers' loss of reproductivity in order to help the queen be more reproductive is enforced, often savagely.*

Fourth, it confuses instinctive animal behaviour with intentional human behaviour.

Fifth, it fails to explain altruistic behaviour in humans. Price's altruism was exceptional only in its magnitude. Many nuns and monks intentionally adopt celibacy and devote their lives to altruistic acts where there is no direct or indirect passing on of genes to the next generation. More typically, when someone in Britain donates £1,000 to the victims of the 2010 earthquake in Haiti, there is similarly no direct or indirect passing on of genes to the next generation.

Sixth, even recognizing that the vast majority of such acts in animals constitute enforced or instinctive collaboration rather than altruism, the genetic relatedness of second cousins is 1/128 (one part in one hundred and twenty-eight).

---

* See page 326.

At this point the indirect reproduction of identical genes is minute, and beyond this degree of relatedness it is negligible. Yet evidence abounds of collaboration between distantly related members of the same species and also between members of different species.*

"Reciprocal altruism"

Harvard theoretical sociobiologist Robert Trivers proposed "reciprocal altruism" to solve the last two problems. In essence he claims animals and humans behave altruistically because a selfless act will be reciprocated in the future.

His seminal paper "The Evolution of Reciprocal Altruism",[9] published in 1971, elaborates Hamilton's Rule with mathematical formulations of net benefits and costs for numbers of each individual's different altruistic acts, controlled by different genes with different population frequencies at different times. By setting the costs to altruists as low and the benefits to recipients as high in his model, Trivers argues that cheats (individuals who benefit from altruism but do not return it) might initially be at an advantage; however, natural selection would subsequently favour individuals who discriminate against cheats and who collaborate with individuals who have assisted them in the past. He employs three cases to support his hypothesis: (1) cleaning symbioses, (2) warning cries in birds, (3) human reciprocal altruism.

The hypothesis, however, is based on a fundamental conceptual flaw. The universally accepted definintion of altruism may be summarized as

> **altruism**   Behaviour characterized by unselfish concern for the welfare of others; selflessness.

"Reciprocal altruism" is a contradiction in terms. If an act is reciprocated it is not a selfless act. It is what Kropotkin described some 70 years previously as mutual aid.

Cleaning symbioses, noted in Chapter 19, exemplify mutual aid between different species: small cleaner fish, such as wrasse, feed on parasites in the mouth of a large host fish, like a grouper. The benefits to each party are immediately realized and not delayed as in the Trivers hypothesis; he offers no evidence of discrimination against cheats or of selective collaboration with individuals who have assisted them in the past.

The second case was documented by Kropotkin: a few birds act as sentries for the rest of the feeding flock when their warning cries attract the attention of predators and make them more vulnerable. Again, Trivers provides no evidence that a bird does not act as sentry for those birds that have refused to act as sentries in the past, or that it only acts for those that have provided this service to it in the past.

---

* See page 328.

In the third case Trivers fails to distinguish between mutual aid between humans, as with mutual and cooperative societies, and altruism. The example of an altruist in Britain donating £1,000 to victims of the 2010 earthquake in Haiti not only refutes the kin-related altruism hypothesis it also refutes "reciprocal altruism": no donor anticipates that earthquake victims will reciprocate by sending him £1,000 in the future, nor are they likely to aid him in any other way.

Despite these conceptual flaws Trivers's paper proved influential among sociobiologists. His ideas, together with those of Hamilton, Price, and Williams, consolidated a gene-centric approach and stimulated a raft of theoretical papers encapsulated in Figure 23.1 as a matrix of four basic behaviours based on gains and losses in evolutionary fitness to the actor and the recipient. Behaviour in which both actor and recipient gain is cooperative, in which the actor loses and the recipient gains is altruistic, in which both actor and recipient lose is spiteful, and in which the actor gains and the recipient loses is selfish.

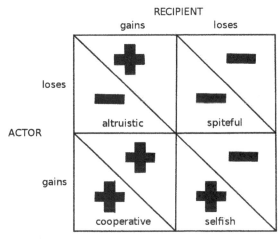

Figure 23.1 Sociobiology's matrix of four basic behaviours viewed as gains (+) and losses (−) in evolutionary fitness of the actor and the recipient

Game theory

In order to support their hypotheses many sociobiologists invoke game theory. This was developed principally by mathematician John von Neumann and economist Oskar Morgenstern in their book *The Theory of Games and Economic Behavior* (1944) in order to predict economic behaviour; it was also employed to devise military strategies after the Second World War. A game consists of between two and *n* players in a competitive situation. Essentially each player is given options, chosen by the game designer, and plays within rules, also chosen by the game designer, with the objective of maximising his own rewards or minimizing his opponents' rewards.

With their arbitrary number of players, choices, rules, rewards, and penalties, such theoretical games can only be evaluated by the success of their predictions. Their track record in Pentagon war games used in Korea, Vietnam, Cambodia, Iraq, and Afghanistan, and in economics where they failed to predict the collapse of the sub-prime mortgage market, the loss of liquidity of major banks, the global financial crisis of 2007–2008, and the rise of China as an economic power, inspire less than total confidence.

The game adopted by most sociobiologists is the Prisoner's Dilemma, which Hamilton, together with political scientist Robert Alexrod, used to try to show how cooperation is selected for in a social group. In the classic case two prisoners are charged with, say, armed robbery for which the sentence is, say, 10 years. The rules are that the prosecutor does not have enough evidence to convict without a confession, and so he keeps them apart and gives each prisoner the choice of either confessing or not confessing without knowing the other prisoner's choice. If each confesses each will be convicted but given a reduced sentence of, say, 6 years; if neither confesses each will be convicted of the lesser crime of illegally possessing a firearm and serve, say, 2 years; if Prisoner A confesses and the other does not, Prisoner A will be rewarded by being released, while his evidence will convict Prisoner B who will serve the full sentence of 10 years; the same applies to Prisoner B.

Prisoner A will be better off if he confesses, irrespective of what choices Prisoner B makes; the same applies to Prisoner B. Therefore a rational prisoner acting in his own self-interest will confess. Yet in a society of two prisoners the social result of each prisoner following this self-interest is worst of all (12 person-prison years); by cooperating with each other and both refusing to confess the social result is best of all (4 person-prison years). Hence, it is claimed, cooperative behaviour benefits the society and so is selected for.

The clear implication here is that cooperation is selected for when the prisoners do not act rationally in their own self-interest. This contradicts the basic NeoDarwinian tenet of competitive self-interest as the driving force of evolution.

Alexrod later devised a more sophisticated version, the Iterated Prisoner's Dilemma, in which the game is played repeatedly on a computer with, for example, each player's choice depending on the other player's previous choice.

Even in this case the prisoners are not permitted to communicate with each other when deciding how to act. But cooperation means working together, and that requires communication between the cooperators. To prevent communication is to prevent cooperation, which makes a nonsense of the game.

In general, by designing the game rules and setting specific numerical values for the rewards and penalties it is possible to demonstrate that it can be in both players' selfish interests to cooperate or that it can be in one player's selfish interest to act altruistically because he will gain in the long term. Sociobiologists adapted such games to explain not only human but also animal behaviour.

In a group of animals, of course, there are not just two members, still less two genes, that are playing a game according to simplistic rules designed by some transcendental Rules Designer who gives each player only two choices, determines the rewards and penalties, and permits no communication between the players and no interaction with other members, other groups, other species, or their environment. Such games, devised in the 1970s, bear as much relationship to reality as did the other games played on the ZX Spectrum 16 kilobyte or Commodore 64 computers of that era. Yet evolutionary biologists are still using them today.

Empirical proof

As for empirical rather than game model proof, the iconic example of "reciprocal altruism" is vampire bats. In the early 1980s animal behaviourist Gerald S Wilkinson, then at the University of California at San Diego, made a study of these bats who regurgitate some of their last meal to feed other bats who have failed to find, or whom Wilkinson prevented from finding, a blood meal. He concluded that giving food depends equally and independently on degree of relatedness and an index of opportunity for reciprocation, and that reciprocity operates within groups containing both kin and nonkin.[10]

However, his data show no evidence of a vampire bat refusing to feed a bat who had previously refused to feed another bat, or how a vampire bat knows that the favour will be returned in the future. Rather they suggest that food sharing takes place between parents and young offspring, and in other cases is determined to a large degree by the need of the underfed, who will likely die without a blood meal for two nights.

Psychologist and evolutionary biologist Marc Hauser and colleagues at Harvard University reviewed more recent attempts at training jays and chimpanzees to play games of the Prisoner's Dilemma, plus Hauser's own experiments with genetically unrelated cotton-top tamarin monkeys, to provide empirical proof of "reciprocal altruism".

One problem here is that in 2010 Harvard University confirmed that a three-year investigation had found Hauser guilty of eight counts of scientific misconduct. The charges included data falsification of cotton-top tamarin behaviour. The US Office of Research Integrity also concluded in 2012 that Hauser was guilty of research misconduct and that he fabricated data for one study and falsely described methodology in other studies. Hauser neither admitted nor denied the charges, but resigned from his post.[11]

In their 2009 paper Hauser and his colleagues concluded

> Darwin knew that altruism, and morality more generally, represent genuine puzzles in light of his theory of natural selection. Adopting the gene's eye view, these altruistic actions no longer represent a challenge to Darwin's logic. The costs of altruism are either neutralized by kinship or by the

prospects of a reciprocally altruistic relationship. *As well as this theoretical perspective works, it largely fails to account for the virtual absence among social vertebrates of reciprocal altruism and spite* [my italics].[12]

That is, the theoretical model fails to account for what is observed.

Stuart West and colleagues confirm this in their 2011 review of the literature. "Overall, after 40 years of enthusiasm, there is a lack of a clear example of reciprocity in non-human species."[13]

As for spiteful behaviour (see Figure 23.1), Andy Gardener of Queen's University, Ontario and colleagues developed a theoretical model for polyembryonic (genetically identical) wasps. They produce soldier wasps who do not defend them but who mediate (actor loses) and who are sterile (recipient loses). With a semantic dexterity of which George Orwell would have been proud, they argue that this is "a prime candidate for spiteful behaviour" but that "other interpretations such as altruism or indirect altruism are valid".[14]

Edward O Wilson's later conclusion on kin selection applies equally to "reciprocal altruism" and any of these mathematical models designed to explain four basic models of animal and human behaviour. It is "constructed to arrive at almost any imaginable result, and as a result is largely empty of content. Its abstract parameters can be jury-rigged to fit any set of empirical data, but not built to predict them in any detail."[15]

Selfish gene

Richard Dawkins popularized and developed the gene-centred approach principally in two books, *The Selfish Gene* (1976) and *The Extended Phenotype* (1982).

Dawkins explains that "The argument in [*The Selfish Gene*] is that we, and all other animals, are machines created by our genes." The competition in nature is not between organisms but between genes for their own survival and replication. "I shall argue that a predominant quality to be expected in a successful gene is ruthless selfishness."[16]

Often dubbed Darwin's rottweiler, Dawkins is honest enough to concede that Darwin would scarcely recognize his own theory in the selfish gene hypothesis, as the quote beneath the chapter heading shows. However, he maintains it is "a logical outgrowth of orthodox neo-Darwinism". I think it is best described as

> **UltraDarwinism**   Any hypothesis that employs the concept of the evolution of things other than organisms by means of natural selection in which the cumulative effect of small random variations in the characteristics of those things, or else characteristics caused by those things, over numerous generations makes them increasingly better fitted to compete for survival and reproduction in their environment.

A single selfish gene "is trying to get more numerous in the gene pool".[17] This gene selfishness "will usually give rise to selfishness in individual behaviour. However... there are special circumstances in which a gene can achieve its own selfish goals best by fostering a limited form of altruism at the level of the individual animal. 'Special' and 'limited' are important words in the last sentence."[18]

These special and limited exceptions are kin-related altruism and "reciprocal altruism", discussed above, in which genes may also achieve their selfish objective by aiding the survival and replication of identical genes in other bodies.

Moreover "not all the phenotypic effects of a gene are bound up in the individual body in which it sits. Certainly in principle, and also in fact, the gene reaches out through the individual body wall and manipulates objects in the world outside, some of them inanimate, some of them other living beings, some of them a long way away....The long reach of the gene knows no obvious boundaries."[19]

For example, although it "is not entirely clear what [a beaver dam's] Darwinian purpose is, but it certainly *must* have one for the beavers expend so much time and energy to build it...a beaver lake...is a phenotype, no less than the beaver's teeth and tail, and it has evolved under the influence of Darwinian selection. Here the choice *must* have been between good lakes and less good lakes....Beaver lakes are extended phenotypic effects of beaver genes [my italics]."[20] I confess it is not entirely clear why these things must be so; unsupported assertions constitute neither good science nor good logic.

I am using more direct quotes than normal because Dawkins frequently complains that sceptics never read beyond the title of *The Selfish Gene*. And lest it be thought that I am quoting 1976 ideas (updated in the 1986 edition I am using) that have been superseded, Dawkins declared in 2006 "there is little in it that I would rush to unwrite now or apologize for".[21]

Despite this declaration, the selfish gene hypothesis is not without some theoretical and empirical problems.

The selfish genes are lengths of DNA that code for proteins. Dawkins does not dismiss the other 98 per cent of the human genome as junk, but says "The simplest way to explain the surplus DNA is to suppose that it is a parasite, or at best a harmless or useless passenger hitching a ride in the survival machine created by the other DNA."[22] Simplest, maybe, but wrong. Most of it is implicated in networks that regulate genes, and such regulation determines an organism's phenotype as much as do its genes.*

Dawkins does say "If we allow ourselves the licence of talking about genes as if they had conscious aims, always reassuring ourselves that we could translate our sloppy language back into respectable terms if we wanted to..."[23] But he never does translate this sloppy language into respectable terms, or explain what the selfish gene is a metaphor for, if indeed it is a metaphor, or explain what causes genes to behave as though they were selfish. In fact he asserts that "The fundamental unit,

---

*   See page 317.

the prime mover of all life, is the replicator...the individual body...did not have to exist."[24] Philosopher Peter Koslowski concludes that Dawkins "concedes a faculty for aspiration, intentionality and consciousness to genes. In so doing he falls into a genetic animism, which apportions perception and decision to the genes."[25]

Gould makes a similar criticism. "The misidentification of replicators as causal agents of selection—the foundation of the gene-centred approach—rests upon a logical error best characterized as a confusion of bookkeeping with causality."[26]

Setting aside the conceptual error of ascribing intentions to segments of an acid (DNA, deoxyribonucleic acid), the scientific test of the selfish gene hypothesis is whether it is supported by evidence.

Using Dawkins's language, cancer genes are the most successful selfish genes. They cause the cells they occupy to replicate uncontrollably, producing a vast number of copies of themselves. They can be so successful that the body, their survival machine, dies, taking them with it. It is difficult to see how this is real success.

At one point in Chapter 3 of *The Selfish Gene* Dawkins concedes that in embryonic development, genes "collaborate and interact in inextricably complex ways, both with each other and with their external environment".[27] Yet the rest of the book reverts to the thesis that the "gene is the basic unit of selfishness"[28] and examples cited assume that each gene codes for a trait. However, most traits arise from the collaboration of very many genes.* This empirical evidence of gene collaboration (not one of the special and limited exceptions) contradicts the hypothesis's axiom of gene competition.

Before he became a popularizer of science, Dawkins was an ethologist (his PhD thesis, supervised by Niko Tinbergen, was "Selective Pecking in the Domestic Chick"), and so it might be assumed that he is on safer ground when seeking evidential support from behaviour. He says that "If *C* is my identical twin [and thus has identical genes to me], then I should care for him twice as much as I care for any of my children [who have half my genes], indeed I should value his life no less than my own."[29] This prediction of the selfish gene theory is contradicted by behavioural evidence. Dawkins attempts to explain it away by saying that animals may not be certain of their relationships. How does this explanation apply to human behaviour?

Dawkins uses Trivers's argument that, because a male produces a large number of small sperms while a female produces a small number of relatively large eggs, the male will want to leave the rearing of offspring to the female so that he can go and have sex with as many other females as possible in order to produce as many copies of his genes as possible. "As we shall see, this state of affairs is achieved by the males of a number of species but in other species the males are obliged to share an equal part of the burden of bringing up children."[30] That is, some evidence supports this assertion for males while other evidence contradicts it. In

---

*   See page 324.

fact, apart from monogamous species, in several groups of birds, like the phalaropes (small wading birds), the female leaves the male to incubate the egg and rear the offspring.

Dawkins concedes that in species living in herds or troops an unrelated female may adopt an orphaned youngster. "In most cases we should probably regard adoption...as a misfiring of a built-in rule. This is because the generous female is doing her own genes no good by caring for the orphan....It is *presumably* a mistake that happens too seldom for natural selection to have 'bothered' to change the rule by making the maternal instinct more selective [my italics]".[31] This seems rather thoughtless of natural selection, whoever she or he is.

Other evidence conceded by Dawkins in a television programme he made for Britain's Channel 4 in 2009 also contradicts the selfish gene hypothesis, but instead of changing the hypothesis to account for such contradictions, Dawkins also attributes these to the "misfiring" of genes without explaining what causes such misfiring.[32]

Genial gene

Stanford University evolutionary biologist Joan Roughgarden also used game theory, but to challenge the selfish assumption of Trivers's parental investment hypothesis and the selfish gene hypothesis. She borrowed a different economic game, one whose rules provide for players communicating, bargaining, and allocating side-payments to each other. In 2009 she published *The Genial Gene: Deconstructing Darwinian Selfishness*, which concludes that the only way that either parent can successfully rear its offspring, and thus achieve evolutionary fitness, is through a significant degree of cooperation.[33]

It is tempting to conclude that Roughgarden's outcome from her model illustrates a Newtonian Third Law of Biology: For every mathematical model in biology there is an equal and opposite model.

Multilevel selection

In 2010 Edward O Wilson had the perspicuity and courage to say he was wrong and sociobiology had taken a wrong path in the 1960s. He and David Sloan Wilson (no relation) had independently arrived at the idea of multilevel selection. This hypothesis claims there is no privileged level—gene or cell or organism or group or ecosystem—at which natural selection operates. In the complex world of biology one level may be more significant, but that level changes for each species with time and with the environment. Moreover, major evolutionary transitions occur when there is a change in selection level, for example when individual eukaryotic cells collaborated to form multicellular organisms. Similarly for eusocial insects, where natural selection has shifted from individual insects to the society of collaborating parts consisting of a head (the queen) and groups of insects each with specialized functions; the insect society behaves like an individual multicellular organism.

NeoDarwinists and UltraDarwinists have attacked this proposal—Richard Dawkins with particular acerbity[34]—which provides an account of biological evolution consistent with the mutual aid and symbiogenesis hypotheses in the following section.

Sociobiologists' mathematical models that extend the NeoDarwinian principles constitute the current orthodox explanation of social behaviours that conflict with that competitive paradigm. Mathematical modelling can be a powerful tool to indicate deeper natural laws when those models map patterns in actual behaviour, and mathematical equations can elegantly express proven natural laws and predict the future behaviour of a system when the parameters of that system are known. However, biology is ill served by employing equations or adopting simplistic, unsuccessful 1950s game theories from economics that are divorced from biological reality, can be designed to produce any outcome, have no predictive value, and are contradicted by observations of animals in the wild and of humans.

Collaboration

Stuart West's view that collaboration is one the greatest problems for the biological and social sciences to explain is a self-imposed problem caused by the adoption of the NeoDarwinian model that is rooted in competition. Two hypotheses solve this problem by the simple device of recognizing that collaboration is a more significant cause of biological evolution than is competition.

Mutual aid

In their paper introducing 15 contributions to a 2009 two-day Royal Society meeting "The Evolution of Society", four current leaders in the field cite passages from Darwin that point out many animals live in groups and cooperate with each other.[35] Chapter 16 of this book notes that Darwin makes these observations in one chapter of *The Descent of Man*, where he says those communities which include the greatest number of sympathetic members would flourish best and leave the greatest number of offspring. But Chapter 16 also notes that by far the greater part of Darwin's books contradict this view in favour of all nature being at war, with the severest competitive struggle being between members of the same or similar species. It is this that provides the foundation of both the Darwinian and NeoDarwinian hypotheses.

"The Evolution of Society" introductory paper claims Darwin "commonly anticipates theoretical developments [in the evolution of animal and human societies] that only occurred 100 years later" and that no comparable work was undertaken until the early1960s. It omits, however, any mention of Peter Kropotkin.

As we saw in Chapter 16, Kropotkin's book, *Mutual Aid: A Factor of Evolution* (first published in England in 1902 with a final edition in 1914) provides extensive evidence from a wide range of animal species to support his findings that the animals which are naturally selected—those which survive longest and breed the most—adopt strategies to *avoid* competition: they collaborate to obtain sustenance, to protect themselves from predators, to migrate temporarily or permanently to a more favourable habitat, and to rear and sometimes train offspring.* Kropotkin concludes "The fittest are thus the most sociable animals, and sociability appears as the chief factor of evolution, both directly, by securing the well-being of the species while diminishing the waste of energy, and indirectly, by favouring the growth of intelligence."[36]

After examining current evidence of species' behaviour in Chapter 19 I concluded that it provides further support for Kropotkin's hypothesis:

> Within each class increased social learning correlates with increased brain complexity, together with increased intelligence measured by the invention of more novel solutions to challenging problems.†

And that overall

> Collaboration plays a more significant role than competition in the survival and propagation of life, from the level of genes through unicellular organisms, organelles within eukaryotic cells, eukaryotic cells within multicellular organisms, and the behaviour of plants and animals from insects to primates.‡

Symbiogenesis
Chapter 16 outlines the symbiogenesis hypothesis devised by Konstantin Mereschkovsky, Ivan Wallin, and others in the early twentieth century that followed symbiotic ideas advanced in previous decades.§ In essence this evolutionary process begins with symbiosis.

> **symbiosis**   The physical association of two or more different kinds of organisms through most of the life history of one of them.

In some cases their metabolic interaction leads to endosymbiosis,

> **endosymbiosis**   An association in which a smaller organism lives inside a larger one, usually collaboratively with each organism feeding on the other's metabolic excretion.

which may evolve into symbiogenesis.

---

*   See pages 267 to 270.
†   See page 330.
‡   See page 329.
§   See page 270.

symbiogenesis The merging of two different kinds of organism to form a single new kind of organism.

Some 40 years later an adjunct assistant of biology at Boston University, Lynn Margulis (then known as Lynn Sagan), developed this hypothesis. Her paper was rejected by at least fifteen journals before the *Journal of Theoretical Biology* published it in 1967. She advanced her ideas in a 1970 book, *Origin of Eukaryotic Cells*.[37] Like the hypotheses of Kropotkin and the early proponents of symbiogenesis, Margulis's proposals were dismissed or ignored by the NeoDarwinists who dominated evolutionary biology.

Based on studies of existing bacteria and protoctists (species of unicellular eukaryotes and their multicellular descendants lacking specialized tissues, like giant kelp) living in an anaerobic, or oxygen-free, environment, Margulis claimed that, towards the end of the Archaean Eon (approximately 3.8 to 2.5 billion years ago) when very little free oxygen existed in the oceans and atmosphere, swimming bacteria ancestors of existing spirochetes attached themselves to archaeabacteria (generally referred to as archaea) in a sulphur-rich environment to form a symbiotic association in which each fed on the other's metabolic waste products. Some archaic spirochetes penetrated the membrane of archaeabacteria in search of sustenance. Parasitism by these spirochete ancestors probably caused the death of many archaeabacteria, but in other cases the association developed into endosymbiosis and, over millions of years, into symbiogenesis in which the DNA of the two organisms merged.

During the Proterozoic Eon (approximately 2.5 billion to 540 million years ago) this generated several evolutionary consequences: a membrane grew around the combined genome to form a cell nucleus; organelles for cell motility (independent movement), like cilia and flagella, developed; and the process of mitosis evolved, whereby the chromosome inside the cell nucleus is duplicated and the nucleus and cytoplasm divide to produce two new cells genetically identical to the original. In some cells the chromosomes duplicated once but the nucleus and cytoplasm divided twice to form four sex cells—eggs or sperm—each possessing half the number of chromosomes of the original cell, thereby enabling sexual reproduction in which a sex cell merges with a complementary cell from another organism to create a new individual that has genes from both parents.

These new, larger, nucleated organisms were anaerobic protoctists, the root of all eukaryotes. A second and a third symbiotic merger in the Proterozoic Eon produced the kingdoms of animals, fungi, and plants (see Figure 23.2).

The second merger was between the nucleated anaerobic protoctists and bacteria that had evolved to respire oxygen in their metabolism and had become aerobic; it produced unicellular aerobic protoctists in which the formerly independent bacteria had become mitochondria. The failure of some of these reproducing cells to separate led to many kinds of multicellular eukaryotes. Some lost the ability to move independently and formed the root of the fungus kingdom. Others retained

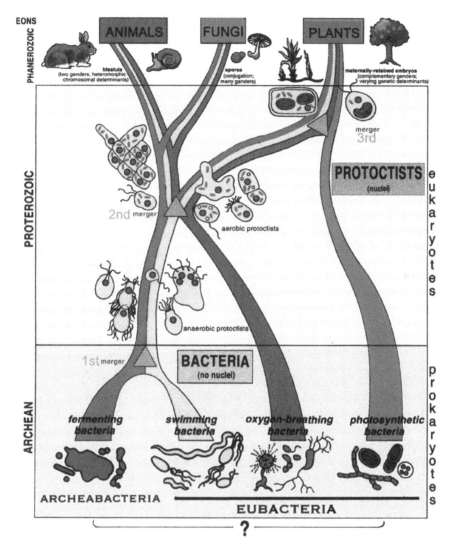

Figure 23.2 Serial symbiogeneses producing the roots of the taxonomic kingdoms of life

motility and formed the root of the animal kingdom. The merger of a swimming sex cell (a sperm) with a sex cell that that had retained the ability to divide but not swim (an egg) gave rise to two-gender sexual reproduction.

The third merger developed when aerobic protoctists ingested photosynthesizing bacteria, which subsequently lost their capacity for independent existence and became chloroplasts. This produced a new type of eukaryote that could use sunlight as a source of energy, green algal protoctists. Of these some lost the ability to move independently and formed the root of the plant kingdom.[38]

Genetic analyses in the 1970s and 1980s revealed that chloroplast genes of some species of algae bore little resemblance to genes in the algae's nuclei but did resemble the genes of photosynthesizing cyanobacteria, while other DNA evidence suggested that mitochondria derive from alpha-proteobacteria similar to existing bacteria known as rickettsiales. This provided genetic support for the symbiogenetic origin of chloroplasts and mitochondria in eukaryotic cells.

While most evolutionary biologists now accept that symbiogenesis provides the most plausible explanation for the origin of mitochondria and chloroplasts in eukaryotic cells, they do not accept Margulis's hypothesis for the origin of eukaryotic cells and the evolution of the taxonomic kingdoms.

Margulis counters by arguing that, outside of mathematical models, a NeoDarwinian gradual accumulation of random genetic mutations has never been proven to produce a new organ or a new species, while symbiosis, endo-symbiosis, and symbiogenesis are widespread in nature. Some species of ants live in symbiotic relationships with aphids: they eat sugary honeydew excreted by the aphids after sucking sap from plants, while in return they protect aphids from predators and transfer them from wilted to healthy plants. Endosymbiosis is the norm: for example, while a human body consists of about 100 trillion cells, it hosts about ten times that number of bacteria in its intestines; different species of these bacteria carry out useful functions, like synthesizing vitamins B and K. A familiar example of symbiogenesis is the more than 15,000 species of lichen. Although they look like plants, they have no plant ancestors; they are formed by symbiogenesis between members of two separate kingdoms: the fungi and either photosynthetic bacteria (cyanobacteria) or protoctists (algae).

According to Margulis, although symbiogenesis is not the only cause of evolutionary innovation, it is the major one. When two or more different kinds of organism merge their identities, the process generates novel behaviours and morphologies: new tissues, new organs or metabolic pathways, and new groups of organisms, including new species.

Margulis found it difficult to get her work published in mainstream scientific and biology journals because, she said in 2011, evolutionary biology has been taken over by the Anglo-American theoretical population geneticists, experimental zoologists, and molecular biologists of the NeoDarwinian school who "block out four-fifths of the information in biology [by ignoring bacteria, protoctists, fungi, and plants] and all the information in geology". They "know nothing about biological systems like physiology, ecology, and biochemistry....They are reductionists ad absurdum."[39]

Leading NeoDarwinist Jerry Coyne was quick to respond to these comments on his website in an article entitled "Lynn Margulis disses evolution in *Discover* magazine, embarrasses both herself and the field". After acknowledging Margulis's contribution to explaining the origin of mitochondria and chloroplasts, he points out that she agrees with the creationists about the inefficacy of the NeoDarwinian paradigm, commenting that "at least she's not crazy enough to accept god as a scientific explanation. But she *is* crazy enough to proffer her 'alternative' theory, which of course is symbiosis [his italics]." He reprises the NeoDarwinian model and concludes that "[w]hen discussing evolutionary biology, then, Margulis is dogmatic, wilfully ignorant, and intellectually dishonest".[40]

The title Coyne chose for this response is indicative. Margulis was a committed evolutionist; if she was showing disrespect it was not to evolution but to the NeoDarwinian model of biological evolution. By equating evolution with NeoDarwinism Coyne reveals part of the problem in contemporary evolutionary biology by which questioning of the current orthodoxy frequently elicits a dismissive response that falls somewhat short of the scientific ethos.

The current consensus in evolutionary biology is that life diverged first into bacteria and archaea, and eukaryotes then evolved from archaea or archaea-like predecessors. Subsequently eukaryotes gained genes from bacteria twice, obtaining mitochondria from alpha-proteobacteria and chloroplasts from photosynthesizing bacteria as Margulis argued, and the three domains of Bacteria, Archaea, and Eukarya branched into a Darwinian descending tree of life caused principally by the sexual transmission and accumulation of random gene mutations within species populations.

It follows from this hypothesis that no characteristically bacterial genes should be found in archaea, and the only bacterial genes in eukaryotes should be those in mitochondrial or chloroplast DNA or those transferred to the nucleus from the bacterial precursors of those organelles, and these should be involved in respiration or photosynthesis. However, recent sequencing of whole genomes confounds this hypothesis by revealing rampant horizontal transfer of multiple genes. Many archaea possess a substantial store of bacterial genes. Nuclear genes in eukaryotes code for nonrespiratory and nonphotosynthetic processes that are critical for cell survival. Moreover, many eukaryotic genes are unlike those of any known archaea or bacteria: for example, genes implicated in two defining eukaryotic features, the cytoskeleton and the system of internal membranes.[41]

This genomic evidence strongly suggests that for the first 2 billion years of biological evolution on Earth genome fusions and horizontal gene transfers resulting from collaborations and symbiogeneses between different species played a far more important role in biological evolution than a NeoDarwinian vertical inheritance of randomly mutated genes.

In the subsequent one and a half billion years horizontal gene transfer very probably remained the most significant mechanism of gene transmission and speciation for prokaryotes. Within the taxonomic kingdom of animals, the mutual aid hypothesis—collaboration within species and between members of different species (symbiosis)—provides a more plausible explanation than does NeoDarwinian competition for the evolutionary pattern of species within each class successfully surviving and reproducing and demonstrating increased cognitive abilities through social learning.

My conclusions at the end of Part 2 will include conclusions drawn from Chapters 21 to 23 plus the next chapter.

CHAPTER TWENTY-FOUR

The Evolution of Consciousness

> Consciousness is so much a total mystery for our own species that we cannot begin to guess about its existence in others.
>
> —LEWIS THOMAS, 1984

Humans are more than the phenotypical characteristics summarized in the taxonomic class *Homo sapiens*. We are conscious beings. Philosophers, psychologists, anthropologists, and neuroscientists disagree markedly among themselves as to what human consciousness is. Most study it as a thing of itself, and I shall consider human consciousness further in Part 3. I suggest, however, that just as the human phenotype evolved from the earliest lifeforms on Earth, so too did human consciousness, and we may gain an understanding of it, or at least its physical correlates, if we can trace its evolutionary trajectory along the human lineage.

To do so I shall use a broad definition of consciousness that applies in however incipient or rudimentary form to all living things and that distinguishes them from inanimate matter.

> **consciousness**   Awareness of the environment, other organisms, and self that can lead to action.

Consciousness thus defined may be mapped by its consequent actions, that is, by an organism's behaviour, and so we need to trace the evolution of organismal behaviour along the human lineage.

The evolution of behaviour

Because fossils rarely indicate behaviour, often the only way to trace this evolution is to examine the behaviour of living species that most resemble the fossils dated by their estimated first appearance in the narrowing taxonomic categories from the domains of *Bacteria* and *Archaea* to the species *Homo sapiens* shown in Table 20.1.

Bacteria and archaea

The species most closely resembling the earliest lifeforms on Earth are prokaryotes: bacteria and archaea. These simplest of organisms act by responding directly to external and internal stimuli in order to survive.

External stimuli include heat, light, and chemicals in their environment, and the presence of other organisms; these may be sources of sustenance or danger. Bacteria and archaea respond by moving towards chemicals and organisms they can use for their own maintenance and away from those that are damaging to them. We saw in Chapter 19 an example of bacteria responding to other bacteria by releasing signalling molecules into their immediate environment and having receptor sites for such molecules. When their receptor sites register a sufficient number—a measure of local population density—this triggers a direct and collaborative response such as the switching on of genes that lead to the production and excretion of polysaccharides for a communal biofilm or enzymes to digest prey.*

Bacteria also show a rudimentary awareness of self by directly responding to internal stimuli. Significant DNA damage triggers what is known as the SOS response: the switching on of normally silent SOS genes that activate a DNA repair mechanism.

Eukaryotes: single-celled

A simple single-celled eukaryote like an amoeba also acts by a simple direct response to an external stimulus, for example by forming temporary extensions of its cytoplasm (known as pseudopods, or false feet) in order to propel itself towards chemicals given off by foods or away from noxious chemicals.

More complex single-celled eukaryotes have a more complex direct response system, using organelles, or specialized cell parts, to serve as receptors of stimulus and as effectors of response. Receptors include stiff sensory bristles in ciliates and light-sensitive eyespots in flagellates. Effectors include cilia (multiple, slender, hairlike projections from the cell surface capable of beating in unison), and flagella (elongated, whiplike cilia) that propel the cell, together with other organelles that draw in food.

Eukaryotes: animals

As life evolved on Earth, producing a more complex world with more and different predators, competitors, and collaborators, the multicellular organisms of the animal kingdom developed a second response system that enables a far more rapid, varied, and flexible response to external stimuli. Their behaviour may be categorized into five overlapping types: direct response, innate, learned, social, and innovative.

---

* See page 322.

a. *Direct response*

Self-preserving direct responses to stimuli constitute most of the behaviours of simple animals, like invertebrates, but they persist as an important component of the behavioural repertoire of more complex and recently evolved animals, exemplified by a human jerking his hand out of a fire.

Direct response behaviour is now taken to include reflexes, which are involuntary responses to stimuli. As animals become more complex along the lineage so too do their reflex responses, which usually involve several muscle groups. For example, in humans the knee-jerk reflex occurs when a tap on the patellar tendon causes in the muscles at the front of the thigh a reflex twitch sufficiently powerful to extend the lower leg at the knee.

b. *Innate*

Direct response and innate behaviours overlap in what may be grouped as instinctive behaviours, which are unlearned responses to specific stimuli. These usually include such things as the urge to have sex stimulated by specific sights or smells, hitting back or fleeing if attacked, and so on.

I think it useful to distinguish an individual's self-preserving direct response to a stimulus from heritable, predictable, fixed actions performed in different environments (which provide different external stimuli) by a member of a species that need not confer a benefit to that individual—indeed often involve a cost—but which may preserve the species.

Some of the most striking examples of such innate behaviours are found in insects, descendants of ancient arthropods. In some 47 species of leaf-cutter ants across Central and South America, for instance, the foraging caste in each colony innately collaborate to cut down a leaf, cut it into small pieces, and transport it across forest or, for some species, desert terrains to their nest, where the pieces are shredded, pulped, and used as mulch for the nest's fungus garden.[1] This behaviour is not a direct response to an individual ant's hunger. Performed by all foraging leaf-cutter ants in their numerous colonies, it helps preserve each species by making and storing sustenance for a colony's future needs.

The genetic, or possibly epigenetic, mechanism for heritable, fixed-action behaviour that may be of no benefit to the individual is poorly understood. We saw in the last chapter that sociobiological hypotheses fail to provide a convincing account. This is underlined by the fact that many innate behaviours are not collaborative actions of a group, whether kin or otherwise, but are performed by individuals.

Such individual innate behaviour is widespread among descendants of the earliest and most primitive vertebrates, like fish from the Cambrian period. Most current species of salmon, after spending between one and nine years in the sea, migrate to the high-level freshwater pool where they hatched in order to spawn. To do so a salmon must find that particular river outlet and battle against the downflowing river and, sometimes, waterfalls, in a gruelling journey of up to a thousand kilometres and a climb of up to two thousand metres. It is difficult to see this as a self-preserving response for the individual; indeed, Pacific salmon die soon after spawning. A stickleback's complicated series of actions that form its courtship and reproductive behaviour is similarly neither a simple direct response nor a learned response.

Which human behaviours are innate is a matter of debate. Linguistic philosopher Noam Chomsky claims that a child's rapid learning of a language would be impossible without a biologically innate language faculty, or innate rules of universal grammar, unlike, say, learning to play chess.[2] On the other hand developmental psychologist Michael Tomasello rejects this and proposes a usage-based theory that claims children learn linguistic structures by reading intentions and finding patterns in their discourse interactions with others.[3]

As species become more complex along the human lineage, so too do their behaviours, with learned, social, and innovative behaviours playing an increasing role.

c. *Learned*

These actions modify direct response or innate behaviours. In rudimentary form they are observed in the simplest and most primitive animals, like molluscs, through habituation. Here, if repeated responses to a stimulus no longer produce the previous outcome, the animal ceases to respond to the stimulus. The converse is sensitization: responses to a stimulus produce heightened outcomes, and the animal subsequently responds to a reduced stimulus.

As biological evolution results in species of greater morphological complexity, copying plays an increasing role in acquiring learned behaviour. Chapter 19 gave examples of fishes learning by copying; the evidence indicates that such behaviour is heritable, suggesting that learned behaviour may become innate.

Chapter 19 also cited a study showing not only how meerkats, small nonprimate mammals, copy but also how adults teach hunting skills to their young.* Copying is observed in nonhuman primates, principally the young copying individual survival skills practised by parents or close kin.

I think a distinction should be made between animals copying, or even being taught by, members of their own species—for which more fieldwork is needed—and animals being taught or trained by humans. The latter gave rise to many animal behaviour hypotheses on which there is little current agreement. Ivan Pavlov's experiments on dogs in the early part of the twentieth century led him to propose that "conditioned reflex", associated with specific areas in the cerebral cortex, explained all behaviour, including

---

* See page 327.

human, and his views became influential in psychology and psychiatry. The experiments, however, simply tell us that animals can be trained to give the same response when the original stimulus is substituted by another; it gives no great insight into the evolution of behaviour in nature.

Psychologist B F Skinner's experiments on rats and pigeons aimed to go further by explaining the learning of new behaviours. His hypothesis of operant conditioning maintains that an animal responds to its environment with a series of random, varied actions. One of these (for example, pressing a bar in his experimental box) eventually leads to a reward (a pellet of food). Such positive reinforcement results in the animal learning to repeat this particular action in future. Skinner claimed that this is Darwinian natural selection of behaviours during an animal's lifetime comparable to Darwinian natural selection of physical traits over very many generations and is equally applicable to human behaviour; he rejected unobservable phenomena like mind and intention. The hypothesis gained considerable acceptance in the twentieth century. However, a more appropriate analogy is with artificial selection rather than natural selection. It is the human experimenter who designs the experiment and provides the reward. The hypothesis amounts to little more than animals learning by trained sensitization and habituation, and Skinner's claim that it explains human behaviour lacks any convincing evidential basis.

While individual habituation and sensitization may make a small contribution to human learning (if you put your hand in a fire and it gets burnt, you learn not to do it again), copying and being taught play a far more significant role. I shall consider this further in Part 3.

d. *Social*

These actions include direct response, innate, and learned behaviours, but extend beyond individual survival and offspring copying individual survival skills from their parents. They include all the ways in which animals living in a group interact with other members of their group.

As we saw in Chapter 19, animals form communities beyond parents and offspring when they collaborate for their survival—defence against the elements (shelter and nest building), defence against predators or competitors, and hunting—and for rearing offspring. Such social living also provides opportunities for mating. Animal social groups are mainly, but not exclusively, based on kin, and many such groups themselves collaborate to form temporary larger groups for survival purposes, like seasonal or permanent migration to a habitat better suited for survival.

The interactions of members of an animal social group are sensory: touch, smell, taste, sight, and hearing.

As the morphological complexity of species increases over evolutionary time along the human lineage, these sensory interactions change. In particular, smell, taste, and touch, which have short ranges, gradually decrease in importance. With the increased complexity of vocalization in primates to communicate a greater variety of messages, like threats, pleasure, warning cries, and so on, to other group members, hearing develops by the evolution of an inner ear that allows the resolution of complex patterning of frequency and timing information. Sight increases in complexity with the development of stereoscopic vision as eyes become forward facing in primates, which also enhances spatial awareness. These developments enable group members to interact in more complex ways over greater distances.

The survival value of social interactions can be direct or indirect. Mammals, like beavers, engaging in play, and primates, like chimpanzees, undertaking mutual grooming, serve to reinforce cohesion of their only social group and so have an indirect survival value.

e. *Innovative*

These actions depart from the typical direct response, innate, or learned actions of its species and are usually shown as responses to new and challenging circumstances. The degree of innovation may be used as a measure of intelligence, sometimes referred to as cognitive ability.

> **intelligence**   The capacity to acquire and successfully apply knowledge for a purpose, especially in new or challenging situations.

Little innovation is shown in the human lineage until primates emerge. (I am considering here only the lineage that leads to humans, and not those that lead to other species showing intelligence, like the crow family and marine mammals like dolphins.) Again, I think it useful to distinguish between primates like chimpanzees being set "intelligence tests" devised by humans, and primates behaving innovatively in the wild. The latter is a more accurate measure of the evolution of innovative behaviour. It is, however, more difficult to ascertain unambiguously from research studies that use different criteria, methodologies, and underlying assumptions.

In 2002 behavioural biologists Simon Reader and Kevin Laland searched the major primate journals and found 533 cases of innovation (defined as apparently novel solutions to environmental or social problems), 445 observations of social learning (the acquisition of information from others), and 607 episodes of tool use. After correcting for such factors as differences in research effort on different species, they concluded that innovation, social learning, and tool use correlated with species' relative and absolute executive brain volumes. The executive brain is essentially the neocortex.[4] I shall consider this correlation later.

Evolutionary pattern

As species become morphologically more complex along the human lineage, so too do their behaviours. While direct response and innate behaviours persist, learned, social, and innovative behaviours play an increasingly significant role. Their greater variety and flexibility mark a rising consciousness.

Physical correlates of rising consciousness

Rudimentary consciousness emerges in prokaryotes and single-celled eukaryotes shown by the actions they take for survival in response to external and internal stimuli. Such simple actions are effected by a physico-chemical direct response system.

The nervous system

The evolution of multicellular animals led to the development of a second response system that enabled a far more rapid and varied response to external stimuli. This is the electrochemical nervous system, which uses specialized electrically excitable cells called neurons for high-speed one-way transmission of electrical impulses from one region of an animal's body to another. The nervous system also coordinates the slower but longer lasting physico-chemical responses to internal stimuli.

> **nervous system**   An organized group of cells, called neurons, specialized for the conduction of electrochemical stimuli from a sensory receptor through a nerve network to an effector, the site at which a response occurs.

> **neuron**   A eukaryotic cell specialized for responding to stimulation and for conducting electrochemical impulses.

Although they come in a variety of sizes and shapes, Figure 24.1 shows the essential structure of most neurons. Dendrites are branched projections from the cell wall that receive electrochemical impulses and conduct them to the cell body. The cell body contains the nucleus and its surrounding cytoplasm. The axon is a relatively long projection from the cell wall that transmits electrical impulses from the cell body to terminals that branch from the end of the axon. Dendrites and axons are commonly called nerve fibres.

In larger and more complex animals like vertebrates the axon can be very long and insulated by a sheath of myelin, a fat-like outgrowth from a series of glial, or supporting, cells called Schwann cells. At gaps between the Schwann cells, known as nodes of Ranvier, the axon is uninsulated; this arrangement allows an electrical pulse to be transmitted even more rapidly.

Most neurons may be grouped into one of three functional categories:

   a. *Sensory neuron*
   Receives stimulation from sensory receptor cells or organs, like cells forming the animal's outer layer or an eye, and transmits an electrical impulse

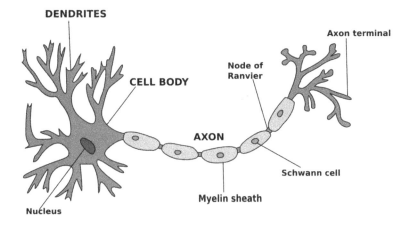

Figure 24.1 Structure of a typical neuron in a vertebrate

either directly to motor neurons or else to coordinating interneurons. In the latter case it is also termed an afferent neuron.

b. *Motor neuron*
Transmits an electrical impulse to an effector cell or organ, like a muscle. It is also termed an efferent neuron.

c. *Interneuron*
Receives impulses from sensory neurons and controls, coordinates, and transmits responses to motor neurons.

Transmission of electrical impulses from one neuron to another or to an effector cell usually occurs at microscopic gaps between them called synapses via chemicals known as neurotransmitters. Direct electrical communication also occurs when membranes are fused; electrical synapses are found mainly in invertebrates and primitive vertebrates.

Through their dendrites and axon terminals each neuron can receive impulses from, and send impulses to, many others, forming a network. Figure 24.2 shows schematically the way these different types of neuron connect with each other and with sensory and effector cells in an animal with a central nervous system.

Interneurons also connect to each other, and together they integrate, coordinate, and control many individual reflexes so that the response of the animal is more than the simple sum of individual reflexes: its behaviour is characterized by flexibility and adaptability to changing circumstances.

The rise in consciousness along the human lineage correlates with the evolution of the nervous system. This does not mean that all nervous systems evolved in the same, linear way, as implied by neuroscientist and psychiatrist Paul MacLean's triune brain model that was influential in the 1970s and 1980s. Evidence accumulated from comparative neuroanatomy during the last thirty

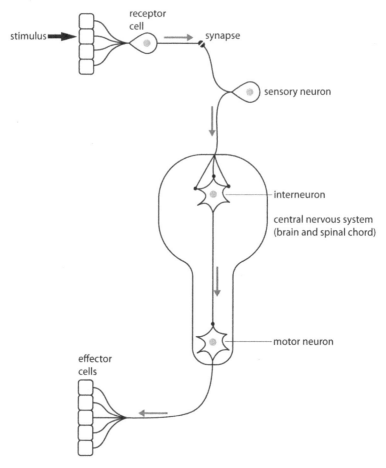

Figure 24.2 Schematic representation of neuronal connections in an animal with a central nervous system

years or so suggests that different nervous systems evolved along different animal lineages leading, for example, to some contemporary octopuses, birds, and marine mammals having structurally different nervous systems from nonhuman primates but with claimed comparable cognitive abilities. However, this quest focuses on the evolution of humans, and so I shall try to follow the evolution of the nervous system that leads to humans.

Nervous systems in the human lineage
Neurons leave no fossils. The only way to trace the evolution of the nervous system in the human lineage is to examine the nervous systems of living species that are morphologically as similar as possible to fossils dated by their estimated first appearance in the narrowing taxonomic categories from the kingdom *Animalia* to the species *Homo sapiens* shown in Table 20.1.

Diffuse nervous system
The earliest animal fossils are radially symmetrical, like the cnidarians of the Ediacaran fauna.* The most primitive nervous systems known today are found in radially symmetrical, very simple animals, like cnidarians (hydroids, jellyfish, sea anemones, corals), and are known as diffuse nervous systems because nerve cells are distributed throughout the organism, usually beneath the outer epidermal layer from which they probably evolved. Their connections form a neural network, or nerve net, illustrated by that of hydra in Figure 24.3 (see next page).

Central and peripheral nervous systems
Flatworms were among the first invertebrates to exhibit bilateral symmetry, with left and right sides that are approximate mirror images of each other.† They were also among the first to show a rudimentary central nervous system and a peripheral nervous system, illustrated in Figure 24.4 (see next page).

The flatworm's nervous system reflects its morphological bilateralism. Two groups of interconnected interneurons form a rudimentary brain from which two cords of nerve fibres run the along the left and right lengths of its body; these cords are connected to each other by transverse nerves much like the rungs of a ladder. This constitutes a primitive central nervous system. Small nerves extend from the cords to the sides, giving rise to interlacing networks of peripheral nerves to form a peripheral nervous system. These peripheral nerves are associated with sensory organs, which are scattered over the body. Eyespots that respond to light are near the brain, while the mouth is halfway along the body. Motor neurons are correspondingly scattered.

The evolution of bilateral animals shows a pattern of an increase in the number of neurons and also centration by which groups of sensory neurons are progressively closer to an enlarged brain.

The human lineage follows an evolutionary path that leads to vertebrates. Their central nervous system comprises a brain protected by a skull and nerve cords protected by a segmented spinal column. The pattern of centration continues with cephalization: groups of sensory neurons—for example those stimulated by smells or sounds—are concentrated near the brain at the head end, while the nerve cords consist of bundles of nerve fibres transmitting signals to and from a brain of two larger, interconnected groups of controlling and coordinating interneurons.

The vertebrate peripheral nervous system is divided into two. The autonomic nervous system comprises motor nerve fibres signalling to internal effector organs like the heart, lungs, and endocrine glands (which produce and secrete chemicals known as hormones into the blood circulation system to distant target cells where

---

* See page 288.
† Why animals should have evolved bilaterally rather than, say, trilaterally or quadrilaterally is a question that is rarely asked and never answered satisfactorily. The "adaptive principle" will not do: the environment has three spatial dimensions.

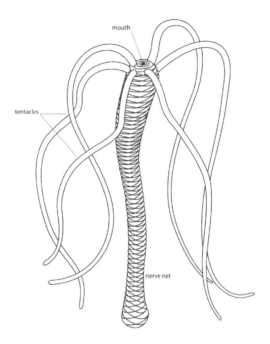

Figure 24.3 Diffuse nervous system of hydra

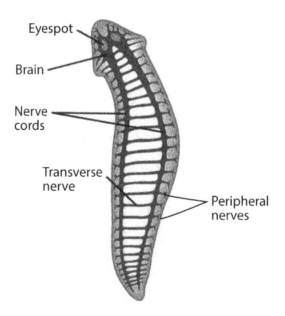

Figure 24.4 Nervous system of the bilateral flatworm

they regulate cellular metabolism) principally to maintain homeostasis, or regular functioning of the adult animal. It is regulated by the most ancient part of the central nervous system, the brain stem (which humans share with descendants of the earliest reptiles) responding automatically to stimuli. The somatic nervous system comprises motor nerve fibres signalling to effector organs, like muscles, principally on or near the outer layers of the animal; it is activated by the central nervous system responding to external, sensory stimuli and is under voluntary control.

The evolution of vertebrates also shows a pattern of complexification. An increase in the number of interneurons controlling responses in morphologically more complex species leads to the brain developing three interconnected outgrowths, each associated with a particular sense: a forebrain associated with smell, a midbrain with vision, and a hindbrain with sound and balance. Further evolution along the human lineage results in interconnected paired outgrowths from the forebrain to form cerebral hemispheres, an outgrowth from the midbrain to form a midbrain roof, or tectum, and an outgrowth of a corrugated ball of nervous tissue from the hindbrain to form the cerebellum (Latin for little brain). Figure 24.5 (see next page) illustrates these structures together with others described below that evolved to form the human brain.

The addition of these nerve centres to the primitive brain stem allows greater coordination and association between the sensory and motor fibres. Evolution from ancient to more recent vertebrates sees a gradual shift of function from the lower brain stem to the higher cerebral cortex. This is the outer layer of the cerebral hemispheres, or grey matter, and consists primarily of cell bodies of neurons above the white matter that mainly comprises axons covered by a whitish myelin sheath.

The emergence of the taxonomic class of mammals is marked by the addition of two new brain structures. The neocerebellum, looking much like a fungal growth, develops from the cerebellum. The neocortex develops from the cortex as a thin—about 2mm in depth—but increasingly convoluted new outer layer.

Figure 24.6 (see next page) shows these developments. For most mammals, like a mouse, the neocortex is small relative to the rest of the brain, leaving the brainstem clearly visible; for a nonhuman primate, like a monkey, the neocortex is relatively larger and grows within the skull by folding, resulting in a convoluted appearance while leaving part of the brainstem visible; for a human the neocortex is much larger, covers all the brainstem, and is very convoluted.

When viewed on the same scale, the striking evolutionary change shown is the growth of the neocortex. In humans it is only twice as thick as in a mouse but has about a thousand times more surface area; while it is only 15 per cent thicker than in a monkey it has at least ten times more surface area.

The addition of new, interconnected specialist regions of the brain, all concentrated in the head, increased both the centration and the complexity of the nervous system. Moreover, the overall growth of the brain along the human lineage meant that not only did the number of neurons increase but so too did

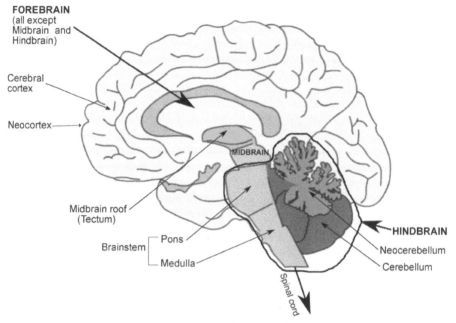

Figure 24.5 **Human brain**

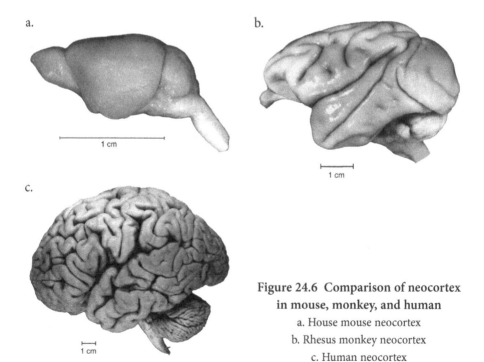

Figure 24.6 **Comparison of neocortex in mouse, monkey, and human**

a. House mouse neocortex
b. Rhesus monkey neocortex
c. Human neocortex

the number of their connections through their dendrites and axon terminals. According to a 2009 estimate the adult male human brain contains approximately 86 billion neurons[5] (which is less than the traditional figure of around 100 billion) passing signals to each other via some 500 trillion (0.5×10^{15}) synaptic connections,[6] making it the most complex thing in the known universe. Furthermore, these neural connections are not fixed as once believed. Studies since the 1970s show the capacity of neurons to change their connections in response to new information and sensory stimulation as well as to damage or dysfunction. This phenomenon, known as neuroplasticity, increases behavioural flexibility and is associated with learning and innovation.

Learning and innovation are generally taken as measures of cognitive ability, or intelligence, characteristic of higher levels of consciousness. Different researchers use different definitions, different methods, and (frequently unstated) different underlying assumptions to correlate these behaviours with a species' nervous system. Four main approaches have been taken.

1. *Brain size*
   On the assumptions that cognitive ability is a function of the number of neurons and their connections, and that these scale with brain size, the cranial capacity has been used as a parameter of intelligence. It is also the most readily measurable in fossils and dead animals.

   Such a simplistic approach, however, is fraught with problems. How do you account for the fact that larger mammals have a larger body with a greater surface area and so need more controlling interneurons processing more sensory and motor neurons rather than neurons processing innovative behaviour? It does not explain why the behavioural flexibility and innovation of, say, macaque monkeys is so much greater than ungulates, like cows, that have 4–5 times the brain size, or why human inventiveness considerably exceeds that of elephants with brains up to 6 times larger.

   If the assumed correlation between absolute brain size and intelligence breaks down across taxonomic orders, does it hold within an order like primates? In 2007 Robert Deaner of the Department of Psychology, Grand Valley State University, Michigan, USA and colleagues analysed the literature covering cognition in primate species and concluded that "absolute brain size measures were the best predictors of primate cognitive ability".[7]

   However, in 2011 Cambridge University biological anthropologist Marta Mirazón Lahr pointed out that the fossil record, albeit incomplete, suggests that ancient humans had larger brains than humans today. For example, an adult male Cro-Magnon human living 30,000 to 25,000 years ago was tall and muscular, with a body weight of 80–90kg and an average brain volume of 1,500cc (approximately the same volume as that of

Neanderthals who became extinct at about the same time). This brain size remained roughly the same until about 10,000 years ago, when it began to reduce to that of a human male today, who has an average brain volume of 1,350cc, a 10 per cent decrease (with a corresponding 10 per cent decrease in body weight).[8]

I don't think anyone suggests that humans of 10,000 years ago showed 10 per cent more innovative behaviour than current humans. Or that women are less innovative and intelligent than men because an adult female brain is approximately 10 per cent smaller than that of an adult male.

2. *Ratio of brain to body weight*
In 1973 neuroanatomist Hartwig Kuhlenbeck compared different species by their ratio of brain weight to body weight.

However, this doesn't overcome other problems. Land-based animals require neural activation of muscle to overcome the gravitational force opposing their movements, whereas this force is largely offset by the buoyancy of water for fishes and marine-based mammals. Moreover, much of the mass of whales, for example, is neurally inert blubber that insulates against deep sea cold.

3. *Encephalization quotient*
In the same year Harry Jerison of the Department of Psychiatry & Behavioral Sciences, UCLA, who specialized in studying fossils and cranial capacities, proposed encephalization quotient, or EQ, as a better measure of effective brain size for all species of mammal. It is based on brain weight to body weight for different mammalian species corrected by a power function derived from measurements of different mammals showing that brain size increases less than body size as the body size increases; it allows the predicted brain mass to be calculated for any species.

EQ indicates how much the observed brain mass of a species deviates from the predicted value for its body mass: an EQ of 1 shows the observed brain mass equals the predicted mass; an EQ greater than 1 means the brain size of the species is larger than expected for its body mass. Compared with the class of mammals, humans have by far the largest EQ, between 7 and 8. Compared with the family of hominids, humans have an EQ of more than 3; gorillas and orangutans may be larger than humans but their brains are about one third of the size.

EQ is the most widely used standard. However, the formula for predicting brain mass depends on the combination of species measured. Moreover, its correlation with intelligence assumes that the larger the brain is relative to the body, the more brain weight is available for more complex cognitive tasks: small-brained animals with large EQs are predicted to have more intelligence than large-brained animals with smaller EQs. But, while capuchin monkeys have much larger EQs than gorillas, they are outranked by them in cognitive performance.

In summary, although an improvement on a simple brain to body ratio, EQ is still too crude a measure that doesn't take into account which particular parts of the brain are responsible for innovative behaviour.

4. *Neocortex*
In 1969 Heinz Stephan of the Max Planck Institute for Brain Research and O J Andy of the Department of Neurosurgery, University of Mississippi Medical Center concluded that all primates as well as many, if not all, recent orders of placental mammals, are descended from ancient insectivores, small insect-eating mammals.

Fifteen years later Stephan, together with Pierre Jolicoeur and colleagues of the Département de Sciences biologiques, Université de Montréal, examined the differences in the brains of 28 current species of insectivores, 63 species of chiropterans (the mammalian order of bats), and 48 species of primates. They found that the brain component which undergoes the greatest relative volumetric increase from basal insectivores to higher primates is the neocortex, which is 150 times more developed in *Homo sapiens* than a hypothetical basal insectivore of the same body weight.[9]

Thus the characteristic that most distinguishes the evolution of the human brain from other mammalian and primate species is the growth of the neocortex, the most recently evolved part of the brain, which processes higher cognitive functions like short-term memory, association, innovation, language, reasoning, and self-reflection, which are all features of higher levels of consciousness.

The convoluted nature of the neocortex is significant for three reasons. First, this folding of neural tissue provides even greater centration: more neurons can fit into the same volume, achieving a higher neuronal density. Second, as a consequence, the lengths of fibres connecting these neurons are shorter than if the neurons were spaced more widely apart, thus reducing interneuron transmission and response times in the brain. Third, it means the head does not have to grow to accommodate more brain. As a result of this third factor, the size of the animal does not have to increase; hence the number and lengths of the nerve fibres connecting to sensory tissue, like skin, and to effector tissue, like muscles, at distant parts of the body do not have to increase. This reduces neural transmission times; it also results in a higher ratio of neurons associated with advanced cognitive functions to neurons associated with motor or maintenance functions. Overall, it optimizes the brain-body size for speed and flexibility of response and for advanced cognitive functions like innovation. All this results not from natural selection from almost limitless random possibilities but from the laws of physics.

CHAPTER TWENTY-FIVE

Reflections and Conclusions on the Emergence and Evolution of Life

It is evident in the post-Copernican era of human history, no well-informed and rational person can imagine that the Earth occupies a unique position in the universe.

—MICHAEL ROWAN-ROBINSON, 2004

Our own existence once presented the greatest of all mysteries, but that is a mystery no longer because it is solved. Darwin and Wallace solved it, though we shall continue to add footnotes to their solution for a while yet.

—RICHARD DAWKINS, 1986

Nullius in verba (Accept nothing on authority).

—MOTTO OF THE ROYAL SOCIETY

Reflections

One of the joys of pursuing this quest is to discover things that I hadn't expected. Cosmologists view the planet Earth as one of the most insignificant things in a galaxy of some hundred billion stars, many of which have their own planetary systems, while that galaxy is only one of more than a hundred billion galaxies in the visible universe, which is, according to current cosmological orthodoxy, an incredibly minute part of the whole universe.

From the relatively new science of astrobiology, however, it is possible to deduce six conditions necessary for the most complex molecules of some 13 atoms found in interstellar space and on asteroids to evolve into things as complex as a human. Those six conditions dramatically reduce the locations in a galaxy and in a solar system where such an evolution can take place in conformity with known physical and chemical laws. The surface of the Earth provides those conditions, but—and here is the surprise—it only does so because of a concurrence of highly unusual stellar and planetary events. If not unique, the Earth is almost certainly a rare location for intelligent life in this galaxy.

Another surprise is the absence of a generally agreed definition of life. Most proposed definitions are neither sufficient nor necessary to distinguish living things from non-living things, while many exclude living things like mules that do not reproduce. I hope the definition suggested in Chapter 13 offers a viable distinction and shows there is an irreversible boundary, however imprecisely delineated, that marks a major evolutionary transition from inanimate matter to life.

I'm most grateful to those astrobiologists and geologists who corrected errors of fact or omission and commented on my conclusions in draft chapters. They did so with an open-mindedness and without dogmatism in the best tradition of science. Perhaps this is because astrobiology is a new, emerging science and geology underwent a paradigm shift in the 1960s when fifty years of geological orthodoxy stubbornly dismissing the hypothesis of continental drift collapsed under the weight of evidence supporting the mechanism of plate tectonics, which is still being worked out in detail.

Sadly, when it came to investigating the evolution of life, a majority of responses to questions and drafts evoked responses similar to those of establishment cosmologists and theoretical physicists outlined in Chapter 11 at the end of Part 1. They may be grouped into five:

a. Peremptorily dismissive
b. Politely dismissive
c. You don't know the facts
d. You appear to know the facts but you don't understand them
e. NeoDarwinism is compatible with all the evidence and all the different hypotheses.

A peremptorily dismissive reply to a request to check a draft chapter beginning with the statement that very few studies of living species in the last forty or so years have been designed to investigate whether or not biological evolution takes place began "I did not get passed [sic] the first sentence of your article.... This claim could not be further from the truth." It went on to say there had been many thousands of experimental studies designed to investigate whether or not biological evolution takes place, adding "I can't help but feel that anyone who was objective and informed would know of these." It went on to recommend a book published in 1986 or any undergraduate evolution textbook.

It was precisely because I recognize my ignorance in his field that I asked him to correct errors of fact or omission in the draft chapter. However, that 1986 book examines not whether biological evolution takes place but methods of investigating its NeoDarwinian mechanism, which is precisely what the second and third sentences of the draft said; the rest of the draft discussed the examples given in current undergraduate textbooks. This response was a more extreme example of evolutionary biologists who equate the phenomenon of biological evolution with the NeoDarwinian mechanism, which is just one of several claimed mechanisms.

One eminent member of the biological establishment in the USA responded within hours of my emailed request and draft chapter to say "I have had a quick read of the text you sent me....What you have written is naive at times, incomplete and biased throughout, not up-to-date in many cases, and full of errors of fact and attribution, as well as of theory."

39 of the 60 references cited, and correctly attributed, in the draft were articles published within the previous six years, including a long encyclopaedia entry by him published online that same year. I replied to say I was extremely concerned by his comments and I'd be indebted if he clarified what was biased and indicated the major errors and misattributions. Despite a reminder, I'm still awaiting a reply.

The politely dismissive said they disagreed with my conclusions but wished me well. Some implied that my attack on evolutionary theory—a misrepresentation of clearly stated intentions—meant I was a closet creationist. I was reduced to prefacing subsequent requests by stating that I was not a believer in creationism or Intelligent Design and previous chapters had concluded both were outside the realm of science.

In the book I briefly trace the development of scientific theory in each field to illustrate how science continually evolves in response to new evidence and new thinking before examining the current orthodox theory to which most researchers in the field subscribe.

One of those consulted objected to the use of the word "orthodox" on the grounds that it implies belief. Two examples of current orthodoxy in biology are the inclusive fitness hypothesis and the "reciprocal altruism" hypothesis, which are invoked by sociobiologists to explain how the widespread phenomenon of collaboration is compatible with NeoDarwinian competition.

Edward O Wilson, generally regarded as its founder, had the perspicuity to say that sociobiology had taken a wrong turn in the 1960s and 1970s. In 2010, together with two mathematical evolutionary biologists, he published an article in *Nature* arguing that inclusive fitness theory has been of little value in explaining the natural world. Five months later *Nature* published a comment signed by 138 evolutionary biologists proclaiming "We *believe* that their arguments are based upon a misunderstanding of evolutionary theory and a misrepresentation of the empirical literature [my italics]." A footnote said that several others who contributed significantly to the comment were not listed because they were named on separate comments. Four other comments attacking different aspects of Wilson's article were signed by a total of 17 other academics. When 155 evolutionary biologists subscribe to an article declaring that they *believe* the founder of sociobiology doesn't understand its theories, I think this illustrates what I mean by orthodoxy and the groupthink mentality it frequently engenders.

Chapter 23 notes that Lynn Margulis found it difficult to get published in the peer-reviewed mainstream science and biology journals, despite genetic evidence proving beyond reasonable doubt that at least several of her hypotheses are valid.

The way in which the biology establishment treats dissenters from within and questioners from without is all too reminiscent of that shown by the cosmology establishment. I suspect it is due largely to the same reasons. In biology the establishment is dominated by a school comprising theorists who use statistical population genetics and experimental zoologists who work mainly on laboratory fruit flies and who extrapolate their findings not only to all other animals but also to plants, fungi, protoctists, and bacteria. That school provided many valuable insights and contributions to biology; its hypotheses explained much of what was known at the time.

More than 60 years later, however, it has ossified into a culture of equating biological evolution with this NeoDarwinian model and of conflating mathematical proof with empirical proof. Contradictory evidence, particularly from the fossil record and from observations of a much wider range of species in the wild, is either ignored or countered by unsupported assertions, logical fallacies, or mathematical models whose arbitrary parameters can be adjusted to produce any desired outcome. Other hypotheses are paid no heed or their proponents are accused of attacking biological evolution.

One factor particular to biologists, especially in the USA, is their response to creationism and attempts to have Intelligent Design taught alongside biological evolution in schools. Paradoxically, ignoring or denying evidence that contradicts NeoDarwinism provides ammunition to creationists and the Intelligent Design movement.

I'm indebted to those who did engage in a sometimes lengthy dialogue while holding to the orthodox view. They referred me to some articles of which I was unaware and drew my attention to loose reasoning and lack of clarity in parts of drafts. Some eventually conceded a draft had made valid points.

The fifth type of response came from evolutionary biologists who did recognize contradictions to NeoDarwinism in evidence from nature plus the plausibility of alternative hypotheses. They took the view that science progresses and so NeoDarwinism is compatible with all the evidence and all the different hypotheses. To quote one, "There are neo-Darwinists who believe that change must be gradual. If there is punctuated equilibrium, or even saltation, that demonstrates that these particular views of these particular scientists are wrong, not that neo-Darwinism is wrong." But gradualism is a core principle of both Darwinism and NeoDarwinism.

Such a broad-church approach, embracing mutually contradictory hypotheses and hypotheses contradicted by evidence, renders a theory meaningless. Any scientific theory is defined by its core principles, and Jerry Coyne should be given credit for defining the core principles of contemporary NeoDarwinism.

The historical pattern of scientific theories shows small, stepwise progress in refining and applying core principles, followed by the growth of new evidence and/ or new thinking that challenges those core principles, defensive resistance to those challenges, and eventual acceptance of a new set of core principles, frequently by a later generation of researchers. This happened in geology and also in physics: Neils Bohr was discouraged from studying physics because classical Newtonianism was inarguable and there were no major discoveries left to be made. Such a view is echoed in the quote by Richard Dawkins below this chapter's title.

Evolutionary biology has yet to undergo its paradigm shift. Humility in the face of what we don't yet know and understand is an admirable quality not universally present in biologists.

More robust evidence revealed in the last fifteen years or so by the rapid sequencing of whole genomes of many different kinds of species has generated models that better reflect the evidence, new ideas, and a new look at previously ignored or rejected ideas compatible with that evidence. I hope it will eventually produce a new theory of biological evolution in which NeoDarwinism, or parts of it, will be seen as a special or limiting case.

Not all evolutionary biologists responded to my drafts in the ways outlined above. One eminent researcher commented "I read your book section, by far the most complete case for complexity increases I've seen," while another wrote "I applaud you for your effort to present as many alternative theories etc. as you have, especially since…many have gone ignored or have just not been made available to the general public.…Overall I think you have kept a long-needed tone of neutrality in your presentation."

Conclusions

Drawing together the findings from Part 2 produces the following conclusions.

1. Six conditions are necessary for the organic molecules of some 13 atoms found in interstellar space and on asteroids to evolve into things as complex as humans: a planet with essential elements and molecules, sources of energy, a minimum and probably a maximum mass, protection from hazardous radiation and impacts, a narrow temperature range just below, at, and just above its surface, and stability of this biosphere over billions of years. (Chapter 12)

2. A concurrence of unusual galactic, stellar, and planetary factors provided these six conditions on Earth, producing a changing flow of energy through a physico-chemical system that has remained stable but far from thermodynamic equilibrium over some 4 billion years. It suggests that the Earth, if not unique, is a rare planet in the galaxy if not the universe in

possessing the conditions necessary for the emergence and evolution of carbon-based lifeforms as complex as humans. (Chapter 12)

3. Life may be defined as the ability of an enclosed entity to respond to changes within itself and in its environment, to extract energy and matter from its environment, and to convert that energy and matter into internally directed activity that includes maintaining its own existence. Its emergence marks a change of kind, not merely degree, from inanimate matter. (Chapter 13)

4. Conclusive fossil evidence of life has been identified in rock dated to 2 billion years ago, some 2.5 billion years after the Earth formed. Assuming that these microorganisms evolved from simpler lifeforms, life must have existed before then. Disputed claims have been made that life dates from 3.5 billion, or even 3.8 billion and 3.85 billion years ago in the Hadean Eon when Earth was bombarded by asteroids and other debris. The current best estimate is probably 3.5 billion years ago. (Chapter 14)

5. The discovery of extremophiles, organisms currently living in extreme conditions similar to those thought to have obtained in the Hadean Eon, suggests that life *could* have existed at that time, during the first 700 million or so years after the Earth formed, but does not prove it. (Chapter 14)

6. Because of the paucity of the fossil record in general, and because nearly all sedimentary rock from the first two billion years has been subducted or metamorphosed, it is almost certain that we will never find firm evidence of the first lifeforms and when they emerged on Earth. (Chapter 14)

7. Genetic analysis of a wide range of cells strongly suggests, but does not prove, that life emerged naturally only once on Earth, and that all current lifeforms—from the numerous types of bacteria to humans—evolved from a single common ancestor. (Chapter 14)

8. No scientific hypothesis explains why proteins—molecules essential for the functioning of every known cell, and hence of the earliest independent cells—form from combinations of up to only 20 different amino acids out of some 500 known amino acids and why only the left-handed out of two possible stereo isomers of each amino acid is used. (Chapter 14)

9. No scientific hypothesis convincingly explains how the number of atoms and molecules comprising the essential complex components, together with their changing configurations and functioning, that constitute the simplest—and presumably the most primitive—independent lifeform emerged from the interactions of atoms and molecules consisting of up to 13 atoms on the surface of the newly formed Earth. (Chapter 15)

    9.1. Biochemistry's orthodox account of how life emerged from a primordial soup of such chemicals lacks experimental support and is invalid because, among other reasons, there is an overwhelming statistical improbability that random reactions in an aqueous solution could have produced self-replicating RNA molecules or even self-replicating peptides, still less ones with only the left-handed amino acid isomers.

9.2. Other hypotheses—like self-organizing complexity, replicating clay or iron pyrites crystal precursors, or a quantum mechanical rather than thermodynamical mechanism—offer intriguing possibilities but so far lack explanations for key steps and also lack empirical support.

9.3. Various proposals that life came to Earth from outer space simply postpone the origin question, and most have been disproven or else rest on highly questionable and unsupported assumptions.

9.4. The weak anthropic principle explains nothing, while interpretations of the strong anthropic principle and proposals for intelligent design of the first cell are not falsifiable and so are outside the scientific domain.

10. As with the emergence of matter, it is very probably beyond the ability of science to explain the emergence of life. (Chapter 15)

11. Evidence from fossils, or from any one of homologous structures, vestigiality, embryology, biogeography, biochemistry, genetics, and genomics of living species, by itself does not offer indisputable evidence for the phenomenon of biological evolution. However, taken together, they provide overwhelming evidence that humans evolved from the earliest lifeforms on Earth that spread across the surface of the planet to form a biosphere or, more accurately, a biolayer. Moreover, the evidence shows a clear pattern. (Chapters 17 and 18)

12. The pattern of the fossil record over time is from simple to more complex: prokaryotes appear before eukaryotes, unicellular eukaryotes before multicellular, radially symmetrical before bilateral and cephalized, invertebrates before vertebrates, fishes before amphibians, reptiles before birds, mammals before primates, and apes before humans. (Chapter 17)

13. This is not a linear progression but a networking, fusing, and branching evolution from a universal common ancestor into very many lineages that, for the vast majority, end in extinction. (Chapters 17 and 18)

14. Although the boundary marking the emergence of a new species is usually indistinct, as we have found for all emergences, it is a boundary nonetheless: beyond it no reversal takes place. (Chapter 17)

15. Lineages reach stasis in which the last species shows little change, either before extinction or, less commonly, continuing for some tens or even hundreds of millions of years to the present day. (Chapter 17)

16. Current evidence indicates that complexity generally increases along each lineage; this is clearly the case for the lineage leading to humans. (Chapters 17 and 18)

17. Living species overall form a pattern of increasing complexity from prokaryotes to humans. (Chapter 18)

18. Collaboration plays a more significant role than competition in the survival and propagation of life, from genes acting in concert, through prokaryotes collaborating in colonies to sustain and replicate themselves, organelles collaborating within a unicellular eukaryote to sustain and replicate the cell, specialized cell groups collaborating within multicellular organisms, and animals collaborating in societies like colonies of insects, shoals of fishes, flocks of birds, and families, clans, troops, and other social groups of many mammals. Collaboration is also widespread between members of different species. (Chapter 19)

19. For unicellular and animal species collaboration principally takes the form of a group within a species communicating and working together to construct a communal shelter, produce and rear offspring, forage, defend themselves, attack prey, and migrate to a better environment for their survival and reproduction. In some cases several groups within a species, and even within related species, also collaborate for mutual benefit, particularly for migration. (Chapter 19)

20. Collaboration in unicellular species is innate. Among simple animals, like many species of insect, collaboration is also innate but more complex and usually is enforced; such behaviour is better described as collectivism to distinguish it from cooperativism in which the benefits are mutual or the collaboration is voluntary. (Chapter 19)

21. As the morphological complexity of species increases from fishes through to primates, social learning, which is more efficient than individual learning by trial and error, increases to supplement innate behaviour; such socially learned skills can be inherited. Moreover within each class, increased social learning correlates with increased brain complexity together with increased intelligence measured by the invention of more novel solutions to challenging problems. (Chapter 19)

22. The current orthodox explanation of the phenomenon of biological evolution is an updated synthesis of a Darwinian natural selection of randomly generated heritable characteristics that make individuals in a species population better adapted to compete for limited resources in their environment and so survive longer and reproduce more, statistically based population genetics, and the central dogma of molecular biology that says information is a one-way flow from a gene to a protein in a cell. (Chapter 16)

23. Many proponents of this NeoDarwinian paradigm say that natural selection causes biological evolution, but it cannot be a cause unless nature chooses because to select is to choose. Neither can it be a metaphor for a natural law because its advocates nowhere invoke, still less offer proof of, a gene-based natural law of biological evolution that applies to all living things. If natural selection is to be more than a circular argument (the organisms that survive and reproduce more are the ones naturally selected to survive

and reproduce more) then it is a passive record of effects caused by other things. (Chapter 21)

24. According to the NeoDarwinian model the three principal causes of the evolution of species are (Chapter 21):

    24.1. randomly generated genetic mutations coding for traits that give a member of a species population an advantage in the competition for limited resources and protection against predators in that population's environment;

    24.2. competition, primarily between members of the increasing population for finite resources in their environment, that enables the winners to survive longer and produce more offspring;

    24.3. sexual reproduction that spreads such advantageous genetic mutations throughout the population's gene pool over tens or hundreds of thousands of generations.

25. This NeoDarwinian orthodox hypothesis plausibly explains the most dominant pattern in the evidence—extensive species extinctions—by one, or a combination, of the following: mortal competition between members of a growing species population for limited resources in that population's environment, predation by other species, the spreading of harmful genetic mutations in a population gene pool because as few as one mutation can prove fatal whereas very many genetic mutations are needed to produce morphological innovation, and the inability of a species to adapt to a change in its environment. (Chapters 16, 17, 18, 19, 21, and 22)

26. Many species undergo reversible variations in response to environmental changes, most likely due to changes in the relative amounts of genetic variations already present in a population's gene pool rather than by a gradual accumulation of new genetic mutations. However, reversible change does not constitute the evolution of species. To distinguish reversible adaptive change from irreversible change, biological evolution is best defined as a process of change in organisms that results in new species. (Chapters 18 and 21)

27. No evidence supports the NeoDarwinian explanation of the irreversible evolution of new species. (Chapter 18)

28. The NeoDarwinian orthodox hypothesis fails to explain

    28.1. why the fossil record for animals shows morphological stasis with minor and often oscillating changes punctuated by the geologically sudden (tens of thousands of years) appearance of fully formed new species that remain substantially unchanged for some tens or even hundreds of millions of years until they become extinct or, more rarely, continue to the present day;

    28.2. why the accumulation of genetic mutations is approximately the same in different lineages, but in some lineages this results in no morphological or species change over tens if not hundreds of millions of years despite considerable changes in the environment while other lineages undergo a series of major evolutionary transitions in the same period;

28.3. why some species with very similar genes have very different phenotypes, while other species differing substantially in their genes have similar phenotypes;

28.4. immediate, rather than gradual, speciation through polyploidization, which is an increase in the normal number of chromosomes per cell that occurs naturally in many plants, and in fishes, amphibians, reptiles, and even some mammals, and which is probably more extensive than previously realized because so few attempts have been made to investigate it;

28.5. why or how acquired characteristics are inherited without any underlying genetic change;

28.6. the function of some 98 per cent of the human genome that does not consist of genes defined as protein-coding lengths of DNA and which was dismissed for some 50 years as "junk DNA" before genomic analyses showed a majority of this is implicated in regulating the expression of genetic networks, a function that determines the phenotype as much as do the protein-coding genes;

28.7. why a single fertilized cell caused by the fusion of a male and a female sex cell replicates into identical stem cells that then differentiate into specialized cells with different functions that subsequently evolve into either a male or female independent complex adult;

28.8. why or how the most numerous species on the planet evolve, namely prokaryotes that do not reproduce sexually but replicate by cloning themselves and engage in horizontal gene transfer, sometimes between very different species;

28.9. why or how two major and related patterns characterize the evidence of biological evolution, namely complexification and collaboration. (Chapters 20, 21, and 22)

29. The claim by advocates of Intelligent Design that the NeoDarwinian hypothesis cannot explain significant biological innovations and their pattern of increasing complexity is supported by substantial evidence. However, to conclude that these biological innovations must therefore have been directed by an intelligent agent is a logical fallacy. If the NeoDarwinian hypothesis cannot explain these phenomena it does not follow that other scientific, testable hypotheses now, or in the future, cannot offer an explanation. Advocates of Intelligent Design fail to offer any testable explanations of their beliefs, which puts Intelligent Design outside the realm of science. (Chapter 22)

30. As for other hypotheses that complement or challenge the NeoDarwinian paradigm, the punctuated equilibrium hypothesis doesn't depart radically

from NeoDarwinism but challenges its method of speciation on the grounds that genetic mutations in a population spread over a wide geographical area are diluted by the population's sheer size and constantly changing environments leading to minor morphological changes that fluctuate about a mean, evidenced by the fossil record. It proposes that genetic mutations spread rapidly in small subpopulations isolated on the periphery leading to a new species unable to breed successfully with the main group. (Chapter 22)

31. The sudden origins hypothesis likewise challenges NeoDarwinian gradualism. Supported by biochemical evidence, it claims persuasively that an organism's normal state is one of homeostasis because cellular repair mechanisms constantly correct randomly generated mutations and thus prevent change. It is only relatively sudden and severe stress that overcomes such mechanisms and allows mutations in sex cells to go unrepaired, causing significant genomic change that effects organismal development. However, the hypothesis is vague about how this leads to viable new species or produces evolutionary transitions that show a pattern of increased complexity. (Chapter 22)

32. Neutral theory challenges the NeoDarwinian gradual accumulation of beneficial genetic mutations as the only mechanism of speciation. It argues that, at the molecular level, by far the majority of accumulated genetic changes confer no adaptive or competitive advantage, and speciation may take place by the chance accumulation of "selectively neutral" mutations. While disproving the single mechanism idea, neutral theory is also vague about how this leads to viable new species or produces evolutionary transitions that show a pattern of increased complexity. (Chapter 22)

33. The whole genome duplication, or 2R, hypothesis first advanced in 1970 offers an explanation of how significant evolutionary change has taken place in vertebrates by the doubling of the normal number of two sets of chromosomes per cell followed by degeneration to two sets of chromosomes per cell but with twice as many genes. Although the hypothesis was ignored or dismissed at the time, recent sequencing of whole genomes provides support by showing, for example, that a large part of the human genome comprises four sets of multiple genes with a similar structure at different chromosomal locations.

This hypothesis offers a plausible explanation for the relatively rapid speciation shown in the fossil record. As currently developed, however, it does not explain what caused such genome duplication or which of several possible mechanisms was involved. Further studies of known polyploidization in plants and animals may reveal answers to this and also how extensive this mechanism has been in biological evolution. (Chapter 22)

34. Epigenetic theory seeks to extend the current paradigm beyond NeoDarwinism by proposing non-genetic mechanisms for both

morphogenesis (differentiation of stem cells into specialized cells that subsequently evolve into a complex independent adult) and also the inheritance of acquired characteristics, but it doesn't explain why such phenomena occur. (Chapter 22)

35. Convergence theory challenges the claim of most current NeoDarwinists that the causes of genetic change are random and hence their pheno-typical effects necessarily will be random. It claims that inherency at the biochemical level and the adaptive principle constrain biological evolution to a limited number of phenotypical outcomes that show, among other things, a pattern of progressive complexification, with the emergence of humans a near-inevitability.

    As currently developed this hypothesis gives inadequate recognition to the evidence of divergence in biological evolution (progressive complexification occurs along diverging lineages). The distinction between convergent evolution and parallel evolution (organs or body plans or behaviours are similar in different species because they, or early versions of them, are found in a common ancestor) is blurred because developmental regulatory genes are very similar across a wide range of species; on the other hand, body plans resulting from such similar genes are very different.

    Patterns of divergence and convergence occur in biological evolution. Attributing the cause of convergence to inherency, however, describes a pattern, not the cause of the pattern, while invoking the adaptive principle suffers from a failure to define a law applicable to all organisms. (Chapter 22)

36. The emergence hypothesis seeks to explain why a limited number of major emergences of increasing complexity, out of innumerable possibilities, occur in the evolution of life. While the idea of one or more underlying selection rules to explain them is attractive, none has yet been identified. Like its explanation of the emergence of life it has not yet progressed beyond a description of the phenomena. (Chapter 22)

37. Self-organizing complexity claims to explain both the differentiation of stem cells into specialized cells that develop into an interdependent adult (morphogenesis) and also the evolution of species by treating each as a self-regulating network governed by a particular class of Boolean logic switching rules that produces stability at the edge of chaos. However, since its introduction in 1991 its mathematical predictions have been contradicted by genetic and genomic evidence, and insufficient progress has been made in grounding its various computer models in biological data. Consequently, as with its explanation of the emergence of life, this hypothesis currently offers no more than hope that corrections to its models will lead to deeper new laws of biology. (Chapter 22)

38. The hypothesis that natural laws of genome evolution exist was advanced in 2011 following the huge increase in data generated by the rapid

sequencing of whole genomes. It claims the available data show several universal patterns in genome and phenome evolution in lineages from such diverse groups as bacteria, archaea, and eukaryotes; moreover these patterns can be accounted for by simple mathematical models, similar to those used in statistical physics. Since the models do not incorporate natural selection, these patterns are shaped by emergent properties of gene ensembles. Consequently, the hypothesis suggests, laws of evolutionary biology comparable in status to laws of physics might be attainable.

This approach offers a potentially fruitful avenue of research to try to determine how outcomes in biological evolution are constrained to a very few out of near limitless possibilities. (Chapter 22)

39. The natural genetic engineering hypothesis was also advanced in 2011 as a challenge to the NeoDarwinian paradigm that biological evolution is caused by randomly generated genetic changes in otherwise static genomes. It claims that cells have an innate ability to re-organize their genomes in response to hundreds of kinds of inputs; moreover, genomes are not read-only memory systems subject to accidental change but are read-write information storage organelles at all time scales, from the single cell cycle to evolutionary eons. As evolution proceeds, so does evolvability: living organisms are self-modifying beings and are intrinsically teleological. It argues that the current paradigm fails to explain the rapidity and variety of biological complexification, and that integrating biological science with information- and systems-based approaches will produce a new paradigm for the twenty-first century.

This hypothesis doesn't provide that paradigm, but rather an agenda for one based on the concept of organismal self-modification that is more consistent with the evidence than is the current paradigm. (Chapter 22)

40. Systems biology challenges the reductionism of the NeoDarwinian model by taking a holistic view of natural phenomena and aims to discover emergent properties. It studies living things as an integrated and interacting network of genes, proteins, and biochemical reactions that give rise to life and are responsible for an organism's form and functions. These emergent properties cannot be attributed to any single part of a living system, which is therefore an irreducible entity.

Such a concept reflects the reality of the biological world far better than does the simplistic NeoDarwinian model and offers a potentially rewarding approach to a more complete understanding of the causes and mechanisms of biological evolution. The principal danger in the future is that, in relying on bioinformatics to cope with an exponentially increasing volume of molecular sequencing data, and in a desire to be as rigorous as the reductionists, systems biologists will become so focused on the specialist investigation of the parts as to lose focus on the whole. (Chapter 22)

41. Broadly similar in concept but wider in its application to the whole Earth, the Gaia hypothesis proposes that the Earth's biosphere, atmosphere, oceans, and soil constitute an interactive, self-regulating system; so closely coupled is the evolution of living organisms to the evolution of their environment that together they constitute a single evolutionary process. While this approach makes intuitive sense, and has some empirical support at a global level, the hypothesis does not yet provide a sufficiently tangible explanation of the evolution of species. (Chapter 22)

42. The hypothesis of formative causation proposes not only how morphogenesis and biological evolution take place but also why they do. According to this view there are no immutable universal laws, which are Newtonian deterministic concepts that fail to answer what caused these laws to exist. Rather it proposes that nature is habitual. All species draw upon and contribute to an evolving universal collective memory of their species that takes the form of evolving universal morphic fields. By a process of resonance each morphic field imposes its behavioural pattern on otherwise random or indeterminate activity to cause, for example, identical human stem cells to differentiate into specialized cells that develop into independent adults. Some experimental evidence on rat behaviour is inconsistent with both the NeoDarwinian hypothesis and the epigenetic inheritance of acquired characteristics but is consistent with formative causation. The lack of experimental testing of the hypothesis since it was introduced in 1981 may be attributed to a lack of open-mindedness by the biology establishment. (Chapter 22)

43. Attempts to explain non-competitive behaviour of all species, including human, according to the NeoDarwinian model resulted in a theoretical gene-centred view of biological evolution adopted by sociobiology. Its principal hypotheses are as follows. (Chapter 23)

   43.1. *Kin-related altruism or inclusive fitness*

   On the basis of mathematical theory, this hypothesis claims that altruistic behaviour results from individuals acting to pass on copies of their genes to the next generation not only directly through sexual reproduction but also indirectly by aiding the reproductive success of close kin sharing some of its genes, even at a cost to itself. However, it

   a. fails to account for the behaviour of the majority of species on the planet, prokaryotes, that replicate by cloning and can pass on genes immediately to other prokaryotes belonging to very different species;

   b. assumes that members of a group will behave the same way if they share the same genes when the evidence shows the relationship between genes and behaviour is far more complex;

c. uses the term "altruism", a selfless act, to describe enforced collaboration;

d. confuses innate and enforced animal behaviour with intentional human behaviour;

e. fails to explain altruism in humans;

f. fails to explain collaboration between animals more distantly related than first cousins.

43.2. *"Reciprocal altruism"*

This hypothesis claims animals and humans behave altruistically because a selfless act will be reciprocated in the future. It is based on a contradiction in terms: if an act is reciprocated it is not altruism; rather it is mutual aid, or cooperation. It suffers from the first five flaws of the kin-related altruism hypothesis. Furthermore, no evidence of untrained species in the wild supports its claims. Its "proof" is theoretical, as are extensions to explain cooperative, selfish, and spiteful behaviours. These are based either on mathematical models that do not map patterns in actual behaviour or else on over-simplistic and unsuccessful 1950s game theories from economics that are divorced from biological realities, can be designed to produce any desired outcome, have no predictive value, and change if not invert the generally accepted meaning of words.

43.3. *Selfish gene*

This hypothesis asserts that genes are the prime movers of all life, that individual bodies do not have to exist, and that genes selfishly compete with each other for their survival and replication. It is based on a conceptual error of ascribing intentions to segments of an acid and is contradicted by substantial evidence.

43.4. *Genial gene*

This illustrates the conclusion that game theories can be devised to produce any desired outcome.

43.5. *Multilevel selection*

This hypothesis concedes that the preceding sociobiological hypotheses are misconceived, offers a more realistic account of the complex world of organisms and their interactions with each other and their environment, and its description of how major evolutionary transitions arise is consistent with collaboration as the principal cause of biological evolution.

44. The view that collaboration is one of the greatest problems for the biological and social sciences to explain is a self-imposed problem caused by the adoption of the NeoDarwinian model that is rooted in competition. This problem is solved by recognizing that collaboration is extensive at every level of life and is the prime cause of organisms developing and surviving (Chapter 19):

44.1. genes work together for the development of an organism, often regulated by other genes that themselves work together;

44.2. a unicellular organism consists of parts performing specific functions that work together to maintain and replicate the organism;

44.3. a eukaryotic cell consists of a nucleus controlling the collaboration of cellular parts, or organelles, that perform specific functions for the maintenance and replication of the cell, while a multicellular organism consists of a hierarchy of specialized cell groups working together with low conflict to maintain the organism's existence and produce offspring;

44.4. organisms from simple prokaryotes to humans collaborate for their survival and reproduction.

45. Moreover, collaboration rather than NeoDarwinian competition more plausibly causes biological evolution characterized by complexification, innovation, and diversification of lifeforms from the earliest simple prokaryotes. Collaboration between different kinds of archaic bacteria for their mutual survival (symbiosis) led to small organisms living inside larger ones (endosymbiosis) followed by their merger (symbiogenesis) to form the first unicellular eukaryotes (larger and more complex with nucleated cells). Subsequent symbiogeneses with other archaic bacteria, followed by gene transfer from endosymbionts to nuclei, formed still larger and more complex eukaryotic cells containing collaborating organelles characteristic of the taxonomic kingdoms of animals, plants, and fungi. Collaborating members of unicellular eukaryotic colonies merged to form still more complex multicellular eukaryotes within these kingdoms. (Chapter 23)

46. Rather than a NeoDarwinian gradual accumulation of randomly generated genetic mutations within a species population over thousands of generations, further evolutionary complexification and innovation in the animal kingdom more plausibly results from (Chapters 17, 18, 22, and 23):

46.1. transposons (jumping genes) implicated in regulating gene expression and generating different cell types and different biological strctures,

46.2. horizontal transfers of suites of genes between species,

46.3. the merging of genomes through hybridization, and

46.4. duplication of whole genomes (possibly through hybridization or polyploidy).

47. Collaboration between members of a species group to construct a communal shelter, produce and rear offspring, defend themselves, forage, attack prey, and migrate to an environment with better resources (the mutual aid hypothesis), together with collaboration between members of different species, is a more important factor than NeoDarwinian competition for increasing the probability of those members surviving and successfully reproducing. (Chapters 16 and 23)

48. Like competition, collaboration can be either innate—behaviour that proved successful for survival and reproduction and was inherited genetically or epigenetically, or perhaps resulted from a morphic field—or it can be intentional as in human mutual and cooperative societies. (Chapter 23)

49. Collaboration, moreover, provides a more plausible explanation than does NeoDarwinian competition for the pattern within taxonomic classes of species demonstrating increased cognitive abilities through social learning. (Chapters 19 and 23)

50. The evolution of life is more than biological evolution. Consciousness, as the awareness of the environment, other organisms, and self that leads to action, emerges in rudimentary form in prokaryotes, the most primitive and ancient lifeforms on Earth, evidenced by their actions taken for survival in direct response to external and internal stimuli. As individual prokaryotes collaborated in communities, collaborative direct responses for collective survival evolved. (Chapter 24)

51. Consciousness rises along animal lineages, demonstrated by the evolution of more complex and varied actions, notably the addition of innate, learned, social, and innovative behaviours. As progressively more complex species evolved along the human lineage, learned, social, and innovative behaviours increased in significance. (Chapter 24)

52. The physical correlate of this rise in consciousness is the evolution in animals of an electrochemical nervous system enabling progressively more rapid, varied, and flexible responses to external stimuli. It is characterized by four interrelated trends. (Chapter 24)

   52.1. *Growth*

   A net increase in the number of neurons, or nerve cells, and their connections develops.

   52.2. *Complexification*

   The nervous system complexifies not only by increasing the number of neurons and their modifiable connections until the human brain emerges as the most complex thing in the known universe, but also by the growth in number and size of specialized interconnected groups of neurons processing and controlling responses to an increasing range of external and internal stimuli. The neocortex—the most recently evolved and most complex part of the brain—shows the greatest growth from early mammals to humans, correlating with an increase in innovative behaviour—a mark of intelligence—from mammals to primates, followed by a vastly greater increase in innovative behaviour and intelligence in humans.

   52.3. *Centration*

   Growth and complexification are accompanied by a progressive centration, from a diffuse network of neurons in simple primitive animals to a centralized nervous system in vertebrates in which most neuron groups become concentrated in the head with the rest in the spinal column.

52.4. *Optimization*

Centration and complexification reduce the length of neural connections, hence transmission times, and hence response times, while increasing the types of response. While animals have generally increased in size as their lineages have evolved, the brain and body size of current humans are smaller than those of early humans, consistent with optimization of the brain-body size for maximum efficiency of the electrochemical nervous system in providing greater speed and flexibility of response.

53. Individual organisms collaborating for survival in their environment most probably caused the natural selection of genetic networks promoting the growth, complexification, and centration of the nervous system that constitute the physical correlates of rising consciousness that so far has reached its highest level in the human species. Optimization of the electrochemical nervous system for speed and flexibility of responses derives from the laws of physics. (Chapter 24)

Overall, four qualitative laws of biological evolution may be inferred from the evidence.

First Law of Biological Evolution   Competition and rapid environmental change cause the extinction of species.

Second Law of Biological Evolution   Collaboration causes the evolution of species.

Third Law of Biological Evolution   Living things evolve by progressive complexification and centration along fusing and diverging lineages that lead to stasis in all but one lineage.

Fourth Law of Biological Evolution   A rise in consciousness correlates with increasing collaboration, complexification, and centration.

In Part 3 I shall examine whether the emergence of humans marks a change of degree or of kind in the evolution of life.

PART THREE

The Emergence and Evolution of Humans

CHAPTER TWENTY-SIX

The Emergence of Humans

The essential quality of man…is conceptual thought.
—JULIAN HUXLEY, 1941

The origin and development of human culture—articulate spoken language and symbolically mediated ideas, beliefs, and behaviour—are among the greatest unsolved puzzles in the study of human evolution. Such questions cannot be resolved by skeletal or archaeological data.
—RUSSELL HOWARD TUTTLE, 2005

Any explanation of where we came from—when, how, and why we evolved from primates—depends on understanding what it means to be human. Hence I shall begin Part 3 by considering the various definitions of human, formulate the definition I shall use, examine the evidence for the emergence of humans, and then evaluate its current scientific explanations.

What is a human?

Like the definition of life there is no agreement as to what is a human. Scientists define human from the perspective of their own specialized disciplines, and even within a discipline significant differences arise.

Palaeoanthropologists draw on skills from archaeology, anthropology, and anatomy to study the emergence of humans, and so it is instructive to begin by considering how they identify humans. According to Donald Johanson, who discovered the fossilized partial skeleton of a creature nicknamed Lucy that he concluded had walked upright on two feet, "bipedalism is the most distinctive, apparently earliest, defining characteristic of humans". But penguins, emus, ostriches, and other birds are bipedal. Among the primates, chimpanzees sometimes walk upright on the ground, but with bent knees. When orangutans move along the branches of trees in which they live they usually walk upright with humanlike straight legs while raising their arms for balance or grasping. And even Johanson assigned Lucy to the prehuman taxon of *Australopithecus afarensis*.

Teeth are frequently the only hominin fossils found in a prehistoric site, and as few as one or two teeth have been used to define a fossil as human or other species. The reasoning is that living chimpanzees, genetically the closest species to living humans, have large, pointed and projecting canines, single-cusped lower premolars that sharpen the upper canines, and thin enamel on their teeth; living humans have smaller canines that resemble their central incisors, bicuspid lower premolars, and thick tooth enamel; hence the degree to which fossilized teeth resemble human compared with chimpanzee teeth indicates the creature's position on the phylogenetic tree from the common ancestor of humans and chimps to living humans. However, the underlying assumption is that teeth changed in the branching lineage leading to living humans but did not change in the branching lineage leading to living chimpanzees. As we shall see later, the discoverers of *Ardipithecus ramidus* claim that the common ancestor of chimpanzees and humans did not resemble living chimpanzees. Moreover, living orangutans have thick tooth enamel.

Other palaeoanthropologists use cranial capacity as the defining characteristic of humans. But as we saw in Chapter 24, elephants have a cranial capacity some six times that of humans, while the cranial capacity of humans some 30,000 years ago is similar to that of Neanderthals of the same period and is estimated to be 10 per cent larger than that of living humans.*

Chris Stringer of Britain's Natural History Museum attempts to cover all bases by using a suite of characteristics to define a human: a large brain volume, a high and domed skull, a vertical forehead, a small and flat face, reduced brow ridges, a lower jaw with a bony chin, smaller and simpler teeth, a lightly built tympanic bone containing the ear bones, a bone in front of the pelvis that is short and nearly circular in cross section, no iliac pillar reinforcing the pelvis above the hip socket, and thigh bones that are oval in cross section and thickened most at the front and back.[1] However, even if such bones were all found in one prehistoric specimen, which has never been the case, they indicate relatively small degrees of difference from other hominins rather than characteristics unique to humans.

Zoologist Desmond Morris claimed the defining characteristic of humans is hairlessness, which distinguishes them from 193 other species of apes and monkeys.[2] However, it doesn't distinguish humans from whales, dolphins, and many other mammals. Even if humans were defined as hairless apes, it would not help identify the emergence of humans because fossilized hairs are extremely rare.

Geneticists define humans by their genes. However, we saw in Chapter 18 that humans share around 94 per cent of their genes with chimpanzees and possibly 90 per cent with mice.† The initial draft analysis of the Neanderthal genome published in 2010 suggests that human and Neanderthal genes differ by only 0.3 per cent;[3] while this is less than the 6 per cent difference from chimpanzees it is

*   See page 403.
†   See page 318.

not much more than the estimated 0.1 to 0.15 per cent variation of genes within living humans.[4]

Some geneticists are currently attempting to identify particular genes, or variations, that are unique to humans, but I think this research is unlikely to produce a unique characteristic, not least because of the genetic variations within humans, the fact that most genes act in concert with many other genes through networks, genes are often implicated in different functions, and their effects depend crucially on their regulation: the degree to which they are activated, when, and for how long.

Few anthropologists today claim that culture rather than genes defines a human; the current orthodoxy is gene-culture co-evolution. But within this framework most claim that culture is unique to humans. For example, in their book, *Not by Genes Alone: How Culture Transformed Human Evolution*, Peter Richerson and Robert Boyd define such culture as "information capable of affecting individuals' behaviour that they acquire from other members of their species through teaching, imitation, and other forms of social transmission". However, as we saw in Chapter 19, in different degrees all these features except teaching are observed in animals from fishes to nonhuman primates, and even teaching has been observed in meerkats.*

Many neuroscientists now claim they can define a human by identifying where capacities unique to humans—like language, use of symbols, and sense of selfhood—are located in the brain and the neural networks and cognitive mechanisms that support them. Whether such capacities are nothing but neural activity in parts of the brain or neural activity is a correlate of these capacities is a major metaphysical, rather than scientific, question that I shall examine later. For the purpose of identifying the emergence of humans from primate ancestors, however, neurons leave no fossils and so other evidence is needed.

Proposed definition

What makes humans unique is not bipedalism, teeth, cranial capacity, skeletal structure, hairlessness, genes, culture as described above, or neural activity in areas of the brain. Such characteristics are either exhibited by other organisms or else the human variant differs only by a small degree from that of some other organisms.

We do possess, however, one characteristic that, as far as we know, is unique.

The pattern of evidence identified in Part 2 gave a clue. The evolution of life overall is marked by rising consciousness, defined as

> **consciousness** Awareness of the environment, other organisms, and self that can lead to action; a property shared by all organisms in differing degrees, from rudimentary levels in very simple organisms to more sophisticated levels in organisms with complex cerebral systems.

*   See page 327.

With increasing complexity, centration and, in the case of humans, optimization of the nervous system, consciousness rises along the human lineage to the point *where it becomes conscious of itself.* Like water heated to 100° Celsius, a phase change occurs: consciousness reflects on itself.

> **reflective consciousness**   The property of an organism by which it is conscious of its own consciousness, that is, not only does it know but also it knows that it knows.

Humans may thus be defined as

> ***Homo sapiens***   The only species known to possess reflective consciousness.

This faculty enables a human to think about itself and its relationship with the rest of the universe of which it knows it is a part.

Evidence for reflective consciousness

The most compelling evidence for this faculty is humans asking, and attempting to answer, such questions as What are we? Where did we come from? What is the universe in which we exist? and so on: the stuff of religion, of philosophy, and subsequently also of science.

This is the full flowering of reflective consciousness. It did not, however, spring suddenly into existence fully formed, just as a mass of heated water does not suddenly all turn into steam when the temperature reaches 100° Celsius.

How then do we detect its emergence, its first glimmerings in prehistoric times? The answer, I suggest, is evidence of those primate secondary faculties it transforms and new secondary faculties it generates. Chief among the secondary faculties that reflective consciousness radically transforms are comprehension, memory, foresight, cognition, learning, invention, intention, and communication. The new secondary faculties it generates include thought, reasoning, insight, imagination, creativity, abstraction, will, language, belief, and morality. Its possession demarcates cooperation (rational, voluntary collaboration) from collectivism (instinctive, conditioned, or coercive collaboration).

These secondary faculties tend to act synergistically, and the results of such actions leave evidence. For example, comprehension combined with invention, foresight, and imagination gives rise to specialized, composite tools; cognition combined with imagination produces beliefs in supernatural powers that result in religious rituals; cognition combined with communication produces representational art that results in paintings and sculptures; add imagination and the art extends to images that are not observed in life; add abstraction and the art extends to symbols.

We saw in Part 2 that rising consciousness in species was also associated with species members living in groups that facilitate, among other things, social learning. Individual faculties transformed or generated by reflective consciousness that are shared in human societies constitute its culture.

> **human culture** A society's knowledge, beliefs, values, organization, customs, creativity expressed in its arts, and innovation expressed in its science and technology, that are learned and developed by its members and transmitted to each other and to succeeding generations.

This definition distinguishes characteristics unique to human societies from Richerson and Boyd's definition of culture that applies to many nonhuman societies.

Human predecessors

To detect the emergence of humans—those first glimmerings of reflective consciousness evidenced by signs of the secondary faculties it transforms or generates—we need to know what humans emerged from.

Table 20.1 (pages 336–340) shows the majority, but by no means unanimous, view of the taxonomy of the human species: *Homo sapiens* is the only surviving species of the genus of humans, *Homo*, within a tribe of hominins that form part of the hominid family within the superfamily of hominoids that comprises a branch of the order of primates. Most researchers in the field consider these classifications reflect descent in a branching evolutionary tree from an ape-like primate genus like the tailless *Proconsul* that existed from around 25 to possibly 5 million years ago in East Africa and is tentatively assigned to the superfamily Hominoidea. Figure 26.1 outlines the broad branches of this evolutionary tree, with examples of existing species. It incorporates an additional subfamily of Homininae that includes the tribe of hominins, the two chimpanzee species, and gorillas, but excludes organutans.

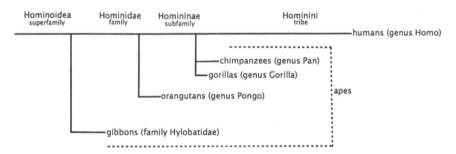

Figure 26.1 Phylogenetic tree showing human descent from hominoid ancestor

However, tracing the lineage to modern humans (species *Homo sapiens*) is highly problematic; it is bedevilled by a paucity of evidence that gives rise to disputed interpretations and disagreements about taxonomic classifications and ancestral relationships, reflected in Figure 26.2.

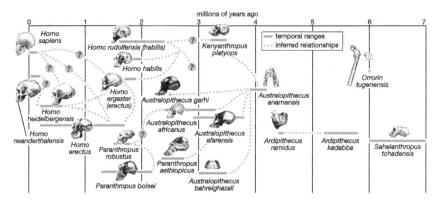

Figure 26.2 Possible pathways for a lineage leading to the emergence of *Homo sapiens* from hominid predecessors

The evidence and its problems

About once a year the popular press, and often the science journals, announce the discovery of a new fossil, or the re-dating of an existing fossil using new techniques, that rewrites the story of human emergence. In their academic papers rather than their press conferences the scientists tend to be more cautious. Understandably so.

As archaeological historian Robin Derricourt of the University of New South Wales, Australia puts it

> Both palaeoanthropology and the archaeology of early humans operate in a framework where the science is not experimental, data are sparse and hypotheses are not easily refutable and replicable. There are few other sciences where an isolated piece of evidence can support or change a broad scale interpretative model without itself being readily testable.
>
> In this framework, personality plays an important role—the conservative thinker or the innovator, the dogmatist or the sceptic, the taxonomic splitter or the lumper.[5]

It provides fertile ground for the Law of Data Interpretation and rare, but significant, cases of fraud.*

Five kinds of evidence are used to infer the evolutionary relationships and behaviour of the species preceding modern humans: fossils, other remains, relative and "absolute" (radiometric) dating, genetics, and living species. Each has its own limitations.

---

*   See page 285.

Fossils

The principal problem is the sheer scarcity of fossils. In the period approximately 2.5 to 1 million years ago only around 50 individuals have been classified as a species of *Homo* and many of these comprise one or two teeth or a bone fragment.

Complete skeletons from prehistoric times are almost never found. Even the celebrated Lucy skeleton dated to 3.2 million years ago consisted of fragments fitting together to comprise 47 bones compared with a total of 206 bones in an adult human. Fragments of teeth, lower jaws, facial and upper cranial bones are the most common fossils, followed by fragments of thighbones, while remains of the feet, hands, pelvis, and spine are extremely rare.

It is easy to be seduced by modern sophisticated 3-D computer-generated images of a skull, but their accuracy depends on underlying assumptions about how crushed fragments should be reconstructed and where they should be placed in one or more skulls, and the assumptions built into the computer program that creates a whole skull from the input data.

Estimated cranial capacities are often used to assign a fossil to a particular species. Table 26.1 lists these brain sizes for different hominins; it also shows how few specimens they are based on.

Table 26.1 Estimated brain size of hominin fossils and their number of examples

Hominin	Average capacity of braincase (cc)	Number of fossil examples
Australopithecus	440	6
Paranthropus	519	4
Homo habilis	640	4
Javanese *Homo erectus* (Trinil and Sangiran)	930	6
Chinese *Homo erectus* (Peking man)	1,029	7
Homo sapiens	1,350	7

Source: Tuttle, Russell Howard (2014)

Even if the estimated cranial capacities of the samples were accurate and the sample sizes much larger, we saw in Chapter 24 that neither brain size nor ratio of brain size to body weight is a reliable criterion of cognitive ability and hence of level of consciousness.*

For skeletons generally, adults are larger than juveniles, and most hominid males are larger than females, but the degree of gender dimorphism varies

---

* See page 403.

between species, making interpretation of a few, often scattered, fragments, problematic. For many years *Ramapithecus*, a group of species thought to have lived 15 to 5 million years ago, was considered ancestral to *Australopithecus* and therefore to modern humans. This view was based on reconstructed jaw and dental fragments found as far apart as Fort Ternan in Kenya and the Siwalik Hills in India. Molecular genetics and later fossils led to the hypothesis that these were actually female members of a previously described genus, *Sivapithecus*, which was reconstructed to be an ancestor of orangutans.[6]

It is impossible from the fossil record to determine whether a specimen is a hybrid. As we saw, this is a difficult question even among living species, when hybridization among related mammals produces fertile offspring (see pages 282 and 308); it occurs also between chimpanzees and bonobos and many species of monkey. Yet successful hybridization can produce significant physiological change in descendants.

The paucity of the fossil record also makes reliable inference of behaviour particularly difficult.

Other remains

Specimens found either with hominin fossils or very often on their own are used to infer the behaviour—and hence the level of consciousness—of hominins.

The commonest ones are tools, and in particular stone tools because they are preserved, which gave rise to the term Stone Age. And therein lies one limitation: we don't know whether early hominins used perishable tools and other implements, say sharpened wooden spears and arrows or poisonous darts, or shelters made from bamboo and hide.

When stone tools are the only remains found in a geological layer, interpreting their function, degree of sophistication, and similarity to stone tools associated with sparse fossil specimens discovered in another part of a continent, or even another continent, in order to deduce which species made them is, at best, an art rather than a science.

Charcoal remains may be from a hearth, showing controlled use of fire for warmth, protection against predators, or food preparation, but they may also be from the stump of a tree burnt in a forest fire. Other remains include fossils of animals, like fragments of lion or wolf bones. These and hominin fossils are usually scattered over a site, and it is frequently far from clear whether the hominin killed or scavenged the animal or vice-versa, or one brought part of the carcass of the other from another location, which affects its dating.

Dating

Relative dating assumes that specimens found in lower geological layers are older than those found in higher layers. But erosions, landslides, earthquakes, and other geological phenomena can confound chronological layering. In rare cases ash from a volcanic eruption can spread widely and specimens found in such a layer can confidently said to be of the same age, but even so this does not tell us the actual age.

Radiocarbon dating introduced in the 1950s held out this promise.* It was later recognized, however, that dating beyond about 30,000 years ago becomes increasingly unreliable because the amount of carbon-14 remaining in a sample after this time is too small to measure accurately and contamination can significantly distort the result. Still later discoveries about cosmic ray fluctuations and changes in the Earth's atmospheric circulation undermined the assumption that the percentage of carbon-14 in the atmosphere has been constant.

Other radiometric techniques, summarized in Table 26.2, were developed to extend the dating range and provide crosschecks. However, each has its limitations and problems, expertly summarized by Chris Stringer.

Table 26.2 Radiometric techniques and their date ranges

Method	Approx. date range
Radiocarbon	0–40,000 years
Uranium series	0–500,000 years
Luminescence	0–750,000 years
Electron spin resonance	0–3,000,000 years
Argon-40/argon-39	0–1,000,000,000 years

Source: Stringer, Chris (2011) pp. 33–44

Techniques are constantly being refined and improved, and this has led to re-dating specimens discovered many years ago, sometimes with startling results. For example, part of a primitive human skull discovered in 1932 in Florisbad, South Africa had been dated by radiocarbon analysis of the peat bog in which it had been found as 40,000 years old. In 1996 electron spin resonance dating of the enamel of an upper molar tooth put its age at 260,000 years.[7]

Genetics
Genetic analysis was embraced as a technique that could bring a more precise, scientific approach to tracing human emergence. After it was discovered that a mother's, but not a father's, mitochondrial DNA was cloned in the mitochondrion of the fertilized cell from which the offspring developed, Rebecca Cann, Mark Stoneking, and Allan Wilson of the Department of Biochemistry, University of California, Berkeley, devised a means of tracing back matrilineal descent from changes in mitochondrial DNA (mtDNA) over time. In their landmark paper of 1987 published in *Nature* they analysed the mtDNA of 147 women from five geographical regions, fed the results into the computer program they had

---

*   See page 170 for a more detailed description of this technique.

created, and concluded that "All these mitochondrial DNAs stem from one woman who is postulated to have lived about 200,000 years ago, probably in Africa."[8]

The scientific as well as the popular press lauded this achievement as identifying the ancestor of all modern humans, the so-called Mitochondrial Eve. The result gave strong support to the Recent African origins hypothesis (see below). However, their work was subsequently criticized on many grounds. For example, their computer program could produce many thousands of ancestral trees and not all were rooted in Africa, their time calibration was questionable, and their samples were disputed (many African samples were actually from African-Americans).[9]

Similar studies were subsequently undertaken on changes in the Y chromosome possessed only by males. Different analyses tracing patrilineal descent back to Y-chromosome Adam range from 142,000 years ago[10] to as recently as 100,000 to 60,000 years ago in Africa.[11] Quite how members of a small population of Y-chromosome Adams managed to impregnate members of a small population of Mitochondrial Eves who were 58,000 or even 140,000 years older was never explained. Such discrepancies reinforce the unreliability of genetic analyses, and the molecular clock technique in particular, for tracing biological evolution, as discussed in Chapter 20 (see page 332).

Living species
Palaeoanthropologists often invoke two kinds of living species in order to infer the behaviour of our prehistoric predecessors: chimpanzees, on the grounds they are genetically the closest species to modern humans, and human hunter-gatherers.

Many researchers associate the aggressive behaviour of adult male chimpanzees with their long, pointed, projecting canines and assume that similar hominin teeth mean similar behaviour. However, as discussed above, the assumption that chimpanzee teeth have remained unchanged over some 5 to 7 million years while human teeth evolved is highly questionable.

The problem of inferring hominin behaviour from skeletal similarities to living chimpanzees is vividly illustrated by bonobos. Whereas the common chimpanzee, *Pan troglodytes* lives in forests across most of equatorial Africa, the bonobo, *Pan paniscus*, is found only in lowland rainforests to the south of the River Congo in the Democratic Republic of the Congo. Although bonobos are sometimes called pygmy chimpanzees, there is considerable overlap in size among individuals of the two species, which are also genetically very similar. If a complete skeleton were unearthed it would be difficult to say whether it was that of a chimpanzee or a bonobo, and impossible if only fragmentary fossil bones were found. Yet their behaviours are significantly different.

Chimpanzees eat mainly fruit and leaves, but occasionally hunt small mammals and monkeys. They live in communities of between 15 and 100 members, but tend to forage and hunt in flexible bands of 6 to 10 individuals. An alpha male dominates each community with the other adult males forming a linear hierarchy beneath in which dominance is maintained, or overthrown, either by shows of aggression or actual aggression; all adult males dominate the females, who form their own more complex social hierarchy. Male chimpanzees regularly patrol their territorial boundaries and engage in vicious and often lethal attacks on members of neighbouring communities.[12]

Bonobos, by contrast, are much less aggressive. They too supplement their diet with occasional mammals, but they do not hunt monkeys; instead they play with and groom them. Unlike chimpanzees, bonobos have not been observed practising infanticide and cannibalism. Their communities are much more egalitarian, with females usually taking the lead, and they engage in more frequent sex of all kinds: heterosexual, same gender, oral, and anal. They use sex to defuse tensions within the community and have fewer conflicts with neighbouring bonobo communities.[13]

In the absence of direct behavioural evidence from human predecessors it may seem reasonable to infer their behaviour from that of living hunter-gatherers, like the Aka and Bofi pygmies of the Congo Basin or the Pila Nguru of the Great Victoria Desert in Western Australia. However, two questions should be borne in mind when using such evidence. First, to what extent have their behaviours evolved from those of their ancestors of say, 15 to 10 thousand years ago? Second, how representative are their ancestors of hunter-gatherers of that period? Individual humans show a range of proficiency in different abilities, typically represented by a bell curve. For example, in intellect we find Einstein and a few others at one extreme, most of us clustered around a median level, while a small number have a low intellectual ability. As discussed in the next chapter, we can be reasonably confident that the majority of archaic human hunter-gatherers invented, or adopted the invention of, farming as a more effective way of sustaining themselves. Living hunter-gatherers comprise around 2 per cent of the human population. Are they the descendants of archaic humans who lacked the ability to invent, or adopt, this better way of obtaining food, or lacked the foresight to migrate to regions where they could do so?

With these reservations about the evidence in mind, I shall try to summarize what we know about our hominid predecessors depicted in Figure 26.2.

Tribe of hominins

Palaeontologist Michel Brunet claims the oldest hominin fossil consists of a cracked and distorted cranium, three fragments of jaw, and several isolated teeth he discovered in 2001 in a desert in Chad, 2,600 kilometres west of the East African rift valley. He classified this as a new species *Sahelanthropus tchadensis* and controversially suggested the creature was bipedal. Indirect dating put its age at 7 million years, which is older than the generally accepted genetic dating of 6.3

to 5.4 million years ago for the last common ancestor of humans and chimpanzees from which it is supposedly descended.[14] However, see page 332 for a critique of the molecular dating technique and page 334 for the wide range of dates for the human-chimpanzee split that different researchers have arrived at using this technique.

Another candidate for a bipedal hominin, at around 6 million years ago, is *Orrorin tugenensis*, classified on the basis of three fragments of thighbone, a partial upper arm bone, part of a jaw bone, and several isolated teeth and finger or toe bones found at four sites in the Tugen Hills of Kenya in 2000.[15]

In 2009 a large international team led by Tim White of the University of California, Berkeley announced a more substantial find from the Afar Rift region of northeastern Ethiopia. Fifteen years' reconstruction of crumbling bone fragments and a skull crushed to 4 centimetres in height had produced 110 specimens, including a partial female skeleton and pieces of 35 other individuals found in sediment dated to 4.4 million years ago.

Classified as a new species, *Ardipithecus ramidus* lived in what was then a mixed woodland and forest environment. The female was small—about 120 centimetres high and weighing around 50 kilograms—with an estimated brain size similar to that of a chimpanzee. According to team member C Owen Lovejoy of Kent State University, Ohio, unlike a chimpanzee whose short legs and long arms are structured for swinging between branches and quadripedal knuckle-walking on the ground, *A ramidus* moved along branches using hands and feet and walked upright on the ground, although not as effectively as later hominins. From its skeleton and its teeth the team concluded that *A. ramidus* demonstrates that the common ancestor of humans and chimpanzees did not resemble a chimpanzee, and that the latter's features evolved along a different trajectory from that of humans.[16] David Pilbream of Harvard University is one of several palaeoanthropologists who remain sceptical about this conclusion.

The team also discovered other fossils in the same area that they dated at around 5.5 million years ago, concluded they represented an earlier species of the same genus, and classified as *Ardipithecus kadabba*.

It is unclear how *Ardipithecus* is related to the later genus *Australopithecus*, of which *Australopithecus afarensis* (Lucy) is the best-known species. Australopithecines are characterized by a fully erect posture and a bipedal gait, although the degree to which all the species could comfortably walk upright on the ground is a matter of dispute. However, their braincase is similar in shape and size to that of chimpanzees, although proportionately greater if body size is taken into account. Lumpers and splitters disagree as to whether there are five or eight or more species of *Australopithecus*; most palaeoanthropologists now re-assign two robust species, *A. robustus* (South Africa) and *A. bosei* (East Africa), to a different genus, *Paranthropus*, as shown Figure 26.2.

Different palaeontologists claim that one or more of the *Australopithecus* species evolved into the first species of the *Homo* genus around 2 million years ago. The

2011 candidate was *Australopithecus sediba*. Based on two well-preserved and approximately half-complete skeletons that he discovered in Malapa Cave, South Africa, Lee Berger of the University of Witwatersrand, together with colleagues, claimed that, although *A. sediba* has long arms and a small braincase like earlier australopithecines, its small teeth, brain endocast showing a large frontal lobe, projecting nose, pelvis shape, ankle joints, long thumb and short fingers, and long legs, are more humanlike. Consequently they argue that *A. sediba* is the most plausible ancestor of *Homo*, possibly of *Homo erectus*. Uranium isotopic analysis of calcite formed by waterflow along the floor of the cave gave a date of 1.977 million years ago. Hence, Berger's team claims, this invalidates classifying any earlier species as human.[17]

Fred Spoor of University College London and the Max Planck Institute for Evolutionary Anthropology in Leipzig is one of several palaeoanthropologists who challenge this, arguing among other things that Berger and colleagues' morphological analyses and comparisons are insufficient to support their claims, an *A. afarensis* upper jawbone from Hadar in Ethiopia is more humanlike than that of *A. sediba*, which it predates by some 370,000 years, and a maximum of 80,000 years is insufficient for the evolutionary change from *A. sediba* to a species of *Homo*.[18]

Genus of *Homo*
Ancestral relationships are no clearer when we consider the genus *Homo*. The earliest species is thought to be *Homo habilis*, fossils of which are claimed from 1.9 to 1.5 million years ago in Kenya and Tanzania, with some fossils accompanied by extremely primitive stone tools called Oldowan (see later); similar tools have been claimed to date from 2.6 million years ago. A re-reconstructed skull re-dated to 1.9 million years ago is regarded either as a variant of *H. habilis* or a separate species *H. rudolfensis*. The discovery reported in 2012 of a face and two jawbones with teeth dated between 1.78 and 1.95 million years ago led to claims that *H. rudolfensis* was indeed a separate species.[19]

For those who split them into separate species, any one or none of *Homo habilis*, *Homo rudolfensis*, or *Homo ergaster* may have been the ancestor of *Homo erectus*, and any one or none of these may have been the ancestor of *Homo heidelbergensis*, *Homo neanderthalensis*, and *Homo sapiens*, while *H. heidelbergensis* may or may not have been the ancestor of either one or both of *H. neanderthalensis* and *H. sapiens*.

Neanderthals
Anatomically, Neanderthals are distinguished from humans by being shorter but more robust. The adult Neanderthal cranial capacity, averaging more than 1400cc, is larger than that of living *Homo sapiens*, but its brain to body size ratio is slightly smaller; the principal differences in skull structure are listed in Table 26.3.

Table 26.3 Features distinguishing Neanderthal and modern human skulls

Neanderthal	Modern human
large brow ridge	reduced brow ridge
chin and forehead sloped backwards	vertical forehead, bony chin
large crest and dent at back of skull	high, rounded skull

Their backward-sloping forehead has been interpreted as indicating that Neanderthals had a less developed forebrain, where higher cognitive functions are processed, than humans.

Whereas mitochondrial DNA analysis had given 500,000 years ago as the date of the split between the Neanderthal and modern human lineages, analysis of their nuclear DNA suggested the split occurred between 440,000 to 270,000 years ago.[20]

Remains assigned to *Homo neanderthalensis* have been found at sites throughout Europe, in western Asia, as far east as Uzbekistan in central Asia, and as far south as the Middle East. They have been dated from 400,000 to 30,000 years ago, meaning that Neanderthals overlapped in time with *Homo sapiens* for at least 10,000 years. That some also overlapped in place is suggested by the first draft of the Neanderthal genome mentioned previously. Analysis of nuclear DNA extracted from three bones taken from three individuals found in Vindija Cave in Croatia indicates that humans have between 1 and 4 per cent Neanderthal genes, implying there was some limited interbreeding.[21]

Stone tools associated with Neanderthal sites include scrapers, triangular bifacial hand axes, triangular points made on flakes suggestive of spears, most probably for thrusting rather than throwing, and awls, or pointed tools used for piercing holes, but no bone tools, like needles.[22]

In 2010 João Zilhão of the University of Bristol and colleagues claimed that two eroded Neanderthal sites in the Murcia province of southeast Spain yielded perforated and pigment-stained marine shells presumed to have been body ornaments. At any other site these would be taken as a sign of a high level of consciousness, if not reflective consciousness, and assumed to be human artefacts. However, Zilhão claims that these shells are "approximately 50,000 years old". They constitute "secure evidence that…10 millennia before modern humans are first recorded in Europe, the behavior of Neandertals was symbolically organized". In fact the supplementary information to his paper lists radiocarbon dates of five shells ranging from 38,150 ± 360 to 45,150 ± 650 years ago from one site, and of four specimens of wood charcoal from the other site ranging from 98 ± 23 to 39,650 ± 550 years ago.[23] This contradicts the claimed age of 50,000 years.

Most palaeoanthropologists point to the lack of firm evidence of symbols, artwork, and other signs of reflective consciousness in Neanderthal sites generally compared with their widespread presence in human sites and conclude that Neanderthals were cognitively inferior to early humans.

In the absence of reliable evidence, the extinction of Neanderthals around 30,000 years ago is a matter of speculation rather than hypothesis. Suggestions by palaeoanthropologists include the following: unlike humans, they failed to adapt to rapid climate changes at that time; they interbred with and were assimilated by humans; they contracted human diseases for which they lacked immunity; they were killed off by humans.

Hobbits

In 1994 a joint Australian and Indonesian team reconstructed fossilized bones found in a cave on the Indonesian island of Flores to produce a partial skeleton of an adult 1 metre tall with long arms, a low receding forehead, and no chin. They estimated its brain capacity as 380 cubic centimetres, around that of the smallest-known australopithecine and less than that of an average chimpanzee; yet the site was associated with stone tools and possibly a hearth together with the remains of a dwarf elephant-like stegodon that apparently had been hunted.

The team concluded that the creature was descended from *Homo erectus* individuals thought to have reached the region about 1 million years ago. On this isolated island, with no competition and limited food, these hominins had dwarfed, just as had the stegodon. (In fact Flores also has remains of enormous rats, huge lizards, and giant tortoises, and it is unclear why some species should have dwarfed while others became unusually large.) They classified the find as a new human species, *Homo floresiensis*, nicknamed the Hobbit, generating worldwide publicity plus controversy within palaeontology.[24] More remains were found, representing six to nine individuals depending on how the remains are reconstructed, and dated from 17,000 to 95,000 years ago.

In 2009 Peter Brown, the University of New England team member who had initially designated the creature as *Homo floresiensis*, questioned his original classification. He claimed further research showed that Asian *Homo erectus* were not ancestors of the Hobbits, whose characteristics had more in common with australopithecines than the 1.75 million-year-old fossils at Dmanisi in Georgia, thought to represent the first *Homo* species outside Africa.[25]

Denisovans

In 2010 an international team led by Svante Pääbo of the Max Planck Institute for Evolutionary Anthropology announced they had identified a new hominin population based not on fossil but uniquely on genetic evidence only. They had extracted DNA from a finger bone unearthed in Denisova Cave in southern Siberia, which stone tool remains indicated had been a site of *Homo* occupation beginning up to 280,000 years ago. The bone was found in a layer dated 50,000 to 30,000 years ago that had previously yielded stone microblades and polished stones typical of the Upper Palaeolithic and also cruder tools more typical of the

Middle Palaeolithic, leading to speculation that the cave had been occupied by both Neanderthals and modern humans at one time or another.

Mitochondrial DNA (mtDNA) analysis indicated this individual was from a group they labelled Denisovan that had diverged from the common lineage leading to modern humans and Neanderthals about one million years ago, that is, about twice as far back in time as the divergence between Neanderthal and modern humans, and so was an earlier species. It also showed that the Denisovan genome differed from the modern human genome by twice as much as did the Neanderthal genome.

However, subsequent analyses of nuclear DNA, based on several assumptions including the date of the human-chimpanzee divergence, indicate that the individual shared a common origin with Neanderthals about 640,000 years ago, and that the Denisovan genome differed from that of modern humans by about the same as did the Neanderthal genome. Pääbo's team used mtDNA analysis of a tooth that doesn't resemble either a human or Neanderthal molar to suggest that Denisovans had a distinct population history. In particular, Denisovans made no genetic contribution to Eurasian humans as did the Neanderthals, but did contribute 4–6 per cent of their genome to that of present-day Melanesians, although not to other East Asian populations closer to Denisova, like Han Chinese or Mongolians.

The team refrain from assigning either Denisovans or Neanderthals to a specific taxonomic classification. They interpret these genetic results from a finger and a tooth in Denisova to mean that at least two forms of archaic hominins existed on the Eurasian mainland in the Late Pleistocene epoch (around 125,000 to 10,000 years ago): Neanderthals, who were widespread in western Eurasia, and Denisovans, who were widespread in east Asia; each had a limited but separate interbreeding with *Homo sapiens*.[26]

Earliest signs of Homo sapiens

Tim White of the University of California, Berkeley and colleagues claimed in 2003 that the earliest fossils most closely resembling modern humans are three crania discovered at Herto Bouri in Ethiopia's Afar Triangle radioisotopically dated to between 160,000 and 154,000 years old. They stopped short of classifying the specimens as fully developed modern human and designated them as a subspecies *Homo sapiens idaltus* that, in their view, most probably represents our immediate ancestors.[27]

In 2008 John Fleagle of Stonybrook University, New York re-dated from 130,000 to around 196,000 years old two specimens discovered by Richard Leakey in 1967 at Kibish, near Ethiopia's Omo River. He claimed controversially that the first, Omo 1, consisting of a partial skeleton and fragments of a face and jawbone, was the earliest fully human fossil. The second, Omo 2, a skull with no face, was more primitive.[28]

Whether or not these claim are valid, they relate only to relative anatomy and not the unique characteristic of reflective consciousness. For this we need to find evidence of those secondary faculties it transforms or generates.

Tools

Earlier I suggested that comprehension combined with invention, foresight, and imagination produced specialized composite tools. Archaeologists broadly classified stone tools by tradition, or industry, named after the site where an assemblage of such tools was first studied, although other classifications such as regional or Modes I–V are used. Not only has the classification of tools changed and continues to change, so too have the cultural periods characterized by different tool types. The Stone Age in Europe was initially divided into two, the Palaeolithic, or Old Stone Age, and the Neolithic, or New Stone Age. The Palaeolithic was later divided into three major phases, Lower, Middle, and Upper, and the Mesolithic subsequently inserted between the Upper Palaeolithic and the Neolithic.

The Lower Palaeolithic (called the Early Stone Age in sub-Saharan Africa) starts with the earliest recognizable stone tools, called Oldowan after the Olduvai Gorge in Tanzania, which consist of crude flakes struck from large pebbles. The flakes are assumed to have been used for scraping, and the chipped pebbles for pounding. Very likely hominins previously used unchipped pebbles for pounding, as some chimpanzees have been observed to do, but these would be extremely difficult to distinguish from any other pebbles. The earliest Oldowan tools currently reported are from Gona in Ethiopia dated to around 2.6 million years ago.[29] The next significant development, from about 1.7 million years ago, is Acheulean tools, named after Saint-Acheul in northern France. Here flakes have been chipped away from two sides of stones to produce typically pear-shaped handaxes and picks. Such bifaced tools have been found at sites in Europe, Africa, the Middle East, India, and Asia throughout the rest of the Lower Palaeolithic that ends around 180,000 years ago in Europe and 150,000 years ago in Africa,[30] accounting for more than 93 per cent of archaeology's time span.

It is difficult to see evidence of reflective consciousness in such primitive tools as Oldowan and Acheulean, which developed over nearly 2.5 million years. Thereafter the rate of improvement, reflecting the makers' level of consciousness, increases.

When specialized composite tools first appeared is unclear. Claims have been made that stone segments and points found at Twin Rivers in Zambia were designed for mounting on wooden handles and date from 260,000 years ago.[31] University of Cape Town PhD student Kyle Brown claimed in 2009 that silcrete flake tools could only have been made by first heating silcrete (a hard, cement-like siliceous crust) to improve its flaking properties and give a gloss to the flakes. Such silcrete tools are found at many sites from about 72,000 years ago and appear as early as 164,000 years ago at Pinnacle Point on South Africa's southern coast.[32] If silcrete had been baked for several hours before chipping, this would signal a high level of cognition and invention.

Stone tools of increasing refinement characterize the Middle Palaeolithic and the transition to the Upper Palaeolithic. They include long, thin bifaced flint flakes, assumed to be crude knives, together with differently sized thin, triangular sharp flakes, assumed to form the heads of arrows and spears: composite tools used for killing at a distance. Upper Palaeolithic sites in Europe have revealed such tools as barbed harpoons and sharply pointed awls together with needles made from mammoth or reindeer ivory, which suggest the use of animal skins sewn together by, say, lengths of gut, as clothing and/or tents. The Upper Palaeolithic traditionally ends in in Europe around 10,000 years ago with the emergence of the Mesolithic, typified by small composite tools known as microliths, prior to tools characteristic of farming and metal-working societies.

In sub-Saharan Africa the approximate equivalent of the Upper Palaeolithic and the Mesolithic is labelled the Later Stone Age, the duration of which varies considerably in different regions. Its tool-making persists in hunter-gatherer communities in some areas today, as it does in some South American rainforests.

Controlled use of fire

When the controlled use of fire first occurred is notoriously difficult to establish, and claims range from 0.2 to 1.7 million years ago. Evidence for the creativity and invention of methods to generate and use fire for warmth, cooking, and warding off predators was found in South Africa in the form of burned bones within a date range of about 200,000 to 700,000 years ago.[33]

Symbols and ornaments

The earliest evidence for the use of symbols and the capacity for artistic expression appears in Africa. Different levels of the Blombos Cave on South Africa's southern Cape coast dated from at least 100,000 to 75,000 years ago have yielded not only new forms of carefully shaped stone points and bone tools but also pieces of ochre engraved with abstract designs plus perforated seashells, some of which were daubed with ochre.[34]

The latter, assumed to have been used as personal ornaments such as necklaces, have also been found at Qafzeh and Skhul in Israel, Oued Djebbana in Algeria, and Taforalt in Morocco. Dating is problematical, but the available estimates are that the Moroccan site is from 85,000 to 80,000 years old, while the layer at the Skhul site in Israel is claimed to be 135,000 to 100,000 years old.[35]

Such symbolically mediated artefacts demonstrate a capacity for thought, imagination, creativity, and abstraction.

Trading?

Three of the Moroccan sites yielding pierced seashells are 40–60km inland, while the Algerian site is 190km from the sea. Francesco d'Errico and colleagues concluded that these ornaments were made and traded by tribes living on the coast.[36] The location of the pierced seashells may also be explained by the migration of tribes from the coast to more sheltered locations or ones that have better and more easily accessible food supplies, and these tribes took their ornaments with them. *If* they were traded goods, then this would be evidence of foresight, creativity, and will; it would also demonstrate cooperation, as distinct from collectivization, and infer language, however rudimentary.

Sea crossings

Humans appear to have arrived in Australia at Deaf Adder Gorge, Northern Territory, between 60,000 and 53,000 years ago if not earlier.[37] In none of the glaciation periods did ocean levels drop sufficiently to provide land bridges from Asia to Australia, and it has been estimated this migration involved at least 100km of open sea.[38] Such a feat requires a high level of comprehension, foresight, and will to plan and navigate the sea crossing, together with correspondingly high levels of creativity and invention to design and build a suitable raft or boat. Success of the whole enterprise almost certainly depended on cooperation, which is supported by what we know of the social organization of descendent Aboriginal tribes, and this in turn most probably required the use of language.

Ceremonial burials and cremations

According to the British Museum of Natural History's Chris Stringer, the oldest known symbolic burial dates to 115,000 years ago at Skhul Cave in Israel and comprises an early modern man clasping the lower jaw of a massive wild boar.[39]

James Bowler of the University of Melbourne and colleagues claim that new dating techniques put the date of the world's oldest ritual ochre burials and the first recorded cremations at 40,000 ± 2,000 years ago at Lake Mungo in western New South Wales.[40]

Such rituals suggest high levels of comprehension, imagination, and very probably belief.

Paintings, figurines, and flutes

In 1994 Jean-Marie Chauvet and two colleagues discovered a series of spectacular images on the walls of a cave in a cliff overlooking the Ardèche River in southern France. They include four horse heads drawn with shading and perspective, together with sophisticated representations of two rhinoceroses locking horns, lions chasing a herd of bison, cave bears, and other animals. Most were drawn with a piece of charcoal or painted with a brush or finger covered in red pigment.

No human remains were detected, but the bones of more than 190 cave bears were found, including a skull on top of a large flat rock. These finds prompted suggestions that the cave was for ceremonial rather than occupational use, perhaps to invoke the spirits of powerful animals for success in hunting.

More than 80 radiocarbon dates have been taken from the torch marks and paintings on the walls, as well as the animal bones and charcoal that litter the floor, providing a detailed chronology of what is now called the Chauvet Cave. The dates show that the artwork was made in two separate periods, one 30,000 years ago and one 35,000 years ago, making them the oldest known cave paintings.[41]

Some palaeoanthropologists challenged these datings, not least because of the sophistication of the paintings, and suggested the samples were contaminated or that the charcoal was from a later period, but the dating techniques were vigorously defended.[42] These paintings may not be the oldest. An Italian team working at Fumane Cave, near Verona, has dated stone fragments bearing images of an animal and one depicting a half-human, half-beast figure to between 32,000 and 36,500 years ago, although these fragments were detached from the cave wall.[43]

Nearly 350 caves have been discovered in France and Spain that contain art from prehistoric times. The most famous are at Lascaux on a hillside in the Dordogne region of southern France. This large complex contains around 600 images on cave roofs and walls, created with charcoal and a range of mineral pigments; some have been incised into the stone. The height of many of them suggests that scaffolding must have been erected for the painters, implying cooperation and very probably language.

In common with other Western European cave art most Lascaux paintings are of large wild animals, such as bison, horses, aurochs, and deer; one has been labelled a unicorn, but it could also be a horse with a spear through its head. The cave walls are also decorated with imprints of human hands as well as abstract patterns called finger flutings. Recent radiocarbon dating gives a range of dates for different images, with the oldest estimated at between 18,900 and 18,600 years.[44]

Images of humans are rare in cave art, and where they do occur they appear to be part-human part-beast, which is evidence of imagination and not just representation; possibly they are of humans wearing headdresses of lions or eagles or other powerful hunters. In either case they constitute evidence of belief, probably in animal spirits whose intercession is sought by shamans.

Evidence of such beliefs and of imagination is reinforced by the discovery of fragments of mammoth tusk at the back of Stadel Cave on the Hohlenstein cliff in southwestern Germany. These fragments, recently radiocarbon-dated to about 40,000 calendar years ago, have been fitted together to produce a spectacularly detailed 30cm-high man with the head of a cave lion. A much smaller (2.5cm-high) lion-headed human also carved from mammoth ivory was found in the Hohle Fels Cave, nearly 40 kilometres southwest of Hohlenstein-Stadel.

The same region of Germany has yielded many more figurines dating from 30,000 to 35,000 years ago in cave layers rich in stone tools. They include a mammoth-ivory figurine of a headless and legless woman with exaggeratedly large breasts, buttocks, and vulva. The carver had deliberately omitted the head and replaced it by a small ivory ring from which the figurine had probably been suspended. Such figurines, less than 10 centimetres high, have been found at other sites in Europe. Some are of animals; many depict women in all stages of fertile life: nubile, pregnant, and giving birth. Although some of these women wear beads or belts, nearly all are naked in an Ice Age when clothing made from animal skins was normally worn. Many portray large breasts and buttocks when such morphology was probably rare (skeletal remains indicate slim bodies that have walked long distances, carried heavy loads, and often suffered malnutrition). The details focus on the naked body rather than limbs or face that are often missing, which suggests they were fertility symbols. Many contain a hole and were probably designed as talismans to be worn as pendants. In hunter-gatherer societies today men take no part in the birthing process and so it is possible that women may have carved many of these figurines.[45]

Three of the German sites within the same dating range also yielded the oldest known musical instruments. Four flutes were made from the wing bones of swans and vultures, with holes pierced at regular intervals along the length; the most complete was found 70 centimetres from the carved headless and legless woman.

Demonstrating even greater invention, imagination, creativity, and reasoning is a flute made from a mammoth tusk, pieced together from 31 fragments found at the Geißenklösterlen cave near Ulm. Instead of starting from the hollow wing bone of a bird, the maker had to split the curved tusk without damaging it, hollow out each half, make three holes along the length, and glue the ivory halves together with no air leaks along the seam.[46]

Many archaeologists still consider such evidence in the Upper Palaeolithic of representational, imaginative, and symbolic art in personal ornaments, sculptures, paintings and decorated tools, together with musical instruments, as a Western European phenomenon. As Colin Renfrew of Cambridge University's McDonald Institute for Archaeological Research wrote in 2008

> It is important to remember that what is often termed cave art—the painted caves, the beautifully carved 'Venus' figurines—was during the Palaeolithic (i.e. the Pleistocene climatic period) effectively restricted to one developmental trajectory, localized in western Europe.[47]

This, however, is by no means the case. As mentioned above, different levels in South Africa's Blombos Cave dated from at least 100,000 to 75,000 years ago have yielded pieces of ochre carved with symbols. Paintings of sophistication comparable to Western European examples have been found on seven slabs of rock brought to the Apollo 11 Cave in the Huns Mountains of southwestern Namibia and dated at 25,500 to 23,500 years ago. Painting on exposed rock faces using similar materials continues to this day in some tribal areas, but exposure to the weather suggests their longevity is limited.[48] Rock paintings have been found in Australia. For example, a red ochre painting discovered in 2010 at the centre of the

Arnhem Land plateau depicts two emu-like birds with their necks outstretched; these have been claimed to belong to the genus *Genyornis*, giant birds thought to have become extinct more than 40,000 years ago.[49] Five clusters of rock shelters on the southern edge of the central Indian plateau display paintings similar to many of the European examples. The most recent dating techniques haven't been employed, but it is thought they are around 12,000 years old; members of local tribes still paint on rock faces.[50]

These findings leave the impression that the much larger number of prehistoric artworks reported in Western Europe may be due more to the effort put into discovering and analysing them, together with more favourable conditions for their preservation, than their being a phenomenon unique to archaic European humans. More rigorous dating techniques are needed to determine whether these African, Asian, and Australian artworks are older, contemporaneous with, or later than comparable European ones.

Language

Members of many species with high levels of consciousness use sounds and gestures to communicate fear, warning, threat, pleasure, and other emotions both to other members and to other species. Reflective consciousness, however, transforms communication into language, which I define as

> **language**   The communication of feelings, narratives, explanations, or ideas by a complex structure of learned spoken or written or signed symbols that convey meaning in the culture within which it is employed.

Language's unique ability to transmit experiences and ideas not only to other living members of the human species but also to succeeding generations, who thereby benefit from accumulated wisdom, enabled the increasingly rapid flowering of reflective consciousness.

It is difficult to determine when language first arose. Some researchers claim the Broca's area of the brain is evidence for speech on the grounds that if this area is damaged a person loses the ability to speak, and possession of a Broca's area can be inferred from an endocast, or impression of the brain on the inside of a skull. However, some chimpanzees have a prominent Broca's area and none has been able to utter a simple sentence despite considerable effort to teach them.

Others think that the FOXP2 gene indicates a capacity for speech because mutations of this regulatory gene are associated with articulation difficulties accompanied by linguistic and grammatical impairment. The Neanderthal FOXP2 protein is identical to the human one, that in chimpanzees differs by only two amino acids, and that in mice by only three amino acids. Svante Pääbo and colleagues from the Max Planck Institute for Evolutionary Anthropology speculate that the human version conferring articulate speech was selected for after the human-chimpanzee split.[51] However, the FOXP2 gene regulates a large number of genes and is implicated in the development not only of neural tissues but also of the lung and the gut.[52] I do not think possession of the human variant of FOXP2 is adequate evidence of speech.

The only unambiguous evidence of language is written records, but how do we detect the use of language in prehistoric times? I suggest it can be traced in those things from which writing evolved.

Most linguistics scholars consider the earliest writing systems were cuneiform engraved on clay tablets from around 5,000 years ago in Sumer, Mesopotamia and Egyptian hieroglyphs engraved on stone from roughly the same period. These evolved from proto-writing—the distinction is blurred—comprising abstract symbols that recorded quantities of goods, like wheat, that had been traded or given as a tribute or tax. These in turn had developed from pictorial symbols, the origin of which probably lies in the painted artworks and symbols together with symbols carved in ochre mentioned previously.

Many early writings record the feats of rulers and their ancestors or of gods or spirits that presumably had been transmitted orally by an elder or storyteller in earlier times. It also seems reasonable to hypothesize that painted or engraved symbols were paralleled by speech, however basic, that subsequently developed more sophistication. I've also suggested that particular achievements, like a group of humans migrating across 100 kilometres of ocean or trading in goods, require speech.

Although the evidence of language in prehistoric times is necessarily indirect, it suggests that spoken language emerged in the Upper Palaeolithic.

Completion of human emergence

A combination of the available evidence indicates the emergence of humans was complete by the end of the Upper Palaeolithic Age (Later Stone Age in Africa), and most probably earlier than that. The first glimmerings of reflective consciousness appear as different manifestations of new or radically transformed secondary faculties at various places and periods, like bubbles of steam bursting out at different times from different parts of boiling water. At the surface some bubbles may condense back to liquid. But above this transitional surface, while water molecules may have the same 100° Celsius temperature as those forming the liquid below, they are indisputably gas: a phase change has occurred. Likewise, consciousness has entered a new, reflective phase that signals the presence of *Homo sapiens*.

Explanatory hypotheses[53]

The paucity of evidence and problems of its interpretation have given rise to many hypotheses that seek to explain this phenomenon. They may be grouped into six, some of which complement, while some contradict, the others. All agree that a wave of emigration of *Homo erectus* from Africa to the Middle East, Asia, and Europe occurred about 1.7 million years ago.

Multiregional model

Advocated by Milford Wolpoff and others, the multiregional model claims that *Homo erectus* groups reached China, Indonesia, and perhaps Europe by 1 million years ago. They adapted to the different environmental conditions in these different global regions and evolved eventually into modern humans whose racial differences reflect these different environments. The unity of humans as a species was maintained by interbreeding between members of regional groups.

Replacement or Recent African origins model

Chris Stringer and others initially claimed that regionally dispersed members of *Homo erectus* died out or evolved into successor species, like *Homo neanderthalensis* in western Eurasia; only in Africa, however, very probably in a small, favoured area like East Africa, did they evolve into *Homo sapiens* by around 130,000 years ago.

Some *Homo sapiens* groups migrated to Israel and the Middle East about 100,000 years ago and some reached Australia about 60,000 years ago. The rapid development of more advanced Later Stone Age tools and complex behaviours by *Homo sapiens* in Africa by about 50,000 years ago enabled some of these to migrate to Europe, which they reached by 35,000 years ago (although, apparently, *Homo erectus* or its successor had reached Asia and Europe about a million years previously without such advanced tools and complex behaviour).

The cognitively more advanced *Homo sapiens* replaced indigenous species of *Homo* that became extinct. Different racial features developed by adaptation to regional environments only after this relatively recent emigration from Africa.

The model's advocates claimed support of the landmark 1987 mitochondrial DNA analysis ("Mitochondrial Eve"), but that analysis was subsequently criticized (see page 436).

Assimilation model

This compromise model advanced by Fred Smith and Erik Trinkaus accepts the claim of the Recent African origins model for the origin of *Homo sapiens* but argues that rather than immigrant modern humans replacing indigenous *Homo* species in different regions, interbreeding and local natural selection led to the assimilation of indigenous species, which explains the racial characteristics of modern humans.

Recent African origins with hybridization model

Stringer and colleagues modified their original model in the light of the genomic analyses mentioned earlier that indicated some limited interbreeding took place between *Homo sapiens* and *Homo neanderthalensis*. The Neanderthal genome is more similar to those of current Asian and Western European humans than to those of African humans, suggesting this interbreeding took place in the Middle East after humans left Africa but before their expansion into Europe and Asia.

Words like interbreeding and assimilation may give an impression of harmony, like contemporary inter-racial marriages, but rape has long been a weapon of war. It is mentioned on numerous occasions in the bible, such as Zechariah 14:

> For I will gather all the nations against Jerusalem to battle, and the city shall be taken and the houses plundered and the women raped.

It recurs constantly throughout historic times up to the present. A United Nations report documented the systematic rape of Tutsi women by the Hutu in their 1994 genocidal war for control of Rwanda and estimated it resulted in between 2,000 and 5,000 pregnancies.

Human revolution model

The discovery of the spectacularly sophisticated cave paintings in France and Spain, in contrast to the two-and-a-half-million-year development of tools comprising stones chipped on one side to stones chipped on two sides, prompted the hypothesis by Richard Klein of Stanford University that modern humans emerged suddenly about 35,000 years ago; the most probable cause was a genetic mutation in the brain that transformed neural processing to produce a dramatic rise in consciousness.

Subsequent finds of comparable art and sculptures on other continents changed the place and time of this hypothesized mutation. The place was Africa and the time was before the emigration of *Homo sapiens*. According to Klein this event occurred about 50,000 to 40,000 years ago, while Paul Mellars of Cambridge University puts it at about 80,000 to 60,000 years ago. It was an event that equipped *Homo sapiens* with the intellectual powers to migrate successfully from Africa (although, as noted above, members of the less intellectually endowed *Homo erectus* had apparently left 1.7 million years before) and replace cognitively inferior *Homo* species across the globe, thus supporting the replacement model.

Gradualist model

Sally McBrearty of the University of Connecticut and Alison Brooks of George Washington University scathingly dismiss the human revolution model as Eurocentric. They point out that evidence of stone tool refinements and composite tools, specialized hunting, use of aquatic resources, ornaments, engravings, and use of pigments in art and decorations did not appear all at the same place and time but in different locations across the African continent at different times beginning some 250,000 years ago. This, they argue, show that cognitive powers developed gradually.

Proposed causes of human emergence

With the exception of the human revolution model the above hypotheses deal with *how* and *when* humans emerged rather than *why*.

Genetic mutation

A genetic mutation causing such a dramatic development in behaviour in so short a time is incompatible with what we know about how genes function. While a single mutation can prove destructive and inhibit an existing behaviour or even result in death (although even here several genetic mutations are usually needed), the development of a new or significantly enhanced behaviour almost always requires many genes operating as a network. We saw on page 324, for example, how a 2010 study showed that human intelligence is controlled by a network of thousands of genes rather than a few powerful genes as previously thought.

If genetic change were the sole or major contributory cause of a relatively sudden revolution in behaviour reflecting a relatively sudden rise in consciousness—in this case to the point of reflective consciousness—then this is most likely to be caused not by a genetic mutation but by a significant genomic change involving many genes and regulatory mechanisms such as that resulting from hybridization or genome duplication.*

Climate change in East Africa

For many decades the orthodox explanation was that humans emerged in East Africa because climate change resulted in forest turning to savannah east of the African Rift Valley, newly created by tectonic movement. Bipedalism was selected for because it was an adaptive advantage that enabled humans to walk more efficiently than the knuckle-walking of chimpanzees and gorillas when travelling long distances in search of sustenance in these open grasslands, and/or it enabled them to see prey and predators at a greater distance, and/or it enabled them to run faster towards prey and away from predators, and/or it exposed less of the body to the heat of a Sun no longer shielded by a forest canopy, and/or it freed hands for using tools.

At the beginning of this chapter I observed that bipedalism was not unique to humans. Moreover, undermining the savannah explanation, *Ardipithecus ramidus* lived in a mixed woodland and forest environment and is claimed to have been bipedal, while chimpanzees have been observed using and even making simple tools. Furthermore, anthropologists Sally McBrearty and Nina Jablonski reported in 2005 the discovery of a 545,000-year-old chimpanzee fossil† in Kenya, contra-

---

* See page 421.
† The fossil consists of three teeth that McBrearty claims are chimp-like incisors and all three have chimp-like thin enamel.

dicting the view that only hominins migrated to and survived in savannah east of the Rift Valley[54] (although, logically, that one chimpanzee may have failed to survive in those conditions).

Climate changes globally

The evolution of hominins took place in the geological epoch known as the Pleistocene, roughly 2.6 million years ago to 12,000 years ago, characterized by repeated major glaciations interspersed with shorter, warm interglacial periods of ten to twenty thousand years.

This prompted Stringer to claim that a key climatic event occurred 450,000 years ago when *Homo heidelbergensis* (a claimed successor of *Homo erectus*), which had spread across Eurasia, including Britain, experienced severe glaciation, cutting it into three main groups that then evolved separately: those in Europe evolved into Neanderthals, those in Asia evolved into Denisovans, while those in Africa evolved into modern humans.

Other palaeontologists have invoked these alternations of glacial and interglacial periods to explain human evolution. According to this view, repeated warm interglacial periods caused desertification and drought in East Africa, forcing humans to migrate. One such period about 60,000 years ago drove humans north and out of Africa, while a major glaciation phase peaking around 25,000 to 15,000 years ago caused a drop of some 120 metres in sea level to create a land bridge at what is now the Bering Straight enabling animals, pursued by humans, to move from Eurasia into North America.

Migration of itself, however, doesn't explain human emergence: many other species have migrated.

Moreover, climate changes are more complex than this. We saw in Chapter 12 that climate is affected by changes in solar radiation like solar flares, the Earth spinning around its axis, the precession of the Earth's axis, seasonal changes due to the tilt of the Earth's axis, while this tilt itself varies over 41,000-year cycles, tectonic movements that affect global ocean and air currents, and so on.* Such causes of climate change operate on different timescales and interact producing, for example, periods of heavy rainfall that are not always predictable. After studying lake sediments, palaeoclimatologists Martin Trauth and Mark Maslin conclude that even in the supposed drought periods there were large deep lakes in the East African Rift Valley.[55]

Rick Potts, director of the Smithsonian's Human Origins Program and curator of anthropology at the National Museum of Natural History, suggests that it wasn't a change from forest to savannah that caused the emergence of humans but rather an environment fluctuating between rainforests, grassland, and desert caused an increase in hominin brain size and cognition enabling them to figure out how to survive in fluctuating circumstances.[56]

---

* See page 100 and following for planetary factors affecting climate.

We do not know how quickly these fluctuations occurred. Ice-core records from Greenland have shown that marked climate change can occur over as little as 20 years or less. As Response of Humans to Abrupt Environmental Transitions (a multidisciplinary research consortium funded by the UK's Natural Environment Research Council) points out, studies correlating human evolution with climate change are compromised because of the inability to synchronize archaeological and geological records with sufficient precision.

Most of the factors determining climate change are global and I see no reason to privilege East Africa as the "the cradle of humanity". Signs of reflective consciousness—specialized composite tools, ornaments, and symbols—have also been discovered in western and southern Africa. Perhaps the fact that more hominin fossils—and even these are few—have been found in East Africa is due to the rapidly eroding highlands filling the Rift Valley with sediment providing more favourable conditions for fossilization rather than to a unique habitat promoting human emergence.

Conclusions

1. The one characteristic that distinguishes humans from all other species is reflective consciousness.

2. Its emergence may be traced by the rise in consciousness from primates to the point that consciousness becomes conscious of itself. Signs of the first glimmerings of this reflective consciousness lie in indisputable evidence of those faculties possessed by primates that it radically transforms in humans together with new secondary faculties that it generates.

3. Insufficient evidence, however, makes it impossible to trace the particular lineage from primates that results in modern humans. Insufficient evidence also makes it impossible to choose between the different hypotheses claiming to explain where and how this process occurred. As for when, the available evidence is compatible to a limited degree with both the human revolution and the gradualist models. Reflective consciousness did not suddenly appear from nowhere. Consciousness rose gradually in different hominins from about 2.5 million years ago; however, its rate of change increased, and continued to increase, over the last 250,000 years or so when different signs of reflective consciousness appeared relatively quickly in different places until all the signs are present during the Later Stone Age in Africa and the Upper Palaeolithic in different continents, roughly 40,000 to 10,000 years ago.

4. The same lack of evidence makes it impossible to identify the specific cause or causes of human emergence. Possibly the root cause was an instinct for survival against predators and a need for sustenance in a fluctuating environment induced by a fluctuating climate, which contributed to a recognition that these were best achieved through cooperation rather than

competition. Hybridization or genome duplication may have played a role in this evolutionary change.

5. Overall, it appears to be a case of systems emergence* in which the interaction of faculties at a lower level of complexity—like comprehension, invention, learning and communication—generates a novel faculty at a higher level of complexity—in this case reflective consciousness—and that higher level faculty causally interacts with the lower level faculties to transform them and generate new ones—like imagination, language, abstraction, and belief.

6. Like the emergence of matter and the emergence of life, because of the inherent paucity of evidence it is almost certain that science will never be able to identify when, how, and why humans emerged. But that it not to deny the emergence has taken place. To extend an analogy used in Part 2, reflective consciousness is like a flower grown from the seed of consciousness. In some environments the seed never grows into a flower. In others it grows into a shoot that produces a bud that eventually withers away. In a different environment that bud blossoms into a flower. A bud is not a flower, but it is impossible to say at what particular time the bud turns into a flower.

---

* See Glossary for the definition of *emergence* and its three principal types.

CHAPTER TWENTY-SEVEN

Human Evolution 1: Primeval Thinking

> Outside and above the biosphere there is the noosphere...with hominisation, in spite of the insignificance of the anatomical leap, we have the beginning of a new age.
>
> —PIERRE TEILHARD DE CHARDIN, 1955

I shall begin by considering how humans evolved, suggest that this process can be divided into three overlapping phases, and examine the first of these phases in more detail.

How humans evolved

Physically

The fossil evidence suggests that, apart from a gradual and small reduction in average size of skulls and skeletons, humans ceased evolving morphologically by at least 10,000 years ago when they inhabited every continent on the globe.

Physical anthropologists used to divide the human species into three broad variations or races arising from climatic adaptation: Caucasoid, Mongoloid, and Negroid. Classification of different peoples into these racial groups was contentious because of lack of agreement on defining these racial characteristics and their causes, while interbreeding resulting from invasions and migrations blurred distinctions between peoples. Mating between racial groups has increased significantly in the last fifty years or so for reasons discussed in the next section, accelerating the trend towards less physical differentiation among the human species.

While some reversible physical changes are taking place—principally a reduction in size due to malnutrition in countries affected by famine, or obesity due to overeating and lack of exercise in some rich countries—humans overall are not evolving physically.

Genetically

In a study published in 2007 anthropologist John Hawks of the University of Wisconsin, Madison and colleagues claim that the natural selection of adaptively advantageous genetic variants has produced an extraordinarily rapid genetic evolution of the human species.[1] Their modelling of data shows that genetic variants, or mutations, increased from 40,000 years ago. As NeoDarwinists they attribute this to the spreading of variants that enable populations successfully to adapt to different environments, such as genetic variants that code for the production of the dark skin pigment melanin in populations near the equator because melanin protects against the harmful ultraviolet radiation of strong sunlight.

However, their modelling also shows that genetic variants peaked around 8,000 to 5,000 years ago, when human populations began a massive expansion, and drop to zero at present. Hawks and colleagues attribute this drop to an inability to detect adaptive variants that "*should* exist…occurring at a faster and faster rate [my italics]".

Even according to NeoDarwinian criteria there are good reasons for challenging their unsupported extrapolation of a past trend and their conclusions that rapid genetic change is increasing in the human species.

One reason is that the genetic variation within the human species today, estimated at between 0.1 and 0.15 per cent, is low compared with that of many other species from fruit flies to chimpanzees.[2]

Another reason, according to UCL geneticist Steve Jones, is that natural selection requires (a) variations in genes randomly generated by mutation, and (b) humans surviving long enough to reproduce and pass on those genetic variations that have aided their survival. He maintains that in 1850s London only about half of newly borns survived to puberty, whereas that figure is now 99 per cent. Thus people with advantageous sets of genes are no longer surviving longer than people with less advantageous genes. Hence, Jones argues, nature has lost its power to select and human evolution has ended, at least in the developed world.[3] To be more precise, it means there is no change in London's population gene pool (and by extension gene pools in the rest of the developed world) arising from the differential accumulation of advantageous genes recorded by natural selection.

The principal causes of child mortality in nineteenth century London were malnutrition and disease. Their removal was due to faculties transformed or generated by reflective consciousness. Human comprehension, creativity, invention, and communication produced more effective methods of food production and distribution; they also produced medical interventions to prevent and cure what had been fatal diseases to those without a natural genetic immunity.

Moreover these faculties, combined with intention, prevent genetic variations spreading when humans, uniquely, have sex but choose to use birth control methods that prevent the passing on of their genes.

Furthermore, unlike other species, different populations of humans have not become genetically isolated in different habitats. The last fifty years of so in particular have seen an increasing expansion and mixing of human population gene pools due to two main factors. First, human reasoning and education are leading to the dismantling of social barriers to mating with those of a different race or class. Many of these barriers were reinforced by legislation, as with the segregation laws in the southern states of the USA until the 1960s and the apartheid laws in South Africa until the 1990s. The process of dismantling these barriers is by no means complete: the caste system that has been entrenched for at least 4,000 years still has a powerful influence separating classes in rural India and Pakistan, while Israelis are isolating themselves from other Semitic races by their government's policies. However, the *overall* trend is towards greater integration of groups previously separated racially or socially.

Second, it used to be the case that you only married within your extended family, and then your village, and then your town. This applied even in the most industrially advanced nations until the invention from the 1950s of inexpensive mass transport systems, like commercial air travel. Consequently, in parallel with the globalization of trade, industry, and higher education, people are increasingly migrating, temporarily or permanently, where they are meeting and marrying people from a different region or country.

The overall result of both these factors is, in NeoDarwinian terms, a trend towards mixing and expanding gene pools, thus maintaining *Homo sapiens* as a single interbreeding species. While minor genetic changes arising from mutations are inevitable (and either neutral or harmful in their effects), irreversible genetic change within geographically or socially isolated human populations to produce new species is not occurring.

But if humans began ceasing to evolve genetically from around 5,000 years or so ago, that does not mean they are not evolving in other ways.

Noetically

The rate of change of human behaviours, particularly social and innovative, since reflective consciousness marked the full emergence of humans during the Upper Palaeolithic some 40,000 to 10,000 years ago (hereinafter referred to as roughly 25,000 years ago) tells us how humans have evolved and are still evolving. What we detect is not physical or genetic but noetic: the evolution of reflective consciousness.

Thus far it may be grouped into three overlapping phases: primeval thinking, philosophical thinking, and scientific thinking.

The evolution of primeval thinking

I define the term as follows:

> **primeval thinking** The first phase of reflective consciousness when reflection on self and its relationship with the rest of the universe is rooted principally in survival and superstition.

This phase was the only kind of reflective thinking for some 90 per cent of modern human existence. It produced six things that had profound consequences for human evolution:

1. the transition from humans adapting to the environment to adapting the environment to their needs;
2. the transition from a nomadic hunter-gatherer existence to settled communities;
3. the invention and dissemination of technologies that drove the growth of agricultural villages to city-states and then empires, each with a social hierarchy based on specialized functions;
4. the invention and development of writing;
5. the foundations of astronomy and mathematics;
6. the development of belief systems and religions.

The evidence is principally threefold: the remains of early humans and their artefacts; contemporary hunter-gatherers, pastoral groups, and subsistence farmers; and written accounts of prehistoric oral narratives and of contemporary events and beliefs.

Reservations about the first two kinds of evidence were detailed in Chapter 26. Written accounts, too, need treating with caution. Rarely were they intended as historical records. Rather their purpose was usually political or religious: to persuade a population to accept a dynastic ruler's commands by extolling his feats and those of his ancestors on their behalf or to inculcate and propagate belief in a god or gods. In some cases the purposes combine when rulers were deified, as were the kings of Egypt.

Oral accounts transmitted down scores of generations were prone to embellishment and the assimilation of other myths at each re-telling. Many warriors became heroes fathered by a god, as was the Greek Heracles (adopted by the Romans as Hercules) or became the incarnation of a god, as was Krishna by the time his story was committed to writing in the Indian epic poem the Mahabharata at least 900 years after his birth.

Bearing in mind these reservations about the evidence, I shall attempt to show how the evolution of primeval thinking produced these six consequences.

Nomadic hunter-gatherer bands to settled farming communities

The emergence of reflective consciousness did not suddenly transform humans. To the contrary those first glimmerings of reflective consciousness struggled with powerful instincts that had been ingrained over millions of years of prehuman ancestry.

Research in the last twenty years or so has dispelled previous romantic notions of Edenic peaceful hunter-gatherers living in harmony with nature or who were only provoked into violence after being persecuted by so-called civilized Westerners (which did occur).

In his book *War Before Civilisation: The Myth of the Peaceful Savage*[4] University of Illinois palaeoanthropologist Lawrence Keeley compiles evidence from prehistoric remains and modern tribal peoples to show that death from raids and massacres in prehistoric times was far more frequent and proportionately on a vastly greater scale than death from warfare conducted by modern nations. Among many examples, ethnographic studies of tribal peoples of highland New Guinea, the Yanomama of the Venezuelan rainforest, and the Murngin of Australia show that some 25 per cent of adult males died from warfare. A century and a half before the arrival of Europeans in South Dakota, a mass grave near a burnt-out village at Crow Creek revealed the remains of more than 500 men, women, and children—an estimated 60 per cent of the village population—who had been slaughtered, scalped, and mutilated; underrepresented among the skeletons are those of young women, suggesting that these had been taken as captives. One half of the people found in a Nubian cemetery dating to some 12,000 years ago had died of violence.

These findings are echoed in *Constant Battles*[5] by Harvard archaeologist Steven LeBlanc who admits that for 25 years he and his colleagues routinely ignored clear evidence of warfare. He now points to hunter-gatherer sites showing evidence of violent deaths, artefacts specialized for warfare, trophy-taking, especially heads, and warfare imagery in rock art.

Homicide is not restricted to members of other hunter-gatherer bands. From the late 1970s Arizona State University anthropologists Kim Hill and Magdalena Hurtado studied Aché bands who pursued a Stone Age lifestyle in the Amazonian rainforest. Among adult males external warfare accounted for 36 per cent of all deaths. However, Hill and Hurtado found that death at the hands of another Aché accounted for 22 per cent of all deaths in their sample of 843, and included infanticide, geronticide, and club fights. Infanticide, particularly for orphans, and the killing of the aged and sick were common, so much so that the Aché often buried alive such members of their band before moving on. It was an accepted part of their culture. Life is hard for hunter-gatherers; most days they are hungry, and they cannot support those who are unable to contribute in their nomadic search for food.[6]

Evidence for the thinking of prehistoric hunter-gatherers is scarce and open to many different interpretations; inferences are no more than conjectures. In such circumstances a thought experiment can illuminate the conjecture. What follows in italics is a thought experiment interpolated with such evidence as does exist.

Imagine that you know nothing of microbial diseases, meteorology, astronomy, or any other science. In fact 25,000 years ago you know nothing other than what you have seen, heard, touched, smelled, or tasted in the territory in which your extended family forage for edible plants and fruits and attempt to kill animals for their meat.

What do you think when one of your sons becomes hot and weak and dies? What do you think of a drought when those plants shrivel before ripening? What do you think of a crack of thunder, a flash of lightning, torrential rain, or a snowstorm?

Generations later you realize that some of these phenomena follow a cyclical pattern. Day follows night in a predictable sequence and provides the basic rhythm of life. Day is when a yellow disc crosses a blue sky and provides warmth and light for hunting and foraging before it sinks below the horizon. Night is when the sky is black, while a white disc the same size as the yellow one crosses the sky for a similar period of time. It provides no warmth and little light; it is the time to sleep. Unlike the yellow disc, though, each night the white disc wanes a little until it becomes a sliver and then waxes until it grows into a full disc once more before repeating the cycle. Many generations later you realize that the number of days between each full Moon is the same as the menstrual cycle associated with the fertility of your women. Hence, you reason, the Moon is associated with fertility.

The Moon was a symbol of fertility in many early human societies, as we shall see later.

Your survival depends on the fertility of your women producing healthy sons to replace those who die as children, or who die in battle defending your food sources and your women, or who leave your band. Around your camp fire you or your women make a symbol of what you want by carving a small piece of mammoth tusk that shows the vulva you penetrate to plant your seed and from which a child emerges, together with full breasts to nourish the child. Carried in a pouch or hung around your woman's neck by a strip of hide, it will bring good fortune and produce healthy babies.

Such figurines, dating from as early as 35,000 years ago, are described in Chapter 26.*

You understand that the opposite of birth is death. This is another thing you reflect upon. You know that you, like everyone else, will die. Either in battle or from wounds suffered in battle, or after becoming weak if you fail to find food, or becoming hot and unable to eat, or, rarely, after living a long time and shrivelling in stature and strength. But death is certain. It raises the profoundly disturbing question: what happens to me when I die?

You recognize the one thing that distinguishes the living from the dead is that the living breathe and the dead do not; their bodies decay.

The idea that breath is the animating force or spirit appears in nearly all the earliest written languages.†

You cope with the psychologically traumatising foreknowledge of death by concluding that your breath or life force goes somewhere else after it has left your body to decay. In your memory and also in your dreams when you sleep—the state between being dead and being alive—you see things and people, including your dead parents. Their life force, or spirit, lives on in another world, a spirit or dream world.

---

*   See page 446.
†   See page 192.

Similar reflection leads you to conclude that all living things—like the animals you hunt or who hunt you, and the plants and fruit you eat, the trees whose branches you use to make shelters, and even the mountains from which life-giving water flows— possess a life force or spirit.

Such thinking constitutes animism, which belief persists today in many forms such as in Shinto and in the traditions of many Native American tribes like the Lakota Sioux and in the Dreaming of Australian Aboriginals.

Because you know nowhere else, this spirit world extends only as far as the terrain and the sky above it that you have seen. You and your extended family feel an identity with one particular spirit in your habitat on which your survival most depends. You may need to have the hunting and killing ability of the lion, which you admire and fear if your territory is in inter-glacial Europe, or of the jaguar if it is in Central America. Or it may be a particular prey, like a herd of bison that regularly migrates across your territory. Or a particular edible plant, like wild wheat, that reliably ripens. It becomes sacred and your group imposes a taboo on how such a spirit is to be treated. You represent it by adorning yourself with the head of the animal, or sheaves of the plant, or by carving an image of it.

Such beliefs and customs, which reinforce group solidarity, constitute totemism, which is practised in various forms today by some tribal peoples in Africa, Australia, Indonesia, Melanesia, and the Americas.

A member of your group becomes adept at inducing a dream state or trance— what we now call a state of altered consciousness—by means of long sessions of repetitive dancing or drumming, or fasting, or ingesting hallucinogenic plants or their extracts. You believe his spirit leaves his body to enter the spirit world where he seeks the help of powerful spirits to cast out the evil spirits who have entered your body and are making you hot and weak; he summons back your true spirit to make you strong. Or he beseeches the sky spirit to provide rain. Or he asks your ancestors what will happen in the future. You respect this member as a wise person, or shaman.

Shamanism in various forms is still practised today by tribal peoples from Siberia and the Arctic regions, like the Evenki, to the Americas where shamans have been called medicine men, and in Africa, where they have been called witch doctors.

You figure out how to make and use pigments to draw on rock walls representations of what you think and imagine, while you develop an increasing ability to express your ideas and feelings in sounds that convey your meaning to others in your group. Thinking, spoken language, and art evolve synergistically to enable you to share your experiences, imagination, and thinking not only with your contemporaries but also with future generations.

In the fluctuating climate of the Pleistocene you experience a hurricane that devastates your territory, but you have also learned from stories passed down generations that there have been times when the sky has sent down no rain and the Sun has burned and shrivelled the plants. The spirits are angry and need to be placated by offerings if you are to survive. They are evolving into gods. Your shaman intercedes with them on behalf of your group. He is evolving into a priest.

Generations later you realize that, as your territory becomes parched with a lack of rain, you stand a better chance of survival if your group moves away in search of more fertile territory, which you discover by a spring or in the alluvial flood plain of a great river. Here, not only can you capture fish, you recognize more cyclical patterns in nature. The wild wheat sheds seeds when day and night are of roughly equal length. The next time this occurs the plants have been nourished by the rains and warmed by the Sun, grown and ripened, and are ready to be gathered and eaten. The river floods at roughly the same time each year, irrigating the surrounding land and leaving a rich mud. But some years those floods are huge and ruin the crops.

Some 15,000 years, or 600 generations, after those first glimmerings of reflective consciousness, you reason that, rather than be constantly on the move in search of food, it is more effective to take those seeds, plant them in the richest soil, harvest them when they ripen, and store the harvest in pots you have made by moulding them from the river mud and heating them. You use the same creativity and invention to make clay bricks.

Downriver from the hills, where rainfall is much less, you reason that you can increase your crop by digging canals from the river to irrigate the dry land before planting the seeds. You also protect your plants from destructive flooding by building river embankments.

Irrigation, planting, and harvesting require many people. You reflect that, rather than fight for food, with the defeated males moving off in search of new territories and new females, it makes sense for your extended family to stay together and cooperate at these times for such labour-intensive tasks and for building permanent dwellings from the clay bricks. Your settlement grows. Later you realize that, rather than stalk migrating animals and kill one of them to eat, it is more effective to capture some, raise them for breeding, and kill them for your food when their offspring are ready to replace them.

Your agrarian community gains cohesion not only through cooperation but also by the rituals you perform. In spring you invoke the goddess of fertility to turn newly planted seeds into abundant crops, in the autumn you celebrate and give thanks for the crops you have harvested, while the winter solstice is the time to persuade the Sun god not to leave you.

This transition to a settled community occurred in different places at different times and reflected different environments and climates. It is evidenced by the remains of bones, crops, tools, pottery, and buildings in what is classified as the Neolithic phase of these different regions. In some places, however, it never occurred. Survival International estimates that today some 150 million hunter-gatherers (about 2 per cent of the world's population) live in small groups in more than 60 countries in regions like equatorial Africa and South American rainforests. Of these, some 100 tribes have not been contacted, though some have been spotted from afar or from aeroplanes.[7]

Agricultural villages to city-states and empires

No longer constantly hungry and on the move, you have more opportunity to reflect and to exchange ideas with other members of your village. Now your tools are not limited by what you can carry and so, in cooperation with others, you invent more effective ones, like ploughs made from antlers to help plant seeds, shaped millstones for grinding grain, and large clay pots for storage; you have the time and creativity to decorate these with images and symbols. These tools, combined with wider-ranging irrigation systems, enable you to extend your farmland.

You revere your parents and their parents in turn for teaching you how to survive and so you make offerings at their graves; you may even put in a place of honour that part of them from which their spirits departed, their skulls.

One of the settlements at Jericho, dated to about 6500 BCE, revealed plastered skulls with cowrie shell eyes, which has been interpreted as evidence of ancestor worship, while excavations at Çatal Höyük in Anatolia (now part of Turkey) also unearthed skulls plastered and painted with ochre to recreate faces. Veneration of ancestor spirits endures in many religions today, from Shinto in Japan and Hinduism in India to Catholicism in Central America where it assimilated many indigenous traditions.

You realize that that even your extended farmland lacks some resources that other peoples possess and you reason that it would be beneficial to exchange your surplus products for these other resources. You make dugout canoes or rafts and then boats with sails in order to travel up and down the river and trade with other communities. In so doing you exchange not only materials and artefacts but also ideas.

You build stone walls around your settlement to protect against nomadic pastoral groups who try to raid your stores or other tribes forced by over-farming or drought or tectonic activity or other natural disaster to leave their territory and attempt to take over yours. Your settlement has become a walled city.

Until now your priests have used the position of the Sun in the sky relative to a mountain in order to predict the most propitious times to make offerings to the gods for planting and harvesting. Under their direction you help build stone mountains—single stones or menhirs, or stone circles—to enable them to make more accurate predictions for these and other rituals associated with celestial events.

No longer are you in awe of the hunting prowess of the lion or the speed of the horse or the ability of any other animal or bird. You are the only species that can kill at a distance; you have tamed the horse and ride it; you have made chariots. You imagine your gods as the most powerful things you know: humans with extraordinary abilities, like a man who possesses the killing prowess of a lion, a man who possesses the speed of a horse, a man who possesses the vision of a hawk, a woman who commands the fertility power of the Moon, a woman who possesses the healing power of the snake that sloughs off its skin each year and renews itself.

In a later generation you figure out, or learn from other groups, that if you find gold-coloured shiny stones in a stream bed and heat them over a fire they form a liquid that you can pour into a ceramic mould where it solidifies into a lustrous metal. The same thing happens when you heat certain coloured rocks. Such metals as gold, copper, and tin, aren't hard enough to make useful tools and weapons, but you mould and engrave them to form prestigious jewellery and other ornaments.

Subsequently your experienced craftsmen discover or learn from other peoples that if they mix some of these liquid metals, like copper and tin, the mixture solidifies into an alloy with superior properties, like bronze that is hard and can be moulded to form sharp edges or points. Your city uses this technology to produce more effective farming implements and weapons.

The craftsmen later discover or learn how to make a furnace hot enough to produce from rust-coloured rock a silvery metal that they can hammer into shape and re-heat, and so forge iron tools and weapons with the properties of today's steel.

Your city uses these new technologies to expand your farmland. Your population grows accordingly but also becomes more complex. The large majority farm and pass on their skills to their offspring. But there are also craftsmen and merchants who pass on their skills to their offspring, as do those skilled in the art of warfare to defend your farmland and to conquer other cities in order to capture their natural resources and also their people as slaves or sacrifices to your gods, and the priests who predict the most auspicious times to make sacrifices to the gods so they will help you survive and prosper.

Rituals are not enough to keep such a large population united. A ruler seizes power and imposes laws to prevent chaos. Whether he comes from the priest class or the warrior class depends on whether the greatest threat to your survival comes from the forces of nature controlled by the gods or from other peoples. Usually it is the latter. Once established, the ruler, like members of the other classes, trains his children in his skills, and the skills here are how to exercise power; they usually include brutal suppression of dissent.

Evidence for this broad evolutionary pattern, within which are individual variations, comes from six regions: what is now the Middle East, Egypt, Europe, the Indian subcontinent, China, and Central America. It is being uncovered continually and in the future will doubtless include other regions like sub-Saharan Africa.

The Middle East

To date the earliest agricultural settlements have been found in the Levant, the region bordering the eastern Mediterranean Sea comprising what is now Israel, the Palestinian Territories, Jordan, Lebanon, Syria, and southern Turkey.

Very likely they developed from the preceding semi-nomadic Natufian peoples of that region. Fresh water springs at Jericho, in what is now the Palestinian Territories, made an attractive camping site for Natufian hunter-gatherers from perhaps 9000 BCE. Archaeological and other remains indicate that this developed into a permanent agricultural settlement consisting of circular dwellings about 5 metres in diameter built of clay bricks, with the cultivation of cereals, like wheat and barley, and the raising of sheep and goats.

About a thousand years later it had apparently grown into a town surrounded by a massive stone wall with a stone tower; population estimates vary from 200–300 to 2,000–3,000. Evidence is missing until about the end of the fourth millennium BCE when Jericho became a walled town again, with its walls rebuilt many times.

Around 1900 BCE Canaanites developed a town that was destroyed sometime between 1550 BCE and 1400 BCE.[8] When this event was recorded several hundred years later, the Hebrew scribes credited their God with responsibility. He had instructed the Israelite king Joshua to order warriors to march round the walls and shout while seven priests blew their horns, whereupon the walls collapsed. Joshua entered the city and took everything made of gold, silver, copper, and iron, while his warriors slaughtered every living thing—men, women, children, and animals—as an offering to their all-powerful God.[9] The city was later rebuilt at a nearby site.

This pattern of conquest, destruction and rebuilding, usually into a more complex settlement, appears to have been repeated several times, although we have no written accounts claiming divine intervention for the other destructions.

A similar pattern is found in Mesopotamia, the area between the Tigris and Euphrates rivers comprising most of modern Iraq and bordered on the north by the mountains of Armenia (modern Turkey), on the east by the Kurdish mountains of Iran, and on the west by the deserts of Syria and Saudi Arabia. It was the biblical Garden of Eden. However, conditions in this area were far from idyllic. One hypothesis, which has divided scholars, is that a transformation was caused by a global climate change that drastically reduced rainfall except in the mountainous northeast,[10] leaving long, hot, dry summers for most of the region, a fast-flowing Tigris causing destructive floods, and marshlands in the delta around the confluence of the Tigris and Euphrates that flows into the Persian Gulf.

The earliest settlements, like Jarmo in the north, probably date from about 7000 BCE. Inconclusive evidence suggests that small agricultural villages formed on the alluvial floodplain of the Euphrates in the south of Mesopotamia from about 5800 BCE.[11] Peoples in the south invented or developed technologies like dams and irrigation canals to enable them to extend their farmland. The Sumerian people invented the potter's wheel that enabled a more effective production of large ceramic storage pots, and later they invented wheeled vehicles. The south lacked many natural resources, like trees producing timber suitable for large constructions, metal ores, and sufficient amounts of stone, and so the Sumerians raided other areas and also developed extensive trading to obtain these materials or their products.

Such technologies and trading drove the growth of these settlements from about 4000 to 3000 BCE into socially stratified cities and then city-states, such as Uruk and Ur, each centred on a temple dedicated to a patron god or goddess believed to protect the city. Initially they were governed as a theocracy, led by a chief priest or priestess, *en*, with a secular leader, *lú-gal* (strong man), elected by citizens for a year. Increasing warfare between these expanding cities meant the greater threat to survival was other people rather than natural forces, resulting in a shift in power to the *lú-gal* who established a permanent and then dynastic rule so that *lú-gal* came to mean king, who assumed some priestly functions.[12]

There followed conquests and assimilations by dynastic rulers of different peoples who extended their control over the whole region, creating successive empires, like the Akkadian, Babylonian, and Assyrian, until Mesopotamia was conquered by Alexander the Great in 332 BCE.

Egypt[13]

Pre-dynastic remains are rare, open to different interpretations, and datings are uncertain, but it appears that Neolithic agricultural settlements developed in the rich alluvial floodplain of the River Nile at places like Merimde in the western delta of the Lower Nile from 4750 to 4250 BCE and at El-Badari and Asyut in the south, or Upper Nile, from about 4400 BCE to 4000 BCE, and might have existed as early as 5000 BCE. Cemeteries in the El-Badari area show remains of wheat, barley, lentils, and tubers, plus cattle, dogs, and sheep together with pottery.

The earliest remains of the Naqada people from further upriver are variously dated at 4400 BCE or 4000 BCE, and their culture spread along the Nile until about 3000 BCE. During this time they engaged in trade with Nubia to the south, desert oases to the west, and Mediterranean peoples to the east. At Hierakonpolis, south of Thebes, excavations revealed a 30-metre-long oval courtyard. Running along one side of this is what appears to be a drainage trench filled with bones. Nearby is a cemetery of mutilated bodies; some had been scalped, others bludgeoned to death.

The Greco-Egyptian Manetho of Sebennytos, writing in the early third century BCE, claims that the kingdoms of the Upper and Lower Nile were united around 3100 BCE by the legendary King Menes, of whom no evidence exists. Some archaeologists think he was a king called Narmer. This view is based on the discovery of a siltstone palette dating from the thirty-first century BCE. It is engraved on one side with the image of a king holding the hair of a man whom he is about to strike with a mace, while the other side shows the king in ceremonial procession plus ten decapitated corpses with their heads at their feet. Other engravings include images contentiously interpreted as symbols of the Upper and the Lower Nile and crude hieroglyphics interpreted as Narmer.

The tombs of Narmer's dynastic successors include victims with dried blood on their teeth or whose tortured poses suggest they were buried alive. Tombs of the kings became ever larger and grander, later developing into massive stone pyramids.

The united kingdom grew rich by trading its surplus grain and developing a powerful army to defend its territory and subjugate other territories and nations to form an empire. Occupation of Nubia to the south, for instance, gave it control over the region's richest gold mines and a virtual monopoly of the metal in the eastern Mediterranean. Its craftsmen gilded monuments like obelisks and created sumptuous grave goods (Tutankhamen's inner gold coffin weighed more than 110 kilograms).[14]

It was ruled by a dynastic king, later called pharaoh, who was worshipped as a god because he was deemed responsible for the annual flooding of the Nile on which the fertility of the crops, and Egypt's wealth, depended; it was governed by an administrative class, many of who were priests. Next in the descending hierarchy were traders, craftsmen, and engineers who designed the increasingly large stone temples and pyramid tombs. Lower still in the hierarchy were the vast majority, the farmers, followed by slaves captured from subjugated nations. Whether Egypt's massive stone structures were built by people cooperating to honour their gods and kings or because they were forced as citizens or slaves is a matter of speculation.

The pattern over 3,000 years and some 30 dynasties shows fluctuating fortunes in defending and extending the kingdom, internecine struggles for power, collapse of central government and rule by local leaders, battles between local leaders, reunification by war, seizing of power by settlers like Canaanites and Libyans or by invaders like Assyrians and Persians who either became dynastic rulers as pharaohs or installed puppet pharaohs, until Alexander the Great conquered Egypt. After Alexander's death one of his generals, Ptolemy, a Macedonian Greek who was governor of Egypt, declared himself king in 305 BCE and instituted the last dynasty before Egypt's conquest and annexation by Rome in 30 BCE.

Europe
Neolithic culture spread north to the mainly cooler and wetter Europe, where the Levant's mud-brick dwellings were replaced by great long-houses, together with long barrows, or burial mounds, made of earth piled over timber mortuary structures.

A notable feature is the four types of megaliths made of large, roughly hewn stones: the passage tomb, the single standing stone, or menhir, the stone row, and the stone circle.

The best known is Stonehenge in southern Britain, developed in three different phases: a circular earthwork enclosure constructed around 2950 BCE, a series of timber settings during 2900 to 2400 BCE when it was used as a cremation cemetery, and a third phase 2550 to 1600 BCE that resulted in concentric stone circles, with the outermost connected by stone lintels, surrounding an altar stone and approached by a long avenue.[15]

Perhaps more revealing of Neolithic life in the British Isles, however, is a recently discovered site, the Ness of Brodgar, covering 2.5 hectares in Orkney, a group of islands lying at the northeastern tip of Scotland, that may have been occupied from 3500 BCE. Only about a tenth has been excavated, but already it has revealed slate-roofed stone dwellings, walls painted with coloured pigments, and what excavation coordinator Nick Card has likened to a temple: a stone building 25 metres long by 19 metres wide with walls nearly 5 metres thick that would have dominated the nearby Standing Stones of Stenness and the Ring of Brodgar stone circle. The whole site is enclosed by a thick stone wall and appears to have been in use for about 1,500 years.[16]

Also predating Stonehenge are the many passage tombs in Ireland.[17] Situated on a ridge above fertile agricultural land in the valley of the River Boyne, 50 kilometres north of Dublin, the massive Newgrange passage tomb has been dated to about 3200 BCE. Richly carved with symbols, particularly spirals, it took up to 30 years to build according to its discoverer, Michael J O'Kelly of University College, Cork. Human remains were cremated, but Newgrange appears to have been the tomb of only a few people, like the royal stone pyramids built 500 years later in Egypt, and its orientation and design suggests it was a ritual centre. I will discuss these aspects in more detail when considering the development of beliefs and religions in a later section.

Evidence of alloying metals in Europe dates from around 3200 BCE in the Mediterranean island of Cyprus, whose craftsmen mixed copper from mines on the island with scarcer tin imported from distant lands to produce bronze artefacts. The Minoans of Crete, named after their legendary King Minos, appear to have controlled much of the trade in bronze, using the proceeds to build lavish palaces like that at Knossos.

Indian subcontinent[18]

Recent research has overturned previous ideas and datings, largely stemming from dates estimated by the German Max Müller in the 1860s and interpretations of remains by archaeologists from the British colonial era. Although dating on the Indian subcontinent is still contentious, it seems that residents of mud-brick houses farmed wheat and barley and herded sheep, goats, and cattle at Mehrgarh, in what is now the Kachi plain of Baluchistan in Pakistan, from 7000 BCE, with evidence of copper smelting from 5000 BCE. Successive settlements occupied the area until around 2600 BCE.

It seems reasonable to suppose it was Mehrgarh people spreading south and then southwest and northeast along the rivers of the Indus Valley that gave rise to what is known as the Indus Valley civilization, which developed from about 3300 BCE. At its height it covered an area of some 1,250,000 square kilometres of present day Pakistan and northwest India and comprised more than a thousand settlements. Chief among these were the cities of Harappa and Mohenjo-daro, with evidence of hilltop citadels, sophisticated urban planning, plumbing and sanitation systems, dockyards, and granaries. Their peoples engaged in trade by land and sea as far as Mesopotamia, Persia, and central Asia.

No writing has so far been deciphered, and we do not know for certain how these cities were governed. The epic Sanskrit poem, the Mahabharata ("Great Story of the Baharatas", the legendary first Indians), probably compiled in its final, expanded version between 400 BCE and 400 CE, describes a historic battle between cousins, each supported by allied kings, for the right to succeed the throne of the Kuru. Such a battle probably did place near or in the Indus Valley, although estimates of dates vary from 950 BCE (mainly European scholars) to as early as 5561 BCE (mainly Indian scholars).[19]

Apart from a dynastic king supported by his prince-brothers, the rest of the social hierarchy may be inferred from the caste system that endures in many parts of rural India and Pakistan today; it reflects a social stratification based on function similar to those of other ancient civilizations examined so far, with a priestly caste (Brahmins), a warrior caste (Kshatriyas), a merchant caste (Vaishyas), a peasant caste (Shudras), and finally the outcastes or untouchables.

The cities and villages were abandoned by about 1900 BCE. Speculations as to the cause include natural disasters like tectonic activity, the drying up or change of course of rivers, over-farming, or conquest and destruction. Plausibly their populations moved east into the Ganges valley.

China[20]

Evidence of early agricultural settlements is scarce, but pottery and graves suggest these developed from 8000 BCE to 2000 BCE in the alluvial flood plains of the Huang, or Yellow, River (northern China) and the Yangtze River (central China).[21] Remains from later periods imply that their peoples cultivated rice and millet, and domesticated pigs, chickens, and water buffalo.

Bronze works probably date from around 2000 BCE and develop through to the later Han period, or 221 BCE, producing weapons like spear heads or ritualistic objects like tripods and vessels marked with motifs such as demons, symbolic animals, and abstract symbols.

The earliest firm evidence of government comes from Yin, near modern Anyang, capital city of the late Shang dynasty in northern China. It follows the now familiar pattern. The Shang dynasty of 31 kings ruled from about 1600 BCE to 1046 BCE during which it moved its capital nine times. Yin was occupied from the thirteenth century BCE and comprised an oval area about 9.75 kilometres by 3.75 kilometres that includes a temple complex with remains of human sacrifice, residential buildings, clusters of aristocratic houses, and workshops plus cemeteries incorporating 11 royal tombs that include the bodies of soldiers and horses.

In 1046 BCE Wu of Zhou from northwest China conquered the Shang kingdom, which he incorporated into his own, centred on Fenghao near present-day Xi'an (or Sian) in the Wei River valley. On Wu's death his younger brother, the Duke of Zhou, became regent. He was later mythologized and credited among other things with formulating the Mandate of Heaven doctrine, according to which Heaven blesses the authority of a dynastic ruler who is just but withdraws it from an unjust one. This was seen as legitimizing the overthrow of the Shang who claimed a divine mandate to rule.

The Zhou dynasty employed a feudal system to administer its empire that expanded to incorporate most of China. The dynasty saw the introduction of iron-working, ox-drawn ploughs, crossbows, and horseback riding.

Central America[22]

Dating in prehistoric Americas is also problematic. It appears that from 1500 BCE if not earlier corn, beans, squash, and other plants were cultivated by the Olmec peoples in the tropical lowlands of southeastern Mexico, possibly beginning in the Coatzacoalcos River valley that runs into the Gulf of Mexico. Here seasonal flooding provided a fertile alluvial soil that attracted large populations, leading to the development of cities.

Characteristic artefacts include giant heads carved from single blocks of volcanic basalt, assumed to be of rulers, engravings interpreted as were-jaguars (humans that turn into jaguars), baby were-jaguars on altar pieces suggestive of sacrifice, plus a ball game and images of ritualistic bloodletting, principally by rulers, that are seen in subsequent Central American civilizations. It lasted for some thousand years before disappearing between 400 and 350 BCE. The cause or causes of its demise are speculative and similar to those of the Indus Valley civilization.

More evidence is available from the Maya civilization that developed in what are now Belize, Guatemala, southeastern Mexico, and the western parts of Honduras and El Salvador.

In their Classic Period, from about 250 CE to about 900 CE, the Maya built ceremonial centres that developed into more than 40 cities typified by a central plaza overlooked by a temple atop a massive stepped pyramid, together with a palace and a large ball court; these cities were constructed mainly from limestone, quarried and carved using harder stone like chert. Despite their advances in astronomy and numbering (considered later), there is no evidence that the Maya worked metal or used the wheel, either for pottery or for transport although they did build white roads (Yucatec Maya *sacbeob*) coated with limestone stucco over a stone and rubble fill, which seem to have been used for religious, military, and trade purposes. Much of this civilization also collapsed, leaving many cities overgrown by jungle. Speculative causes include over-farming, drought, conquest and destruction, or war-related disruption of trade routes.

However cities like Chichén Itzá, Uxmal, and Mayapán in the flat rainforests of the northern Yucatán peninsula continued to flourish for several centuries, possibly having been settled or conquered by Toltecs from the north. Later buildings at Chichén Itzá are no longer engraved with motifs of a serpent, representing Ixchel, Maya goddess of fertility and medicine, but of a plumed serpent, representing the legendary Toltec god-king known to the Maya as Kukulkán and also to the Aztecs as Quetzalcóatl. Sculptures, murals, and written codices depict warriors victorious in battle and a skull rack, typically used by ancient Mesoamerican cultures to display the heads of war captives or other sacrificial victims. They portray a social stratification similar to other ancient cultures of a hereditary king, priest-astronomers, nobles, farmers, and slaves.

The development of writing

The visible recording of language on materials like stone, clay, or paper is unique to the human species. It enables the transmission of ideas over vast distances of space and lengths of time. Without it no large and complex societies could have developed.

The arrangement of letters or symbols in groups or sequences to express defined and understood complex meanings evolved from pictorial and then symbolic representations of things. It developed independently in at least three different places at different times,[23] typically to record amounts of agricultural products or artefacts traded or else taken as a tax or tribute to a ruler, or to promulgate laws, or to record stories venerating gods and rulers, or to record instructions for rituals.

As noted in Chapter 26, most linguistics scholars consider the earliest writing systems were the Sumerian cuneiform script engraved on clay tablets from around 3000 BCE and Egyptian hieroglyphs engraved on stone from roughly the same period. Whether Egyptian writing developed independently or under the influence of Sumerian writing is contentious.

Cretan scribes used an indigenous hieroglyphic script on clay tablets from about 2200 BCE, which developed from around 1900–1800 BCE into what is termed Linear A, not yet deciphered, presumably for the Minoan language. Linear B appears in the Peloponnese from about 1500 BCE and also in Cretan palaces; it was used for writing Mycenaean Greek and predates the classical Greek alphabet by several centuries.

What appears to be a script inscribed on rectangular seals made of soft stone, and on tools, miniature tablets, copper plates, and pottery was found at Harappa in the Indus Valley and dated from about 2500 BCE, but it has not yet been deciphered and it is unclear whether this is proto-writing or writing. Sanskrit, the classical language of India, has its roots in Vedic Sanskrit, in which ancient hymns to the gods were written, but Western and some Indian scholars disagree about the extent to which it developed from local languages or was imported into India by Aryans invading from what is now Iran. However, the Aryan invasion hypothesis has little or no evidential support, and the origin of Sanskrit is unclear.

Writing was invented independently in China, where the earliest evidence so far discovered consists of inscriptions on cattle shoulder-blades and tortoise shells used in divination rituals dating from around 1250 BCE in the Shang dynasty. The Chinese writing that subsequently developed does not employ an alphabet with a limited number of characters, like Western languages; instead each character generally represents one syllable of spoken Chinese and may be a word on its own or part of a polysyllabic word, leading to more than 5,000 characters.

Writing almost certainly developed independently too in Central America. A stone slab discovered in the Mexican state of Veracruz and dated to around 900

BCE has been claimed as the oldest writing system on the continent because of the sequencing patterns of its engraved 62 signs. As yet undeciphered, it has been attributed to the Olmecs.[24] The best understood ancient language in that region is Mayan of the Classic Period (about 200–900 CE), frequently called a glyphic script. It has been found carved in wood and stone, painted on pottery and walls, and written on bark-paper parchments known as codices. Most of the latter were burnt on the orders of Diego de Landa, the Franciscan charged with converting the Maya to Catholicism, but a few survived. The most extensive is the Dresden Codex, named after the library in Germany where it is kept, that dates from about the eleventh or twelfth century and is thought to be a copy of an original text of some 300 to 400 years earlier.[25]

The foundations of astronomy and mathematics

What we now deem the scientific or intellectual disciplines of astronomy and mathematics were founded for practical purposes and intimately bound up with the beliefs of early humans.

Many of the megalithic monuments in Europe demonstrate a knowledge of astronomy. In 1963 astronomer Gerald Hawkins published an article in *Nature* that used computer analyses showing celestial phenomena viewed from Stonehenge in 1500 BCE to claim that features of Stonehenge were oriented with 13 solar and 11 lunar events and that its design enabled the prediction of lunar eclipses.[26] The latest view, advocated by University of Sheffield archaeologist Mike Parker Pearson, is that Stonehenge was a place of ritual at the winter solstice rather than the summer solstice as previously thought.

Undisputed is the design of the earlier Newgrange passage tomb in Ireland, radiocarbon dated to 3200 BCE. Shortly after sunrise on the winter solstice sunlight enters through a slit cut into the rock above the tomb door and traces a pencil-thin beam along the floor of the 18-metre-long passage to reach the front edge of the basin stone at the end chamber, in which cremated remains were presumably kept. Over the next few minutes it expands into a 17-centimetre-wide beam, bathing the basin stone, together with the intricate spirals and other carvings that decorate the chamber, side chambers and corbelled roof, in a golden glow before narrowing and plunging the tomb back into darkness.[27]

Several European megalithic "circles", including the inner ring of standing stones at Stonehenge, are actually ovoid, suggesting at least a practical knowledge of the Pythagorean theorem.

In order to count large numbers the Sumerians invented a sexagesimal, or base 60, system that we still use today to divide a minute into 60 seconds, an hour into 60 minutes, a day into 24 hours, a circle into 360 degrees of arc, a foot into 12 inches, and so on. Their successors, the Babylonians, adopted this system and developed arithmetic, algebra, trigonometry, and geometry for the practical purposes of trade, surveying, designing buildings, and dividing fields. Thousands of clay tablets recovered from the Old Babylonian period (2000–1600 BCE) show the results of computing linear and quadratic equations, and geometrical constructions; they also list multiplication tables and tables of squares, square roots, and reciprocals. The Babylonians probably knew the theorem that the West attributes to Pythagoras. The Plimpton 322 tablet dating from 1700 BCE, more than a thousand years before Pythagoras was born, lists the values of squaring the hypotenuse of a right-angled triangle as the sum of squaring the other two sides $(a^2 + b^2 = c^2)$.

From about the twelfth century BCE the Babylonians started to organize their earlier celestial observations and divinations into a series of cuneiform tablets known as the *Enūma Anu Enlil*, which refer to the sky god and king of the gods Anu and god of the air Enlil. They also produced star catalogues, such as *Mul. Apin*, containing descriptions of the Sun, Moon, constellations, individual stars, and planets associated with their gods. Their purpose was astrology: to predict celestial events, like eclipses and the movement of the planets across the heavens, that they interpreted as omens signalling the intention of those gods.

Their star catalogues were subsequently adopted and developed by the Greeks and then the Romans, giving modern astronomy the legacy of planets named after Roman gods—Mercury, Venus, Mars, Jupiter, Saturn, Uranus, and Neptune—and constellations named with the Latin equivalent of the Babylonian ones, like Taurus (the bull) marking the spring equinox, Leo (the lion) marking the summer solstice, Scorpius (the scorpion) marking the autumn equinox, and Capricornus (goat-horned), marking the winter solstice.[28]

Excavations at Harappa, Mohenjo-daro, and other sites of the Indus Valley civilization (probably 3300 BCE to 1700 BCE, see above) have revealed the use of mathematics for practical and religious purposes. For example, the Harappans had a system of weights and linear measures based on a decimal series.

The Sulba Sutras, instructions for designing fire altars most pleasing to the Vedic gods of India, show how to make a square equal in area to a given rectangle, which involves the results of the Pythagorean theorem, making a square equal in area to a given circle, which involves ratios that give an approximate value of π (where we calculate πr^2 as the area of a circle of radius r), and the addition of ratios that give a remarkably accurate value of the square root of 2.[29]

Many Western scholars consider they were transmitted from Babylon before reaching India a thousand years later. However, Subhash Kak, now Regents Professor of Computer Science at Oklahoma State University, argues that archaeo-astronomy—computer analyses showing the positions of stellar constellations viewed from different places at different times in the past—correlated with archaeological remains and the Sanskrit texts shows the Sulba Sutras are the same age as the Old Babylonian calculations. Similarly he maintains that the *Vedanga Jyotisa*, a Vedic astronomical manual, dates from around 1300 BCE.[30]

An inscription on an oracle bone found in Yin, capital city of the late Shang dynasty in northern China, records a lunar eclipse on the fifteenth day of the twelfth moon of the twenty-ninth year of King Wu-Ding, which corresponds with 23 November 1311 BCE.[31] For each successive dynasty astrologers in ancient China prepared calendars recording celestial events and indicating the phases of the Moon and the time of solar year in order to plan the agricultural seasons and provide omens.

Detailed records of astronomical observations were made during the Warring States period (about 475–221 BCE) and include star catalogues attributed to astronomer/astrologers Shi Shen and Gan De. Most were lost, but copies were included in the *Great Tang Treatise on Astrology of the Kaiyuan Era*, compiled from 714 to 724 CE. Stellar charts and lists of astronomical events like eclipses, comets, and novae expanded from the Han period (206 BCE–220 CE) onwards.[32]

The Maya preoccupation with predicting the natural cycles on which they believed their survival depended led them to calculate the solar year, most probably at Copan somewhere between 300 BCE and 900 CE, as 365.2420 days, which is more accurate than that of the Gregorian calendar used in the West today. They used three calendars: a civil one called Haab of 365 days, a sacred one called Tzolkin of 260 days (which is the period between the two zenith passages of the sun at Copan), and a long calendar calculated as the smallest common multiple of the two (the smallest common multiple of 365 and 260 gives a number of days equivalent to approximately 52 years).[33]

Priest-astronomers used these calendars to determine propitious times for such things as planting and rituals. In order to make such calculations and long-term predictions the Maya devised a sophisticated numbering system of base 20 (compared with the current European base 10, or decimal, system) that uses the concept of zero, which neither the Greeks nor the Romans invented for their numbering systems.

Surviving codices show the Maya had star catalogues and tables listing, among other things, tables predicting eclipses, the appearance of Venus as the bright morning star (a good augury for going to war), and the retrograde motion of Mars.[34]

The development of beliefs and religions

In Chapter 2 I summarized nine principal themes evident in the creation myths of different peoples during this phase of primeval thinking and suggested most were best explained by a lack of understanding of natural phenomena together with political and cultural needs. These conclusions are reinforced what we know about the development of other beliefs and religions that occurred in different regions at different times in this phase.

By the time of the cave paintings and figurines dating from some 35,000 years ago in Europe described in Chapter 26, these beliefs most probably involved animism, shamanism, totemism, and ancestor worship. After humans adopted a settled existence these beliefs evolved. As agricultural villages grew into cities with social hierarchies and eventually empires, the spirits evolved into a pantheon of gods reflecting the needs and hierarchies of the cities: the fertility goddess gradually lost her dominance to a male sky god whose power to unleash claps of thunder and bolts of lightning reflected the power of the warrior-king who defended the city and extended the farmland.

In Europe from around 3200 BCE the alignment of megaliths to heavenly events like solstices, together with evidence of cremation or ritual burial, grave goods, and artwork like the finely engraved, elaborate spirals that recur throughout many of the passage tombs, indicate belief in an afterlife and the existence of supernatural beings or gods. The design of the Newgrange passage tomb, for instance, suggests belief in a communion between the spirit of a cremated ruler and the Sun god. Such beliefs are explicit in the funerary literature of ancient Egypt.

The earliest, known as the Pyramid Texts, date from about 2400 to 2300 BCE and were carved on pyramid walls and sarcophagi of kings of Egypt's Old Kingdom; they include spells and incantations to ensure the safe passage of the king to the afterlife, where Ra, the Sun god, rules over the heavens and was sometimes referred to as creator and father of all things.

They developed into the Coffin Texts of the Middle Kingdom (about 2000 BCE), which included those of wealthy individuals as well as kings, and then into texts written on papyrus and placed inside the mummy case dating from the eighteenth dynasty (1580–1350 BCE) that incorporated extracts from the Pyramid and Coffin texts and became known collectively as the Book of the Dead.

Amenhotep IV, who became king sometime between 1370 and 1358 BCE, became the originator of monotheism in the fifth year of his reign by rejecting as superstition the pantheon of traditional gods based on natural forces and depicted as half man and half beast or bird. The chief god then was Amun-Ra, local god of the capital Thebes, Amun, who had became ruler of the gods and identified with the Sun god Ra after the ruler of Thebes defeated the last Hyksos king of Egypt and founded the eighteenth dynasty. Amenhotep IV declared there was only one god, the Sun god, supreme source of light and life, who was to be depicted only as the disc of the Sun, Aten, and that he was the son of Aten, hence-forth to be known as Akhenaten. To outmanoeuvre opposition from the powerful and wealthy priests of Amun-Ra, Akhenaten moved the capital from Thebes to halfway between Thebes in the Upper Nile and Memphis in the Lower Nile where he built a new city dedicated only to Aten.

Monotheism did not survive his death and neither did his city. The priests of Amun-Ra gained their revenge by declaring Akhenaten a heretic and then attempted to remove all traces of his existence. The Egyptians reverted to a pantheon of gods in which the god who protected the city of whoever gained the throne became elevated to sky god.[35]

A similar pattern occurs when a city-state conquers another and develops an empire. Often the conquerors assimilate gods of the vanquished people but elevate their local god to chief god, whom they identify with the sky god who is also the god of war. Worshipping the sky god as their chief protector and as ruler of the other gods was common in many societies, with examples like the Indian Indra, the Greek Zeus, the Roman Jupiter, the Norse Thor, and many more.

The oldest religious texts in the world are the Vedas of India, possibly compiled over two thousand years beginning around 4000 BCE.* The first part of the texts, known as *karma-kanda*, are hymns and commentaries on rituals that portray the devas—gods and goddesses—as elemental powers, led by Indra, and include Agni, god of fire, Usha, goddess of the dawn, Surya, god of the Sun, and Soma, god of intoxicating liquor (probably hallucinogenic and drunk by priests to obtain visions).[36]

Several Vedantic schools today claim an occult meaning to the hymns, saying for instance that Indra actually represents the human organs of sense and action. Such claims are unconvincing; there is no evidence that the hymns were intended, or understood at the time, to be anything other than what they are. The same is not the case for the later Upanishads, insights of seers that are traditionally appended to the Vedas. Many are explicitly allegorical, and I shall consider them in the next chapter.

In China 60 per cent of all recovered oracle bones date to the reign of Wu-ding (about 1250–1192 BCE). The pantheon Wu-ding sacrificed to was dominated by his lineal ancestors and some of their consorts, plus nature gods like River and Mountain. The most important god was Di (lord) who, like the king, commanded; he could determine the outcome of warfare, the weather, the harvest, and the fate of the capital city. Although princes had the right to interpret the oracle bones, only the king could communicate with Di.

The size and layout of the royal tombs, their wealth of furnishings including traces of numerous human sacrificial victims, suggest the king was the point of contact between humans and the world of the gods, and that tombs were places of worship. John Lagerwey of the Chinese University of Hong Kong and Marc Kalinowski of École Pratique des Hautes Études, Paris, argue that royal ancestor worship, with its mythologizing and rituals, was used to legitimize the rule of the dynastic king and regulate the behaviour of the masses.[37]

We still see such practices. In 2012 military chiefs of the so-called communist republic of North Korea arranged three days of mass rituals to mark the succession of Kim Jong Un, the Great Successor, after the death of his father, Kim Jong II, the Dear Leader, who is said to have been born on a sacred mountain. He in turn had succeeded his father, Kim Il Sung, who had ruled for 46 years after being installed by the Soviet Union as premier and then gaining absolute power as Eternal President and Great Leader. The tomb of his mummified body was the centre for one of those mass rituals.

An ancestor cult persists among many present-day Maya. As for the ancient Maya, their beliefs, inferred from surviving buildings, engravings, books, and codices plus accounts by their Spanish conquerors, vary over time, location, and dialects that give different names to similar things. They indicate that every

* See page 9.

natural phenomenon on which their survival depended possesses a spirit that evolved into a god or goddess whose importance varied in different environments.

After their Spanish conquerors forced them to accept Catholicism, many Maya identified their gods as Catholic angels and saints, and Catholic feast days today take on a distinct Maya form. On All Saints Day and on the anniversary of an ancestor's death it is a common practice to leave food and drink on an altar in the house dedicated to a Catholic saint, or at the ancestor's tomb, for the returning spirit to consume.[38]

As Chapter 2 noted, monotheism re-emerged some 700 years after Akhenaten when Josiah, king of Judah, wanted to legitimize and sanctify the union of Judah with the fallen kingdom of Israel under his sole and absolute rule.*

Like their contemporaries, the different Hebrew tribes worshipped many gods, such as Asherah, Baal, Anath, El, and Dagon. Their bible was written after the official religion became a monotheism that reflected a patriarchal ruler who defended his people but exacted vengeance if his laws were disobeyed. It is instructive to see how it deals with these other gods. In some books, like the Second Book of Kings, their worship is condemned and punished by their Lord God.[39] In others their God is shown to be superior to other gods, as in Exodus, when Aaron and Moses ask the king of Egypt to release their people from bondage because their God wills it. On God's instructions Aaron throws his staff on the ground and it turns into a snake. The king's magicians throw their staffs on the ground and they too turn into snakes. However, Aaron's snake devours the others, proving his God is more powerful. After the king still refuses to free the Israelites, their God wins further magic contests with those of the Egyptian priests before bringing down plagues and death to the first-born of the Egyptians.[40]

In some books other gods are subsumed into the one God. Ezekiel has a vision of him glowing like brass in the heart of flames and having four faces: those of a lion, a bull, an eagle, and a human (the lion was the totemic symbol of the god of the tribe of Judah, the bull that the god of the northern tribes, and the eagle that of the tribe of Dan).[41]

Other books demote the gods to a heavenly court of angels who surround the Lord and praise him. Isaiah describes them as flaming creatures with six wings.[42] Other angels are named and have specific tasks, like Gabriel the messenger who explains to the prophet Daniel the meaning of his visions,[43] Michael, Israel's guardian angel, who helps another angel fight the guardian angel of Persia,[44] and Satan who questions God's commands.[45]

Conclusions

1. For the majority of the time since they fully emerged from hominins roughly 25,000 years ago modern humans pursued a hunter-gather

---

* See page 14.

existence in small, extended family groups where competition for survival with other such groups and with predators resulted in a high death rate.

2. The early glimmerings of reflective consciousness evolved slowly until some 10,000 years ago when humans invented farming as a more effective way of providing sustenance and understood the benefits of cooperation by establishing larger agricultural village communities. This development occurred in different places at different times and in some places it never occurred.

3. With more opportunities to reflect and to transmit ideas by drawing, speech and writing, humans in these agricultural village communities cooperated to invent technologies that improved and extended their farmlands, and also cooperated with other settlements to trade in both goods and ideas, enabling their settlements to grow in size and complexity.

4. Although cooperation had begun to evolve it struggled against the instinct for competition, ingrained over millions of years of prehuman ancestry, that produced battles for control of these settlements and their agricultural and other resources both from within and from other settlements. It resulted in centralization and enforced collaboration.

5. As these settlements grew they developed a social hierarchy that reflected classes of skills passed on from parents to offspring, typically ruler, priests, warriors, merchants and craftsmen, farmers, and slaves. They expanded into autocratically ruled cities, city-states, and then empires that rose and fell. The overall pattern across the globe was an increase in size, complexity, and centration of human societies.

6. The evolution of primeval thinking was intimately bound up with the evolution of superstitious beliefs that arose from imagination combined with a lack of understanding of natural phenomena and fear of the unknown. From the animism, totemism, and ancestor veneration of hunter-gatherers, beliefs gave rise to organized religions* that reflected the growth in size, complexity, structure, and specializations of human societies. They developed from worship of a fertility goddess through polytheism to a pantheon ruled by a powerful male sky and war god, and lastly to a patriarchal monotheism with other gods subsumed into one God or demoted to angels.

7. In applying reflective consciousness to devise technologies for survival and reproduction, and to influence the supernatural forces believed to determine these factors, primeval thinking produced the foundations of art, music, spoken and written language, mathematics, and astronomy.

8. Although many ideas and inventions spread and developed through cultural transmission, cases of convergent or parallel evolution are evident,

---

* See Glossary for the meaning of *religion* used in this book.

instanced by the independent evolution of writing and the independent evolution of astronomy and numbering systems.

9. This first phase of human evolution, primeval thinking, was thus characterized by the imperative to survive and reproduce, a domination of the deeply ingrained competitive instinct over a newly evolving cooperation, and superstition.

Figure 27.1 gives a simplified representation of the evolution of some of the main strands of primeval thinking following the emergence of reflective consciousness. It is not a conventional tree diagram, like a genealogical tree with fixed relationships, but a two-dimensional snapshot of a four-dimensional dynamic process in which its branches and sub-branches not only change over time—developing, ramifying still further or withering, becoming moribund, or dying—but also interact with other branches to hybridize or mutate or generate a new branch.

For example, imagination—the ability of the mind to form images, sensations, and ideas not seen or otherwise experienced at the time—not only gave rise to superstitious beliefs, such as a spirit with the body of a man and the head of a lion, but it also interacted with creativity to produce practical new things, such as the wheel, while creativity also gave rise to non-practical things, like abstract art.

Invention—the ability to make new things—may have resulted from the interaction of creativity and imagination, but it could also result from experimentation, such as trial and error when mixing liquid copper and tin in different quantities to in order to produce an alloy, bronze, with superior properties to either for use as tools and weapons.

Figure 27.1 also shows instinct, which was ingrained for several million years in our hominid ancestors.

> **instinct**   An innate, impulsive response to stimuli, usually determined by biological necessities such as survival and reproduction.

It did not vanish when reflective consciousness emerged from consciousness to distinguish *Homo sapiens* from other species. To the contrary, even today instinct remains a powerful force and must be taken into account when trying to understand human behaviour.

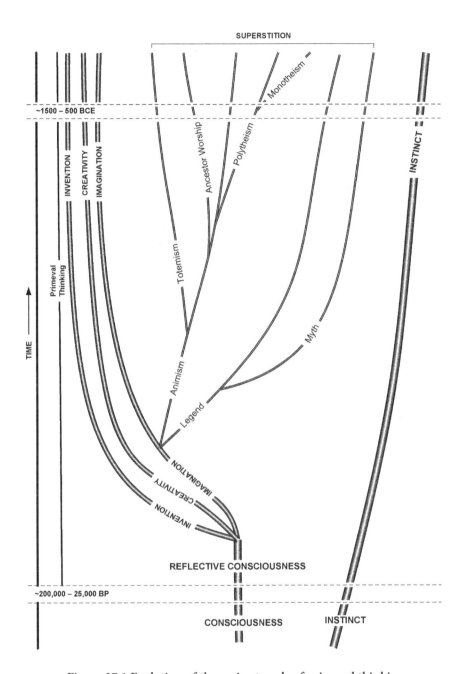

Figure 27.1 Evolution of the main strands of primeval thinking

Unlike a conventional tree diagram with fixed relationships, this is a simplified snapshot of a four-dimensional dynamic and interactive process. It is not drawn to scale and dates are best estimates.

CHAPTER TWENTY-EIGHT

Human Evolution 2: Philosophical Thinking

There is only one thing that a philosopher can be relied on to do, and that is to contradict other philosophers.

—WILLIAM JAMES, 1904

Disputation is a proof of not seeing clearly.

—ATTRIBUTED TO ZUANGZI, ABOUT FOURTH CENTURY BCE

The second phase of human evolution begins when reflection on self and its relationship with the rest of the universe branches off from superstition into philosophy.

Rather than adopt one of many definitions of philosophy currently employed in the West, I shall use one more in keeping with the original sense of the word.

> **philosophy**   Love of wisdom; thinking about ultimate reality, the essence and causes of things, the natural world, human behaviour, and thinking itself.

One book, still less one chapter, cannot cover the evolution of all philosophical thinking. I shall concentrate on those parts relevant to answering the central questions of this inquiry: what are we? and where did we come from?

The emergence of philosophical thinking

Bertrand Russell had no doubts about the origin of philosophy.

> [The Greeks] invented mathematics and science and philosophy.... Philosophy begins with Thales, who, fortunately, can be dated by the fact that he predicted an eclipse which, according to the astronomers, occurred in the year 585 BC. Philosophy and science—which were not originally

separate—were therefore born together at the beginning of the sixth century.[1]

As Chapter 27 shows, Russell was wrong about the origin of mathematics and astronomy. The fact that they were developed for practical purposes or to advance superstitious beliefs like astrology in no way detracts from their existence more than a thousand years before the Greek Thales.

If Thales did predict a solar eclipse—and this is highly questionable—he very probably used Babylonian celestial charts if not Babylonian predictions, as we shall see later. Yet Russell's view of the origin of philosophy persists today in the West.

Since Russell was writing *A History of Western Philosophy* perhaps he may be excused. My copy of *The Oxford Companion to Philosophy* states on the jacket "Covers the whole history of philosophy, worldwide." Yet of its 1,922 entries, 22 (1 per cent) relate to India, 10 (0.5 per cent) to China, and 10 (0.5 per cent) to Japan, that is, some 2 per cent devoted to the major non-Western centres of philosophy. It too asserts

> The first three universally recognized philosophers—Thales, Anaximander, and Anaximenes—all came from Miletus, a prosperous Greek city.[2]

To discover when, where, and how philosophical thinking emerged we need to abandon Eurocentrism.

India[3]

The earliest known examples of philosophical thinking are contained in the ancient Indian Upanishads, of which the German philosopher Arthur Schopenhauer said "From every sentence deep, original, and sublime thoughts arise."[4]

Traditionally they were attached to the end of the Vedas, hymns to the gods and associated rituals described in Chapter 27. They record the insights of illumined seers, and their purpose, according to Eknath Easwaran, a modern translator, is

> to discover for certain who we are, what the universe is, and what is the significance of the brief drama of life and death we play out against the backdrop of eternity[5]

which is directly relevant to the question asked in this inquiry.

Upanishads continued to be written, but the oldest, particularly the ten commented on by the philosopher and teacher Adi Shankara (c. 788–820 CE), are generally recognized as the most important. Although each is different they share two fundamental insights: *sarvam idam brahma* (all is Brahman) and *ayam atma brahma* (the Self—translated with a capital S to mean the essential Self as distinct from the phenomenal self or individual personality—is Brahman). This identity—the Upanishadic answer to the pivotal self-reflective question what am

I?—is expressed succinctly as *tat tavm asi* (You are That, where You is the pure consciousness "which makes the eye see and the mind think", namely the Self, and That is the indescribable ultimate reality, Brahman).

The Upanishads variously employ metaphor, simile, allegory, dialectic, and thought experiment to convey the meaning of the otherwise ineffable Brahman. It may be expressed as the Cosmic Consciousness from which all things spring and of which all things consist; it is transcendent, existing formlessly out of space and time, and also immanent in the phenomena perceived by our five physical senses and by our mind, our chief instrument of perception.

According to the Upanishads, training the mind by disciplined meditation to focus on the interior object of contemplation until it becomes absorbed in it enables us to gain insight into this ultimate reality.

By contrast with that for the ancient Greek philosophers, the evidence for the Upanishads is first-hand. Estimates of the age of these Sanskrit texts, however, vary. On the basis of their literary form, Michael Nagler, professor emeritus of Classics and Comparative Literature at the University of California, Berkeley, considers they were not written down before about 600–400 BCE, but notes that oral tradition holds the Upanishads are much older than the texts.[6] Jitendra Nath Mohanty, Emeritus Professor of Philosophy at Temple University, estimates the texts range from 1000–500 BCE,[7] while Arindam Chakrabarti, Professor of Philosophy at the University of Hawaii, says they were compiled between about 1000 and 900 BCE.[8] David Frawley, co-director of the Institute of Vedic Studies in the USA, agrees with those Indian scholars who claim that Sanskrit developed from one or more indigenous languages rather than being imported by Aryans invading from the north; he dates the texts from 2000–1500 BCE.[9]

D S Sarma, author of an anthology of the Upanishads, comments that the seers who taught in the forest ashrams were

> apparently engaged in the mighty task of transforming a rather low type of sacrificial religion prevalent at the time into a great mystical religion true for all time.[10]

This illustrates a general pattern. Emergence is a gradual process, and the earliest philosophical thinking is inevitably entangled in the roots of superstition from which it branched.

China[11]

Philosophical thinking emerged in China from around the sixth century BCE during the disintegration of the Eastern Zhou Empire and flourished in large part as a response to the chaos and violence of the Warring States period.

All the ancient Chinese philosophers believed in, and made sacrifices to, Tian, variously translated as the Sky, Heaven, or Nature. However, they used reasoning and insight rather than claimed divine revelation to arrive at their ethical (which included political) and metaphysical ideas. In this sense they began to depart from the primeval thinking of their culture.

Most taught the Dao (previously transliterated as Tao) as the Way, or Path, to be followed, but their understanding and interpretation of the Dao differed. Historians of the later Han dynasty classified the many philosophical schools into different traditions. The three principal ones are Confucianism, Mohism, and Daoism.

The first takes its name from Confucius, the Latinized name of K'ung Fu-tzu, (now transliterated as Kongzi), which means Master K'ung. Tradition has it that he was born about 551 BCE and was a former minister in the court of the Duke of Lu, a state in the Eastern Zhou Empire. Ideas attributed to him are contained in the *Analects*, a short collection of discussions with his followers, which seems to have been compiled over a period of thirty to forty years by first- and second-generation disciples. Other strands of Confucianism developed, most notably in *The Book of Mencius*, the Latinization of Meng-tzu (371–289 BCE), who studied under the grandson of Confucius and was later recognized as the true transmitter of Confucian philosophy.

For Confucianism the Way to restore social stability and just government was through traditional values. Its thinking focused on ethics, and in particular the relationships between ruler and subject, parent and child, older and younger brother, husband and wife, and friend and friend. These five relationships should be based on reciprocity, "What you do not want done to yourself, do not do to others,"[12] a version of what later came to be known as the Golden Rule advocated by many thinkers in the ancient world. This virtuous prescription should be cultivated by adherence to etiquette and ritual.

Mohism was the school that followed the thinking of Mozi (c. 490–403 BCE), who opposed using traditional values as the Dao because they changed and were unreliable. Ethical values are constants of Tian, or Nature, that show an impartial concern for all. Mozi's approach to determining the ethical Way was pragmatic. He argued that acts should be measured to assess the balance of their consequential benefit against harm. His conclusions included opposition to military aggression and injuring others, and advocacy of a state governed by a benevolent sovereign and administered by a meritocracy. Mohist views stimulated the thinking of others, like Mencius, and many were absorbed into Confucianism.

The early Daoist thinkers, like the seers of the Upanishads, focused by contrast on ultimate reality, and their prescription for human behaviour comes as a consequence of that.

The first of the principal Daoist texts is *Laozi*, or Old Master, after its legendary author said to have been an older contemporary of Confucius. In the West this collection of aphorisms is known as *The Book of the Way and Its Power*. The second, and longer, key book is the treatise *Zuangzi*, named after its author who, it is claimed, lived about two hundred years after Laozi. Modern scholarship suggests that *Laozi* and *Zuangzi* probably constitute collections of ideas compiled by different writers at different times; it was common practice in several ancient cultures for disciples of a teacher to attribute their views to the founder of their school.

In these texts the Dao is the Way everything in the universe was brought into existence and which sustains everything. It is the ultimate, ineffable reality, like Brahman, but is manifest in the Way the natural world functions. This dynamic quality demonstrates continuous flow and change in an unending cycle between two opposites that constitute a whole, symbolized by the dark *yin* and the bright *yang* in the ancient Chinese Diagram of the Supreme Ultimate (see Figure 28.1).

Figure 28.1 The Supreme Ultimate, symbolized by the dark *yin* and the bright *yang*

A dynamic rotational symmetry in which each dot indicates that one force contains within itself the seed of its opposite; when one force reaches its extreme the seed of the opposite grows while the force diminishes until the opposite force reaches its extreme, and so on for ever.

Most likely originating from the cyclical patterns in nature, like night and day, *yin* and *yang* give rise to all the complementary opposites we perceive: winter and summer, female and male, Earth—dark, receptive, female—below and Heaven—bright, strong, male—above, and so on.

To achieve the ideal state of being we should harmonize our lives with the Dao. Following this Way, or universal law, is similar to the Dharma of Hinduism and Buddhism discussed later.

These early Daoist ideas were gained by insight. The *Zuangzi* explicitly rejects reasoning:

> The most extensive knowledge does not necessarily know it; reasoning will not make men wise to it. The sages have decided against both these methods.[13]

Europe[14]

In Europe philosophical thinking began in the sixth century BCE with the ancient Greeks. I shall trace its emergence by following the arbitrary convention of considering those thinkers before Socrates made his mark a century later. The place to begin is Ionia, a narrow strip of what is now Turkey's western coast that had been colonized by the Greeks.

As for why it should begin here we can only speculate. The most plausible explanation is that Miletus was an Ionian seaport, one of the most prosperous Greek cities of its time, that had extensive trading and cultural links with Egypt, Babylonia, and other regions to the east. Most likely Egyptian and Babylonian astronomy and mathematics reached Miletus, and thence the rest of Ionia, as a by-product of trading. These stimulated Ionians, distant from the religious orthodoxy of mainland Greece, to develop their own ideas.

Russell gives no evidence to support his assertion that the first philosopher was Thales of Miletus. This is unsurprising. Nothing of what Thales may have written survives. We have to rely on Aristotle two centuries later saying that Thales, Anaximander, and Anaximenes of Miletus were the first *physici*, who gave natural explanations of the world, as distinct from the *theologi*, who believed everything was due to the actions of impetuous gods. It seems, however, that Aristotle had no written material on which to base his claim that Thales was the founder of natural philosophy.

According to his account Thales said that everything was made of water, and that magnets are alive (or have a soul, *psyche*, which meant much the same thing) because they made iron move. Later sources, like the unreliable Roman Diogenes Laertius writing some 800 years after Thales and claiming secondary sources of which no trace remains, credit Thales not only with being the first to predict a solar eclipse but also with other achievements that are implausible or plainly wrong.

One part of one sentence of Anaximander survives in the form of a quotation nearly a thousand years later by the NeoPlatonist Simplicius that cites Theophrastus, Aristotle's chief assistant and successor in 322 BCE as head of the Athens Lyceum, and it is unclear if this is a paraphrase or a direct quotation from a book. According to Simplicius, Anaximander expressed his views in the poetic language of Greek mythology. The view attributed to Anaximander by Simplicius and other tertiary sources is that the *arche*, or fundamental stuff of the universe, is not Thales's water but *apeiron*, usually translated as boundless or indefinite, which has no observable qualities of its own but from which all observable phenomena arise.

The first to arise are the opposites of hot and cold, and these are engaged in a struggle, overseen by Time, in which each encroaches on the other and thus repays the other's "injustice". This has been conjectured to mean that fundamental opposites follow an indefinite cycle, with the dominant one giving way to the other in an alternating pattern, and out of which all opposites are generated, similar to the Chinese *yin* and *yang* of the Dao.

Nothing survives of the work of the third of the Melesian philosophers, Anaximenes. Again we have to rely on tertiary sources, principally Simplicius, that his idea of the *arche*, or fundamental stuff of the universe, was not Thales's water or Anaximander's unobservable boundlessness, but air, which supports the Earth. It appears to be air in the sense of *psyche*, which meant both breath and soul, or the life-giving force, as we saw earlier in the Homeric myths. In its most rarefied form it is fire, its ordinary form is the air we experience, but progressive condensation results in wind, cloud, water, earth, and stones. Constant motion causes these rarefactions and condensations.

Cicero, writing four centuries later, claimed that Pythagoras (c. 570–c. 500 BCE) was the first to use the term philosopher (*philosophos*, lover of wisdom) when describing himself to Leon, the ruler of Phlius. Pythagoras was an Ionian, but founded a sect in Croton, a Greek colony in southern Italy. According to Russell

> This mystical element entered into Greek philosophy with Pythagoras, who was a reformer of Orphism as Orpheus was a reformer of the religion of Dionysus. From Pythagoras Orphic elements entered into the philosophy of Plato, and from Plato into any philosophy that was in any degree religious.[15]

The problem in disentangling truth from myth about Pythagoras is that nothing of what he wrote or said survives, and Pythagoreans, the secretive, ascetic cults founded, or inspired, by Pythagoras, attributed all their ideas to him, these very probably including the geometric theorem that bears his name.

Pythagoreans believed the soul is eternal and is reincarnated in all living forms. Release from reincarnation and reunion with the god from whom all life had sprung can be achieved by a life of contemplation and inquiry. Pythagoreans maintained the Orphic distinction between what can be learned of the world through the sense organs and pure knowledge attained by the soul, that is, insight. Examining geometry, music, and the celestial bodies will reveal the divine principles that order the universe, and Pythagoreans saw those expressed in numbers and their relationships, although there is no evidence of their making detailed observations of the stars or deducing mathematical relationships from them as they did with musical scales.

Heraclitus (c. 540–475 BCE) was a native of Ephesus, one of the twelve cities of the Ionian League conquered by the Persians in the middle of the sixth century BCE. The 130 or so surviving fragments of his works reveal the enigmatic aphorisms of a mystic who believed that what he had to say went beyond the limits of human language. His key to understanding the world was introspection, and his insights show considerable congruence with those of Eastern mystics. His universe is a whole that has always existed. It consists of opposites perceived by the senses to be in a state of constant change and strife, but these opposites are somehow the same because all things are one, which is the *logos*, or guiding principle of the universe. This is similar to the Dao giving rise to the interplay of *yin* and *yang* and subsequent opposites.

Underneath this flux of the phenomenal world is *aiezoon pyr* (ever-living fire) that transforms. This is similar to *prana*, the Sanskrit word used in the Upanishads to denote vital energy, the essential substrate of all forms of energy.

Parmenides, born around 510 BCE in the Greek colony of Elea in southern Italy, and his followers opposed this idea of constant change. From the 150 or so surviving lines of his poem *On Nature* he appears to have had an insight of reality ("what-is") that is one, timeless, complete, and unchanging. There is no such thing as "what-is-not" (nothingness). Consequently, what-is could not have come from what-is-not, and so there could not have been a creation event. Since there is no what-is-not for what-is to move into, no motion or change is possible. What we perceive as change is an illusion created by our five senses.

This view was defended by his disciple Zeno in a series of provocative paradoxes. For example, at any instant in time after an arrow has been fired it is stationary. Its flight consists of a series of such stationary instants. Hence the arrow is stationary.

Of the next generation, Empedocles (c. 492–433 BCE), a citizen of Acragas on the south coast of Sicily, appears from secondary and tertiary sources to have been a colourful individual who claimed to be divine after many reincarnations. He seems to have pronounced that everything is made of up of a mixture of four irreducible elements—earth, air, fire, and water—and that the force of love causes them to combine and the force of strife drives them apart.

Anaxagoras (c. 500–428 BCE) brought the tradition of Ionian thinkers to Athens around 460 BCE. He stayed there as a friend of the democratic statesman Pericles for some 30 years before being convicted of impiety for denying that heavenly bodies were gods to be worshipped; stars, he said, were made of glowing stones. He was opposed to superstition and was reputed to have claimed that a universal Mind (*nous*, which may also be translated as reason) brought order out of primordial undifferentiated matter and somehow controls all natural processes.

Leupicius is thought to be the founder of the atomist school, and apart from that we know precious little about him. Democritus (c. 460–370 BCE) developed the idea that everything in the cosmos consists of different types of irreducible entities called atoms. The meaning of this Greek word is "uncuttable"; this answers Zeno, whose paradoxes depend on infinite divisibility.

Innumerable atoms are in ceaseless chaotic motion in the void of empty space. Random collisions result in different types of atoms adhering to each other to produce all the phenomena—both physical and mental—we perceive. But these adhesions are only temporary; eventually they decompose into atoms once more. According to this view there is no place for a creator god or for a guiding principle of nature or for the immortality of the soul: everything ultimately returns to atoms and death is a total annihilation. In an infinite universe our world is one of many possibilities, one in which the conditions are right for life. This prefigures the modern multiverse speculation.

For Democritus the atomist idea had ethical and medical implications. Atoms constituting the mind or soul become disordered and unsettled by passions, and so an individual's happiness is produced by moderation, equanimity, and the absence of disruptive desires. Likewise this prescription is necessary for a civilization to flourish and prevent disintegration into barbarism. Physical health too is an equable balance of atoms in the body.

The Middle East[16]

The German psychiatrist and philosopher Karl Jaspers (1883–1969) coined the term "Axial Age" to describe the period 900 to 200 BCE during which similar revolutionary thinking appeared independently in India, China, Greece, and the Middle East. While listing several of the thinkers considered in this section and the following one on the evolution of philosophy, Jaspers adds the Hebrew prophets Elijah, Isaiah, Jeremiah, and the Persian Zoroaster.

Elijah's purpose was to lead the Israelites away from worshipping false gods like Baal and persuade them to worship only Yahweh. According to the Hebrew bible he argued his case by raising the dead to life, calling fire down from heaven, and ascending to heaven in a chariot of fire. While Isaiah and Jeremiah each advocated peace and justice rather than warfare, they maintained this could only be achieved by seeking the intercession of the God of Judah.

Zoroaster was a priest and prophet and founded a religion in ancient Persia. He preached a dualism of two primeval powers engaged in conflict in this world: Good, personified by the god Ahura Mazda, and Evil, personified by Angra Mainyu, who may or may not have been his twin brother (Zoroastrian texts differ on this point). Zoroaster advocated the worship only of Ahura Mazda and the practice of good thoughts, words, and deeds to achieve the triumph of Good.

While their advocacy of righteous behaviour, justice, and peace opposed the predominant aggressive struggle for dominance and territory rooted in competition, the thinking of each of these four is circumscribed by a belief in, and worship of, an interventionist god or God. None represents an attempt to move beyond primeval thinking.

Central America

I'm not aware of any attempt by indigenous populations in Central America, or the rest of the Americas, to go beyond primeval thinking until centuries after their conquest by Europeans.

The evolution of philosophical thinking

India[17]

Even as it struggled to emerge from its primeval roots, philosophical thinking in India began to evolve. It diverged principally into six orthodox Hindu traditions and a movement that rejected religious beliefs and rituals and from which emerged two more traditions.

Hindu orthodox philosophical traditions
The word "Hindu" derives from the ancient Persian for Indus and probably referred to the peoples of the Indus Valley and their descendants. I use the word "orthodox" to mean the various interacting traditions that, broadly speaking, share a reverence for the Vedas together with three beliefs: ultimate reality is Brahman, or Cosmic Consciousness, which is identical with the individual soul or Self (Atman); all creatures are trapped in *samsara*, an endless cycle of birth, suffering, death, and rebirth according to *karma*, the principle by which an individual's moral deeds in past lives determine his or her present life; and that *moksha*, or liberation from this cycle of reincarnation, can be attained.

Each tradition branched into many schools, arguing contradictory positions such as dualism and monism, whether or not Brahman is God, and the existence of gods as realities or aspects of God.

Shramana
A loose movement known as Shramana consisted of individual seers who sought direct insight, most commonly by renouncing their life in society and practising asceticism. Of the very many such seers, two subsequently developed large followings and became traditions in themselves.

Jainism
The first of these was Jainism, whose leading figure, Vardhamana, was known as Mahavira, or Great Hero. According to tradition he was born in 599 BCE as a prince in what is now the state of Bihar in eastern India through which the Ganges flows. At the age of 30 Mahavira renounced the world and became a wandering holy man. Twelve years later he achieved the insight that led to his recognition as a Tirthankara, literally "Fordmaker", an enlightened being who shows how to cross the river of reincarnations to the state of eternal liberation of the soul.

The Jain canon, preserved in an ancient Prakrit dialect but only committed to writing several hundred years later, claims that Mahavira was the twenty-fourth and final Tirthankara of this particular half-cycle of the eternal cosmos.* He advocated the application of his ethical insight, which requires extreme asceticism and non-violence, in order to achieve liberation from reincarnation.

Like other seers Mahavira attracted disciples, but it was only after Chandragupta, founder of the Maurya Empire, converted to Jainism, as did several kings of Gujara, that it developed into a major tradition which today has some six million

---

* See page 10 for the Jain belief in the cycles of the cosmos.

followers, mainly in India. Jain monks and nuns adopt the extreme lifestyle taught by Mahavira, while laypeople vow to follow the practices as best they can in their occupations.

To reconcile endlessly disputing schools of philosophy, followers of Jainism developed the idea of non-exclusivism, a theory of logic that recognizes seven truth values: true, false, undecidable, and four combinations of these three. They also developed a metaphysical theory, according to which things have infinite aspects so that no description is wholly true and none is wholly false.

A major division developed over centuries between the sky-clad, or naked, and the white-clad. Each of these communities branched further, while reverence for Tirthankaras, real and mythical, developed into worship, with some sects incorporating Hindu deities such as Rama and Krishna into their pantheon.

Buddhism

The second of these major Shramana traditions was Buddhism, founded by Siddhartha Gautama. Separating the facts from the myths that grow after the death of a charismatic leader is always difficult, and especially so in this case.

We may reasonably conclude that Siddhartha was the son of the ruler of the Shakyas, who occupied the foothills of the Himalayas in present-day Nepal. Different traditions give different dates for his birth; Western scholars used to give a date of 566 or 563 BCE, but more recent research indicates sometime between 490 and 480 BCE.

At the age of 29 he forsook his life of closeted luxury, his wife and son, and lived as a wandering mendicant to try to understand the physical and mental suffering he had witnessed and how it could be alleviated.

It seems he studied meditation under different teachers and for a time practised Jainism, or at least a life of severe asceticism, but decided it was too extreme. According to Buddhist scriptures, at the age of 35 he sat beneath a pipal tree and vowed not to move until he had found the answer to his question. His meditation gave him the insight now referred to as the Four Noble Truths.

1. All human life is suffering (*duhkha*).
2. Suffering is caused by craving, especially for pleasurable sensations.
3. Suffering can be ended by the cessation of craving.
4. Craving can be ended by following the Eightfold Noble Path: right view (understanding), right resolve, right speech, right action, right livelihood, right effort, right mindfulness, right meditation.

Following this path leads to release from suffering and the cycle of reincarnation. Thereafter he became known as the Buddha, or Enlightened One, who taught his insight of this Middle Way (between the extremes of worldly indulgence and severe asceticism) to a small group of disciples who formed a *sangha*, or community, committed to practising his insight.

He wrote nothing and taught in his local dialect. Several versions of his teaching survive as scraps of text in Sanskrit and Prakrit dialects. The earliest complete set of scriptures was written in Pali, one of the Prakrit dialects, most probably between 29 and 17 BCE in Sri Lanka. This Buddhist canon comprises the *Sutra* (Pali, *Sutta*) or discourses intended to be memorized, the *Vinaya* or monastic rules, and the *Abhidharma* (Pali *Abhidhamma*), a metaphysics that includes the idea of co-dependent origination. According to the latter, all phenomena—physical, mental, and emotional—arise from, and depend on, a temporary network of contingent causes; consequently nothing is permanent. This means there is no eternal Self or soul; on the death of an enlightened one who has achieved release from the cycle of reincarnation no personal identity or boundary of mind remains (in contrast to the Jain belief). Whether this was the reasoning of Siddhartha or of disciples who formulated it in his name is a matter of speculation.

The body of Buddhist teaching is called the Dharma. This Sanskrit word was used in Hinduism to mean the natural law of the cosmos and also the conduct of individuals in conformity with this law. Both Buddhist and Hindu meanings are similar in concept to the Chinese Dao.

Siddhartha Gautama and his small *sangha* remained just one of many such Shramana groups until the adoption of his teachings by a powerful ruler. In the third century BCE the Emperor Ashoka, distressed by the suffering caused by his conquest of the east Indian kingdom of Kalinga, embraced Buddhism. By this time he controlled most of India, and he sent Buddhist missionaries south to what is now Sri Lanka and as far afield as kingdoms in the Hellenistic world.

But even by the third century BCE Buddhism had begun to diverge. Theravada Buddhism, or the Doctrine of the Elders, claimed to preserve Siddhartha's original insight and teaching.[18]

Other schools took the view that membership of a *sangha*, or monastery as it had become, was too exclusive and Buddhism should be available to all. By about the first century BCE this movement had developed into Mahayana Buddhism, or the Greater Path. This held that seeking individual liberation from suffering and reincarnation was selfish, and the ideal path was that of the *bodhisattva*, one who postponed his merited liberation until he had helped others achieve this state. Siddhartha Gautama was one of several such *bodhisattvas*, temporary Earthly manifestations of an eternal, omniscient Buddhahood. Mahayana schools revered relics and images of *bodhisattvas*, and appealed to them for aid and assistance, in a way that was indistinguishable from the worship of gods.

While this was occurring Buddhism spread south and east, and subsequently north, adapting to local conditions as it progressed. Missionaries had taken Theravada south to Sri Lanka, where it became the dominant religion, and from there and also from India to Thailand, Burma, and then Indonesia.

From about the first or second century CE varieties of Mahayana spread east along the Silk Route towards China, assimilating indigenous traditions and beliefs. In China it encountered periods of opposition and support, diverging into several forms in response to local conditions, including a meditative school called Ch'an. In about the sixth century Prince Shotoku, regent in Japan, adopted Buddhist ideas and practices, and in Japan Ch'an developed into Zen.

In India, however, Buddhism as a distinct philosophy had disappeared by about 1000 CE due to two main factors. First, Mahayana's adoption of rituals and beliefs made it less distinguishable from the Brahmanic religion, commonly referred to as Hinduism, while Hinduism's eclectic nature assimilated Buddhist thinking and practices. Second, invading Muslims from the west destroyed most Buddhist monasteries because they deemed images of Siddhartha and other *bodhisattvas* blasphemous.

In the eighth century CE versions of Buddhism had spread north into Nepal and thence to Tibet, but only in the eleventh century when rulers in west Tibet adopted Buddhism did it really take hold there. The dominant version adopted many Hindu religious rituals, like repeated recitation of mantras, or sacred sounds to invoke the power of deities, and mandalas, or sacred diagrams, considered to play a similar role. Consorts of *bodhisattvas* took on great importance as deities and became versions of the ancient Mother Goddess. This branch of Buddhism became known as Vajrayana, or the Diamond Way. It assimilated several elements of the Tibetan Bon religion, which was rooted in animism and shamanism, and developed into a religion comprising different schools, each governed by the abbot, known as the grand lama, of its principal monastery.

The authority of the grand lamas increased after it was agreed that each was a successively reincarnated *bodhisattva*. The monasteries competed for political control—sometimes by armed conflict—after conquering Mongols from China appointed the grand lama of the Sakya monastery as viceroy. By the seventeenth century this control had passed to the grand lama of the Dge-lugs-pa order, known as the Dalai Lama and regarded as the fifth such reincarnation.

Tibet became if not a theocracy then a Buddhocracy. It remained so until Chinese Communist troops invaded in 1950 and seized control. In 1959 Tenzin Gyatso (born Lhamo Dondrub), recognized as the fourteenth reincarnation of the first Dalai Lama, fled to northern India where he established a government in exile.

While a few Buddhist schools across the world today maintain a non-theistic philosophical tradition, most became religions in which Siddhartha and other mystics are paradoxically worshipped as gods.

China[19]

A broadly similar pattern occurred in China.

Following his brutal unification of China in the third century BCE, the self-declared First Emperor, Qin Shi Huang, adopted Legalism as official government policy and suppressed other philosophical schools. Mohism declined and withered away. The fortunes of Confucianism fluctuated according to support given by emperors. The Han dynasty from the first century CE saw its precepts as endorsing the established hierarchy, which largely they did, and so emperors made sacrifices to Heaven and venerated Confucius.

However, Confucianism had to compete with Daoism, and then Buddhism, and suffered a decline from the third to the seventh centuries. It was not until the Sung Dynasty (962–1279) and the development of NeoConfucianism that it became the dominant philosophy among educated Chinese. Drawing on Daoist and Buddhist ideas, NeoConfucian thinkers formulated a system of metaphysics, which had not been part of older Confucianism, but they retained the hierarchical political and social vision of early Confucian teachings. The philosopher Chu Hsi (1130–1200) unified different strands, and his system dominated subsequent Chinese intellectual life until 1911 and the overthrow of the monarchy, with which it had been associated. Its decline accelerated after the Communist revolution of 1949.

The current Chinese Communist Politburo, having introduced capitalism in order to increase prosperity in the country, is now promoting Confucianism at the expense of Marxist-Leninism in order to support its own authority and establish social stability.

Although Daoism emerged as a mystical insight into ultimate reality, seeing a natural law or pattern to the way the cosmos functions, it underwent a profound change in 142 CE, some 600 years after the death of its legendary founder Laozi. Zhang Daoling, a hermit, announced that Laozi had appeared to him in a vision and appointed him Heavenly Master. He founded a religion that worshipped Laozi as the incarnation of the Supreme God, and which taught that the essence of divinity also resides in each individual; longevity and immortality could be cultivated by meditation, diet, and alchemical treatments.

Daoist religious sects adopted the institutions of Buddhism. In the following centuries they became linked to popular movements opposing Confucian support for the established social and political order, to local religions with their nature gods, to alchemy, and to martial arts. Consequently, after about the fifth century Daoism ceased to exist as a system of philosophical thinking.

Europe[20]

Socrates

Socrates (469–399 BCE) is generally regarded as the first of three great thinkers who shaped the evolution of philosophy in the West. He wrote nothing, and we are left with four sources for his ideas. The playwright Aristophanes was the only one to write about Socrates during his lifetime, and he set out to discredit Socrates. Xenophon, a former pupil of Socrates, became an historian after a military career but lacked the intellectual grasp to act as a reliable guide to Socrates's philosophy. Aristotle, born fifteen years after Socrates's death, learned of Socrates's views second-hand from Plato. The most extensive source is Plato, a disciple who idolized Socrates to the extent that he attributed everything he thought wise to Socrates, and what Plato thought wise changed a great deal over the course of forty years.

What we do know is that Socrates was an Athenian of noble stock who forsook wealth to devote himself to thinking, that he employed a particular kind of reasoning, and that he died for his beliefs. Unlike most of his Greek predecessors he was not concerned with ultimate reality or what was the basic stuff of the universe. Socrates focused on ethics: how humans should behave and why. His method was to interrogate the beliefs of others, exposing the flaws in their arguments, in an endeavour to reach the truth.

As usually presented, Socrates and the Buddha, who lived about the same time, could not differ more. Socrates professed that his whole life was an unfulfilled quest for wisdom whereas the Buddha claimed enlightenment. Socrates employed dialectical reasoning whereas the Buddha gained insight through meditation.

Deeper investigation, however, reveals some striking similarities in addition to their common focus on ethics. According to Plato, during his military service Socrates

> started wrestling with some problem or other about sunrise one morning, and stood there lost in thought, and when the answer came he still stood there thinking....And at last, toward nightfall, some of the Ionians brought out their bedding after supper...partly to see whether he was going to stay there all night. Well, he stood there till morning, and then at sunrise he said his prayers to the sun and went away.[21]

Substitute sitting for standing, add a pipal tree, and this description is comparable to Siddhartha Gautama's meditation to gain insight.

Socrates reasoned that our conduct should be determined not by expectation of heavenly reward or fear of punishment in any afterlife but because it gives happiness now. This, however, is not selfish because the only way to achieve happiness is by acting justly and behaving virtuously towards everyone including—by contrast to Greek convention—one's enemies as well as one's friends. His prescription is similar to the Buddha's eightfold Noble Path, and his description of happiness is a state of serenity like the Buddha claimed for those who followed his Noble Path.

Plato

The second great thinker was Plato (c. 429–374 BCE), an Athenian nobleman who left Athens after his teacher and friend Socrates was convicted and sentenced to death by a jury of 500 citizens for refusing to recognize the state gods. It confirmed in Plato a deep distrust of democracy to which he later added a deep distrust of tyrants (a tyrant then meant one who seized power, often by leading a popular revolt against an aristocratic or plutocratic government, and ruled with absolute authority but not necessarily cruelly or oppressively).

He travelled for 12 years, spending time at, among other places, the Greek cities in southern Italy where he was influenced by Pythagoreans, and the Greek city of Syracuse in Sicily, where he encountered the tyrant Dionysius I.

He concluded that only philosophers should become rulers, or else rulers should become philosophers, because only philosophers possessed the knowledge and wisdom to act justly. On his return to Athens he founded an academy to educate the sons of noblemen in a wide range of philosophical inquiry. His written output is traditionally divided into three phases, although the dating and authenticity of some texts are questionable.

His most influential conjecture is his idea of Forms. A Form is a transcendental and eternal ideal that our senses perceive only through imperfect, temporal examples. For instance, a circular plate is only an imperfect, impermanent specimen of the ideal Circle that exists above and apart from the material world. Similarly, Wisdom, Justice, and Goodness are the Forms of qualities only imperfectly found in this world but to which we should aspire.

Aristotle

Most of the early Greek thinkers reputedly were polymaths. If the secondary and tertiary sources are to be believed, the third great philosopher Aristotle (384–322 BCE) surpasses them all put together in the sheer range and volume of his output. Yet only about a fifth to a quarter of his work is estimated to have survived, and he wrote none for publication. It appears to consist of notes he prepared for lectures or discussion, notes made by his pupils, and research notes, all of which were later edited and compiled into books by his pupils, particularly Eudemus and Theophrastus, at the Lyceum, the academy in Athens where Aristotle taught for 12 years.

The version that reaches us was edited by Andronicus of Rhodes in the first century BCE. It was discovered in the eighth century CE by Muslim scholars who translated it into Arabic and Syriac, after which commentaries were written by Muslim philosophers, notably Averroës in the twelfth century. Translations from Arabic and Syriac into Latin began to appear in the West later in the twelfth century.

This body of work includes rhetoric, logic, ethics, political theory, literary theory, metaphysics, theology, and what are now classified as the sciences of biology, anatomy, physics, astronomy, and cosmology. Lauded in the mediaeval West as "The Philosopher", this former pupil of Plato is credited with inventing science and the scientific method. Because the purpose of this quest is to discover what science can tell us about human evolution from the birth of the universe, I shall focus on this aspect of Aristotle's work.

His scientific method derives from his system of logic that formalized deductive reasoning and his search for the causes of things and their essence, or form. Unlike Plato's transcendental Form, Aristotle's form is intrinsic to a thing.

He classified things on a scale of perfection expressed in their form, which can be inferred from their internal organization that he called their *psyche*, or soul. Plants are the lowest form of life, possessing a vegetative soul responsible for reproduction and growth, animals a vegetative plus a sensitive soul responsible for mobility and sensation, and humans a vegetative, a sensitive, and a rational soul capable of thought and reflection.

Aristotle applied his method most successfully to the physiology and behaviour of some 540 zoological specimens based on his own dissections and observations and the reports of others. Although he made several mistakes, parts of his account of the heart and vascular system, for example, were not bettered until the seventeenth century.

He divided animals into those with blood and those without blood (at least without red blood), corresponding to the modern distinction between vertebrates and invertebrates, and further grouped species with common forms into a genus. His classification was not surpassed until Linnaeus's in the eighteenth century.

In his other investigations of nature, however, Aristotle dispensed with empiricism. In anatomy he reasoned that men had more teeth than women. In physics he reasoned that bodies fell to Earth with a speed proportional to their mass, which Philoponus in the sixth century CE disproved by a simple experiment. Aristotle didn't measure. His work is better described as natural philosophy rather than science as now understood and defined in Chapter 1.

In cosmology he reasoned that the Moon, Sun, and constellations are embedded in a series of concentric crystalline perfect spheres that rotate forever at different constant speeds around the Earth. And here he moves into metaphysics. Everything has a cause. In one book Aristotle reasons that the spheres are connected, so that the innermost one is caused to move by the adjacent one, which in turn is caused to move by the next farthest from Earth, and so on. But the chain of causality has to stop somewhere. What causes the outermost sphere to move? It is the Unmoved Prime Mover. Aristotle reasoned that this Uncaused First Cause must be simple, unchanging, perfect, eternal, and without physical size or shape; it consists entirely of intellectual contemplation, the purest exercise of reason: it is Divine.

Unlike most of his predecessors in the Greek-speaking world and philosophers from other countries, Aristotle shows no sign of using insight.

Hellenistic philosophy
The death of Aristotle a year after that of his former pupil Alexander the Great conventionally marks the end of the Hellenic (Greek) phase of philosophical thinking and the beginning of the Hellenistic (Greekish) phase. Alexander's empire fragmented, with his governors seizing power in their regions to create their own monarchies. Other intellectual centres developed to rival Athens, most notably Alexandria in Egypt, Antioch in Syria, Pergamon in Asia Minor, and later the Aegean island of Rhodes.

Aristotle's Lyceum continued in Athens, but fell into decline after the middle of the third century BCE. Three other principal schools of thinking arose in this period. The Epicureans, Stoics, and Sceptics agreed that philosophy should be practical and its objective was to achieve *ataraxia*, a state of tranquillity. This suggests they were influenced more by Socrates than Aristotle.

Epicurus (341–271 BCE) followed the atomistic metaphysics of Leucippus and Democritus, and argued that, since the mind was nothing but a collection of atoms that would disperse when life departed, this removed the fear of death and we should seek pleasure in this life. Detractors and, later, Christians performed an effective spin job in portraying this as overindulgence in food and sex. However, Epicurus's meaning of pleasure was "freedom from pain in the body and from disturbance of the soul", and he seems to have led an abstemious life, cultivating friendships and showing generosity to slaves.

The Stoics, founded by Zeno of Citium in the early third century BCE, believed that *ataraxia* would be achieved when man lived in accord with nature. Hence they inquired into nature, concluded that everything is determined by fate, and so taught that an attitude of acceptance of fate would lead to tranquillity.

The Sceptics held that we have no certain knowledge of the nature of things, demonstrated by arguing every question both ways, and so we achieve peace of mind by suspending judgement.

Greco-Roman philosophy

These strands of thinking continued and evolved during the ascendancy of the Roman Empire when Romans adopted and developed Greek thinking. Aristotelianism revived during this era, but its followers concentrated on preserving and commentating on Aristotle's works rather than advancing them, and the school eventually died out in the third century CE.

It had failed to transform philosophy into a purely rational way of thinking. In the second century, for example, the Roman emperor and Stoic Marcus Aurelius advised withdrawing into oneself, and the self into which one should withdraw was "like the universal intelligence that permeated and organized the world; indeed it was part of that intelligence".[22] This is similar to the central insight of the Upanishads: not only everything is Brahman, but also the Self is Brahman.

Arguably the leading thinker of this phase was Plotinus (c. 205–270 CE), considered to be the founder of NeoPlatonism. From the evidence of his *The Six Enneads* he was a mystic whose purpose was union with the One, which he claimed to have experienced on several occasions. To enjoy blissful union with the One he advised that we must abandon

> proof by evidence, by the reasoning process of the mental habit. Such logic is not to be confounded with that act of ours in the vision; it is not our reason that has seen it; it is something greater than reason.[23]

Through contemplation and moral virtue we can rise in our minds towards the One. This mental journey is not only a return to the source of our existence but also a discovery of our true selves. "When the soul begins again to mount, it comes not to something alien but to its very self."[24] This insight, too, mirrors the Upanishadic insight.

Philosophical thinking, however, whether by reasoning or insight as distinct from belief in supernatural forces like gods or a God, was stifled in the West after 529 CE when the Roman emperor Justinian closed down the Athenian schools because they undermined Christianity.

Scholasticism

Christian scholars learned of Aristotelianism in the twelfth century through Arabic and Syriac translations and commentaries. It had been of little or no significance in the West for nearly a thousand years, which is nearly twice as long as it had existed (though it had been revived in the Muslim world after five centuries).

Christian theologians who became known as the Scholastics adopted Aristotelian reasoning to prove the existence of God. The most notable of these, the thirteenth century Dominican Thomas Aquinas, used among other arguments the Uncaused First Cause and the Unmoved First Mover.

Paradoxically, Christianity had been founded on insight. The gospels of Matthew, Mark, and Luke each tell of the Spirit of God descending on Jesus of Nazareth who spent 40 days and nights in the desert without food. He emerged to teach his insight that to love God and love our neighbours as we love ourselves (a version of the Golden Rule) was more important than blind adherence to the laws proclaimed in the Hebrew bible.

Spiritual insight continued with the Desert Fathers of the third century, subsequent monastic orders, and a series of contemplatives like Meister Eckhart, Saint Teresa of Avila, and Saint John of the Cross through to the sixteenth century.

From the end of the twelfth century, however, Scholastic reasoning became dominant and formed virtually the only method of thinking in Western schools and universities, which were controlled by the Roman Catholic Church.

Modern philosophy

Reasoning became so embedded in the West that Protestant reformers in the sixteenth century used it to attack the Roman Church, while the Enlightenment thinkers of the eighteenth century used it to argue against any Church. Today, departments of philosophy in Western universities, the vast majority of which are secular, teach only reasoning. Of nine American and British encyclopaedias and dictionaries I consulted, seven define philosophy solely in terms of reasoning. That, I think, conflates essence with method.

The ramification of philosophical thinking

As philosophical thinking evolved in different parts of the globe it branched into innumerable traditions and schools that continued to ramify.

The basic division commonly presented is between the mystical East and the rational West from the time of the ancient Greeks. I think a more fundamental branching is between insight and reasoning, defined as

insight   Seeing clearly the essence of a thing, usually suddenly after disciplined meditation or following an unsuccessful attempt to arrive at an understanding through reasoning.

reasoning   An attempt to understand the essence of a thing by a logical process, based either on evidence or on assumptions taken as self-evident.

In addition to disciplined meditation to gain insight, reasoning was employed in ancient India in several of the Upanishads, in the Nyaya logical system that was comparable in many ways to Aristotelian reasoning, in the Jain seven-valued logical system, in the Buddha's discourses and Buddhist metaphysics of co-dependent origination, and so on. In China, while the mystical Daoist *Zuangzi* rejects reasoning as a method of inquiry, *Mencius*, attributed to the fourth century BCE Confucian philosopher, uses reasoning in its discourses, while Mohists used logic to arrive at their ideas.

These are but few of many examples showing that thinkers in the East employed both insight and reasoning, and many schools still employ reasoning, even if it is not their principal means. Moreover, nineteenth century colonialism by Western powers resulted in many Eastern countries adopting Western academic norms. It is mistaken to characterize Eastern philosophy as only mystical.

Insufficient evidence means we cannot say whether many thinkers in the ancient West used insight or reasoning. There are no reliable accounts that pre-Socratic thinkers other than Zeno used reasoning to arrive at their ideas, while such evidence as does exist suggests that Anaximander's boundlessness probably resulted from insight, as did the mystical ideas of Pythagoras and the Pythagoreans. Heraclitus was a mystic, while the young Socrates may have employed insight. Philosophers like Marcus Aurelius in the second century CE and Plotinus in the following century undoubtedly were mystics, while thinkers like Archimedes had mathematical and scientific insights, a topic I shall explore in the next chapter.

The clearest exception to this mix of insight and reasoning is Aristotle, who prized reasoning above all. But Aristotelianism was only adopted in the West from the twelfth century. The subsequent enshrining of reasoning in Western philosophy may be due not only to centuries of Church control of education but also to the view that it is superior to insight. The distinguished British philosopher Anthony Quinton wrote in 1995

> Philosophy is a collaborative pursuit, unlike the meditative activity of sages, which is commonly conceived to flourish best in isolated or even hermetic conditions. The form of collaboration involved, however, is not collaborative but competitive...a business of critical argument. Argument is meant to persuade, and to succeed must overcome counterargument. Sages merely issue pronouncements to those who visit their retreats.[25]

This last sentence is far from the case. Moreover, it is questionable whether one method of philosophical thinking is superior to the other. Reasoning depends upon a prior set of assumptions or the selection and interpretation of evidence. The supreme rationalist Aristotle reasoned that celestial bodies are embedded in crystalline spheres rotating round the Earth, together with other things that are plainly wrong.

On the other hand the mystical insight that an ineffable ultimate reality—whether called Brahman or the Dao or Anaximander's boundlessness or Heraclitus's *logos*—gave rise to all things in the cosmos is close to the modern quantum field hypothesis that the universe we perceive materialized from a fluctuation in a pre-existing cosmic quantum field.*

Likewise the mystical insight of the Upanishads that underlying all matter and energy is a universal energy force (*prana*), like Heraclitus's insight of *pyr*, resembles the modern string conjecture that all matter and energy ultimately consist of strings of energy.†

So too the mystical insight of the early Daoists that the Dao is not only the source of everything but also the Way the natural world functions, like Heraclitus's *logos*, prefigures the physical laws of science that regulated the formation and evolution of the universe we perceive but whose cause science cannot explain.‡

Such insights do not involve an anthropomorphic god or God. I shall call them mystical to differentiate them from spiritual insights that claim revelation from God or a god or his or her messenger who usually exhorts the recipient to advocate a course of action for believers in the deity. In this sense spiritual insight forms part of primeval thinking as defined in the previous chapter. The border between the two blurs, for example when Aristotle attributes divinity to the Uncaused First Cause or when some Hindu schools interpret Brahman as the Supreme Godhead beyond all form. Other Hindu schools, however, cross that boundary by personifying Brahman as the god Vishnu or the god Shiva, each of whom, according to their devotees, has periodically incarnated to intervene in human affairs.

If we take the fundamental branching of philosophy to be that which occurred from the twelfth century onwards by the West's identification of reasoning as the only method of philosophy, how do we map the subsidiary ramifications in the evolution of philosophical thinking?

As philosophical thinking spread and interacted and multiplied, and new things were discovered generating yet more thinking, countless schools formed, but a clear trend to cope with this vast increase in ideas was specialization. The most useful way to map the next level of ramification is by the objects of inquiry, which fall into six main specialist fields. Their boundaries are indistinct, but they may be summarized broadly as follows.

---

* See page 46.
† See page 62.
‡ See page 92.

metaphysics   The branch of reasoning that investigates and attempts to understand ultimate reality or the essence and cause of all things whether material or immaterial.

natural philosophy   The branch of reasoning that investigates and attempts to understand the natural world perceived by our five senses and how it operates.

logic   The branch of reasoning that aims to systematically distinguish valid from invalid inferences by devising rules principally for deductive and inductive reasoning.

epistemology   The branch of reasoning that investigates the nature, sources, validity, limits, and methods of human knowledge.

ethics   The branch of reasoning that evaluates human behaviour and often attempts to produce codes governing good conduct between individuals, between an individual and a group of individuals (like a society or state), and between groups of individuals.

aesthetics   The branch of reasoning that attempts to understand and communicate the essence of beauty in nature and in human creations.

These fields evolved, and continue to evolve. Before Plato there was no distinction between metaphysics and natural philosophy, while natural philosophy gradually evolved into science, as we shall see in the next chapter.

Each of these branches ramified and continues to ramify. For instance, metaphysics divided into three schools of thought: monism, which holds that ultimate reality consists of one thing; dualism, which holds that material and mental things (or consciousness) are fundamentally different; and pluralism, which holds that reality consists of many kinds of things that cannot be reduced to one or two.

Insight tends to be holistic, but what ramified considerably was the interpretation by later-generation disciples of a seer's ineffable insight. Nonetheless, I think it useful to show a branching between the objects of insight while recognizing that, because of its holistic nature, its branches overlap even more than do the branches of reasoning. Thus Siddhartha Gautama's psychological insight of the Four Noble Truths merged into the ethical insight of the Eightfold Noble Path.

I suggest six basic divisions, defined as follows, but acknowledge that these are debatable.

mystical insight   Direct understanding of ultimate reality: the essence and cause of all things.

scientific insight   Direct understanding of the essence or causes of natural phenomena, their interactions or other relationships, and often the rules governing such interactions or relationships.

mathematical insight   Direct understanding of the properties of, or relationships between, numbers, real and abstract shapes, and often the rules governing any such relationships.

psychological insight   Direct understanding of why and how individuals, or groups of individuals, think and behave as they do.

ethical insight   Direct understanding of how, and often why, humans should behave either as individuals or as a group towards other individuals and other groups.

artistic insight   Direct understanding that results in the creation of beautiful or thought-provoking visual, musical, or written works.

Scientific insight frequently arises not from a process of disciplined meditation but after a scientist has failed to solve a problem through reasoning and he or she has entered an altered state of consciousness, often unintentionally, by relaxing, ceasing to reason, and switching off—thinking about very little—when the answer appears as if from nowhere.* The same is true for mathematical and other insights.

Overview of noetic evolution

Figure 28.2 is an attempt to give an overview of the evolution of philosophical thinking, showing its emergence from the superstition of primeval thinking, together with the evolution of the other main strands of primeval thinking discussed in the previous chapter, plus instinct.

Like Figure 27.1 from which it is derived, Figure 28.2 is a two-dimensional snapshot of a four-dimensional dynamic interactive process in which its branches, sub-branches, and sub-sub-branches not only change over time but also interact with other branches to hybridize or mutate or generate a new branch. For example, Siddhartha Gautama's psychological and ethical insights interacted with metaphysics to develop a strand within metaphysics that reasoned—and continues to reason—about such things as the essence of self and not-self, while logic interacted with creativity and invention to generate the new branch of computer languages.

Nor is the diagram drawn to scale. While reflective consciousness only fully emerged from consciousness in the Upper Palaeolithic, between 40,000 to 10,000

---

*   See page 509 for examples.

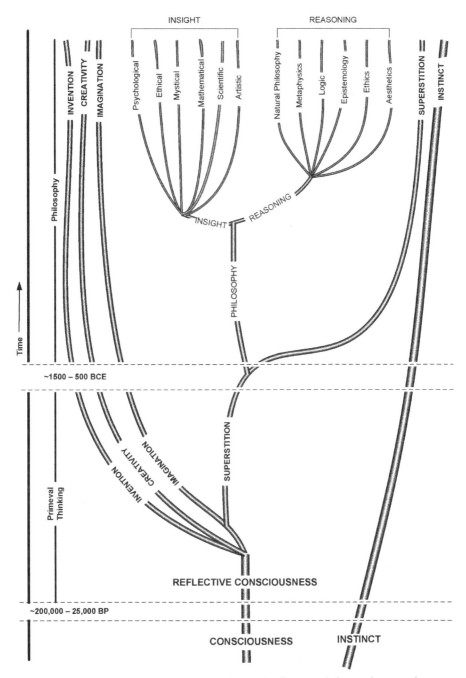

Figure 28.2 Emergence of philosophical thinking and the evolution of its main strands

Like Figure 27.1 from which it is derived, this is a simplified snapshot of a four-dimensional dynamic and interactive process. It is not drawn to scale and dates are best estimates.

years ago (abbreviated to roughly 25,000 years ago), philosophical thinking only emerged from the superstition of primeval thinking about 3,000 years ago, and insight and reasoning diverged some 800 years ago. Put another way, if human existence were represented by a 24-hour clock, philosophy would appear at around 2 hours 53 minutes before midnight, while reasoning as virtually the only method of philosophical thinking taught in the West appears at 45 minutes before midnight.

Superstition—in the form of myths and religious beliefs arising from ignorance of the laws of nature or fear of the unknown and that we now know to be in conflict with firm evidence—has been inculcated over more than a thousand generations. It is unsurprising to find its influence so strong and pervasive today when it remains a formidable competitor with reason and insight in attempts to answer the central questions of human existence.

Conclusions

1. Attempts to answer the pivotal questions of what we are and where we came from entered a new phase roughly 3,000 years ago at different places across the globe when philosophical thinking branched off from the superstitions of primeval thinking. It was characterized by a desire to seek explanations that did not invoke imagined spirits or anthropomorphic gods or God, which had been believed for more than 20,000 years.

2. Philosophical thinking most probably emerged first on the Indian sub-continent, while the other main centres were China and the Greek colony of Ionia.

3. Philosophers employed insight, often resulting from disciplined meditation, and reasoning, based on prior assumptions or interpretations of evidence.

4. Like all emergences considered so far the early groping attempts at philosophical thinking were hard to distinguish from their roots, the primeval superstitions inculcated by religions. As they grew and spread, their evolution showed a pattern similar to that of phyletic evolution in biology in responding to different local environments. Some buds of philosophical thinking failed to develop; some developed and then withered; some were assimilated and changed by other schools of thinking; some interacted with local beliefs and mutated into religions; some were assimilated or destroyed by religions or rulers; some lay fallow and were revived much later; some were promoted by rulers and flourished; surviving ones continued to evolve and ramify.

5. The insights of seers, whether Indian, Chinese, Greek, or Roman, tended to share common features, in particular the underlying unity of all things. In most cases this underlying unity or ultimate reality was experienced as something ineffable, best described as a transcendent cosmic consciousness or intelligence existing formlessly out of space and time but also immanent in that it gives rise to all the phenomena perceived by

our five physical senses and our mind; arising from this underlying unity the essence of each of us is identical to the whole. Moreover, this ultimate reality is manifest in, or regulates, how the cosmos functions, and we should harmonize our lives with it in order to achieve fulfilment.

6. While insight tended to be holistic, it may be divided into branches according to the object of inquiry. What ramified considerably were the schools founded to interpret, practise, and teach the insights of particular seers.

7. Reasoning was employed in teaching insights and also as a method of inquiry. It too may be divided into branches according to the object of inquiry, and these in turn ramified as schools of reasoning spread, interacted, and multiplied in response to new thinking.

8. Where thinking focused on ethics—how we should behave—virtually all ancient philosophers, whether using insight or reasoning, taught that we will only achieve happiness or tranquillity by acting unselfishly and doing to all others as we would have them do onto us. This ran counter to the prevailing inherited instinctive drive in their societies for aggression, warfare, and conquest. At root it is a prescription for cooperation not competition in order to achieve progress for humanity.

9. A fundamental branching between insight and reasoning occurred from late in the twelfth century when the West adopted reasoning as virtually the only method of philosophical thinking, although evidence does not support the superiority of one method over the other.

In the next chapter I shall focus on the emergence of the next phase of human evolution, scientific thinking and its subsequent development.

CHAPTER TWENTY-NINE

Human Evolution 3: Scientific Thinking

A scientist, my dear friends, is a man who foresees; it is because science provides the means to predict that it is useful, and the scientists are superior to all other men.

—HENRI DE SAINT-SIMON (1760–1825)

Science may be described as the art of systematic oversimplification.

—KARL POPPER, 1982

The third phase of human evolution is characterized by scientific thinking, when knowledge and understanding are obtained empirically rather than by philosophical speculation or belief based on claimed supernatural revelation.

As we saw in Chapter 1, the meaning of science has changed over the centuries and so it is helpful to restate the contemporary definition used in this book.

> **science**   The attempt to understand and explain natural phenomena by using systematic, preferably measurable, observation or experiment, and to apply reason to the knowledge thereby obtained in order to infer testable laws and make predictions or retrodictions.

Although it overlaps with technology, I think it useful to make a distinction between them.

> **technology**   The invention, making, and use of tools or machines to solve a problem.

Science is inseparable from its method. This too has changed over time. The current understanding is that it consists of five steps, summarized in the following definition.

scientific method (notional)

1. Data is collected by systematic observation of, or experiment on, the phenomenon being studied;
2. a provisional conclusion, or hypothesis, is inferred from this data;
3. predictions deduced from this hypothesis are tested by further observations or experiments;
4. if these tests confirm the predictions, and the confirmations are reproduced by independent testers, then the hypothesis is accepted as a scientific theory until such time as new data conflict with the theory;
5. if new data conflict with the theory, then the theory is either modified or discarded in favour of a new hypothesis that is consistent with all the data.

In practice these steps are not always followed. The antibiotic properties of penicillin were discovered not by Step 1 but by Alexander Fleming returning to his laboratory after a few weeks to find that bacteria inadvertently left on a culture plate had been destroyed by a mould he called penicillin. Wilhelm Röntgen discovered X-rays in a dark room after covering a cathode ray tube on which he had been experimenting and then noticing a fluorescent screen was illuminated. We saw in Chapter 3 how evidence for the cosmic microwave background was discovered by chance.*

Insight rather than Step 2 has usually been responsible for major scientific advances. It is popularly called the Eureka moment after the ancient Greek mathematician and inventor Archimedes, who was said to have leapt naked from a bath and run through the streets shouting *eureka* (I have found it) after suddenly realizing that that the upward buoyant force exerted on a body immersed in a fluid is equal to the weight of the fluid the body displaces. The German chemist August Kekulé reported that, after failing to work out the molecular structure of benzene, he dozed in front of a fire. In the flames he saw an image from ancient myths of a serpent eating its tail, prompting the sudden realization that benzene had a ring, not a linear, structure.[1] When describing how he arrived at his revolutionary ideas, Albert Einstein said

> A new idea comes suddenly and in a rather intuitive way. That means it is not reached by conscious logical conclusions.[2]

Too often the response of scientists when they find data conflicting with a cherished theory is not to take Step 5 but to question the data and the method by which it was collected, or ignore it,† or interpret it in such a way as to render it consistent with the theory.‡

---

* See page 21.
† See, for example, page 79, page 84, page 271, and page 388.
‡ See, for example, page 83, page 380, and page 409.

What is sometimes called the hypothetico-deductive method is better described as a logical analysis of how scientific theories derive support from evidence, but it tends to be the only scientific method now taught in universities.

The emergence of scientific thinking

The early sciences may be divided into three overlapping groups—physical sciences, life sciences, and medical sciences—the branches of which emerged at different times.

> **physical sciences** The branches of science that study inanimate phenomena; they include astronomy, physics, chemistry, and the Earth sciences.

> **life sciences** The branches of science that study features of living organisms (such as plants, animals, and humans) and also relationships between such features.

> **medical sciences** The branches of science that are applied to maintain health, prevent and treat disease, and treat injuries.

Medical sciences
This last group applies findings from the first two, particularly life sciences, for its purpose of human survival. Its roots go back to ancient times when illness and disease were thought of as things inflicted by spirits or gods. The earliest attempts to understand and treat the physical causes of sickness emerged from, and were entangled with, such superstitions.

Nonetheless the *Edwin Smith Papyrus* dated at around 1600 BCE, but thought to be a copy of several earlier Egyptian works, and the Babylonian *Diagnostic Handbook* dating from around 1050 BCE each detail the examination, diagnosis, treatment, and prognosis of numerous ailments in a rational manner.

The foundational text of Chinese medicine is the *Huangdi neijing*, traditionally attributed to the mythical Huangdi, or Yellow Emperor, and probably dating from the third or second century BCE. It discusses empirical approaches to diseases and treatments based on Daoist philosophy and aims to restore the dynamic balance of vital energies in the body, particularly the *qi* (the life force) and blood flow by harmonizing the *yin* and *yang* features of both behaviour and also bodily organs.[3] Its legacy is the traditional Chinese medicine practised today.

Ancient India's Ayurveda, which means the knowledge of long life, applied rational methods to medical treatments. Its two most influential texts belong to the schools of Charaka and of Sushruta, both born about 600 BCE. The former details eight branches of medicine while the latter incorporates a wide range of surgical techniques including cataract surgery and plastic surgery to reconstruct the nose (which was often cut off as a punishment). Ayurveda applied the theory of five elements from which everything in the universe is composed and stressed the need to balance three elemental energies or humours to maintain good health. This is similar to the approach in ancient Greece (see below). Whether they were arrived at independently or resulted from cultural transmission is an open question.

The ancient Greeks built *asclepieia*, temples dedicated to Asclepius, the god of healing, where patients came for treatment. From such superstitious beginnings came 70 or so medical texts attributed to the fifth century BCE physician Hippocrates of Kos, but most probably written by his students over several decades, which describe and categorize many illnesses, treatments—including surgical—and prognoses. These are based on a view current at the time that everything in the cosmos is comprised of four elements: fire, air, water, and earth.* Through the body run four humours, or fluids—black bile, yellow bile, phlegm, and blood—consisting of different mixtures of the four elements, and these humours need keeping in balance for a healthy body and mental disposition.

Aristotle's dissections of animals stimulated a blossoming of ideas about human anatomy and physiology during the Hellenistic period, culminating in the Greco-Roman period in the large body of works of the Greek Galen (c. 130–c. 200 CE). His synthesis and development of Aristotelian, Platonic, and Hippocratic anatomical and medical theories, including the four humours, was deemed authoritative.

By about 750 CE scholars in the Muslim empire had translated the works attributed to Charaka, Sushruta, Hippocrates, and Galen into Arabic. Their physicians, notably the polymath Ibn Sina (known in the West as Avicenna), used and developed these ideas. After Avicenna's magisterial encyclopaedia, the *Canon of Medicine,* completed in 1025, was translated into Latin in the twelfth century it became the most influential medical textbook not only in the Muslim world but also in Europe until the seventeenth century.

In the mid-sixteenth century, however, investigation of human corpses by the Belgian Andreas Vesalius of the University of Padua showed some 200 errors in Galen's human anatomy (Galen's ideas were based on dissecting animals and drawing conclusions about human anatomy). In 1628 the English physician William Harvey published *Anatomical Exercise on the Motion of the Heart and Blood in Animals*, which describes a series of experiments leading to his discovery of the circulation of blood and the role played by the pumping action of the heart that overturned many Galenic ideas. Such repeatable empirical work marked the emergence of modern medical science from its roots in ancient medicine.

---

* See page 489.

Life sciences

The life sciences can be traced back to the categorization of zoological specimens by Aristotle* and that of botanical specimens by Theophrastus, his successor as head of the Lyceum.

Perhaps the life sciences as now understood most clearly emerged in the seventeenth century with the experiments of Harvey plus the observations made with the newly invented microscope, like those by Antonie van Leeuwenhoek who discovered red blood cells, spermatozoa, and bacteria.

Physical sciences

The oldest of the physical sciences is astronomy.

> **astronomy**   The observational study of moons, planets, stars, galaxies, and other matter beyond the Earth's atmosphere together with their motions.

As we saw in Chapter 27, evidence from the alignment of European megaliths dating from around 5,200 years ago is followed by Babylonian, Indian, Chinese, and Maya star catalogues.† The fact that astronomy was used to serve superstition does not negate its methods of systematic observation of natural phenomena from which accurate measurements and predictions were made.

Islamic scholars used Arabic translations of Greek and Indian works on astronomy to develop the science. They established observatories, most notably at Maragheh in Persia in 1259. The evidence is unclear as to whether it was Islamic astronomers who first proposed that the Earth rotates. Nicolaus Copernicus, the Polish polymath and canon at the Catholic cathedral in Fauenberg, East Prussia, is usually credited with claiming that the Earth rotates on its axis and orbits the Sun, as do the other planets. Most probably he drew on the geometrical arguments of Islamic astronomers like Ala al-Din Ibn al-Shatir and others when he compiled his pivotal book, *On the Revolutions of the Celestial Spheres*, that challenged the Aristotelian and biblical Earth-centred view. He refused to publish it until 1543, the year of his death, for fear of the Catholic Church's response.[4]

Copernicus's theory was supported and refined by astronomers Tycho Brahe and Johannes Kepler, but it was Galileo Galilei in the early seventeenth century who provided not only mathematical but also compelling observational support after he had made one of the earliest telescopes and reported, among other things, four moons orbiting Jupiter and the phases of Venus.

Astronomy's interaction with natural philosophy played a major role in the emergence of physics.

---

*   See page 497.
†   See pages 473 to 475.

physics   The branch of science that investigates matter, energy, force, and motion, and how they relate to each other.

This emergence was barely recognized at the time, and the term "natural philosophy" was still used long after knowledge was acquired by empirical methods, which distinguished it from knowledge gained by the reasoning or insight of philosophy.

In fact physics as currently understood had begun to bud from natural philosophy as early as the sixth century CE. In an experiment usually credited to Galileo more than a thousand years later John Philoponus disproved Aristotle's theory of falling bodies by dropping two very different weights and showing they hit the ground at the same time.[5] This bud, however, was stillborn. Eight centuries later a group of scholars known as the Oxford Calculators introduced measurement and calculation into natural philosophy and produced the concept of velocity; they also distinguished between heat and temperature.[6] This bud, too, never flowered.

Galileo had read of Philoponus's experiment. He expressed in mathematical terms uniform acceleration for bodies in motion and laid much of the groundwork for the mechanics that Isaac Newton published in 1687 with his three laws of motion and his law of universal gravitation that consigned Aristotle's cosmology to the trash bin. Newton's experiments with prisms separating white light into different colours led to his theory that light consisted of tiny corpuscles, or particles, that obeyed his laws of motion.

In a period of about 150 years, from the mid sixteenth to the late seventeenth centuries, scientific thinking emerged principally in Europe not only to disprove many of the ideas of natural philosophy, astronomy, and medicine that had originated in ancient times but also to propose new theories with predictions supported by systematic observation and experiment. It was called the scientific revolution.

Its emergence was due to five synergistic factors. First, the translation into Latin of Arabic translations, and developments, of ancient Greek texts from the twelfth century onwards helped forge the Renaissance. This revival of Classical culture produced an initial acceptance of the authority of thinkers like Aristotle and Galen that gave way to a questioning of their ideas.

Second, the increasing demand for knowledge led to an increase in the number of universities in Renaissance Europe where ideas were disseminated and discussed.

Third, technology for survival was adapted to further knowledge for its own sake. Glass making had been refined in the Italian cities of Venice and Florence in the thirteenth century and used to produce single lenses for magnifying glasses and then spectacles in order to compensate for poor or failing eyesight. It was not until 1608, however, that inventing the telescope was credited to Dutch eyeglass makers. One of those, Hans Lippershey, had accidentally discovered that looking through two aligned lenses of different optical lengths produced magnification of distant objects. He called it a spyglass because he intended it to spy on distant armies in war.

News of the invention spread. The English mathematician and scientist Thomas Harriot made a 6-powered telescope to observe the Moon in August 1609. Galileo made a 20-powered one that he used to observe the Moon, discover four moons of Jupiter, and resolve nebular patches into stars; he published the results in March 1610.[7] With every increase in magnifying power, knowledge of heavenly bodies and their movements multiplied.

What the telescope did for astronomy, the microscope did for the life sciences.

Ways of measuring time, like a sundial, water clock, and hourglass, had been invented for astrological/astronomical, agricultural, religious, and other purposes. However, Dutch scientist Christiaan Huygens's invention in 1665 of the pendulum clock provided much greater accuracy, and it was used to collect data and test theories in astronomy and physics.

Fourth, the dissemination of knowledge increased rapidly due to another technological invention. Printing on paper had been invented in China in the ninth century to copy and distribute sayings of the Buddha, but it was Johannes Guttenberg's invention in 1439 of the moveable metal typeface that transformed the volume and speed of printing, and hence the dissemination of knowledge, new discoveries, and theories.

Fifth, cooperation developed among those stimulated by the dissemination of received knowledge and new discoveries. Competition was still a strong factor, exemplified by attempts to patent the telescope and rivalries between scientists like Newton and Leibniz over the invention of calculus. Increasingly, however, the new breed of scientists took a different approach, cooperating to develop ideas and disseminate information about discoveries. In 1652 a group of physicians in the German city of Schweinfurt founded the Academy of the Curious as to Nature, which was to evolve into the German Academy of Sciences; in 1670 it published the world's first medical and scientific journal. In 1660 a group including Robert Boyle and Christopher Wren decided to meet weekly to exchange information, discuss, and conduct experiments. By the following year it had received a royal charter. The Royal Society of London for Improving Natural Knowledge, which evolved into the national Royal Society, took as its motto *Nullius in verba* (accept nothing on authority) as an expression of its determination to question received authority and seek verification of claims by experiment or observation. It disseminated its findings in its journal, *Proceedings of the Royal Society*. A similar pattern occurred with the founding in 1666 of the French Academy of Sciences, in 1700 with the Prussian Academy of Sciences, and in 1725 with the Russian Academy of Sciences.

The evolution of scientific thinking

Once scientific thinking had emerged it evolved rapidly due to the factors just mentioned, and in particular the widening of access to education, the development of new technologies, now also specifically designed for scientific inquiry rather than just adapted from those made for other purposes, the development of ever more effective means of copying and disseminating information, and increasing cooperation.

These drove an increasingly rapid expansion of scientific knowledge that resulted in the ramification of scientific inquiry into progressively more specialized branches with their own learned societies and publications and their own university departments. Figure 29.1 is an attempt to give an overview of this process. It focuses on the ramification of the branch of natural philosophy shown in Figure 28.2. Like that figure it is a simplified two-dimensional snapshot of a four-dimensional dynamic and interactive process. For the sake of clarity not all branches and sub-branches are shown.

The emergence of scientific thinking did not mean that superstitious thinking ceased among its practitioners. To the contrary. Superstition, shown as a major branch in Figure 28.2 interacting with natural philosophy, continued to interact with science. The acclaimed sixteenth century mathematician, astronomer, geographer, and hydrographer John Dee attempted to communicate with angels using prayers, crystals, mirrors, mystic numbers, and other magical equipment. Kepler's observations and calculations showed that the orbits of planets round the Sun were elliptical rather than circular, but he continued to practise astrology, believing that the motions and positions of the Moon, Sun, and planets influenced the behaviour and fortunes of individuals. Newton, often described as the father of physics if not of science, devoted by far the greatest part of his life to alchemy, the esoteric art of transforming base metals into gold and of discovering the elixir of eternal life. According to philosophy historian Anthony Gottlieb, Newton wrote more than a million words of gibberish on the subject.[8] All this simply reinforces the pattern that new branches of thinking are inevitably entangled with their roots.

Before indicating how and why each of these early branches of scientific thinking ramified as shown in Figure 29.1, I shall mention two technologies used by them all, the progress of which demonstrates the rapid acceleration of this phase of human thinking.

The first is aids to calculating. Mechanical calculating machines were developed in the seventeenth century, and more sophisticated versions remained in use until the twentieth century. In the second half of the twentieth century, however, transistorized computers using digital calculations and storage rapidly developed through microprocessor controllers to computers with 128 bytes of Random Access Memory to the development from the 1970s of personal computers and then supercomputers like the Titan designed for climate studies. Unveiled in 2012 it performs 17.59 thousand trillion calculations per second, ten times more powerful than the world leader of three years before.

The second is the dissemination of knowledge, which stimulates new thinking. Guttenberg's fifteenth century hand-operated moveable metal typeface printing

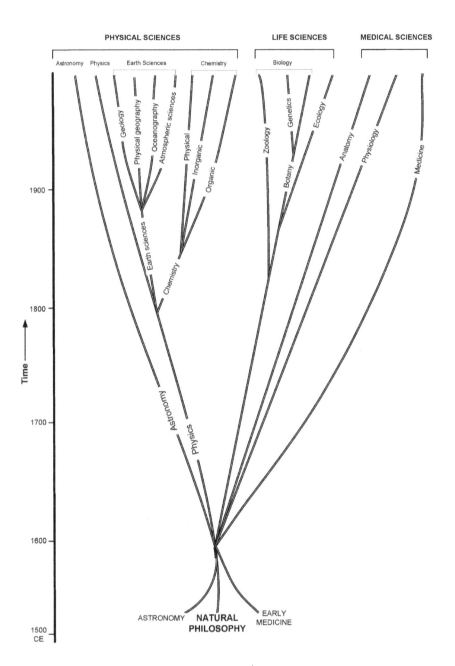

Figure 29.1 Evolution of natural philosophy into the early branches of science

A simplified two-dimensional snapshot of a four-dimensional dynamic and interactive process. For the sake of clarity not all branches and sub-branches are shown. Superstition's interaction with natural philosophy and with physics, for example, played a significant role (see text). When a new shoot emerges as a distinct sub-branch is difficult to determine; accordingly the timescale shown is the best estimate.

machine became much faster with the invention in 1810 of the steam-powered printing press. The speed and the volume of production increased still further, reducing the unit cost of printed material, with the invention of offset lithograph printing at the beginning of the twentieth century. However, digitization and the development from the 1970s of computer networks resulting in the Internet, followed by the World Wide Web (WWW), not only resulted in virtually instant dissemination of new knowledge but also globalized that dissemination.

The WWW was invented by Tim Berners-Lee of CERN (the European Organization for Nuclear Research) in 1991 for the purpose of sharing research information on nuclear physics. After it became popular in 1993 with the introduction of a browser with a graphical interface and CERN making it freely available, it transformed and democratized information access. Prior to that scientific papers were published in the printed journals of a scientific society, like the Royal Society, and circulated to its members and corporate subscribers, such as university libraries, mainly in its home country. Now digital versions of these journals and individual papers could be downloaded by anyone in the world for a price. Moreover, public access sources like arXiv (beginning with physics in 1991 but expanding to other sciences) and PLoS (Public Library of Science, from 2003) allow researchers to make their work available at no cost. Many scientists also provide copies for free download on their homepages, while search engines like Google Scholar make it easy to find what has been published by whom on what topic.

Physical sciences

Astronomy

The telescope was invented for spying on opposing armies. However, once modern astronomy had emerged following the rejection of the Earth-centred universe belief, astronomers developed technologies to serve their needs rather than use those designed for other purposes.

Newton realized that the refractive telescope, which consisted of two glass lenses, would give different refractions of light for different colours and so he developed a reflecting telescope that used mirrors rather than lenses in order to eliminate chromatic aberrations. Reflecting telescopes of increasingly sophisticated designs became the principal type of astronomical telescope. Individuals observing the sky with their own instruments gave way to increasingly larger teams of cooperating astronomers and technologists, leading to the European Extremely Large Telescope planned for operation in the 2020s and sited in Chile. Its main mirror will consist of 798 hexagonal segments stretching 40 metres in diameter and will be capable of gathering more light than all existing telescopes on the planet combined.[9]

The accidental discovery in 1931 of naturally occurring radio frequency emissions from the sky led to American radio engineer Grote Reber building in his backyard a 9-metre parabolic radio telescope that he used in 1940 to identify the Milky Way galaxy as a source. By 1957 Manchester University's astronomy department had constructed a 76-metre radio telescope at Jodrell Bank in northwest England to investigate cosmic rays, meteors, quasars, pulsars, and other sources of radio frequency waves and also to track space probes.

Investigating celestial phenomena by their emissions and absorptions at frequencies outside the visible spectrum resulted in ever more developed technologies and ever wider cooperation beyond members of a single university. The Orbiting Solar Observatory, a series of eight satellites launched between 1962 and 1971 by the USA's National Aeronautics and Space Administration (NASA), detected images at ultraviolet and X-ray wavelengths that would have been filtered out by the Earth's atmosphere. The Hubble Telescope, launched in 1990 by NASA and operated not only by its scientists but also by those from the European Space Agency, probed the visible, ultraviolet, and near infrared wavelengths. By the end of the twentieth century specialist teams of international astronomers were mapping sections of the sky across every part of the electromagnetic spectrum and using computer analyses and digital imaging and reconstructions.

Physics
Physicists are fond of the aphorism "If a biologist doesn't understand he asks a chemist; if a chemist doesn't understand he asks a physicist; and if a physicist doesn't understand he asks God." It captures the reality that physics is the fundamental science because the relationships between matter, energy, force, and motion lie at the heart of all other physical sciences, which branched from physics as the increase in knowledge led to increasingly specialized fields of study.

Although for clarity's sake Figure 29.1 shows physics as one branch, it ramified into sub-branches like mechanics—which itself divided into solid mechanics and fluid mechanics, which includes such strands as hydrostatics (forces on fluids at rest), hydrodynamics (forces on fluids in motion), aerodynamics, and pneumatics—acoustics, optics, thermodynamics, static and current electricity, and magnetism.

From such classical physics, which studied what is normally observable, physicists developed technologies to investigate matter and energy not normally observable. Ernest Rutherford's investigations of radioactive decay at the end of the nineteenth and beginning of the twentieth century led to his postulating the existence of the atomic nucleus. This in turn helped lead to the second scientific revolution begun by quantum theory and relativity theory early in the twentieth century.

This second revolution produced yet more ramification into the increasingly specialized sub-branches of atomic physics, nuclear physics, particle physics, and plasma physics.

From physicists working alone, like Einstein, and small laboratories like Cambridge University's Cavendish Laboratory where Rutherford had worked with J J Thomson, cooperation increased not only between scientists but also between countries. Established in 1954, the 20 member states of CERN cooperate to finance the construction of particle accelerators designed by its physicists and engineers to discover how fundamental particles interact. In 2008 it began operating the largest and most complex piece of technology the world has known, the Large Hadron Collider.

Each of the specialized sub-branches of physics has its own sub-department within university physics departments and its own specialist publications.

Chemistry
One sub-branch that became recognized as a science in its own right is chemistry.

> **chemistry**   The branch of science that investigates the properties, composition, and structure of substances and the changes they undergo when they combine or react under specified conditions.

It has its roots in alchemy, munitions manufacture, and medical remedies. When it emerged as a science in the current meaning of the word is debateable. Some, especially in the English-speaking world, attribute this to the English theologian and physicist Robert Boyle, best known for his laws on the properties of gases. It was, they claim, his publication of *The Skeptical Chymist* in 1661 that separated chemistry from the esoteric art of alchemy. This is a misreading of the book, which criticizes only those alchemists who divorced their work from any theoretical underpinning. Boyle continued practising alchemy during the rest of his life.

Chemistry more clearly escaped from its superstitious roots and branched from physics during the last quarter of the eighteenth century due to the work of a group of French scientists, most notably Antoine Lavoisier. Because of the secrecy of the alchemists and munitions makers, compounds and their interactions were given different names, and comparisons and independent testing were difficult if not impossible. Lavoisier introduced a new chemical nomenclature that enabled chemists to share and expand empirical knowledge, and wrote the first chemistry textbook, which included a research programme and quantitative methodology for future chemists.

Recognition as a separate branch of science came with the establishment of its own learned societies, like the Chemical Society of London in 1841, which became the Royal Chemical Society, followed by similar societies in Germany, the USA, and elsewhere, plus separate departments in universities.

Increasing knowledge and understanding drove this branch of science to ramify. Chemists recognized that the properties of compounds of biological origin were very different from those of mineral origin, and so some specialized in investigating what they called organic chemistry. In 1828 Friedrich Wöhler discovered that an organic molecule could be made from inorganic ones, and so the definition changed. Organic chemistry became simply the chemistry of carbon and its compounds, inorganic chemistry covered all other elements, while physical chemistry was the study of the physical properties of substances, such as their electrical and magnetic behaviour and their interaction with electromagnetic fields.

These branches ramified still further into specialist strands such as organo-metallic chemistry, polymer chemistry, and nanochemistry.

Earth sciences
It is even more difficult to pinpoint when the Earth sciences branched off as a distinct group.

> **Earth sciences**   The branches of science that study the origin, nature, and behaviour of the Earth and its parts, including their interactions.

From the tenth to the twelfth centuries scholars of the Muslim empire wrote on such things. In Part 2 of his compendium *The Book of Healing* published in 1027 the Persian physician and polymath Avicenna proposes on the basis of field studies in what is now Uzbekistan explanations for the formation of mountains, the origin of earthquakes, the formation of minerals and fossils, and other topics now labelled as geology and meteorology.[10] In the late fifteenth century Leonardo da Vinci correctly speculated on the nature of fossils and on the role that rivers play in the erosion of land and the stratification of sedimentary rocks.

Perhaps the Earth sciences most clearly emerged as a separate discipline in 1795 when James Hutton* published his theory of uniformitarianism, which says that processes we observe now that shape the Earth, such as erosion, volcanism, and so on, were also responsible for shaping the Earth in the past. Hence the Earth was far older than had been calculated from the bible, and its features were not due to catastrophic events like the worldwide flood from which Noah was spared.

The Earth sciences developed into four interacting branches: geology, the study of the Earth's rocky crust, or lithosphere, which itself branched into such specialities as mineralogy, petrology, palaeontology, and sedimentology; physical geography, the study of the surface features of the planet; oceanography; and atmospheric sciences.

Life sciences
Chapter 16, which describes the evolution of ideas about biological evolution, indicates how the life sciences evolved.

This development of what was then regarded as natural history pursued mainly by clerics and gentlemen of private means into today's life sciences was brought about not only by physics envy—the desire to put biology on a mathematical footing†—but also by advances in technology. Improvements in the magnifying

---

*   See page 249.
†   See page 273.

power of Leeuwenhoek's single-lens microscope were achieved by the invention and developments of the compound microscope, but its effective limit of 2,000-fold magnification due to the wavelength of visible light was dwarfed in the 1930s by the invention of the electron microscope, which now can produce a magnification of more than a million-fold.

X-ray imaging, which developed in the early twentieth century after Röntgen's 1895 discovery, shows things like skeletal structure beneath the skin and soft tissue, while X-ray diffraction images of DNA led to the identification of its structure in 1953 and began a new era of molecular biology.*

The rate of technological innovation increased still further in the second half of the twentieth century with such techniques as NMR (nuclear magnetic resonance) spectroscopy to study molecular structure, measure metabolic rate, and produce images of internal soft structure, like muscles and tendons, PET (positron emission tomography) to monitor metabolic or biochemical activity by tracking a radioactive tracer injected into the bloodstream, and fMRI (functional magnetic resonance imaging) to map neural activity in the brain.

Biology branched into zoology—which ramified into specialized fields like ethology (animal behaviour), entomology (insects), marine biology, ornithology, and primatology—and botany, from which branched genetics at the beginning of the twentieth century.

The wealth of data generated by the use of increasingly sophisticated technologies, coupled with the widespread dissemination of that data, drove the ramification of the life sciences branches in the second half of the twentieth century into increasingly specialized fields such as neuroscience, cell biology, molecular biology, genomics, and bioinformatics.

Increasing cooperation was epitomised by the Human Genome Project begun in 1990. Funded principally by the USA's National Institutes of Health and a UK charity, the Wellcome Foundation, thousands of geneticists working in more than 100 laboratories in the USA, the UK, Japan, France, Germany, and Spain plus 13 other countries cooperated over 13 years to produce a map of the human genome. It didn't mean, of course, that competition disappeared. Craig Venter's Celera Genomics corporation pursued the same objective with the intention of patenting its results in order to exploit them for profit.

Medical sciences

According to Sherwin B Nuland, Clinical Professor of Surgery at Yale, physicians continued treatments based on the ancient idea of bodily humours for nearly three hundred years after this notion had been disproved empirically.[11] However, medical sciences evolved, particularly from the second quarter of the twentieth century, by applying many of the same technologies used by the life sciences, and they too ramified into specialized branches.

---

*   See page 275.

One of the most significant developments followed from the technology of genetic engineering in which new or modified genes are introduced into an organism's cells. This technique has been used to produce drugs like insulin and human growth hormone. Its most dramatic planned application is human gene therapy to replace defective genes involved in some 3,000 disorders. Although the concept is relatively simple, progress since the heady optimism of the 1990s has been slow due to problems of implementation and the elimination of fatal side effects arising from the use of retroviral vectors to insert genes into the genome.

Psychology

The newer science of psychology, which emerged in the late nineteenth century, has its roots in philosophy.

> **psychology**   The branch of science that investigates the mental processes and behaviours of individuals and groups.

Its name derives from the ancient Greek *psyche*, meaning soul or breath, but it came to mean the study of the mind. In the seventeenth century the French philosopher and mathematician René Descartes claimed a distinction between body and mind: the body was comprised of matter, which was measurable and divisible; the mind was an entirely separate, incorporeal, indivisible, non-spatial thing whose function was to think. This was the philosophical belief of dualism.

> **dualism**   The speculation or belief that there are two fundamental constituents to the universe: matter and mind, or consciousness.

The science of psychology began to emerge from philosophy when physiologist Wilhelm Wundt established the first psychological laboratory in 1879 at the University of Leipzig for the experimental study of sensation, memory, and learning. His work attracted students from all over the world, and in 1881 he founded the first journal to disseminate the results of experimental psychology. In 1890 William James, a philosopher who taught physiology at Harvard University, published *Principles of Psychology*, which embraced the experimental approach but also stressed the subjective experience of the mind. It was not until the twentieth century that professional associations and university departments of philosophy and psychology became distinct.

Thereafter psychology expanded rapidly, diversifying into a tangle of branches according to the purpose, object, and method of study, together with consequent schools of thinking. Overlapping purposes are either to acquire knowledge and understanding for its own sake or to apply such knowledge. Ramifications of applied psychology include the clinical—to cure specific disorders like schizophrenia or depression—through the educational, like child psychology, to the motivational, like sports psychology. Objects of study include intelligence, memory, learning, emotion, personality, and group behaviour, while methods range from the use of drugs through hypnosis, psychoanalysis, questionnaires, and experiments, to neuroscience technologies like PET and fNMR brain scans. Schools range from behaviourist to Freudian and Jungian. In Chapter 24 I considered two behaviourist ideas, Pavlov's conditioned reflex and Skinner's operant conditioning, that were influential in the early and middle parts of the twentieth century respectively, and concluded that they failed to explain why humans think and behave as they do.*

By the early 1960s the locus of most of this protean tangle had shifted from university departments to hospitals, clinics, private consulting rooms, and companies. I shall focus on two of the multiplicity of sub-branches of psychology because they claim to answer the pivotal question of this investigation: what are we?

Neuropsychology
In *The Astonishing Hypothesis* published in 1994 Francis Crick, who had received the Nobel Prize for jointly discovering the structure of DNA, answered the question succinctly:

> The Astonishing Hypothesis is that "You," your joys and your sorrows, your memories and your ambitions, your sense of personal identity and free will, are in fact no more than the behaviour of a vast assembly of nerve cells and their associated molecules.[12]

Like most scientists, Crick rejected Descartian dualism in favour of monism,

> **monism**   The speculation or belief that all existing things are formed from, or are reducible to, the same ultimate reality or principle of being.

and its particular version known as physicalism.

> **physicalism**   The speculation or belief that only physical matter is real and that all other things, such as mind or consciousness or thoughts, will eventually be explained as physical things or their interactions; also called materialism, it incorporates a wider view of physicality than matter, e.g. non-material forces like gravity that arise from matter.

Crick's view is what Julian Huxley called "nothing buttery". It is clearly not the case that we are nothing but the behaviour of our neurons and associated molecules.

---

*   See page 393.

Neurons transmit electrical impulses in response to stimuli, but the neurons and their networks do not understand the informational content of those impulses.

Even if it is assumed that our minds are generated by, or emerge from, our neurons and their interactions, a mind is not the same thing as a brain. For example, your decision to shoot your neighbour causes the activation of neurons in your brain that send signals to activate muscles in your arm and fingers to pick up a gun, point it at your neighbour, and press the trigger. But it is not the neurons of your brain that make that decision.

Moreover, neuropsychology, or at least Crick's claims for it, has not been able to provide independently verified observations or experiments that explain what it is like to have subjective experiences of phenomena, such as a notion of self, a feeling of pride, listening to music, and seeing a colour (referred to as qualia). The problem is described by two leading neuroscientists, V S Ramachandran and Colin Blakemore.

> The riddle of qualia is best illustrated with a thought experiment. Imagine a neuroscientist in some future century, who has complete knowledge of the workings of the brain—including the mechanisms of colour vision—but who happens to be colour blind and cannot herself distinguish between red and green. She uses the latest scanning techniques to generate a total description of all the electrical and chemical events in the brain of a normal human as he looks at a red object. The functional account may seem complete, but how could it be so without an explanation of the nature of the unique experience of red, which the scientist herself has never had? There is a deep epistemological gulf between descriptions of physical events in the brain and the personal, subjective experiences that we presume to be associated with those events.[13]

Until neuropsychology can explain empirically such subjective experiences, Crick's answer to the question must remain in the realm of philosophy rather than science, and even in that realm his argument is flawed.

Evolutionary psychology

The other sub-branch of psychology that claims to answer the question is evolutionary psychology: what we are, in the sense of what we think, what we feel, and how we behave, is caused by the NeoDarwinian accumulation over thousands of generations of random genetic mutations responsible for psychological mechanisms naturally selected to provide an adaptive advantage in the competition for survival of our Stone Age ancestors. As Leda Comides and John Tooby, co-directors of the University of California, Santa Barbara's Center for Evolutionary Psychology put it: "Our modern skulls house a stone age mind."[14]

The assumptions of evolutionary psychologists, for example that the human mind consists of a large collection of computationally distinct modules each shaped by natural selection to solve a particular Stone Age problem, are highly questionable. Their methods are the gene-centred mathematical and games models of sociobiology. I won't repeat here the considerable defects of these models when applied to animals set out in Chapter 23,* which are even more defective when applied to self-reflective humans. The evidence refutes the claims of evolutionary psychologists. For example, if you were born and educated in Tibet you are much more likely to think and behave as a Buddhist than if you were born and educated in Saudi Arabia where you are much more likely to think and behave as a Muslim, whereas if you were born and raised on the Falls Road in Belfast you are much more likely to be a Catholic Christian who hates Protestant Christians born and raised on the Shankill Road on the other side of a 12-metre-high "peace wall".

Faced with such evidence some evolutionary psychologists moved away from the hardline gene machine explanation of human nature that says we are puppets of our genes by proposing the hypothesis of the environment of evolutionary adaptation. This says there is a basic human nature genetically programmed for Stone Age survival and this determines thinking, behaviour, and emotions such as pride and guilt. However, there is also a genetically determined developmental program that absorbs information from the social environment and adapts the maturing mind accordingly: in effect fine-tuning these basic faculties so that, for example, some felt less guilt than others. We have inherited these ranges of Stone Age thinking, behaviour, and emotions.[15] No evidence supports this speculative attempt to rationalize the defects of genetic determinism. Moreover, it still fails to show how human thinking such as devising quantum mechanics to explain the interactions of subatomic particles or composing a symphony resulted from a human nature genetically programmed for Stone Age survival.

A further shift in emphasis towards the environment having an influence on what we are takes the form of the gene-culture co-evolution, or dual inheritance, hypothesis that claims what we are and how we behave results from the interaction of two different Darwinian evolutionary mechanisms: genetic and cultural. This also is not without its problems. For example, various studies disagree on whether or not there is a genetic basis for homosexuality. Those who claim that there is subscribe to NeoDarwinism. Accordingly they need to explain how such a combination of genes is passed on and accumulates in the human gene pool over very many generations when homosexuals are much less likely to breed and pass on their genes. Attempts to do so[16] are, at best, totally unconvincing.

No reasonable person denies that genes played a role in human biological evolution or that they *may* play a role in the evolution of human thinking, emotion, and behaviour. However, as we saw from Part 2, despite the claims of NeoDarwinism, the complexities of their role, their regulation, and their

---

*   See pages 373 to 383.

interaction with the environment in biological evolution are poorly understood at present. This warrants less assertiveness by genetic determinists in the field of psychology.

For the sake of completeness I add that not only does the faculty of reflective consciousness enable humans to transcend their genetic inheritance, it also enables them to transcend their cultural inheritance: many born and raised on the Falls Road have reasoned that it is morally wrong to hate Protestants living on the other side of the "peace wall" and are campaigning for its removal.

Interactivity and hybrid sub-branches

The simplified two-dimensional snapshot of Figure 29.1 does not show the psychological sciences or the other social sciences such as archaeology, anthropology, and sociology that branched from the life sciences during the late nineteenth century.

It also fails to convey not only the profusion of specialist sub-branches but also their interactions to produce new, hybrid sub-branches like astrophysics, astrochemistry, biogeography, and biochemistry.

Convergent trend

Figure 29.1 shows the divergent ramifying pattern of scientific thinking since the late sixteenth century. Counter to this predominant trend, however, is one of convergence, led by the fundamental science of physics.

Figure 29.2 draws on the account of the discovery of four fundamental forces of nature in Chapter 8 to illustrate this trend. It began as far back as the late seventeenth century when Newton recognized that the force that causes objects to accelerate to the ground is the same force that causes planets to orbit the Sun and moons to orbit their planets. Electrical current was discovered in the middle of the eighteenth century and realized to be a manifestation of the power of static electricity, which had been known from ancient times, while from the first quarter of the nineteenth century physicists showed that electrical force and magnetic force are different aspects of an underlying electromagnetic force.

The realization that underlying apparently different physical phenomena was the same cause increased rapidly in the twentieth century. Physicists theorized that the electromagnetic force itself and the weak nuclear force—the interaction between elementary particles that is responsible for nuclear decay and is several orders of magnitude weaker than the electromagnetic force—were particular manifestations of the same force, which they called the electroweak force, or electroweak interaction. The theory predicted the existence of new particles, which experimentalists found.

Theoretical physicists subsequently produced several mathematical models for a Grand Unified Theory (GUT) that unites the electroweak force and the strong nuclear force, the strongest of all known force that acts only over very small distances within an atomic nucleus and binds together particles within the

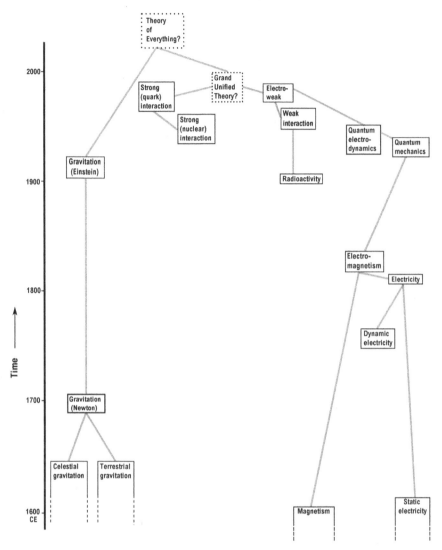

Figure 29.2 Convergence of the principal branches of physics towards a single theory that describes all the interactions between all forms of matter

nucleus, although no experimental data has been found so far to validate any of the models.*

The holy grail of physics, however, is to unite these theories, which seek to explain interactions of phenomena on the scale of the very tiny—the size of an atom or smaller—with relativity theory, which explains interactions on the

---

* See pages 41 and 109.

scale of the very large—the size of a star and above—in a Theory of Everything.* Candidates, like the many versions of the string conjecture and loop quantum gravity, have been advanced. While currently there is no known way to subject these hypotheses to empirical testing,† the thrust of theoretical physics, especially over the last twenty-five years, is to try and show that all physical phenomena in the universe are lower energy manifestations of one fundamental energy at the beginning of the universe.

Results of scientific thinking

A remarkable result of this phase of human evolution is the growth of empirical knowledge. Figure 29.1 fails to convey its sheer volume. According to science historian Derek de Solla Price,

> all crude measures, however arrived at, show to a first approximation that science increases exponentially, at a compound interest of about 7% per annum, thus doubling in size every 10–15 years, growing by a factor of 10 every half century, and by something like a factor of a million in the 300 years which separate us from the seventeenth-century invention of the scientific paper when the process began.[17]

His principal measure was the growth in the number of scientific publications up to 1960. Peder Olesen Larsen and Markus von Ins monitored that growth between 1907 and 2007, and concluded there are no indications that the rate has decreased in the last 50 years.[18] Moreover, as noted earlier in this chapter, since the end of the twentieth century to this must be added the growth in the number of scientific papers published in open access websites like arXiv, PLoS, and scientists' homepages.

Turning to individual scientific branches, one measure of the result of scientific thinking in the medical sciences is life expectancy at birth shown in Table 29.1.

Although these are crude estimates for all but the most recent times, what is striking is the great increase in life expectancy over the last 100 years or so in scientifically developed countries like the USA, demonstrating unequivocally that this cannot be attributed to natural selection. And this has not merely been a prolongation of life disabled by sickness but the extension of an active life.[19] The most reasonable explanation is the correlated evolution of medical sciences that has produced an understanding of diseases and developed measures to combat them. For example, clean drinking water, sewage disposal, garbage disposal, food safety standards, mosquito control programmes, and health education programmes have drastically reduced the spread of many infectious diseases.

---

*   See page 109.
†   See page 69.

Table 29.1 Life expectancy at birth since Neolithic period

Period	Life expectancy at birth (years)
Neolithic (beginning c. 10,500 years ago in the Middle East)	21
Ancient Greece and Rome (c. 2,800 – 1,500 years ago)	28
New England (Colonial America) c. 1800 CE	35
USA c. 1900 CE	47
USA c. 2000 CE	77
Industrialized countries early 21st century	78

Sources: Galor, Oded and Omer Moav (2007) "The Neolithic Revolution and Contemporary Variations in Life Expectancy" in *Working Papers*, Economics: Brown University; "Mortality" *Encyclopædia Britannica Online* (2013); "Life Expectancy" *Gale Encyclopedia of American History* (2006)

Vaccines have almost eliminated formerly common diseases like diphtheria, tetanus, poliomyelitis, smallpox, measles, mumps, and rubella. In addition to the technologies used for diagnosis mentioned earlier, this period has also seen the development of increasingly refined surgical interventions that include kidney, heart, and other organ transplants.

Part 1 of this book and the first chapter of Part 2 outline the major results of scientific thinking in the physical sciences that have produced our rapidly increasing knowledge and understanding of the origin and evolution of the universe and how the Earth became a planet fit for life, together with the current limits of that knowledge and understanding. The rest of Part 2 does likewise for the life sciences in terms of the emergence and evolution of life in general, while Part 3 is an attempt to show the results of our knowledge and understanding of our own emergence and evolution not only through the physical and life sciences but also through the social sciences.

This geometric growth of knowledge—a remarkable achievement for scientific thinking—has necessarily resulted in the increasing specialization of scientific thinking, demonstrated by the growth in the number of scientific publications, each of which focuses on a particular field of research. However, the distinguished entomologist and thinker Edward O Wilson sounds a note of warning.

> The vast majority of scientists have never been more than journeymen prospectors. That is even more the case today. They are professionally focused; their education does not orient them to the wider contours of

the world. They acquire the training they need to travel to the frontier and make discoveries of their own, and as fast as possible, because life at the edge is expensive and chancy. The most productive scientists, installed in million-dollar laboratories, have no time to think about the big picture and see little profit from it....It is therefore not surprising to find physicists who do not know what a gene is, and biologists who guess that string theory has something to do with violins. Grants and honours are given in science for discoveries, not for scholarship and wisdom.[20]

To be a successful scientist today means spending a career on colloid chemistry or palaeoarchaeology or studying chimpanzees, or some equally if not more specialized field. This narrowness of inquiry has produced a depth of knowledge but has also carved out canyons of expertise from which its practitioners find it difficult to engage in meaningful dialogue with other specialists except where canyons intersect in a cross-disciplinary study of the same narrow subject.

Few scientists transcend their specialist field to address fundamental questions of human existence such as what are we? The few who do rarely engage in open-minded debate. Too often they are unable to see the bigger picture and from their canyons they tend to fire a fusillade of views derived from the training, focus, and culture of the narrow academic discipline in which they have spent their professional lives.

Specialization also reinforces a belief in scientific reductionism—the taking apart of a thing into its constituent parts in order to understand what it is and how it functions—as the only method of scientific inquiry. This analytical tool has proved extremely successful in increasing our understanding of natural phenomena and may explain satisfactorily the essence and behaviour of relatively simple and isolatable systems, like why a crystal of rock salt dissolves in water. Its explanatory power becomes less effective the more complex and the less isolatable the phenomenon is. As noted in Chapter 13, it has considerable limitations when applied to explaining and defining life.* Its limitations are even more exposed when offering an explanation of what humans are, as we saw above in Crick's simplistic conclusion.

Moreover the question arises as to how far down reductionism goes. Crick may have thought that reducing to the level of the cell provides a complete explanation of consciousness, but why stop there? If we continue reducing to the level of molecules, and then atoms, and subatomic particles, we reach the quantum level. Several interpretations of quantum theory require conscious observation of the quantum field before it collapses into matter.† If these interpretations are correct, then consciousness cannot result from material phenomena like neurons and their interaction.

As far as reductionism is concerned, the quantum theorist David Bohm commented

---

*   See pages 198 to 206.
†   See page 87.

A centrally relevant change in descriptive order required in the quantum theory is thus the dropping of the notion of analysis of the world into relatively autonomous parts, separately existent but in interaction. Rather the primary emphasis is now on undivided wholeness.[21]

Like Bohm, most of the pioneers of quantum theory subscribed to a different version of monism than physicalism called idealism.

> **idealism**   The speculation or belief that material things do not exist independently but exist only as constructions of the mind or consciousness.

When applied to answering the fundamental questions of human existence such an idealist monism converges towards a common mystical insight. Indeed many quantum theorists came to hold views similar to those of mystics. Erwin Schrödinger, who created the wave equation for quantum mechanics, expressed in his last book, *My View of the World*[22] a metaphysical outlook that closely parallels the mysticism of the Vedantic school, Advaita; this sees material phenomena either as an illusion or else as a manifestation of the Cosmic Consciousness that is the source and ground of reality. According to Max Planck, who formulated the revolutionary quantum theory of radiation,

> All matter originates and exists only by virtue of a force....We must assume behind this force the existence of a conscious and intelligent Mind. This Mind is the matrix of all matter.[23]

In *The Tao of Physics*, first published in 1975, the high-energy physicist Fritjof Capra made a systematic comparison between the findings of modern physics and the insights of Eastern mysticism. He concluded

> The two basic themes [of Eastern mystical insights] are the unity and interrelation of all phenomena and the intrinsically dynamic nature of the universe. The further we penetrate into the submicroscopic world, the more we shall realize how the modern physicist, like the Eastern mystic, has come to see the world as a system of inseparable, interacting and ever-moving components with the observer being an integral part of the system.[24]

If the branches of physics have been converging over the last 25 years or so towards an explanation that all we perceive in the universe are lower energy manifestations of a single energy at its beginning, and this also converges with ancient mystical insights, this would indeed be a profound convergent tendency opposing the divergent tendency that has characterized more than 400 years' evolution of scientific thinking.

Uniqueness of Humans

We are just a third species of chimpanzee.

—Jared Diamond, 1991

This is a present from a small, distant world, a token of our sounds, our science, our images, our music, our thoughts and our feelings. We are attempting to survive our time so we may live into yours.

—Message on Voyager spacecraft by US President Jimmy Carter, 1977

Current orthodoxy

A large majority of primatologists, anthropologists, and evolutionary biologists deny that humans differ in kind from other animals. Their arguments may be grouped into three overlapping sets.

Ego-anthropocentricism

According to this argument, while humans may have some characteristics that are superior to those of other species, such as large brains and manual dexterity, other species have their own superior characteristics. Thus bats are superior echo-locators, birds are superior navigators, and so on for other species. To privilege one set of characteristics over another simply because we possess them is anthropocentric, egotistical, unobjective, and unscientific.

This view focuses on physiological differences and fails to recognize the faculties that are transformed by, or result from, the unique higher faculty of reflective consciousness. While humans may not be good echolocators, they have invented radar, which performs that function for them. While humans may not naturally be good navigators, they have invented the Global Positioning System that provides them with a precise navigation capability. It is difficult to think of any superior animal characteristic the function of which has not been enabled for humans by their creativity and invention. No other species has transcended its natural physiological characteristics in this way.

Genetic identity

This argument, articulated by Gerhard Roth of the University of Bremen's Brain Research Institute, claims that humans share about 99 per cent of their genes with the two chimpanzee species. These three species are more closely related genetically to each other than chimpanzees are to any other living primate. Hence there should be a separate new taxon comprising chimpanzees, bonobos, and humans.[1]

In fact humans and chimpanzees do not share 98.5 per cent of their genes as frequently asserted. This 1.5 per cent difference is a measure of the difference between equivalent genes; the actual difference in complements of genes is more like 6 per cent.* Moreover, a series of mutations in human DNA since the common ancestor of chimpanzees and humans produces, among other things, RNA molecules that are implicated in the development of the brain in human foetuses.[2]

More significantly, this narrow focus on the 2 per cent of the human genome that consists of protein-producing genes ignores the much more extensive regulatory sequences that determine when, to what degree, and how long genes are switched on, and consequently the organism's observable characteristics. It also ignores the qualitative behavioural and other differences considered in response to the next claim.

Behavioural difference in degree only

This view claims that human behavioural characteristics are not unique. They are all possessed by our genetically closest species, the chimpanzee. Human behaviour differs only in degree, not in kind.

Its advocates support this claim by examples of chimpanzees in the wild. This first occurred in the 1960s when a chimpanzee was observed stripping a twig of leaves in order to poke into a termite nest and fish out termites to eat. It prompted anthropologist and palaeontologist Louis Leakey to conclude "Now we must redefine tool, redefine man, or accept chimpanzees as human."[3] Other observations of tool use include some chimpanzees putting a nut on a stone and using another stone to break open the nut, described rather anthropomorphically as manufacturing and using a hammer and anvil. Moreover chimpanzees learn such skills. Chimpanzees also make different sounds to communicate different things, demonstrating a simple vocal language.

The dominant intraspecies behaviour of great apes is aggression. Male gorillas fight each other to own a harem, and the victor frequently kills the loser's infants as well as the loser. A chimpanzee raiding party will often enter the territory of another chimpanzee community where they launch a surprise attack to isolate an adult, immobilize him or sometimes her, and beat and tear the victim long after he or she has ceased moving. Within a chimpanzee community a change in the male hierarchy, headed by the alpha male, is usually brought about by a vicious

---

*   See page 318.

fight, while maintaining the hierarchy is often achieved by a display of aggression rather than an actual fight.

Primatologists, however, emphasize behaviours that contrast with such aggression, like mother-infant bonds, mutual grooming, and collaboration in hunting. Examples of humanlike behaviours include living in hierarchical societies where deception, cuckoldry, and submissiveness are common. Moreover, they point to humanlike emotions in great apes. The oft-quoted example of grief is that of a gorilla named Gana in Münster Zoo, northern Germany, who in 2008 refused to give up her dead three-month-old baby; she cradled and stroked it for more than a week.

Furthermore, cognitive tests carried out by primatologists on captive young great apes give rise to more examples of humanlike behaviour (older apes are normally too aggressive to test). Young chimpanzees often outperform human children aged less than three years. Some primatologists claim chimpanzees have a sense of self because they can recognize themselves in a mirror. Claims have also been made that some infant great apes had learned language by pressing buttons with symbols on a computer, and that a chimp and a gorilla had each acquired American Sign Language: they had learned hundreds of words, strung them together in meaningful sentences, and coined new phrases like water bird for a swan.

Steven Pinker, a cognitive psychologist who specializes in language, has debunked these language claims, which were never subject to peer review but made directly to popular science journalists in the press and on television.[4] Essentially, the claimants trained their chimpanzees and gorilla in much the same way as circus trainers do. They no more demonstrate an understanding of phonology, morphology, and syntax than those TV commercials broadcast in the UK of young chimpanzees dressed in scaled-down adult human clothes and drinking from teacups demonstrate their preference for PG Tips tea.

Despite intensive training no chimpanzee (unlike many parrots) has proved capable of mimicking speech. Vocalization of different emotions in the wild, like warning of danger, is qualitatively different from human conversation, and no untrained chimpanzee has written a word, still less a poem or a novel.

Human babies are born with soft and flexible skulls that expand rapidly to the ages of two to three years old to accommodate an increase in brain size and structure.[5] Consequently it is unsurprising that young chimpanzees, who are born with their brains fully formed, perform better at certain cognitive tests than two-year-old children, just as it unsurprising that chimpanzees fail to perform better than older children.

Unique human behaviours

In Chapter 24 I mapped the rise in consciousness along the human lineage by examining five overlapping types of behaviour: direct response, innate, learned, social, and innovative. The emergence of reflective consciousness began a radical transformation of many of these behaviours.

Humans still exhibit self-preserving direct responses. However, unlike other animals, humans can decide to suppress them. Using the example given in Chapter 24, a human can decide to keep his or her hand in a fire, or even burn himself or herself to death: self-immolation has a centuries-long tradition in some cultures and has been used in modern times as a form of political protest.

The overlap of direct response and innate behaviours that may be grouped as instinct still plays a powerful role. For example, the vast majority of humans will respond to an attack on them by fighting back or fleeing. However, if your reflections lead you to believe in the principle of non-violence and to choose to demonstrate this, you can suppress the instinct to fight or flee, as did the followers of Gandhi who allowed themselves to be killed by British colonial troops in pre-independent India. Less dramatically, most human males suppress the instinct to have sex with any female whose appearance stimulates them because they have rationally chosen to respect the rights of women and/or they have agreed to abide by a law determined by the society of which they are a part, while humans who have decided to diet suppress the instinct to eat when stimulated by hunger.

Human learned behaviour is radically different from that of any other primate because their learning process differs qualitatively and quantitatively in four significant ways:

1. nonhuman primates learn almost exclusively by copying, whereas humans learn principally by being taught;
2. with primates the learning relationship is normally between offspring and their parents or close kin, whereas human parents teach their offspring for about the first five years of life but the next ten to twenty years or so of teaching is undertaken by non-kin: different specialists in schools, colleges, universities, business enterprises, and so on who employ diverse means like lectures, books, audiovisual aids, and the internet;
3. primates learn only direct or indirect survival skills, like foraging, hunting, making rudimentary tools for individual use, or mutual grooming, whereas humans also learn a vast variety of things that range beyond survival, like literature, the arts, philosophy, and science;
4. humans are able to teach themselves, using human-created resources like libraries and the World Wide Web.

In terms of social behaviour, nonhuman primates belong to only one social group, mainly kin-related, whose purposes are survival and reproduction. In their so-called fission-fusion societies membership of the group may change, but at any one time an individual is a member of only one group. By contrast humans belong simultaneously to very many social groups whose purposes range beyond survival and reproduction and that operate at family, local, regional, national, supranational, and global levels.

Interactions between members of primate groups are sensory: touch, taste, smell, sight, and hearing. With humans, however, while touch, taste, and smell operate within families, these senses play a relatively small role in their other social interactions. Moreover, sight and hearing are augmented by human inventions such as letters, emails, photographs, films, webcam images, landline and mobile phones, and online conferencing that extend their range from local to global.

Such inventions result from the fifth type of behaviour, innovative.

The 533 cases of innovation, defined as apparently novel solutions to environmental or social problems, reported by primatologists for all primates* can be extended by researchers studying the crow family and marine mammals like dolphins and whales. However, such examples differ from those of human innovations both qualitatively and quantitatively.

The only purpose of nonhuman innovations is survival, such as obtaining sustenance, as when a chimpanzee strips leaves off a twig in order to fish out termites from their nest or uses a stone to crack open a nut.

Moreover, primatologists' claims about the similarity in genes and behaviour, like tool making, of humans and apes illustrates one of the consequences of scientific specialism mentioned in Chapter 29, namely a narrow focus that makes it difficult to see the bigger picture. In this case they fail to recognize that ape behaviours are also practised by many, genetically very different, species. New Caledonian crows trim and sculpt twigs to fashion hooks in order to poke out insect larvae from holes in trees. Scientists studying this species claim the crows' tools are more sophisticated than those of chimpanzees. Researchers examining Florida scrub jays, another corvid species, report that these jays stash perishable and non-perishable foods in caches over a wide area and can remember the locations and contents of each cache. Some jays watch and pilfer caches, but then take precautions to conceal their own caches from other jays.[6]

It appears that chimpanzees using a stone to crack open nuts dates from at least 4,300 years ago.[7] Quite likely they were stripping twigs to poke into termite mounds at that time, but clearly no such evidence would remain today. They have not advanced since then in stone use, not even to the primitive Oldowan stage of flaking a stone to provide a cutting edge practised by prehuman hominins. They have invented very few tools. Each is for the use of one individual and has a survival function.

By contrast, the sheer number, complexity, and size of tools invented by humans, most of which have purposes other than survival, are indicated in Chapter 29. The claim that a single chimpanzee using a stone to crack open a nut is the same kind of thing as a large international team of scientists cooperating to

---

*   See page 395.

invent and construct the Large Hadron Collider in order to discover how fundamental particles interact is, I suggest, somewhat less than valid.

Paradoxically, around the time claims were being made that chimpanzee and human tools showed a difference only in degree between the two species, humans invented two tools called Voyager 1 and Voyager 2, spacecraft designed, and operated remotely, to explore the outer planets before leaving our solar system in September 2013, more than 35 years later. Each carried a gold-plated audio-visual disc that includes a message from US President Jimmy Carter reproduced below the chapter heading.

An ability to recognize a reflection of oneself in a mirror is something quite different from a consciousness that reflects on itself and its place in the universe.

No nonhuman animal behaves in ways other than to acquire food and shelter, escape predators, find a mate, and rear progeny or to reinforce collaboration within its group to increase its own chances of survival and reproduction.

Humans, by contrast, behave in a vast range of ways that have nothing to do with their survival and reproduction. They ask questions about themselves as both physical and thinking entities, about their environment and the universe beyond, and about their own behaviour. No nonhuman animal possesses this capacity for self-reflective thought or the resulting ability to decide to act contrary to its genetically or culturally determined behaviour.

All the valid evidence put forward to support these three overlapping fallacies is consistent with the theory developed in Parts 2 and 3 of this book. Correlating with increasing biological complexity and centration, consciousness rises and intensifies along different evolutionary lineages until stasis is reached or, in the case of one species, a phase change occurs when consciousness becomes conscious of itself. To use an analogy, this intensification is like heat applied to a pan of water: it causes a temperature rise. In the pans representing the lineages of gibbons, gorillas, and chimpanzees, say, this heat, correlating with neural complexity and centration, is insufficient to raise the water temperature above, say, 85°, 90°, and 95° Celsius respectively, representing their level of cognition and behaviour. In the pan representing the human lineage, however, the heat intensity raises the temperature to boiling point. Bubbles of steam form within the liquid and burst through the surface to constitute a gas above the seething water; the heat maintains a temperature of 100° Celsius and the liquid continues to undergo a phase change into gas at an increasing rate as the volume of water decreases.

While humans have been shaped by their genetic inheritance and their cultural environment, this phase change to reflective consciousness has given them the unique capacity to transcend both.

CHAPTER THIRTY-ONE

Conclusions and Reflections on the Emergence and Evolution of Humans

Today the network of relationships linking the human race to itself and to the rest of the biosphere is so complex that all aspects affect all others to an extraordinary degree. Someone should be studying the whole system, however crudely that has to be done, because no gluing together of partial studies of a complex nonlinear system can give a good idea of the behaviour of the whole.

—MURRAY GELL-MANN, 1994

The exploration of space and the planetary character of economic, ecological, and cybernetic complexity are building the foundations of an inevitable global consciousness.

—RICHARD FALK, 1985

In modern scientific man, evolution was at last becoming conscious of itself.

—JULIAN HUXLEY, 1959

Conclusions

Drawing together the findings from Part 3 produces the following conclusions. The evidence and its analysis leading to each conclusion is given in the chapter indicated in brackets.

1. Human anatomical and genetic characteristics differ only in degree from those of other primates. What distinguishes *Homo sapiens* from all other

known species is reflective consciousness, that is, not only does an adult modern human know but also it knows that it knows. (Chapters 26 and 27)

2. Despite the claims of most primatologists, anthropologists, and evolutionary biologists, the emergence of humans with their unique capacity for reflective consciousness marked a change of kind, not merely degree, in the evolution of life, just as the emergence of life marked a change of kind from the evolution of inanimate matter. (Chapter 30)

3. Human emergence may be traced in broad terms to the rise of consciousness in predecessor primates to the point that consciousness became conscious of itself. Lack of evidence, however, means that it is not possible to trace the lineage from a specific prehuman ancestor. (Chapter 26)

4. The first glimmerings of reflective consciousness are shown by its consequences such as specialized composite tools, symbols, ornaments, paintings, sculptures, musical instruments, ceremonial burials and cremations, and sea-crossings. Such things have been found in different locations on the planet. While their dating is often uncertain, most are present during the Later Stone Age in Africa and the Upper Palaeolithic in different continents, spanning roughly 40,000 to 10,000 years ago, although incomplete or questionable evidence suggests that reflective consciousness may have emerged earlier. (Chapter 26)

5. Like the emergence of matter and the emergence of life, because of the paucity of evidence it is almost certain that science will never be able to identify when and how humans emerged. Most likely its root cause was an instinct to survive against predators and a need for sustenance in a fluctuating environment induced by a fluctuating climate which led to a recognition that these were best achieved through cooperation rather than competition. Hybridization or whole genome duplication may have played a role in this evolutionary change. (Chapter 26)

6. Overall it appears to be a case of systems emergence in which the interaction of faculties at a lower level of complexity—like comprehension, learning, and communication—generates a novel faculty at a higher level of complexity—in this case reflective consciousness—and that higher level faculty causally interacts with the lower level faculties to transform them and generate new ones—like imagination, belief, language, abstraction, and morality. (Chapter 26)

7. As with all major emergences, the boundary separating humans from prehumans is indistinct, but it is a boundary nonetheless. Beyond it a process of irreversible change began: the evolution of reflective consciousness. (Chapter 26)

8. This evolution may be divided into three overlapping phases: primeval thinking, philosophical thinking, and scientific thinking. (Chapter 27)

9. For the vast majority of the time since they fully emerged from hominins, humans pursued a hunter-gather existence in small, extended family

groups where competition for survival with other such groups and with predators resulted in a high death rate. (Chapter 27)

10. Primeval thinking evolved slowly until some 10,000 years ago when humans invented farming as a more effective way of providing sustenance and understood the benefits of cooperation by establishing larger agricultural village communities. This development occurred in different places at different times, and in some places it never occurred. (Chapter 27)

11. With more opportunities to reflect and to transmit ideas by drawing, speech and writing, humans in these agricultural village communities cooperated to invent technologies that improved and extended their farmlands, and also cooperated with other settlements to trade in both goods and ideas, enabling their settlements to grow in size and complexity. (Chapter 27)

12. Although cooperation had begun to evolve it struggled against the instinct for competition ingrained over millions of years of prehuman ancestry that produced battles for control of these settlements and their agricultural and other resources both from within and from other settlements, resulting in the rise and fall of dynasties and empires. (Chapter 27)

13. As these settlements grew they developed a social hierarchy that reflected classes of skills passed on from parents to offspring, typically ruler, priests, warriors, merchants and craftsmen, farmers, and slaves. The overall pattern across the globe was an increase in size, complexity, and centration of human societies. (Chapter 27)

14. The evolution of primeval thinking was intimately bound up with the evolution of beliefs that arose from imagination combined with a lack of understanding of natural phenomena or fear of the unknown, that is, superstition. From the animism, totemism, and ancestor veneration of hunter-gatherers, religions developed that reflected the growth in size, complexity, and specializations of human societies, from a fertility goddess through polytheism to a pantheon ruled by a powerful male sky and war god, and lastly to a patriarchal monotheism with other gods subsumed into one God or demoted to angels. (Chapter 27)

15. In applying reflective consciousness to devise technologies for survival and reproduction, and to influence the supernatural forces believed to determine these factors, primeval thinking produced the foundations of art, spoken and written language, mathematics, and astronomy. (Chapter 27)

16. Attempts to answer the pivotal questions of what we are and where we came from entered a new phase roughly 3,000 years ago at different places across the globe when philosophical thinking branched off from the superstitions of primeval thinking. It was characterized by a desire to seek explanations that did not invoke imagined spirits or anthropomorphic gods or God, which had been believed for more than 20,000 years. (Chapter 28)

17. Philosophical thinking most probably emerged first on the Indian

sub-continent, while the other main centres were China and the Greek colony of Ionia. Philosophers used insight, often resulting from disciplined meditation, and reasoning, based on prior assumptions or interpretations of evidence. (Chapter 28)

18. Like all emergences considered so far the early groping attempts at philosophical thinking were difficult to distinguish from their roots, the primeval superstitions inculcated by religions. As they grew and spread, their evolution showed a pattern similar to that of phyletic evolution in biology in responding to different local environments. Some buds of philosophical thinking failed to develop; some developed and then withered; some were assimilated and changed by other schools of thinking; some interacted with local beliefs and mutated into religions; some were assimilated or destroyed by religions or rulers; some lay fallow and were revived much later; some were promoted by rulers and flourished; surviving ones continued to evolve. (Chapter 28)

19. The insights of seers, whether Indian, Chinese, Greek, or Roman, tended to share common features, in particular the underlying unity of all things. In most cases this underlying unity or ultimate reality was experienced as something ineffable, best described as a transcendent cosmic consciousness or intelligence existing formlessly out of space and time but also immanent in that it gives rise to all the phenomena perceived by our five physical senses and our mind; arising from this underlying unity the essence of each of us is identical to the whole. Moreover, this ultimate reality is manifest in, or regulates, how the cosmos functions, and we should harmonize our lives with it in order to achieve fulfilment. (Chapter 28)

20. While insight tended to be holistic, it may be divided into branches according to the object of inquiry. What ramified considerably were the schools founded to interpret, practise, and teach the insights of particular seers. (Chapter 28)

21. Reasoning was employed in teaching insights and also as a method of inquiry. It too may be divided into branches according to the object of inquiry, and these in turn ramified as schools of reasoning spread, interacted, and multiplied in response to new thinking. (Chapter 28)

22. Where thinking focused on ethics—how we should behave—virtually all ancient philosophers, whether using insight or reasoning, taught that we will only achieve tranquillity and fulfilment by acting unselfishly and doing to all others as we would have them do onto us. This ran counter to the prevailing instinctive drive in their societies for aggression, warfare, and conquest. At root it is a prescription for cooperation not competition in order to achieve progress for humanity. (Chapter 28)

23. A fundamental branching between insight and reasoning occurred from late in the twelfth century when the West adopted reasoning as virtually

the only method of philosophical thinking, although evidence does not support the superiority of one method over the other. (Chapter 28)

24. The third phase of human evolution, scientific thinking, is characterized by the attempt to explain natural phenomena by using systematic, preferably measurable, observation or experiment, and to apply reasoning to the knowledge thereby obtained in order to infer testable laws and make predictions or retrodictions. (Chapter 29)

25. Such thinking was initially employed in three areas: physical sciences that study inanimate phenomena, life sciences that study living organisms, and medical sciences that are applied to maintain health and treat illnesses and injuries. (Chapter 29)

26. The medical sciences have their roots in the healing practices of ancient times, which were entangled with superstition and emerged as part of modern science in the seventeenth century, as did the life sciences. The oldest physical science, astronomy, dates from prehistoric times where it developed to serve superstitious beliefs before it emerged as a modern science in the sixteenth century. Physics, the fundamental physical science, emerged from natural philosophy in the sixteenth and seventeenth centuries. During this period, commonly referred to as the scientific revolution, many of the ancient ideas of medicine, astronomy, and natural philosophy were disproved and new, empirically based theories proposed. Most of its practitioners, nonetheless, retained superstitious beliefs. (Chapter 29)

27. By the nineteenth century the study of humans and their social relationships led to the social sciences branching from the life sciences. (Chapter 29)

28. Once it had emerged, scientific thinking evolved with increasing speed mainly due to five synergistic factors (Chapter 29):

    28.1. aids to calculating that developed from mechanical calculating machines to personal computers and supercomputers;

    28.2. the dissemination of knowledge, which stimulates new thinking, that developed from hand-operated moveable metal typeface printing machines to the Internet, followed by the World Wide Web that not only provides virtually instantaneous communication of new knowledge but also globalization of that knowledge;

    28.3. the development of new technologies designed for specific scientific investigations in addition to using technologies designed for other purposes;

    28.4. cooperation among scientists through scientific societies dedicated to sharing knowledge and thinking and also through teamwork in specific scientific investigations that developed from local through national to international;

    28.5. the widening access to education that resulted in the training of more scientists.

29. These factors resulted in (Chapter 29):

    29.1. the invention of technologies to investigate matter and energy not normally observable that led to the second scientific revolution in physics with the development of quantum theory and relativity theory;

    29.2. a geometric growth of empirical knowledge that produced the ramification of scientific thinking into progressively more specialized branches and sub-branches that investigate ever narrowing fields of inquiry;

    29.3. a depth of knowledge in these narrowing fields that resulted in canyons of expertise from which its practitioners find it difficult to engage in meaningful dialogue with other specialists except where canyons intersect in a cross-disciplinary study of the same narrow subject;

    29.4. few scientists equipped to transcend their specialism and address fundamental questions of human existence such as what are we?

    29.5. a belief in scientific reductionism as the only method of scientific inquiry despite the limitations of this powerful analytical tool when applied to complex, interacting, and emergent phenomena like life or humanity;

    29.6. a corresponding belief by many scientists in physicalism, which holds that only physical matter is real and that all other things such as mind or consciousness will eventually be explained as physical things or their interactions.

30. Scientists in two sub-branches of psychology, which investigates the mental processes and the behaviour of individuals and groups, claim to answer the question of what we are. (Chapter 29)

    30.1. Neuropsychologists who believe in physicalism claim that we are no more than the behaviour of a vast assembly of nerve cells and associated molecules. Even if it is assumed that our minds are generated by, or emerge from, our nerve cells and their interactions, a mind is not the same thing as a brain. Moreover, until neuropsychologists provide independently verified observations or experiments that explain what it is like to have subjective experiences of phenomena, such as a notion of self, a feeling of pride, listening to music, and seeing a colour, then such a claim remains in the realm of philosophical speculation rather than science.

    30.2. Evolutionary psychologists claim that what we are in the sense of what we think, feel, and behave is explained by the NeoDarwinian accumulation over thousands of generations of random genetic mutations responsible for psychological mechanisms that are naturally selected to provide an adaptive advantage in the competition for survival of our Stone Age ancestors. Their evidence consists of the simplistic gene-centred mathematical and games models of sociobiology that are divorced from reality, are designed to produce any desired outcome, have no predictive value, and are refuted by evidence of cultural beliefs and behaviours incompatible with a NeoDarwinian gene machine explanation of human nature.

30.3. In order to accommodate such evidence, some evolutionary psychologists propose the hypothesis of the environment of evolutionary adaptation. No evidence supports this speculative attempt to rationalize the defects of genetic determinism.

30.4. Others go further by proposing a gene-culture co-evolution hypothesis. However, despite the claims of NeoDarwinism, the complexities of the functioning of genes, their regulation, and their interaction with the environment in biological evolution are poorly understood at present, and are even less understood in explaining thinking, feeling, and behaving.

31. Counter to the predominant divergent, ramifying pattern of scientific thinking since the late sixteenth century, a convergent trend began in the fundamental science of physics that accelerated from the early twentieth century with the second scientific revolution. This stemmed from the insight that underlying apparently different physical phenomena is the same cause. The main thrust of theoretical physics over the last 25 years or so has been to try to show that all physical phenomena in the universe are lower energy manifestations of one fundamental energy at the beginning of the universe. (Chapter 29)

32. Many pioneers of the second scientific revolution rejected reductionism as the method of explaining physical phenomena in the universe and argued that the focus should be on an undivided wholeness. (Chapter 29)

33. Several interpretations of quantum theory require conscious observation of the quantum field before it collapses into matter, and these conflict with physicalism. Some quantum theorists subscribed to idealism, the conjecture that material things do not exist independently but exist only as constructions of the mind or consciousness. Several expressed a metaphysical view similar to the ancient mystical insight of the underlying unity of all things as a transcendent Cosmic Consciousness that gives rise to, and regulates, all phenomena perceived by our five physical senses and our mind. (Chapter 29)

34. If the branches of physics have been converging over the last 25 years or so towards an explanation that all we perceive in the universe are lower energy manifestations of a single energy at its beginning, and this also converges with ancient mystical insights, this would indeed be a profound convergent tendency opposing the divergent tendency that has characterized more than 400 years' evolution of scientific thinking. (Chapter 29)

35. While humans have been shaped by their genetic inheritance and their cultural environment, their possession of self-reflective consciousness has given them a unique capacity to transcend both. (Chapter 30)

Reflections

Reviewing the phenomenon of human evolution since reflective consciousness marked its full emergence some 40,000 to 10,000 years ago, several broad patterns, some longstanding and some more recent, are evident.

Reduction in aggression

In a century that began with Al-Qaeda's attack on New York's Twin Towers causing 2,750 civilian deaths, the US-led invasion of Iraq and military intervention in Afghanistan, plus civil wars in the Congo, Liberia, Sudan, Darfur, Libya, and Syria, following a century marked by two worlds wars and the use of nuclear weapons, it may seem counterintuitive if not perverse to say that human evolution has been characterized by a reduction in aggression.

Any impression of an increase in aggression is due to three main factors. First, perceptions of what is aggression have changed. In scientifically advanced societies most people no longer regard it as acceptable, still less honourable, to wage war against infidels in the belief that this is what God wants, or to punish a wrong according to the biblical or qur'anic eye for an eye, to settle disputes by duelling, to kill a female member of one's family for an alleged breach of a moral code in order to restore that family's honour, to torture to death, or to execute for a crime of theft.

Between the eleventh and thirteenth centuries, however, Western European kings and nobles, at the behest of popes promising immediate entry to heaven for those killed, conducted holy wars against Muslims while slaughtering Jews and others in the process. For those who are appalled by current public floggings, chopping off of hands, and beheadings in less developed countries for a raft of crimes from theft to adultery, it is salutary to remember that such practices were common in our own societies until relatively recently. The Tower of London contains a museum of instruments of torture, which practice was not abolished in England until the mid-seventeenth century. On 13 October 1660 Samuel Pepys records matter-of-factly in his diary that he saw Major-general Harrison publicly hanged, drawn, and quartered to the cheering of crowds.[1] At the beginning of the nineteenth century in Britain more than 200 offences, ranging from poaching through theft to murder, were punishable by death. Public executions in the UK only ceased in 1868, while the last execution took place in 1964. In France judicial public beheadings continued until 1939, while the death penalty was abolished only in 1981. In 2014 taking a life for a life was a legal punishment by the federal government and by 32 states of the USA.[2]

The second reason is a focus on the numbers of victims of aggression while ignoring the geometric rise in populations. It is more rational to examine aggression relative to the size of population under consideration.

Third, until the invention of global communications most people simply didn't know what was occurring in other countries. Now we are inundated by 24/7 satellite TV broadcasts, together with videos uploaded to YouTube, that show graphic images of warfare, rape, and other violence from around the world.

Yet the evidence for a reduction in aggression is clear. In Chapter 27 I summarized studies showing the high death rates in the hunter-gatherer era of primeval thinking due to chronic inter-group battles and intra-group club fights, infanticide, and geronticide. In his 2011 book *The Better Angels of Our Nature* Harvard psychologist Steven Pinker assesses the available evidence from forensic, archaeological, ethnographic, historical, and statistical investigations. A majority of human existence was spent as hunter-gatherer and hunter-horticulturalist societies. Studies of 27 such groups show a variation in annual death rates from warfare. The highest rates are from earlier times while the lowest are from tribes living in isolated inhospitable environments, like parched deserts or frozen wastelands, where they are not in competition with other tribes, or where they have been pacified by industrially developed nations or empires. The average annual death rate from warfare in these studies is 524 per 100,000.

By comparison, adding up all deaths from organized violence—wars, genocides, purges, and man-made famines—for the world for the twentieth century produces an annual rate of around 60 per 100,000. For the year 2005, with the USA's armed forces embroiled in conflicts in Iraq and Afghanistan, military deaths per 100,000 for the USA and also the entire world are each too small to register on Pinker's chart.[3]

Reasons for reduction in aggression

The underlying cause of this reduction in aggression, I suggest, is the evolution of human thinking as traced in Part 3 of this book. In the phase of philosophical thinking, those from different cultures who reflected on human behaviour, whether by insight or reasoning, advocated the golden rule of do unto others as you would have them do unto you. There was, as we saw, great resistance to such a view, which ran counter to the ingrained competitive instinct. However, it began a long process of changing the understanding of how to conduct relationships between humans that became codified in laws within societies and in agreements between societies.

The First World War was the last major conflict in which fighting was regarded as noble by most people in the scientifically and technologically developed countries of the world. The dissemination of information resulting from scientific and technological developments helped change popular consciousness. By the middle of the twentieth century not only were there mass circulation newspapers but also radio broadcasts and newsreels shown in cinemas where people saw the horrors of warfare, including the destruction of two Japanese cities by atomic bombs. This had one consequence unimagined by the governments and their military arms who had commissioned the bombs: the growing realization that warfare between competing nations and empires is self-defeating.

The initiative was taken by leading thinkers of that era. Philosopher Bertrand Russell together with Albert Einstein contacted distinguished scientists, including Joseph Rotblat who had been recruited by the American government to help design their atomic bomb. In 1955 they published in London what became known as the Russell-Einstein Manifesto. Speaking "not as members of this or that nation, continent, or creed, but as human beings, members of the species Man...", they posed the stark question: "Shall we put an end to the human race; or shall mankind renounce war?"[4]

This led to the series of Pugwash conferences that bring together influential scholars and public figures concerned with reducing the danger of armed conflict and seeking cooperative solutions for global problems. In 1995 Pugwash and its co-founder, Joseph Rotblat, were jointly awarded the Nobel Peace Prize.

The manifesto also prompted peace movements around the world, but their calls for disarmament were resisted by governments and the self-interest of their military-industrial complexes, as US President and former general Dwight Eisenhower warned in his 1961 valedictory address. After noting that the USA spent more on military security than the net income of all US corporations, he said pointedly

> we must guard against the acquisition of unwarranted influence, whether sought or unsought, by the military-industrial complex. The potential for the disastrous rise of misplaced power exists and will persist.[5]

The second half of the twentieth century onwards saw faltering steps away from competing empires and towards the cooperation of human societies on a supra-national and global basis. The victors of the Second World War recruited 51 countries in 1945 to form the United Nations with the objectives of maintaining international peace and security and promoting international cooperation to tackle global problems affecting all humanity. The attainment of this aim, however, was hindered by the Cold War between the USA and the Soviet Union. Nonetheless the UN did make progress on a global scale on economic, cultural, and humanitarian purposes through agencies it established like the World Health Organization, the United Nations Educational, Scientific and Cultural Organization, and the United Nations Children's Fund (UNICEF). Paradoxically the Soviet empire was defeated not by force of arms but by their spiralling costs, which crippled the Soviet economy, combined with demands by the peoples of its subjugated nations for freedom from Soviet oppression.

Europe, the continent where the scientific revolution began, took the lead in pursuing cooperation rather than competition between nation states. In 1951 France, Germany, Italy, Belgium, Holland, and Luxembourg founded the

European Coal and Steel Community (ECSC). Their aim was to integrate their coal and steel industries so that never again could they be used by member states to wage war on each other. The ECSC evolved into the current European Union (EU) of 28 nations, 17 of whom share a common currency. The loss of parts of national sovereignty in exchange for peace and social and economic benefits, coupled with the perceived bureaucratic imposition of budgetary measures in 2013 and 2014, produced an increased vote for ultranationalist factions in the 2014 European Parliamentary elections. A refugee crisis resulting from internal wars in the Middle East and Afghanistan fuelled the rise of nationalism across Europe and led to the 2016 UK referendum decision to leave the EU. But noetic evolution, like biological evolution, is not a smooth process. Fundamentally, however, it is unthinkable that any of its member states would declare war on another, a fact recognized by the award in 2012 of the Nobel Peace Prize to the EU that "for over six decades contributed to the advancement of peace and reconciliation, democracy and human rights in Europe".[6]

The overall reduction in aggression between societies has been paralleled by a reduction in aggression both between individuals and also between a society and individuals, be they members of that society or another society. Broad acceptance of this trend was marked by the UN General Assembly's adoption in 1948 of the Universal Declaration of Human Rights by a vote of 48 for, 0 against, and 8 abstentions by the Soviet bloc, South Africa, and Saudi Arabia. The first article declares

> All human beings are born free and equal in dignity and rights. They are endowed with reason and conscience and should act towards one another in a spirit of brotherhood

while others denounce slavery, torture, and cruel, inhuman or degrading treatment or punishment. It is a mark of the development of human thinking that most people in scientifically advanced societies are appalled when such acts do take place, especially by, or by agencies of, national governments, like torture by the US military in the Abu Ghraib prison in 2003 and 2004 in Iraq, or the rape of political protestors by interrogators and Iranian Revolutionary guards in Iranian prisons following the 2009 elections.[7]

Increase in cooperation

The evolution of thinking that caused a reduction in aggression also caused a rise in cooperation.

Peter Kropotkin, the late nineteenth century naturalist who argued that mutual aid was a greater factor than competition in biological evolution,* extended this argument to human evolution. He maintained that the historical record gives a distorted impression:

> The epicpoems, the inscriptions on monuments, the treaties of peace—nearly all historical documents bear the same character; they deal with

---

* See p. 267.

breaches of peace, not with peace itself. So that the best-intentioned historian unconsciously draws a distorted picture of the times he endeavours to depict.[8]

This is echoed today by the distinguished historian David Cannadine in his 2013 book *The Undivided Past: History Beyond Our Differences*,[9] where he argues that historians have stressed the conflicts within and between various divisions of humanity, like religion, race, and nation, but have ignored countless fruitful interactions and cooperation across these lines.

Evidence, however, does exist from the earliest humans not just of collectivisation,* as with insect societies, but of cooperation defined in Chapter 19 as

> **cooperation** Working together voluntarily for mutual benefit or to achieve commonly agreed aims.

Unlike instinctive or coercive collectivisation, cooperation requires reflective thought, and this is what makes it unique to the human species. Early signs of cooperation are given in Chapter 26, which examines the emergence of humans.† Chapter 27, on the evolution of primeval thinking, shows how it struggled to grow against the dominant competitive instinct that resulted in enforced collectivization and the development of hierarchical empires.‡ As philosophical thinking evolved, the golden rule of do unto others, including your enemies, as you would do unto yourself is a prescription rooted in cooperation that countered the prevailing aggressive competition with other states for resources and territory.

The collapse of the Western Roman Empire in the fifth century ushered in what Western historians called the medieval period. Yet the Goths, Vandals, Angles, Saxons, Lombards, and other tribes disparagingly labelled barbarian that invaded from the east brought with them from their rural societies the tradition of the village *folkmote*, or meeting of all adult men, who made decisions on the farming of land, which was usually owned in common. Most of these *folkmotes* also elected representatives to arbitrate on disputes, decide who was responsible for an injury or injustice, and levy a fine as compensation instead of the injured party taking revenge. Part of this compensation was retained by the village for communal works.[10] Such traditions were the forerunners of modern parliamentary and judicial systems.

As these barbarian villages grew, Europe developed into a fluctuating patchwork of different socio-political systems. The Church became a secular power; the former invaders became Christianized, while citizens of the former Western Roman Empire assimilated barbarian traditions; new waves of invasions occurred;

---

* See Glossary.
† See, for example, page 445.
‡ See, for example, Conclusions 4 and 5, page 479.

bands of warriors, each led by a warlord, offered to the peasants who worked the land protection in return for tributes, resulting in the feudal system. Cooperation was nonetheless evident in most parts of Europe as prosperous rural villages developed into towns and cities, reasserting their cooperative roots as members of different professions formed brotherhoods, or guilds, based on democratic self-government and mutual support. Many of these were permanent, like guilds of merchants, craftsmen, agricultural workers, and teachers, while temporary guilds formed for specific purposes, like an overseas trading expedition or the building of a church or cathedral.

Medieval cities throughout Europe organized themselves as a double feder-ation: of householders grouped in small territorial unions—basically small village communities each with its own parish—and of men united by oath in guilds according to their profession. Residents of some cities cooperated further by forming communes in which citizens swore an oath binding themselves in a bond of mutual support for trade and defence, while communes also developed in some rural areas. These democratic self-governing communes grew in strength, gaining substantial independence from feudal overlords. This independence was reinforced when they cooperated with other communes in federations such as the Lombard League in central and northern Italy.

In the space of about 350 years from the beginning of the eleventh century the face of Europe was transformed by such cooperative activity. Unremarkable villages and small towns grew into rich, walled medieval cities in which flour-ished commerce, the arts, crafts and learning, together with the construction of impressive buildings like the Gothic cathedrals of Cologne and Chartres.

The demise of most of the medieval communes during the fourteenth to sixteenth centuries was due to a variety of factors in different places. Chief among them was a departure from cooperative principles in order to preserve wealth and power. This alienated newcomers and the peasants who worked the land, and paradoxically left the communes vulnerable to usurpation by ruthless tyrants or subjugation by the power of centralizing states. Typically such a state was ruled by a hereditary monarch, a descendant of the most powerful warlord.

While cooperation as a form of government suffered a setback, its traditions influenced the various accommodations made between the competing powers of monarch, nobles, Church, and merchants, particularly in the parliaments that developed to approve, and then make, laws and to levy taxes.

Cooperation re-emerged in mid-nineteenth century industrialized Britain as a response to capitalist exploitation of the workforce. Enlightened thinkers like the successful industrialist Robert Owen founded cooperative communities in which members could work together to meet their economic and social needs, including education for their children. None of these lasted, due largely to their economic structure. However, they paved the way for 28 men to found in 1844 the Rochdale Equitable Pioneers Society on a set of socio-economic principles that proved successful. Cooperative societies organized on these principles grew in number and spread around the world, building on local cooperative traditions.[11] Only 51 years after the declaration of these cooperative principles in the then cotton-manufacturing town of Rochdale, national cooperative federations cooperated among themselves to form a global alliance in 1895.[12] In 2012 the International Cooperative Alliance estimated that more than a billion people—one in seven of the world's population—were members of a consumer cooperative, a worker cooperative, a housing cooperative, an agricultural cooperative, or other cooperative enterprise.[13]

Political and economic organizations rooted in competition were slower to follow. From the middle of the twentieth century onwards, however, the hierarchical organization of empire ruled by an autocrat gave way to democratic nation states cooperating regionally, as in the European Union, and attempting, not always successfully, to cooperate globally through the United Nations and its agencies.

Chapter 29 shows that progress in science was achieved primarily through cooperation. Individuals certainly made significant conceptual breakthroughs, but these were advanced through cooperation with the founding of scientific societies, the publishing of their findings, and the development of international teams to investigate phenomena.

Rate of change
Human evolution has been accelerating at an ever-increasing rate. Humans fully emerged as a self-reflecting species at least 40,000 to 10,000 years ago. If we take the mean of that range, 25,000 years ago, as the starting point, then for 88 per cent of that time human societies were exclusively primeval, when reflections on self and its relationship with the rest of the universe were based on superstition and concerned solely with survival. Only in the last 12 per cent of that time did human reflection include the pursuit of knowledge and understanding for its own sake through insight and reasoning, while the scientific age employing systematic observation and testing of evidence occupies less than 2 per cent of human existence.

To use the 24-hour clock analogy, if humans emerged at time zero then philosophical thinking emerged at 2 hours 53 minutes before midnight, while scientific thinking emerged only at 27 minutes before midnight as illustrated in Figure 31.1.

Globalization
Uniquely, humans spread round the globe, using reflective consciousness to devise ways of surviving in any environment, without diverging into different species. Globalization, rather than global dispersal, began to accelerate following the development from the mid-twentieth century onward of affordable international travel. This in turn increased racial interbreeding and reduced physiological and cultural differences, aided by the invention of global electronic communications.*

*   See page 457.

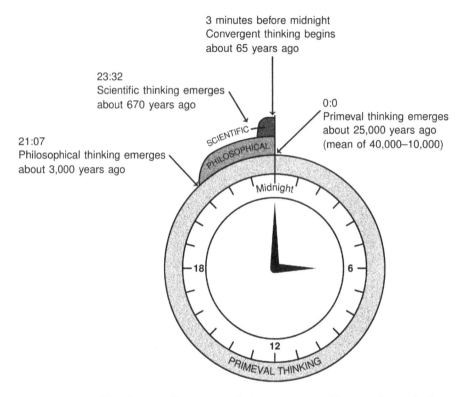

Figure 31.1 The phases of human evolution represented by a 24-hour clock

Trade increasingly became globalized. But more than that. Thinking started to become globalized. Supranational scientific, political, economic, educational, and humanitarian organizations, like the UN and its agencies mentioned earlier, began groping their way towards becoming global thinking networks.

In December 1968, at the height of the Cold War, pictures beamed back to the world by the Apollo 8 astronauts showed for the first time the blue, green, brown, and white planet Earth rising into the cold blackness of space above the grey, lifeless lunar surface. That image played a significant role in creating a global consciousness. No national boundaries were marked on the planet. It was the home of all us, members of one species trying to survive in a small, precariously balanced habitat. Such a vision proved a life-changing experience for the ex-fighter pilots who witnessed it first hand. For the rest it prompted global environmental movements and a growing recognition of the self-defeating nature of war in opposition to the predominant attitude fostered by the military-industrial complex.

Small, local groups rapidly grew into global nongovernment organizations. Médecins Sans Frontières, founded in 1971 by a few doctors and the editor of a medical journal in France, now has offices in 23 countries and provides emergency medical aid based solely on need to people affected by armed conflict, epidemics, and natural disasters. It acts impartially, paying no heed to political or religious affiliations or national borders. In 1999 it was awarded the Nobel Peace Prize. Greenpeace was founded in 1971 by 12 people who protested against the use of nuclear weapons by sailing into the Alaskan islands zone in which the US military was preparing to test a nuclear bomb. It expanded its activities to include global environmental issues and now operates from more than 50 national and regional offices worldwide, retaining its commitment to peaceful protest.

Developments in technology meant that from the early twenty-first century everyone on the planet could, in theory, be connected to everyone else through global multimedia networks. In practice rapidly growing numbers of people were connected globally through social networking sites like Facebook that claimed in 2012 to have more than a billion active users. Overall, in 2011 one third of the world's 7 billion population was online, while there were 6 billion mobile phone subscriptions.[14] By 2020, according to the Swedish telecommunications company Ericsson, a global networked society will be realized by more than 50 billion connected devices, of which 15 billion will be video-enabled.[15]

Complexification

For at least 60 per cent of human existence, humans lived in small, simple groups, mainly extended families, constantly on the move with the sole purpose of survival and with a simple division of labour between hunting and gathering; all adult males participated in warfare. As detailed in Chapter 27, the invention of agriculture with its associated permanent settlements began a process of complexification of human societies and their culture.

The evolution of scientific thinking led to a rapid increase in complexification in every aspect of human societies and their culture, from social, economic, and political organizations through communications, education, occupations, arts, and leisure activities. An individual in a scientifically advanced country in the twenty-first century could live with his or her immediate family and be a member of an extended family dispersed around the globe who communicate with each other electronically. Through education he or she could become socially mobile and pursue a very different occupation from that of his or her parents. This could be in a company or organization that is local, regional, national, or international. He or she could be a member of a neighbourhood, a local community group, a political party, a municipality, a nation, and a supranational organization like the European Community. At the same time he or she could also be a member of a local choir, a football club, and a reading group that is local or global, plus a member of several global social networks. Such complexification only began in the last 65 or so years of 25,000 years of humankind, that is the last 0.25 per cent of human existence.

Leading edge trend

These patterns characterize the leading edge of human evolution. A majority of the human population are not members of scientifically developed societies; their cultures broadly resemble those of European societies at different stages of development. In Bangladesh today, for instance, textile workers are housed in slums, employed for long hours in poor conditions, and paid among the lowest wages in the world; governments in league with factory owners thwart attempts by workers to unionize, arrest labour activists, and ignore factory owners intimidating their workforce and preventing them from forming trade unions.[16] This might have been a description of textile workers in northwest England during the late eighteenth and early nineteenth centuries. In general, developing countries are evolving in a similar way, but more rapidly, to that in which the scientifically advanced nations evolved.

A majority of the population of scientifically developed societies, however, are not immune to some hundred generations' inculcation of beliefs arising from primeval imagination and fear of the unknown. In Figure 31.1 primeval thinking doesn't cease when scientific thinking emerges. It is unsurprising that, say, such a large percentage of the population of the scientifically highly developed USA believe in the biblical account of creation when not only the overwhelming weight of evidence but also logical consistency disproves it.* Or that in scientifically advanced countries so many members of the world's largest religion should believe that God chooses the leader of their Church and inspires him to speak infallibly on matters of faith or morals.

The leading edge is small and barely a hundred years in the making. But it indicates the direction of travel of human evolution. It is noteworthy, for example, that the Roman Catholic Church is rapidly losing members in the scientifically developed regions of Europe and North America, while a majority in these regions who do remain do not believe in the Church's teaching on moral matters like birth control; the Church only retains fully believing members in underdeveloped regions, principally in Africa and South America.

Overall, the instinct for aggressive competition ingrained over millions of years of pre-human ancestry and primeval beliefs inculcated over thousands of years are extremely powerful but gradually declining forces shaping human evolution. Cooperation, as distinct from collectivization imposed by king or emperor or religious leader, struggled against these forces and only began to exert a significant effect on human evolution over less than the last hundred years or so, roughly the same period that discoveries and inventions following the second scientific revolution began to bring about significant changes in human societies. But these two factors—cooperation and scientific thinking—are rapidly increasing forces shaping human evolution.

---

* See page 11.

Convergence

The direction of travel, however gropingly, of human evolution's leading edge is towards convergence.

The appearance of the human species that spread around the Earth created what the palaeontologist, visionary, and Jesuit Pierre Teilhard de Chardin labelled a noosphere (better termed, I suggest, a noetic layer) that evolved from the biosphere (better termed a biolayer) that in turn had evolved from the planet's lithosphere (better termed a geosphere). Teilhard foresaw in an abstract way that this complexifying thinking layer would intensify and converge towards a new stage of the cosmic evolutionary process. The secular humanist Julian Huxley shared this vision but stopped short of Teilhard's identification of this stage as the Christification of the cosmos, which Teilhard did from a deeply held religious belief rather than metaphysical argument.[17]

Neither Teilhard nor Huxley lived to see the last 65 years or so when their shared vision began to be realized. Not only has a thinking layer formed around the Earth but, following the rapidly accelerating rate of scientific discovery and technological innovations from the mid-twentieth century, the increase in cooperation among humans together with the complexification of human societies and the globalization of human activity is producing a trend towards convergence in opposition to the divergence characteristic of competition.

At a biological level signs of this convergence are apparent in the decreasing distinction between subspecies. The term "race" is rarely used today in a scientific sense for the good reason that it is impossible to define any but the most isolated human groups as possessing unique physical characteristics or genomes. The trend towards convergence is most evident in the intercommunicating, inter-acting, and cooperating networks that humans are forming at the individual, family, local, national, international, and global levels. It is producing a shared human consciousness that reflects on such things as the survival of the whole human species, our planetary habitat, other species on the planet, and the future evolution of itself as a species and its relationship with the rest of the cosmos of which it is a part. This does not imply a progression towards uniformity but rather a variety in unity. Just as the self-reflecting consciousness of a single individual correlates with the interaction and collaboration of countless different networks of individual nerve cells, so too the self-reflecting consciousness of the whole species correlates with the interaction and cooperation of countless different networks of different human consciousnesses.

This convergent and accelerating trend is extremely recent in human history. Figure 31.1 shows it beginning at 3 minutes to midnight where 24 hours represents the period of human existence.

Hominization

This evolutionary process of hominization, or what Julian Huxley called progressive psychosocial evolution, is accelerating but is very far from complete; members of the species are better described as human becomings rather than human beings. By analogy with the development of an individual human, the human species was born after a very long gestation, endured a perplexing and often frightening childhood, experienced a disturbing adolescence, and is now groping towards adulthood.

Changing duality of human nature

A basic duality in human nature has been evident from the emergence of humans, manifested in such things as cooperation versus competition, selflessness versus selfishness, and compassion versus aggression. It has been seen as a conflict between good and evil, mythologized in stories throughout the ages from the Hebrew bible's Cain and Abel through the temptation of the Buddha before his enlightenment by Mara and the temptation of Jesus in the desert by the Devil, to the late nineteenth century Robert Louis Stevenson's Jekyll and Hyde that explicitly recognizes the conflict within an individual.

But this conflict is not static. Reflective consciousness has been evolving at an ever-increasing rate from its emergence and is changing the balance in this duality. While the aggressive competitive instinct remains a powerful but gradually decreasing force shaping human evolution, it has been countered by an increasing comprehension that peaceful cooperation is the only way for the human species to survive and continue to evolve.

Integration of patterns in the evidence

Integrating these dynamic patterns in the evidence for the evolution of modern humans may be represented schematically in Figure 31.2 (not to scale).

The evolution of reflective consciousness began very slowly, and its progress was groping in that, countered by a dominating instinct inherited from prehuman ancestors, its consequences, like cooperation, made two steps forward and one step back: cooperation was often subverted by aggression into enforced collaboration, an altogether different thing. Only as recently as 3,000 years ago, with the phase of philosophical thinking, did reflective consciousness begin to impact in a significant way on human societies. Thereafter its acceleration continued to increase in the scientific phase. As of now, its principal consequences—cooperation, altruism, complexification, and convergence—haven't overtaken those of instinct, but a continuation of this increased acceleration implies that they will do so in the not too distant future.

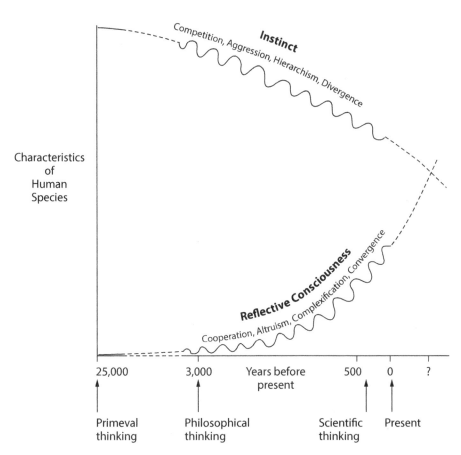

Figure 31.2 Patterns in the evolution of modern humans
The wavy lines indicate groping progression, a series of advances overcoming setbacks.

PART FOUR

A Cosmic Process

CHAPTER THIRTY-TWO

Limitations of Science

The scientific method is the only means of discovering the nature of reality.

—Peter Atkins, 2011

Nominally a great age of scientific inquiry, ours has become an age of superstition about the infallibility of science.

—Louis Kronenberger, 1954

It has become clear in the course of this quest that there are limits to what science can tell us about what we are and how we evolved from the origin of the universe. I shall review two categories of limitations: (1) within the domain of science, and (2) of the domain of science. I shall briefly mention some things outside the domain of science that we need to know in order to answer the questions what are we? where did we come from? how did we evolve? and conclude by suggesting another possible limitation.

Limitations within the domain of science

In Chapter 29 I repeated the definition of science as currently understood and added the generally accepted scientific method from which it is inseparable, while noting that scientists did not always follow all steps of this method, especially for major conceptual breakthroughs. Within this domain are limitations to what science can tell us now but which may be overcome as science advances, while other limitations are insurmountable. It is helpful to consider both within the various, interrelated aspects of science.

Observation and measurement

Science uses systematic, preferably measurable, observations and experiments, but if two current underpinning scientific theories are valid then there are limits to what can be observed and measured.

If quantum theory is valid then it follows from Heisenberg's Uncertainty Principle that the more certain we are in measuring an object's position the less certain we are in measuring its velocity at the same time. This also applies to measuring the energy of an object at a specific time. For visible objects the product of the two uncertainties in each case is so small as to be negligible. However, for objects with the mass of an atom or less, like an electron, the uncertainty is significant.

Furthermore, we cannot measure a time interval of less than 10^{-43} seconds or say anything meaningful about an event occurring within such a time interval. While this is not a practical limitation for most phenomena, many of cosmology's various inflation models speculate that significant events occurred within such a time interval after the beginning of the universe.

If the Special Theory of Relativity is valid then nothing can travel faster than light and so we cannot observe anything further from us than the distance travelled at light speed from the beginning of the universe; this is known as the particle horizon. Moreover, if cosmology's current orthodox model, the Hot Big Bang, is valid, then we cannot observe anything before about 380,000 years after the Big Bang because electromagnetic radiation hadn't decoupled from matter; this is known as the visual horizon.*

Data

Scientific data—the information obtained by systematic observation or experiment—is limited in four main ways. First, there are data that have not yet been obtained but are being actively sought, such as what is the dark matter hypothesized to reconcile gravitational theory with observations,† although some theorists claim that changes to gravitational theory remove the need for dark matter to exist.‡

Second, there are unanticipated data that will be discovered in the future and lead to changes in theory. The discovery of radioactivity eventually resulted in the theory of the weak nuclear interaction as one of the four fundamental forces of the universe.§

The next two are insurmountable limitations. Existing data may be intrinsically unreliable. For example, the earliest written records were rarely, if ever, intended as a factual account but served the propaganda needs of ruler or religion, as with the biblical story of the fall of Jericho, while the accounts of most ancient Greek philosophical thinking are from secondary and tertiary sources written several hundred years later.

Lastly, data may be irretrievably lost. We saw in Chapter 14 that, because fossilization is such a rare occurrence and virtually all sedimentary rock was subducted

---

*   See page 75.
†   See page 41.
‡   See page 119.
§   See page 104.

or metamorphized in the Earth's first billion years, it is almost certain that we will never obtain evidence of what were the first lifeforms on Earth.* The paucity of fossil specimens makes it impossible, among other things, to determine the particular lineage from primates to modern humans.†

Subjectivity

The scientific method is often claimed to be the only way to arrive at an objective explanation of phenomena but, as immunologist and Nobel laureate Peter Medawar observed, "Innocent, unbiased observation is a myth."

Subjectivity can limit science's explanatory power in three related ways: adoption (often unconscious) of underlying assumptions, selection of data, and interpretation of data.

Adoption of assumptions
Underlying any scientific explanation is an explicit or implicit choice of assumptions. For example, we saw in Chapter 3 that, in order to solve the field equations of general relativity when applied to the universe, Einstein and others made two explicit assumptions: the universe is isotropic (it appears the same in all directions) and it is omnicentric (this is true if observed at the same time from any other point). The first of these is not completely valid: the stars in our own galaxy, for instance, form a distinct band of light across the sky, known as the Milky Way. The second can never be tested.

With these two simplifying assumptions, which necessarily result in an homogeneous universe (it is the same at every point), the equations of general relativity produced three solutions: a closed universe whose expansion will slow down, stop, and go into reverse, ending in a Big Crunch; an open universe whose expansion will continue indefinitely at a steady rate producing in effect an empty universe; and a flat universe whose rate of expansion slows but never stops. Cosmologists adopted the flat universe solution as their underlying assumption for what became the orthodox model, the Big Bang and its subsequent modifications. Chapters 3 and 4 show how that model is inconsistent with observations. However, the underlying assumptions of the model are now seldom if ever stated, still less questioned, while the open or closed universe solutions are rarely examined to see if they fit observational data better.

Part 2 shows the many assumptions made when biologists have attempted to explain the emergence and evolution of life. For instance, the orthodox explanation of the evolution of lifeforms rests on the assumption that genes, lengths of DNA coding for the production of proteins, are responsible for the characteristic appearance and behaviour of an organism. This was a reasonable assumption in the mid-twentieth century, but it hardened into a doctrine whose proponents

---

*   See page 208.
†   See page 431.

later resisted or ignored challenges that it was at best a gross oversimplification. For more than 50 years proponents of NeoDarwinism assumed that the 98 per cent of the human genome that didn't consist of genes was junk DNA that had no function. Experimental work in recent years following the mapping of the human genome has shown this to be very far from the case and has vindicated those who challenged the underlying assumption.*

Another example in Chapter 20 is the list of assumptions underpinning the molecular clock technique used to determine when species diverged together with substantial challenges to those assumptions.†

As for the evolution of human thinking, feeling, and behaviour, the underpinning assumption of evolutionary psychology is that such characteristics resulted from the NeoDarwinian accumulation of genetic modifications naturally selected over thousands of generations to give humans a competitive advantage for survival in the Stone Age. I set out the irrationality and conflict with evidence of this assumption in Chapter 29.‡

Assumptions that were reasonable at the time they were made should be explicitly recognized as assumptions that ought to be reviewed, modified, or abandoned in the light of conflicting evidence rather than hardening into dogma that limits the progress of science's explanatory power.

Selection of data
The decision as to which phenomena to observe systematically or measure or which phenomena to experiment on, and what data to collect from such observations or experiments, is rarely disinterested. It is heavily influenced by the conscious or unconscious choice of underlying assumptions and often by the expectation of results. In the molecular clock example quoted above, researchers select which data to use to calibrate the molecular clock and select which nucleotide sequences to compare in different species. This can produce very different conclusions. For instance, estimates for the branching of the chimpanzee genus *Pan* and the human genus *Homo* range from 2.7 million years ago to 13 million years ago.§

The selection of data is frequently bound up with the interpretation of data.

Interpretation of data
Data by itself, like the reading of a thermometer, is meaningless. Science is concerned with the interpretation of data. Objective interpretation of data by scientists is an ideal that is never achieved. Philosophers Thomas Kuhn and Paul Feyerabend each maintained that data interpretation is determined to a large

---

*   See page 317.
†   See page 332.
‡   See page 524.
§   See page 334.

degree by prior theory. This is undoubtedly the case. Part 1 contains very many examples of astronomical data interpreted differently by proponents of orthodox and other theories. These include the very high red shifts of radiation associated with quasars,* radiation from very distant Type 1a supernovae,† ripples in the cosmic microwave background,‡ and the Wilkinson Microwave Anisotropy map of the cosmic microwave background.§

Moreover, other factors are also involved. Scientists aren't automata. They are no different from any other people and have the same motivations, egos, ambitions, and insecurities as the rest of us. Investigating these different interpretations led me to suggest, not entirely mischievously, the Law of Data Interpretation.

> **Law of Data Interpretation**   The degree to which a scientist departs from an objective interpretation of the data from his investigation is a function of four factors: his determination to validate a hypothesis or confirm a theory, the time the investigation has occupied his life, the degree of his emotional investment in the project, and his career need to publish a significant paper or safeguard his reputation.

In the best science this degree is small. However, when the degree is large it shifts into defective science considered later in this chapter.

Part 2 also found considerable problems with data interpretation to explain the evolution of life. Even in those very rare cases that a complete fossil is discovered, its interpretation can be wrong. For example, a fossil found in Canada's Burgess Shale was thought to be of a previously unknown species that walked on spiny stilts and in 1977 was given the name *Hallucigenia sparsa*. 14 years later evidence from China showed that the fossil had been studied upside down: it was a member of an existing phylum that walked on tentacle-like legs and had two rows of spines projecting from its back.¶

Most fossils, however, consist only of small fragments, and Chapter 17 documents the difficulties in interpreting them to ascertain the evolution of life.

Chapter 26 examines the problems of interpreting fossil and other evidence to explain how, where, and when the first humans emerged, and how different interpretations produce different lineages leading from hominins to humans.

Method
Reductionism—the method of breaking a thing down into its constituent parts in order to understand what it consists of and how it functions—has proved an

---

* See page 78.
† See page 45 and page 78.
‡ See page 81.
§ See page 83.
¶ See page 284.

enormously successful tool in increasing our understanding of natural phenomena. It may offer a complete explanation for relatively simple, isolatable systems, such as the formation of crystal structures in a saturated saline solution.

However, reductionism is limited in its explanatory power for complex, open, interactive, and emergent systems. We saw its limitations in Chapter 13 when defining life* and in even more so in Chapter 26 when considering the emergence of humans with reflective consciousness from prehumans. Most quantum theorists consider it inadequate for explaining natural phenomena.

Those scientists who use only this analytical tool limit their explanatory ability by failing to combine it with a holistic approach.

Theory

Several theories advanced as scientific involve infinity. The Friedmann-Lemaître mathematical model of the universe on which the Big Bang idea is based is infinite in extent. As pointed out in Chapter 6, infinity is quite different from a very large number.† Whether a quantity or number that is infinite corresponds with reality, or at least with reality perceived by finite beings like ourselves, is a metaphysical question that science cannot answer. If any such theories cannot be tested by systematic observation or experiment they are outside the realm of science. I shall consider some relevant ones later in this chapter.

As Chapter 6 also noted, each of the two current underpinning theories in physics is limited in its explanatory power. Relativity theory cannot explain phenomena on the scale of an atom or less, while quantum theory cannot explain phenomena with the mass of a star or greater or with a speed approaching that of light.

It is reasonable to suppose that, just as relativity theory and quantum theory were advances on earlier theories of classic mechanics, a further advance may produce a deeper theory that overcomes the limitations of relativity and quantum theories. It may also overcome the measurement and observation limits of those two theories.

Another theory that is limited in its scope is the Principle of Increasing Entropy, which says that in a closed or isolated system disorder either stays the same or, more usually, increases while available energy decreases.‡ While this principle is supported empirically for most systems, it does not explain the evolution of the universe, by definition a closed or at least an isolated system, which shows an increase in complexity over time.§

In the Conclusions to Chapter 10 I suggested one solution to this problem would be the identification of a new form of energy that is involved in energy transformations and associated changes in complexity.

---

* See page 199.
† See page 88.
‡ See Glossary for complete definition.
§ See page 145.

The history of science has shown such cases of the unforeseen that have radically changed our understanding of nature. Before the mid-nineteenth century no one had foreseen that electrical and magnetic phenomena would be explained by the existence of a universal energy field (theoretical physicists currently consider that the range of the electromagnetic field is the whole universe). Before 1932 the existence of the energy field responsible for binding particles together in the atomic nucleus (the strong nuclear force) was unknown. Before 1956 the existence of the energy field responsible for one type of radioactive decay (the weak nuclear force) was unknown. Yet the electromagnetic, strong nuclear, and weak nuclear interactions, together with the gravitational interaction, constitute what science currently recognizes as the four fundamental interactions of nature.*

Psychic energy?
One candidate for a new form of energy is psychic energy. This may be defined broadly as energy associated with the mind that is not reducible to known forms of energy.

Claimed examples of psychic energy in the nineteenth century were entangled with superstitious beliefs like spiritualism and emotions like grief; very many claims were self-delusive or fraudulent. Consequently most scientists dismissed the idea and comparatively little research has been carried out in this field.

More recent areas of study of psychic energy in Europe and north America include extrasensory perception, psychokinesis (the manipulation of physical objects by the power of the mind), out of body experiences, and near death experiences.

Psychologist Raymond Moody's 1975 book *Life after Life* provided the first detailed contemporary accounts of the near death experience (NDE). It prompted the founding of organizations like the International Association for Near-Death Studies that sought, collected, categorized, and investigated such claims. A Gallup Poll in 1992 estimated that some 8 million Americans claimed to have had an NDE. Many of these, like seeing one's life flash by in an instant, a feeling of serenity, going through a tunnel of light, and being greeted by a deceased relative or a deity (Jesus for Christians, Shiva or another god for Hindus, and so on) have perfectly plausible physiological explanations, like anaesthetic and other drugs administered to the patient that may cause hallucinations and memory distortions, excess carbon dioxide, oxygen starvation to neurons as the blood pressure falls, the gradual shutting down of areas of the dying brain, and a transient surge in gamma brain waves immediately after cardiac arrest as demonstrated in rats.[1]

One category, however, appears to defy such explanations. Some patients who were resuscitated after being pronounced clinically dead reported leaving their body, watching the resuscitation effort from near the ceiling, and giving an accurate account of the activities and conversations of the surgeons.

---

*   See page 104.

Sam Parnia, now a critical care physician and director of resuscitation research at the Stony Brook University School of Medicine, New York, together with a UK expert on near death experiences, consultant neuropsychiatrist Peter Fenwick, set up a study in 2008 to investigate whether the mind can separate from the body.

The study was subsequently modified to examine a broad range of mental experiences of 2,060 patients who suffered cardiac arrest in 15 hospitals, and the results were published in the December 2014 edition of *Resuscitation.*[2] 140 of the 330 survivors were able to be interviewed. 39 per cent described a perception of awareness but did not have any explicit recall of events. According to Parnia this suggests that a higher proportion of people may have vivid death experiences but do not recall them due to the effects of brain injury or sedative drugs on memory circuits.

Among the 101 patients who completed further interviews, nine patients (9 per cent) reported experiences compatible with near death experiences while two patients recalled out of body experiences, "seeing" and "hearing" events associated with their resuscitation, though it proved impossible to rule out recall bias and confabulation (the unconscious filling of memory gaps by fabrications believed to be true). One reported validated memories of auditory stimuli during a three-minute period when there was no heartbeat.

The researchers concluded that the results justified more extensive research, while sceptics viewed them as entirely negative.

If information can be received when the lungs have stopped breathing, the heart has ceased pumping blood to the brain, and the brain shows no neural activity, this would suggest that the mind is not bounded by the approximately 1,200 cubic centimetres of 1.5 kilograms of jelly-like matter that constitutes a human brain or even that it depends on the functioning of the brain in order to exist. This in turn would require a new form of energy for the functioning of the mind.

There is no space here to consider in depth the classic mind-body problem that, in the scientific world, is polarized between the majority, who believe in physicalism or materialism, and the minority, who believe either in idealism (mainly quantum theorists) or in dualism, which holds that the mind and the body are not the same thing. Suffice it to say that the brain alone cannot be identical to the mind: an isolated brain is a biological nonsense.

All the systematically and replicably testable neuroscientific studies claimed to show that the mind is nothing but the interactions of neurons in the brain can be interpreted as the immaterial mind directing the material body through the brain. Thus, to use the example in Chapter 29, if you decide to shoot your neighbour, the mind wills the actions that are carried out by the body according to electrochemical signals generated by the brain. It would be somewhat less than convincing to argue in court that you had no choice because Francis Crick, Stephen Hawking and other materialists maintain that free will is an illusion: all your actions are determined by your genes and brain chemistry, a view termed neurogenetic determinism.

Studies of damage to different parts of the brain that impair different mental functions, like memory or sight, are invoked to support the materialist view that the mind is nothing but the brain in action, but neuroscientist Mario Beauregard echoes many dualists by asserting that this "is as illogical as listening to music on a radio, demolishing the radio's receiver, and thereby concluding that the radio was creating the music".[3]

Extrasensory perception (ESP) is the claimed ability to acquire information without the use of the five senses. It includes telepathy (the information comes from another person), remote viewing (the information comes from a distant object or event), and precognition or retrocognition (the information comes from the future or the past).

There is no reason in principle to assume that such a thing is impossible. Before the middle of the nineteenth century no one thought it possible to transmit visual information other than by material means such as the postal service. Within a hundred years moving pictures and sound were being sent and received across vast distances as waves of electromagnetic energy.

A secret $20 million project on remote viewing and psychokinesis was begun in 1975 at the height of the Cold War and lasted 20 years. Funded by US intelligence agencies and NASA, it was directed by two laser physicists, Harold Puthoff and Russell Targ, at the Stanford Research Institute in California. It employed dozens of psychics not only for research but also for operational remote viewing and, according to one participant, psychokinetic disruption of enemy electromagnetics.

In 2002 cognitive neuroscientist Michael Persinger published a study into the remote viewing capabilities of one of the Stargate psychics, Ingo Swann, that concluded not only that they showed positive results but also that these could be correlated with neurophysiological processes and physical events; moreover, remote viewing may be enhanced by applying magnetic fields to the subject's brain.[4]

In 2012 Targ published a book, *The Reality of ESP: A Physicist's Proof of Psychic Phenomena*, that documents his decades of scientific research in this field, including declassified evidence from the Stargate project. His claimed examples include a psychic retired police commissioner, Pat Price, who drew to scale a Soviet weapons factory at Semipalatinsk with great accuracy that was later confirmed by satellite photography, and in 1982 his group making $120,000 by forecasting for nine weeks in a row the changes in the silver commodity futures market.

Targ believes that such psychic phenomena can be explained in part by the quantum mechanical concepts of non-locality and quantum entanglement, and that it is bound up with the ancient mystical insight that the Self of pure consciousness is the same as the ultimate reality of Cosmic Consciousness.* Although a few individuals naturally possess powerful psychic abilities, Targ considers that ESP can be taught.[5]

---

* See page 483.

The Institute of Noetic Sciences chief scientist Dean Radin's 1997 book, *The Noetic Universe: The Scientific Evidence for Psychic Phenomena*[6] (published in the USA as *The Conscious Universe*) gives other examples, as does his 2006 *Entangled Minds*,[7] which shares Targ's belief that quantum entanglement may explain psychic phenomena. Radin argues that most of the scientific community currently refuses to accept strong evidence for psychic energy because of confirmation bias, a psychological term by which evidence supporting prior beliefs is perceived as plausible but evidence challenging prior beliefs is perceived to be implausible and therefore assumed to be flawed or faked.

Certainly most scientists dismiss psychic energy as a phenomenon. However, hypnotism is intentional mind-to-mind communication mediated by vocal communication from the emitter to the receiver. It induces in the receiver a deep physical relaxation and an altered state of consciousness correlated by changes in the functioning of the receiver's brain. In this altered state the receiver's mind responds to suggestions from the emitter that produce mental changes, such as relief from anxiety, and physical changes, such as relief from pain.

Numerous scientific trials have confirmed its effects,[8] and hypnotherapy has been adopted by orthodox medicine in scientifically developed countries since the 1990s as an effective treatment for specific conditions.[9]

Furthermore, in 2014 theoretical physicist Giulio Ruffini and a multidisciplinary international team demonstrated the transmission of information directly from one human brain to another across vast distances without using the senses.

The team developed recent techniques in brain-computer interfaces by which electrodes attached to the scalp record electrical currents when the subject has a specific thought, such as wanting to move a robotic arm or controlling a drone. In this case the output target was another human. The emitter's thoughts were translated into binary code via an electroencephalogram connected to the internet and were received 8,000 kilometres away by robotized transcranial magnetic stimulation headsets on three people, directly stimulating their visual cortex. The blindfolded receivers correctly reported flashes of light in their peripheral vision that corresponded to the message.

Because the transmission required intentional mental activity by emitter and receiver, Ruffini sees this as the early stages of technologically supported telepathy and concludes that in the not so distant future there will be widespread use of human mind-to-mind technologically mediated communication.[10]

Some proponents and practitioners of different kinds of psychic energy, however, eschew the scientific method as inappropriate, arguing that subjective experience, not objective testing, is its essence.

While many nineteenth century practitioners of psychic energy were enmeshed in superstition and their claims subsequently disproved, this was also the case for early practitioners of most scientific disciplines, like chemistry. To dismiss the possibility of psychic energy is no more reasonable than dismissing the possibility of a new, testable theory to replace quantum and relativity theories because attempts to do so over the last thirty years have failed despite the investment of vastly greater resources than in investigating psychic energy.

Moreover, most cosmologists are content to conjecture the existence of a new, mysterious dark energy that accounts for more than two thirds of the universe when other interpretations of one phenomenon, other hypotheses, and other cosmological models disagree.*

New forms of energy may be discovered and validated in the future just as they have been in the past. It is a reasonable conjecture that a candidate for a new form of energy is one associated with the empirically established rise in reflective consciousness and correlated increase in complexity, that is a mental or psychic energy.

Defective science

Science's explanatory power is limited when accounts claiming to be scientific contradict a basic tenet of science or its methods. The border between science and defective science is often indistinct. I suggested above that objectivity in choice of assumptions and in data selection and interpretation is an unachievable ideal. Where the degree of subjectivity is small we have good science; where it is large, and especially where it is also deliberate, we have defective science. This investigation has encountered several other types of defective science.

Inappropriate use of models

A mathematical or computer model can be extremely helpful in constructing or refining a hypothesis that is then tested empirically. Conflating the model with reality, however, is flawed science.

Moreover, assuming that mathematical proof constitutes scientific proof when there is no known way of testing the model empirically is defective science.† This is the case, for example, with the various string "theories" and M-theory,‡ and with the claim that the loop quantum gravity model has established the existence of a prior collapsed universe with specific properties.§

Introducing an arbitrary parameter or scalar field into a mathematical model and altering its value in order to produce a desired outcome or validate a theory also constitutes defective science. The model lacks any predictive function.

As Chapter 3 noted, Einstein introduced into his field equations of general relativity the cosmological constant lambda (Λ) and assigned a value calculated to produce a static universe. He and others then removed Λ from the equations after they accepted observational data that the universe was expanding. When major problems arose with this basic Big Bang model, inflation theorists re-introduced

---

*   See page 78.
†   See page 36.
‡   See page 69.
§   See page 56.

Λ but with a vastly greater value than originally assigned by Einstein and only for an incredibly brief period in order to produce the critical mass density of an incredibly expanded universe that undergoes stable decelerating expansion. In order to conform with their interpretation of radiation from Type 1a supernovae, cosmologists later assigned another arbitrary and very different value to Λ in order to show that, contrary to the universe undergoing a steadily decelerating expansion, its rate of expansion began to accelerate after two-thirds of its lifetime; Λ in this case was said to represent an unknown dark energy for which there is no other empirical support.* Hoyle and colleagues assigned a fourth value to Λ in order to produce a universe expanding overall in a steady state through 50-billion-year cycles of expansion and contraction.†

In order to explain animal and human behaviour that contradicted the NeoDarwinian model, sociobiologists adopted oversimplistic economic games models from the 1970s that are divorced from biological realities. By arbitrarily choosing the rules of the games, the parameters, and the values assigned to those parameters, they produce the outcome they wanted.‡ A different economic game with a different set of choices produced an opposite outcome.§

Claims made beyond, or in contradiction to, the evidence
Belief in assumptions rather than an objective assessment of the evidence produces exaggerated claims. For example, the claim that the space-based Wilkinson Microwave Anistropy Probe (WMAP) enabled scientists to distinguish between different versions of what happened within the first trillionth of a second of the universe's existence is based on a nest of empirically unsubstantiated assumptions.¶

Palaeontology's orthodox account of five major mass extinction events in the last 500 million years, with estimates of percentages of species extinguished in geologically sudden periods, goes far beyond the fossil evidence, quite apart from the uncertainties of what constituted a species at those times.**

Based on the NeoDarwinian assumption of how morphological evolution occurs, most biologists claim that more complete fossil evidence would show gradual change. However, this contradicts the evidence documented by palaeontologists of morphological stasis with minor and often oscillating changes punctuated by the geologically sudden appearance of new species that then remain substantially unchanged until they disappear from the fossil record or continue to the present day as so-called "living fossils".††

---

*   See page 43.
†   See pages 57 to 61.
‡   See pages 377 to 379.
§   See page 383.
¶   See page 82.
**   See page 292.
†† See page 296.

Unlike the founders of NeoDarwinism, most contemporary NeoDarwinists deny that biological evolution shows a pattern of progressive complexity, despite the overwhelming evidence, because it conflicts either with their assumed cause of biological evolution or with an ideology that believes all species are equal.*

Biased data selection
When the selection of data is deliberately biased in order to support a particular theory or hypothesis, it becomes flawed science.

Chapter 6 notes that a leading string theorist's claim to explain why the universe takes the form that it does fails to discuss unfavourable data.†

In the dispute over whether minute traces of carbon found in ancient rock were the oldest fossils of life on Earth or inorganic deposits, one of the claimants was accused by his former research student of ignoring her protests about the selectivity of his evidence.‡

The panel that reviewed 20 years' research on the causes of the mass species extinction deemed to have occurred 65 million years ago (the Cretaceous-Tertiary extinction) supported the orthodox explanation of a massive asteroid hitting Mexico's Yucatán Peninsula. However, some researchers accused the panel of misrepresenting their findings while others accused the panel of ignoring a vast body of evidence inconsistent with their conclusion.§

Flawed methodology
Bernard Kettlewell's experiments in the 1950s and 1960s designed to show natural selection through industrial darkening of the peppered moth contained numerous flaws that undermine the conclusion he sought. Although these were identified in 1998, his work is still widely used as a proof of NeoDarwinism.¶

A landmark paper in 1987 used mitochondrial DNA to trace human descent from one woman living 200,000 years ago in Africa, labelled Mitochondrial Eve. Its methodology was later challenged on several grounds, including the sample selection and the computer program used. Yet its conclusions are still quoted.**

Perpetuation of flawed theory
The perpetuation of a theory that is contradicted by empirical evidence conflicts with a basic tenet of scientific methodology. It usually arises from the choice of assumptions and is reinforced by the subsequent selection and interpretation of

---

*   See pages 348 to 355.
†   See page 84.
‡   See page 210.
§   See page 295.
¶   See pages 311 to 312.
** See page 435.

data that depart to a significant degree from objectivity in order to justify the theory. We saw in Chapter 12 how geology's orthodox crinkling theory to account for the surface features of the Earth was maintained for more than fifty years despite mounting evidence supporting the theory of continental drift (subsequently developed into the theory of plate tectonics).*

For the last thirty or so years the orthodox explanation for the emergence of the first independent lifeforms on Earth says that energy from the Sun and possibly other sources caused random reactions in the aqueous soup of molecules comprising at their largest some 13 atoms that resulted over time in the UltraDarwinian selection of self-replicating RNA molecules. However, not only have experimental attempts to provide support for this idea failed, predictably for chemical reasons according to several scientists, but also the statistical probability of random reactions producing molecules of such size and complexity is close to zero.†

The orthodox theory of biological evolution since the middle of the twentieth century assumes the validity of Darwinian gradualist natural selection, which it incorporates in a mathematical model of population genetics. However, Chapters 17–23 include considerable empirical evidence from the fossil record and from studies of a very much wider range of species in the wild than laboratory fruit flies and nematode worms that conflict with this NeoDarwinian model. Nonetheless one of its leading proponents claimed in 2006 that NeoDarwinism was no longer just a theory but was established fact.‡ Indeed the vast majority of biologists today equate the phenomenon of biological evolution (which has compelling evidential support) with the NeoDarwinian mathematical model, while they tend to treat any questioning of NeoDarwinism as an attack on the reality of biological evolution.§

These and other examples illustrate how theory often hardens into dogma, which is the antithesis of science and limits its progress. The groupthink mentality it engenders is exemplified by 155 biologists signing articles in *Nature* saying they believe that the founder of sociobiology doesn't understand its theories after he had the perspicuity and courage to publish an article conceding that after 40 years those theories had made negligible progress in explaining animal and human behaviour.¶

Suppression of alternative theories
Scientific progress is limited still further when the proponents of a theory that has become dogma hinder and even suppress attempts by others to develop or publish alternative theories.

---

*   See pages 172 to 174.
†   See page 227.
‡   See page 344.
§   See page 409.
¶   See page 408.

In the same year that *Nature* published evidence accumulated over 50 years showing that the continental drift theory accounted for geological phenomena far better than the orthodox crinkling theory, the *Journal of Geophysical Research* dismissed a similar paper from another researcher as "not the sort of thing that ought to be published under serious scientific aegis". *

Chapter 5 notes the complaints that string theorists' domination of committees deciding academic appointments and grants in theoretical physics in the USA has made it very difficult for alternative approaches to obtain funding, despite string conjectures' lack of success over 30 years, while some string theorists engage in dubious practices to suppress dissenting opinions.†

We saw in Chapter 3 how Fred Hoyle's continued espousal of the steady state theory led to his being ostracized by his academic colleagues and to his almost unprecedented resignation of his Cambridge professorship.‡ Several distinguished astronomers protested that academic journals refused to publish their interpretation of astronomical data that contradict the orthodox interpretation which, they claim, was made by mathematical theorists, not observational astronomers. Moreover they allege that there have been many successful attempts to stop them giving invited papers at conferences§ and that they are denied time on telescopes in order to test their hypotheses.¶

Chapter 23 shows that those who offer alternatives to the current orthodox model of biological evolution fare no better. A striking example is Lynn Margulis's development of the theory of symbiogenesis, which she found difficult to get published. Even after NeoDarwinists later conceded that her explanation of the evolution of mitochondria and chloroplasts in eukaryotic cells was supported by genetic evidence and was correct, she complained that the mainstream science and biology journals still refused to publish her work. One prominent NeoDarwinist blogged on his website in 2011 that she was crazy enough to proffer her alternative to the NeoDarwinian model.**

The ossification of the best available theory at the time into dogma results in a defensive response to any questioning that is characteristic of most institutionalized beliefs, and science has inevitably become institutionalized through its specialist societies and university departments that train apprentice scientists. Nearly all human institutions, such as a political party, a government, or a religion, respond in similar ways when challenged from within or without, so a resistance to change by the proponents and teachers of current scientific orthodoxy is unsurprising. However, it behoves the members of a scientific

---

* See page 173.
† See page 71.
‡ See page 21.
§ See page 80.
¶ See page 60.
** See page 388.

institution to remember that what distinguishes science from political parties, governments, or religions is its commitment to empirical testing in as objective a way as possible and to modifying or rejecting a theory when new data conflict with the theory.

Fraud

In the vast majority of cases scientists' significant departure from objectivity in choice of assumptions, selection and interpretation of data, perpetuation of flawed theories, and even suppression of alternative theories is motivated by honest, if often misguided, reasons. However this quest has encountered cases of fraud. The Piltdown Man fake fossil went undetected for 40 years, while Reiner Protsch systematically falsified the datings of his fossils over a period of 30 years.* In 2010 Harvard University found Marc Hauser guilty of falsifying data from animal behaviour experiments.†

These, however, are not just isolated examples of "bad apples" as often portrayed in the media. In 2009 Daniele Fanelli of the Institute for the Study of Science, Technology & Innovation at the University of Edinburgh carried out the first systematic review and meta-analysis of survey data on scientific misconduct. He found that nearly 2 per cent of scientists admitted to having fabricated, falsified or modified data or results at least once—a serious form of misconduct by any standard—and up to one third admitted other questionable research practices. When asked about their colleagues, 14 per cent thought other scientists falsified data and up to 72 per cent thought they engaged in other questionable research practices. Fanelli concludes, "Considering that these surveys ask sensitive questions and have other limitations, it appears likely that this is a conservative estimate of the true prevalence of scientific misconduct."[11]

All this is not to disparage science, which has revolutionized and vastly expanded our knowledge of natural phenomena, but to recognize the limitations within the domain of science of our current understanding of how we emerged and evolved from the origin of the universe.

Limitations of the domain of science

Many scientists and most of those who subscribe to empiricism, positivism, naturalism, physicalism, materialism, or eliminative materialism believe there are no limits to science's domain.

The argument for this belief is essentially

    1. Nothing exists except physical things and their interactions.

---

*   See page 285.

†   See page 379.

2. Science is the empirical study of physical things and their interactions.
3. Therefore there are no limits to the domain of science.

As a corollary,

4. If science cannot explain some phenomenon now, most probably it will do so in the future when new or better data are acquired.

The validity of the conclusion depends on the validity of its two premises. The first is self-evidently invalid because the premise itself is not a physical thing or an interaction of physical things. Put another way, the belief that there are no limits to the domain of science cannot itself be tested by systematic observation or repeatable experiment and so the belief is false.

Science has limited its domain to the empirical in order to differentiate its method of inquiry from that of religion, which seeks answers believed to have been revealed by God or a god, and of philosophy, which seeks answers through insight and reasoning. I shall argue that there are things outside this limited domain of science that we need to know in order to understand what we are and where we came from. These things include subjective experiences, social concepts and values, untestable conjectures, and answers to some metaphysical questions.

Subjective experiences
As individuals we are more than the sum of our cells, molecules, atoms, and their interactions. We are shaped to a large degree by what we see, hear, feel, remember, think, and so on. Such subjective experiences cannot be objectively observed or experimented on by the scientific method. This is epitomized by qualia examined in Chapter 29 in which I quoted the example given by neuroscientists V S Ramachandran and Colin Blakemore: a neuroscientist who has been colour blind from birth might be able to monitor all the physico-chemical processes occurring when a subject looks at a red rose, but she would never be able to experience the quality of redness perceived by the subject.

Neuroscientific reductionists, however, claim they can make objective tests of vision. In 2012 a team at the University of California, Berkeley employed functional magnetic resonance imaging (fMRI) to monitor the neural activity of some 30,000 locations in the cortex of the brain of five male subjects as they watched two hours of movie clips containing 1,705 object and action categories such as animals, buildings, and vehicles. They found the five males shared similar brain maps of overlapping active areas corresponding to the categories.[12] This doesn't, however, inform us which particular animals, buildings, and vehicles people see when these areas are active unless (as in this experiment) the researcher is watching the same movie clip or else the person tells the researcher.

To give another example, if in future a neuroscientist is able to monitor the activity of every single neuron in a subject's brain while the subject is dreaming, the neuroscientist cannot predict or know what particular dream the subject is having or be able to share the subject's experience of that dream.

Similarly, listening to a performance of Mozart's Jupiter symphony cannot be reduced to sound waves vibrating the eardrums and being transmitted along physico-chemical pathways to activate neurons in the brain. The experience of listening involves a vast number of things like a feeling of pleasure, a comparison with the playing of other orchestras, an anticipation of the next bars, memories of when you last heard the piece, and so on. Such experiences cannot be measured or systematically observed by independent testers or be replicated. All these responses combine to provide you with a subjective, holistic experience that is unique to you.

I've concluded that the unique characteristic of the human species is reflective thought. In 2010 neuroscientists at University College London correlated introspective ability with the structure of a small area of the brain's prefrontal cortex. However, the researchers didn't know which was causal: did this area develop as we get better at reflecting on our thoughts, or are people better at introspection if their prefrontal cortex is more developed in the first place?[13] More fundamentally, the objective test didn't show what those introspective thoughts were.

While neuroscience is achieving great success in showing the physical correlates of many of these non-physical things, it is a fundamental error to confuse the physical correlates of subjective experience with the experience itself.

Some subjective experiences, like being inspired by a teacher, falling in love, witnessing a birth or a death, and achieving a major goal can play a large part in determining what we are at the level of an individual.

Social concepts and values

We are not just individuals; each of us is also a member of many societies. Just as reflective consciousness gives an individual human the capability to think and reason, so too it enables a human society to develop social concepts and values. To give just one example, a human society may develop the concept of a just war. Political and social scientists may systematically study the embodiment of the concept (for instance, Britain's 1945 declaration of war on Germany), but the concept itself, and its associated ethical values, lie in the domain of philosophy, not science.

Untestable ideas

As we saw in Chapters 4–7, neither science's current orthodox explanation of the universe, the Inflationary Hot Big Bang, nor any alternative conjecture, provides an empirically testable explanation of the origin of the universe, and hence the origin of the matter and energy of which we consist. It is almost certainly beyond science's ability to do so.

Some explanations of how we evolved in the way we did that are presented as scientific involve the existence of other universes with which we have no contact, like the various multiverse explanations examined in Chapter 7.* If we have no contact with something, then we cannot test it by systematic observation or replicable experiment. Such conjectures may be plausible and, for all we know, one of them may be true. However, they lie outside the domain of science as currently understood.

Metaphysical questions

The natural sciences explain and predict many natural phenomena by applying physical and chemical laws that determine the interactions of physical things. But science cannot explain the essence of these laws. As mathematical cosmologist George Ellis points out, to name some interaction "the force of gravity" helps us to predict what will happen, but does not tell us *how* matter is able to exert an attractive force on other matter a long way away; relabelling it "the effect of the gravitational field" does not alter this situation.[14]

Crucially, science cannot tell us what caused these laws to exist. Newton believed they were created by God, while Einstein too thought they were caused by some transcendental intelligent power.

Another possible limitation

Finally, there is a possible limitation not of science itself but of the human mind. Is there a reality that humans cannot comprehend in a way that rocks cannot comprehend that the Sun provides light and warmth, and chimpanzees cannot comprehend the existence of black holes?

* See page 99.

CHAPTER THIRTY-THREE

Reflections and Conclusions on Human Evolution as a Cosmic Process

[A]ll aspects of reality are subject to evolution, from atoms and stars to fish and flowers, from fish and flowers to human societies and values—indeed…all reality is a single process of evolution.

—Julian Huxley, 1964

Reflections

I began this quest with an open mind to find out what science can tell us about what we humans are and how we evolved not just from the earliest lifeforms on Earth but from the origin of matter and energy. The more I learned the more I realized how much we do not know. Recognition of the extent of our ignorance is often lacking, perhaps understandably, in scientists focused on the latest discovery in their narrow, specialized field. It is also lacking, perhaps less understandably, in scientists whose espousal of a particular theory has hardened into unshakeable belief.

Chapter 32 considers the limitations of our knowledge within the domain of science and also of the domain of science. That domain differentiates scientific from other types of explanation by its attempts to understand and describe natural phenomena by means of repeatable, systematic, preferably measurable, observation or experiment and to infer from this data testable laws used for predictions or retrodictions.

Cosmology's current orthodox account of the origin of the matter and energy of which we consist fits poorly within this domain of science.

The basic Big Bang model is contradicted by observational evidence. Two major modifications made to this model in order to resolve these conflicts produced the Quantum Fluctuation Inflationary Hot Big Bang model, but the central claim of these modifications—that the universe we can observe is an incredibly minute part of the whole universe, the rest of which we can obtain no information about—is untestable. Moreover, these modifications produce a logically inconsistent model if it is held that the Big Bang is the start of everything or else a model that contradicts this basic tenet if a quantum vacuum and an inflation field existed prior to the conjectured Big Bang.

Furthermore, reconciling this modified model with the current orthodox interpretation of observational data requires invoking a mysterious "dark matter" comprising 27 per cent of the observable universe and an even more mysterious anti-gravity "dark energy" comprising 68 per cent of the observable universe. The model has considerable conceptual problems, not the least of which is providing a convincing explanation of the creation of matter and energy out of nothing.

Other hypotheses advanced to modify still further or else supersede the orthodox model are either untested or untestable by any known means. Whether or not the cosmos is eternal cannot be established by the empirical discipline of science.

We can reasonably infer from empirical evidence that we evolved from primordial energy into carbon-based, mobile, behaviourally sophisticated lifeforms possessing the most complex thing in the known universe: a human brain. However, that evolution was only possible because the virtually limitless interactions of matter and energy were constrained by a series of physical and chemical laws plus the fine-tuning of six cosmological parameters, two dimensionless constants, and three parameters of nucleosynthesis. Moreover, this evolution required a planet with a limited mass range that possessed necessary chemicals and was in a limited part of a galaxy and a limited part of a solar system where it received just enough of the right kind of energy for several billion years while being protected from fatal energy radiations and collisions from comets and other space debris.

Science, an empirical discipline, cannot explain what caused these physical and chemical laws to exist, why these parameters have their critical values, and why there was a concurrence of highly improbable and unusual factors on the planet Earth, all of which combined to produce the conditions necessary for our evolution. (The many multiverse explanations are each untestable speculations, and most are based on questionable logic.*) Without this knowledge, however, the chain of causality leading to the emergence and evolution of humans is incomplete.

While science can inform us about the physical correlates of our different subjective experiences, it cannot tell us about the essence of these experiences. They combine holistically to provide each of us with a unique sense that shapes to a considerable degree what we are at the level of the individual. Likewise science cannot tell us about the essence of concepts such as the values that help shape humans at the level of a society.

Within science's domain, limitations broadly comprise two types: permanent and temporary. The latter will be lifted as science progresses with new data

* See page 99.

and new thinking. But some limitations, like the irretrievable loss of data, are permanent.

Nonetheless, what science currently tells us about what we are and where we came from is substantial. Moreover, there is an overall pattern in the evidence.

Conclusions

Reviewing the findings from Parts 1, 2, and 3 leads to the following conclusions.

1. It is almost certain that the empirical discipline of science will never be able to explain the origin of the matter and energy of which we consist.

2. Primordial matter-energy probably comprised a highly energized dense plasma that lost energy (that is, it cooled) as it expanded and matter condensed out. As it did so, matter became increasingly complex. At the scale of the observable universe the chaotic plasma cooled to form a hierarchy of rotating structures of solar systems, galaxies, galactic clusters, superclusters, and sheets of superclusters separated by voids that together constitute a complex whole. At the microscopic scale fundamental particles of matter combined to form more complex nuclei of hydrogen that successively complexified by combining to form the nuclei of some 95 naturally occurring elements. These combined with electrons to complexify further into atoms and then molecules comprising up to 13 atoms found so far in outer space and on meteorites.

3. On the surface of one of the aggregations of matter, planet Earth, which formed over the course of some 4.6 billion to 4 billion years ago from the combination of material debris orbiting the newly ignited Sun, life emerged. Several conjectures have been advanced to explain how, probably within half a billion years of Earth's formation, inanimate molecules comprising up to 13 atoms combined to form the complexity, the size, the changing structures, and the functioning of the simplest possible independent lifeforms. However, because geological processes have irretrievably destroyed the evidence, it is beyond the ability of science to explain the emergence of life just as it cannot account for the emergence of matter.

4. Although the boundary is blurred, the emergence of life—the ability of an enclosed entity to respond to changes within itself and in its environment, to extract energy and matter from its environment, and to convert that energy and matter into internally directed activity that includes maintaining its own existence—marks a change of kind, not simply degree, from inanimate matter.

5. It is highly probable, although not certain, that life emerged only once on Earth, and that all living things on the planet evolved from this one event.

6. Most plausibly the active collaboration of some of the earliest lifeforms for their mutual survival led to their merger and resulted in the divergence of

life into different branches, most of which ramified further into increasingly complex species.

7. The increasingly vast number of species spread across the surface of the planet to occupy habitats conducive to their particular means of maintaining their existence and reproducing themselves. They formed a biolayer above the inanimate geosphere.

8. Competition for resources, damaging genetic mutations, and rapid environmental change resulting in the loss of conducive habitat caused the most dominant pattern among living things, namely the extinction of species.

9. Collaboration at the levels of genes, genomes, cells, tissues, organs, and organisms caused the evolution of more complex species.

10. The animal branch was characterized by mobility, sexual reproduction that increased complexification, and a developing centration of a nervous system for sensing and responding to internal and external stimuli. These three factors caused the animal branch to evolve as diverging and sometimes fusing lineages.

11. The evolution of species with increased morphological complexity and centration of their nervous system correlated with a rise in consciousness of these species.

12. With one exception, surviving lineages ended in stasis of the last species, with some minor reversible morphological changes in response to reversible environmental changes.

13. That one exception was the human species, in which consciousness had risen to the point that it had become conscious of itself: its members not only know but know that they know; uniquely, they possess the ability to reflect on themselves and on the cosmos of which they know they are a part.

14. Because of the irretrievable paucity of the fossil and other evidence it is almost certain that science will never be able to explain with confidence precisely where, when, why, and how the emergence of humans took place. It probably occurred in Africa, though not necessarily in East Africa, and was complete at least 40,000 to 10,000 years ago and possibly even earlier. Most plausibly it was caused by a fluctuating climate producing a fluctuating habitat that prompted greater creativity and invention together with an understanding of the benefits of cooperation over competition in order to survive. Hybridization or whole genome duplication may have played a role in this evolutionary change.

15. It appears to be a case of systems emergence, where the interaction of faculties such as invention, creativity, and communication gave rise to a new, higher level faculty of reflective consciousness that then caused the transformation of the lower level faculties and the generation of new ones such as imagination, belief, language, abstraction, and morality.

16. Like the emergence of life from inanimate matter, although the boundary is indistinct the emergence of reflective consciousness from consciousness marks a change of kind not simply degree. Thereafter the evolution of humans was not primarily morphological or genetic but noetic: the evolution of reflective consciousness.

17. This noetic evolution can be divided into three overlapping phases: primeval, philosophical, and scientific. Primeval thinking, the only kind for some 90 per cent of human existence, evolved when the imperative was to survive and reproduce. It is characterized by creativity, invention, imagination, and beliefs.

18. About 10,000 years ago primeval thinking produced the realization that a more effective way than small, mainly kin-based groups constantly moving and adapting to a changing habitat in order to survive was to adapt a habitat to human needs. The invention of agriculture and corresponding human settlements was associated with the understanding that this was best achieved through cooperation that also resulted in the trading of goods and ideas with other settlements.

19. However, as these settlements developed and expanded, this comprehension struggled with the predominant instinct for aggressive competition ingrained over several million years of prehuman ancestry. This instinct produced battles for control of these settlements and their agricultural and other resources both from within and from without. It resulted in centralization and enforced collaboration.

20. As the settlements grew they developed a social hierarchy that reflected classes of skills passed on from parents to offspring, typically ruler, priests, warriors, merchants and craftsmen, farmers, and slaves. They expanded into autocratically ruled cities, city-states, and then empires that rose and fell. The overall pattern across the globe was an increase in size, complexity, and centration of human societies.

21. The evolution of primeval thinking was intimately bound up with the evolution of beliefs that arose from imagination combined with a lack of understanding of natural phenomena or fear of the unknown, that is, superstition. From the animism, totemism, and ancestor veneration of hunter-gatherers, religions developed that reflected the growth in size, complexity, and specializations of settled human societies. They developed from worship of a fertility goddess through polytheism to a pantheon ruled by a powerful male sky and war god, and lastly to a patriarchal monotheism with other gods subsumed into one God or demoted to angels.

22. Religions expanded when adopted by empires, while autocratic rulers of empires right up to the First World War claimed a divine mandate to legitimize and consolidate their power.

23. Primeval thinking was applied to devise technologies for the survival

of human societies and to influence the supernatural forces believed to determine their fate. It produced the foundations of art, music, spoken and written language, mathematics, and astronomy.

24. Reflective consciousness enabled human societies, uniquely, to adapt any land on the planet as their habitat while remaining a single, interbreeding species.

25. Philosophical thinking emerged around 3,000 years ago—barely the last 10 per cent of human existence—in different places across the globe to seek answers to the self-reflective questions of what we are, where we came from, and how we should behave that did not invoke imagined spirits or anthropomorphic gods or God. These ancient philosophers employed insight, usually resulting from disciplined meditation, and reasoning, based on observations of natural phenomena or assumptions taken as self-evident. This phase marked a quest for knowledge for its own sake, not solely for survival and reproduction.

26. Seers from different cultures tended to experience a similar insight that all things have an underlying unity in an ultimate ineffable reality. This is best described as a transcendent cosmic consciousness or intelligence that exists formlessly out of space and time but which is immanent in the phenomena perceived by our five physical senses and our mind; the essence of each of us, as distinct from our phenomenal self, is identical to this undivided whole. Moreover, this ultimate reality is manifest in, or regulates how, the cosmos functions, and we should harmonize our lives with it in order to achieve fulfilment.

27. When thinking focused on how we should behave towards each other, nearly all ancient philosophers, whether using insight or reasoning, taught that we will only achieve tranquillity and fulfilment by acting unselfishly and treating all others as we would wish to be treated. This ran counter to the prevailing instinctive drive in their societies for warfare and conquest. At root it is a prescription for cooperation and altruism, not aggressive competition, in order to achieve progress for humanity.

28. Philosophical thinking branched fundamentally from the twelfth century when the West adopted reasoning as the only method taught in universities. Reasoning ramified according to the object of inquiry, and these branches diverged still further into different schools of thinking. Insight was principally holistic, but what ramified considerably were schools founded to interpret ineffable insights.

29. Although some branches have roots in ancient times, scientific thinking, the third phase of human evolution, most clearly emerged over the course of some 150 years beginning around the middle of the sixteenth century as attempts to understand natural phenomena by systematic, preferably measurable, observation or experiment that could be used to infer testable laws and make verifiable predictions. Its principal branches were the

physical sciences that studied inanimate phenomena, the life sciences that studied living things, and the medical sciences that sought to promote the survival of humans. The study of humans and their social relationships developed into the social sciences by the nineteenth century.

30. Science disproved many superstitious beliefs and philosophical conjectures about natural phenomena. However, its emergence did not result in the rejection of superstition. This had been inculcated over tens of thousands of years, and most pioneers of scientific thinking retained superstitious beliefs. Moreover, they believed that the natural laws they sought to discover had been created by God.

31. The scientific phase has so far spanned less than the last 2 per cent of human existence, but in this time the rate of human evolution accelerated even more. Enabled by the invention of increasingly sophisticated technology, scientific thinking has produced a geometric increase in our knowledge of what we are physically, where we came from, the universe of which we are a part, and our inter-relationships.

32. It resulted in the further ramification of science into increasingly narrow fields, carving out canyons of expertise from which its practitioners found it difficult to see the whole picture or engage in meaningful dialogue with other specialists except where canyons intersect. This narrow focus fostered a belief in reductionism as the only scientific method and also a belief in physicalism, which is irrational because such a belief is not itself a physical thing or the interaction of physical things.

33. Some hundred years ago technology that enabled the investigation of phenomena not normally observable helped bring about a second scientific revolution in physics. Quantum mechanics revealed a submicroscopic realm of indeterminacy, quantum entanglements, and interdependencies that contrasts with the determinism of classical, or Newtonian, physics of the first scientific revolution. Many of its pioneers interpreted its predictive and empirically validated mathematical models as requiring consciousness to cause the materialization of physical phenomena. Several espoused holistic views similar to the ancient insights of a transcendent cosmic consciousness or intelligence that is manifest in the physical phenomena perceived by our senses and our mind and that regulates the interactions of these interdependent phenomena.

34. From the middle of the twentieth century, physicists increasingly sought to demonstrate that all matter and energy—that is, all physical phenomena—are lower energy manifestations of a single energy at the beginning of the universe. This convergent trend in the fundamental science, in opposition to the hitherto divergent trend in the sciences, is also consonant with the ancient insight of a fundamental energy underlying all things.

35. Having passed the threshold of reflective consciousness, humans developed a dual nature. The ingrained heritage of instinctive, aggressive competition

and enforced collaboration began to be opposed by reflection that resulted in peaceable cooperation and altruism. This new trend very gradually increased over the three overlapping phases of human evolution while instinct gradually decreased from its dominating position.

36. Only since the middle of the twentieth century—barely a quarter of one per cent of human existence—has this rising trend of cooperation and altruism begun to have a planetary impact on human societies, shown by supranational and global organizations cooperating for peaceful, humanitarian, scientific, and educational purposes.

37. Due to scientific and technological developments during the same brief period in human history, a trend of globalization and associated convergence emerged, both physically and noetically, that is, by reflective thinking. From global dispersal, with different human societies occupying different habitats across the planet, humans, at least in scientifically advanced societies at the leading edge of human evolution, increasingly used the whole globe as their habitat. Perhaps more significantly, they increasingly communicated almost instantly with each other through a growing multiplicity of global electronic networks that produced still greater intensification of thinking.

38. A noetic, or mental, layer thus evolved from the biolayer that had evolved from the geosphere that evolved ultimately from primordial energy.

39. The rate of this cosmic evolutionary process has accelerated exponentially, with the inanimate stage measured in 10–20 billion years, the biological stage in around 3.5 billion years, and the human stage in tens of thousands of years; the philosophical phase of hominization has occupied some 3,000 years, the scientific phase about 450 years, while globalization and convergence at the leading edge of human evolution began barely 65 years ago.

The short answer to the question what are we? is that, uniquely as far as we know, we are the unfinished product of an accelerating cosmic evolutionary process characterized by combination, complexification and convergence, and the self-reflective agents of our future evolution.

Notes

Chapter 2: Origin Myths

1. For translations I have used The Rig Veda (1896) and The Upanishads (1987).
2. I have drawn on Sproul (1991) and Long, Charles H "Creation Myth" *Encyclopædia Britannica Online* (2014) for many of the myths summarized in this chapter
3. Graves (1955) p. 27
4. Sproul (1991) pp. 19–20
5. See Kak, Subhash C (1997) "Archaeoastronomy and Literature" *Current Science* 73: 7, 624–627 as an example of a small but growing band of Indian academics who are challenging what they see as the colonial interpretation of Indian history and culture rooted in Victorian scholarship
6. Finkelstein and Silberman (2001)
7. The Revised English Bible (1989) Genesis 1:1
8. The Holy Qur'an (1938) Sura 7:54 and Sura 41:9–12
9. Buddha (1997)
10. According to a survey taken 21–22 April 2005 by Rasmussen Reports
11. Ussher's calculation is given in Gorst (2001). For contemporary endorsements see, for example, the Creation Science Association and its website http://www.csama.org/
12. See, for example, Kitcher (1982); Futuyma (1983)
13. Sproul (1991) p. 17
14. Ibid. p. 6
15. Ibid. p. 4
16. Ibid. p. 10
17. Ibid. p. 29
18. Reanney (1995) p. 99
19. Finkelstein and Silberman (2001)
20. Quoted in Snobelen, Stephen D (2001) "God of Gods and Lord of Lords: The Theology of Isaac Newton's General Scholium to the Principia" *Osiris* 16, 169–208
21. See the appendix dealing with his religious belief in Darwin, Charles (1929)
22. Einstein (1949)
23. Schrödinger (1964)
24. See Krishnamurti and Bohm (1985), (1986), (1999)
25. http://www.cnn.com/2007/US/04/03/collins.commentary/index.html Accessed 6 February 2008

Chapter 3: The Emergence of Matter: Science's Orthodox Theory

1. Burbidge, Geoffrey (2001) "Quasi-Steady State Cosmology" http://arxiv.org/pdf/astro-ph/0108051 Accessed 29 December 2006
2. Assis, Andre K T and Marcos C D Neves (1995) "History of the 2.7 K Temperature Prior to Penzias and Wilson" *Apeiron* 2: 3, 79–87
3. See, for example, Bryson (2004) pp. 29–31
4. Burbidge (2001)
5. Maddox (1998) pp. 33–34
6. Fowler, William A "Autobiography" (1983) *Nobel Foundation.* http://nobelprize.org/nobel_prizes/physics/laureates/1983/fowler-autobio.html Accessed 31 October 2007
7. Magueijo (2003) pp. 79–85
8. Ibid. pp. 109–111; Linde (2001)
9. Guth (1997) pp. 213–214
10. Magueijo (2003) pp. 94–98
11. Guth (1997) p. 186
12. Linde (2001)
13. Guth, Alan and Paul Steinhardt (1984) "The Inflationary Universe" *Scientific American* 250, 116–128
14. Linde (2001)
15. Clowes, Roger G, et al. (2013) "A Structure in the Early Universe at Z ~ 1.3 That Exceeds the Homogeneity Scale of the R-W Concordance Cosmology" *Monthly Notices of the Royal Astronomical Society*, January
16. Horváth, I, et al. (2014) "Possible Structure in the GRB Sky Distribution at Redshift Two" *Astronomy & Astrophysics* 561 id. L12, 4pp. http://arxiv.org/abs/1401.0533v2 Accessed 29 August 2014
17. Hawking (1988) p. 46
18. Guth (1997) p. 87
19. Ibid. pp. 238–239
20. Braibant, et al. (2012) pp. 313–314
21. Rowan-Robinson (2004) pp. 89–92
22. Ibid. p. 99
23. Burbidge, Geoffrey and Fred Hoyle (1998) "The Origin of Helium and the Other Light Elements" *The Astrophysical Journal* 509, L1–L3; Burbidge (2001)
24. Linde (2001)
25. Guth (1997) p. 186
26. Ibid. p. 250
27. Ibid. pp. 278–279
28. Ibid. p. 286
29. Alspach, Kyle (2004) "Guth, Linde Win Gruber Cosmology Prize" *Science & Technology News* 1 May, 1,3
30. Ade, P A R et al. (2014) "Detection of β-Mode Polarization at Degree Angular Scales by BICEP2" http://arxiv.org/pdf/1403.3985 Accessed 18 March 2014
31. Cowan, Ron (2014) "Big Bang Finding Challenged" *Nature* 510, 20
32. Coles, Peter (2007) "Inside Inflation: After the Big Bang" *New Scientist Space.* Special Report, 3 March 2007 http://space.newscientist.com.libproxy.ucl.ac.uk/article/mg193 25931.400;jsessionid=CCHNEIPIDDIE Accessed 2 April 2007
33. Rowan-Robinson (2004) p. 101

34. Ellis (2007) S.5
35. Maddox (1998) p. 55

Chapter 4: What Science's Orthodox Theory Fails to Explain

1. Rodgers, Peter (2001) "Where Did All the Antimatter Go?" *Physics World*. http://physicsweb.org/articles/world/14/8/9 Accessed 12 June 2006
2. Ellis (2007) S.2.3.6
3. Leibundgut, Bruno and Jesper Sollerman (2001) "A Cosmological Surprise: The Universe Accelerates" *Europhysics News*, 32 (4) http://www.eso.org/~bleibund/papers/EPN/epn.html Accessed 10 February 2007; Riess, Adam "Dark Energy" *Encyclopædia Britannica Online* (2014) http://www.britannica.com/EBchecked/topic/1055698/dark-energy Accessed 19 February 2014
4. Ellis, George (2005) "Physics Ain't What It Used to Be" *Nature* 438: 7069, 739–740
5. Kunz, Martin, et al. (2004) "Model-Independent Dark Energy Test with Sigma [Sub 8] Using Results from the Wilkinson Microwave Anisotropy Probe" *Physical Review D (Particles, Fields, Gravitation, and Cosmology)* 70: 4, 041301
6. Shiga, David (2007) "Is Dark Energy an Illusion?" http://www.newscientist.com/article/dn11498-is-dark-energy-an-illusion.html#.U5GjRSj5hhI 30 March
7. Durrer, Ruth (2011) "What Do We Really Know About Dark Energy?" *Philosophical Transactions of the Royal Society A* 369: 1957, 5102–5114
8. Lieu, Richard (2007) "ΛCDM Cosmology: How Much Suppression of Credible Evidence, and Does the Model Really Lead Its Competitors, Using All Evidence?" http://arxiv.org/abs/0705.2462
9. Ellis (2007) S.2.3.5
10. Krauss, Lawrence M (2004) "What Is Dark Energy?" *Nature* 431, 519–520
11. Tolman (1987)
12. Guth (1997) pp. 9–12 and 289–296
13. Tryon, Edward P (1973) "Is the Universe a Vacuum Fluctuation?" *Nature* 246, 396–397
14. Rees (1997) p. 143
15. Hawking (1988) p. 129
16. Maddox, John (1989) "Down with the Big Bang" *Nature* 340, 425–425

Chapter 5: Other Cosmological Conjectures

1. Hawking (1988) pp. 132–141
2. Penrose (2004) pp. 769–772
3. Linde (2001)
4. Quoted in *Science & Technology News* 1 May 2004, p. 3
5. Guth (1997) pp. 250–252
6. Linde (2001)
7. Borde, Arvind and Alexander Vilenkin (1994) "Eternal Inflation and the Initial Singularity" *Physical Review Letters* 72: 21, 3305–3308
8. Magueijo (2003)
9. Barrow, John D (2005) "Einstein and the Universe" *Gresham College Lecture* London, 18 October
10. Smolin (1998) pp. 112–132
11. Ashtekar, Abhay, et al. (2006) "Quantum Nature of the Big Bang: An Analytical

and Numerical Investigation" *Physical Review D (Particles, Fields, Gravitation, and Cosmology)* 73: 12, 124038

12. Narlikar and Burbidge (2008) Chapter 15
13. Ned Wright's Cosmology Tutorial at UCLA (2004) http://www.astro.ucla.edu/~wright/stdystat.htm
14. Lerner (1992) updated on http://www.bigbangneverhappened.org/index.htm
15. Scarpa, Riccardo, et al. (2014) "UV Surface Brightness of Galaxies from the Local Universe to Z ~ 5" *International Journal of Modern Physics D* 23: 6, 1450058
16. Steinhardt, personal communication 24 June 2007
17. Steinhardt, Paul J and Neil Turok (2004) "The Cyclic Model Simplified" Physics Department, Princeton University http://www.phy.princeton.edu/~steinh/dm2004.pdf Accessed 11 March 2007
18. Leake, Jonathan (2006) "Exploding the Big Bang" *The Sunday Times* London, 20 August, News p. 14
19. Steinhardt, personal communication 9 March 2007
20. Steinhardt and Turok (2004), (2007)
21. Steinhardt, personal communication 12 March 2007
22. Ibid.
23. Steinhardt, personal communications 30 April and 7 May 2007
24. Steinhardt, personal communication 20 August 2014
25. Susskind (2005)
26. Smolin (2007) p. 105
27. Magueijo (2003) p. 239
28. Gross, David "Viewpoints on String Theory" (2003) *WGBH* http://www.pbs.org/wgbh/nova/elegant/view-gross.html Accessed 15 August 2006
29. Quoted in Smolin (2007) p. 154
30. Quoted in ibid. p. 197
31. Friedan, D (2003) "A Tentative Theory of Large Distance Physics" *Journal of High Energy Physics* 2003: 10, 1–98
32. Smolin (2007) p. 198
33. Woit (2006)
34. Smolin (2007) p. 176

Chapter 6: Problems Facing Cosmology as an Explanatory Means

1. Ellis (2007) S.2.3.2
2. Maddox (1998) p. 36
3. Ibid. p. 27
4. http://hubblesite.org/newscenter/archive/releases/1994/49/text/ 26 October 1994
5. http://www.nasa.gov/centers/goddard/news/topstory/2003/0206mapresults.html#bctop 11 February 2003
6. http://www.esa.int/Our_Activities/Space_Science/Planck/Planck_reveals_an_almost_perfect_Universe 21 March 2013
7. Rowan-Robinson (2004) p. 163
8. Bonanos, Alceste, et al. (2006) "The First Direct Distance Determination to a Detached Eclipsing Binary in M33" *The Astrophysical Journal* 652, 313–322
9. Rowan-Robinson (2004) p. 164
10. Lerner (1992) with data plus extensive references updated on http://www.bigbangneverhappened.org/ Accessed 16 February 2014

11. Ned Wright's Cosmology Tutorial at UCLA (2003) http://www.astro.ucla.edu/~wright/lerner_errors.html#SC

12. Ellis (2007) S.4.2.2 and S.2.3.5

13 "Quasar" *McGraw-Hill Encyclopedia of Science and Technology*; "Quasar" *The Columbia Electronic Encyclopedia, Sixth Edition* Accessed 29 January 2008

14 Rowan-Robinson, personal communication 21 November 2007

15 Burbidge, personal communication 14 January 2008

16 Arp (1998)

17 Burbidge, personal communication 14 January 2008

18 Arp, Halton and C Fulton (2008) "A Cluster of High Redshift Quasars with Apparent Diameter 2.3 Degrees" http://arxiv.org/pdf/0802.1587v1 Accessed 28 February 2008

19 Arp, personal communications 18 and 25 February 2008

20 Das, P K (2007) "Quasars in Variable Mass Hypothesis" *Journal of Astrophysics and Astronomy* 18: 4, 435–450

21 Singh (2005) pp. 462 and 463 respectively

22 McKie, Robin (1992) "Has Man Mastered the Universe?" *The Observer* London 26 April, News pp. 8–9

23 Editorial (1992) "Big Bang Brouhaha" *Nature* 356: 6372, 731

24 Narlikar, J V, et al. (2003) "Inhomogeneities in the Microwave Background Radiation Interpreted within the Framework of the Quasi–Steady State Cosmology" *The Astrophysical Journal* 585: 1, 1–11

25 http://www.nasa.gov/home/hqnews/2006/mar/HQ_06097_first_trillionth_WMAP.html 16 March 2006

26 Lieu, Richard and Jonathan P D Mittaz (2005) "On the Absence of Gravitational Lensing of the Cosmic Microwave Background" *The Astrophysical Journal* 628, 583-593; Lieu, Richard and Jonathan P D Mittaz (2005) "Are the WMAP Angular Magnification Measurements Consistent with an Inhomogeneous Critical Density Universe?" *The Astrophysical Journal Letters* 623, L1–L4

27 Larson, David L and Benjamin D Wandelt (2005) "A Statistically Robust 3-Sigma Detection of Non-Gaussianity in the WMAP Data Using Hot and Cold Spots" *Physical Review D* http://arxiv.org/abs/astro-ph/0505046 Accessed 25 May 2007

28 Land, Kate and João Magueijo (2005) "Examination of Evidence for a Preferred Axis in the Cosmic Radiation Anisotropy" *Physical Review Letters* 95, 071301

29 Paul Steinhardt, personal communication 20 March 2007

30 http://www.esa.int/Our_Activities/Space_Science/Planck/Planck_reveals_an_almost_perfect_Universe 21 March 2013

31 Ellis, George (2005) "Physics Ain't What It Used to Be" *Nature* 438: 7069, 739–740

32 Tegmark, Max (2003) "Parallel Universes" *Scientific American.* 1 May http://www.sciam.com/article.cfm?articleID=000F1EDD-B48A-1E90-8EA5809EC5880000 Accessed 8 August 2006

33 Quoted in Ellis (2007) S.9.3.2

34 Ibid. S.9.3.2

35 Davies (1990) p. 10

Chapter 7: Reasonableness of Cosmological Conjectures

1. I've drawn on Ellis (2007) S.9.3.3 for these divisions.

2. Tegmark, Max (2003) "Parallel Universes" *Scientific American.* 1 May http://www.

sciam.com/article.cfm?articleID=000F1EDD-B48A-1E90-8EA5809EC5880000
Accessed 8 August 2006
3. Penrose (2004) pp. 17–19 and 1027–1029
4. Ward, Keith (2004) "Cosmology and Creation" *Gresham College lecture*. London, 17 November
5. Elgin (1993) Chapter 13
6. Smolin (1998) p. 242
7. Weinberg (1994) p. 94
8. Rees (2000)
9. Smolin (1998) p. 198
10. Ibid. p. 184
11. http://www.accesstoinsight.org/ptf/dhamma/sagga/loka.html   Accessed   9   June 2014
12. Tegmark (2003)
13. Ellis (2002) S.6.6

Chapter 8: Evolution of Matter on a Large Scale

1. Kak, Subhash C (2003) "Indian Physics: Outline of Early History" http://arxiv.org/abs/physics/0310001v1 Accessed 30 September 2005
2. Griffiths (1987) pp. 37–48
3. Smolin (1998) p. 65
4. Rowan-Robinson (2004) p. 99
5. Lochner, et al. (2005)
6. Rowan-Robinson (2004) pp. 26–42
7. Guth (1997) p. 238
8. Steinhardt, Paul (2014) "Big Bang Blunder Bursts the Multiverse Bubble" *Nature* 510: 7503, 9
9. Eggen, O J, et al. (1962) "Evidence from the Motions of Old Stars That the Galaxy Collapsed" *Reports on Progress in Physics* 136, 748
10. Searle, L and R Zinn (1978) "Compositions of Halo Clusters and the Formation of the Galactic Halo" *Astrophysical Journal* 225 (1), 357–379
11. Rowan-Robinson, Michael (1991) "Dark Doubts for Cosmology" *New Scientist*: 1759, 30
12. Saunders, Will, et al. (1991) "The Density Field of the Local Universe" *Nature* 349: 6304, 32–38
13. Lindley, David (1991) "Cold Dark Matter Makes an Exit" *Nature* 349: 6304, 14
14. Springel, Volker, et al. (2005) "Simulations of the Formation, Evolution and Clustering of Galaxies and Quasars" *Nature* 435: 7042, 629–636
15. Springel et al. (2005)
16. http://www.nasa.gov/home/hqnews/2006/aug/HQ_06297_CHANDRA_Dark_Matter.html 21 August 2006
17. Ellis (2007) S 2.5.1, S 4.2.2, and S 4.3.1
18. Chown, Marcus (2005) "Did the Big Bang Really Happen?" *New Scientist*: 2506, 30
19. Springel et al. (2005)
20. Clowes, Roger G, et al. (2013) "A Structure in the Early Universe at Z ~ 1.3 That Exceeds the Homogeneity Scale of the R-W Concordance Cosmology" *Monthly Notices of the Royal Astronomical Society*, January

21. Schilling, Govert (1999) "Planetary Systems: From a Swirl of Dust, a Planet Is Born" *Science* 286: 5437, 66–68
22. Ward-Thompson, Derek (2002) "Isolated Star Formation: From Cloud Formation to Core Collapse" *Science* 295: 5552, 76–81
23. Kashlinsky, A, et al. (2005) "Tracing the First Stars with Fluctuations of the Cosmic Infrared Background" *Nature* 438: 7064, 45–50
24. Ward-Thompson (2002)
25. Rees, Martin J (2002) "How the Cosmic Dark Age Ended" *Science* 295: 5552, 51–53
26. Rees (2002)
27. Smolin (1998) pp. 144–172
28. Rees (2002) pp. 31–32
29. Adams and Laughlin (1999)
30. Barrow, John D (2006) "The Early History of the Universe" *Gresham College lecture* London, 14 November
31. Krauss, Lawrence M and Michael S Turner (1999) "Geometry and Destiny" *General Relativity and Gravitation* 31: 10, 1453–1459

Chapter 9: Evolution of Matter on a Small Scale

1. http://www.foresight.org/Nanomedicine/Ch03_1.html. Accessed 22 June 2007
2. This section draws principally on Rowan-Robinson (2004) pp. 22–26; Morowitz (2004) pp. 48–53; and Lochner, et al. (2005)
3. Burbidge, E Margaret, et al. (1957) "Synthesis of the Elements in Stars" *Reviews of Modern Physics* 29: 4, 547–650
4. http://www.windows.ucar.edu/tour/link=/earth/geology/crust_elements.html Accessed 22 June 2007
5. Chang (2007) p. 52
6. Barrow and Tipler (1996) pp. 250–253
7. This section draws principally on Ellis (2002) Chapter 3; Barrow and Tipler (1996) pp. 295–305; Morowitz (2004) pp. 51–57
8. Barrow and Tipler (1996) pp. 295–305. The values of these constants are taken from the Physics Laboratory of the National Institute of Standards and Technology http://physics.nist.gov Accessed 15 November 2007
9. Snyder, Lewis E, et al. (2002) "Confirmation of Interstellar Acetone" *Astrophysical Journal* 578: Part 1, 245–255
10. Fuchs, G W, et al. (2005) "Trans-Ethyl Methyl Ether in Space. A New Look at a Complex Molecule in Selected Hot Core Regions" *Astronomy and Astrophysics* 444: 2, 521–530
11. Bell, M B, et al. (1997) "Detection of $HC_{11}N$ in the Cold Dust Cloud TMC-1" *Astrophysical Journal* 483: part 2, L61–L64
12. Lunine (1999) pp. 51–53; "Chondrite" *Cosmic Lexicon*. Planetary Science Research Discovery, 1996

Chapter 10: Pattern to the Evolution of Matter

1. Davies (1990) p. 56
2. Penrose (2004) p. 726
3. Hawking (1998) p. 149

4. Penrose (2004) p. 707
5. Ibid. p. 731
6. Hawking (1998) pp. 49–50
7. Ellis (2002) S.5.4.6
8. Davies (1990) p. 52
9. Ellis (2007) S.2.5

Chapter 12: A Planet Fit for Life

1. JPL Document D-34923 of 12 June 2006: http://exep.jpl.nasa.gov/files/exep/STDT_ Report_Final_Ex2FF86A.pdf Section 1.3.1.1.3 Accessed 21 March 2014
2. Pollack, James B, et al. (1996) "Formation of the Giant Planets by Concurrent Accretion of Solids and Gas" *Icarus* 124: 1, 62–85
3. JPL Document D-34923 of 12 June 2006
4. Morowitz (2004) p. 65
5. Schilling, Govert (1999) "Planetary Systems: From a Swirl of Dust, a Planet Is Born" *Science* 286: 5437, 66–68; Lunine (1999) p. 4
6. Ibid. pp. 124–125
7. http://gsc.nrcan.gc.ca/geomag/nmp/long_mvt_nmp_e.php Accessed 23 May 2008
8. James Kasting, personal communication 30 May 2008
9. Ryskin, Gregory (2009) "Secular Variation of the Earth's Magnetic Field: Induced by the Ocean Flow?" *New Journal of Physics* 11: 6, 063015
10. Bryson (2004) pp. 228–229
11. http://www.geolsoc.org.uk/gsl/geoscientist/features/page856.html Accessed 23 May 2008
12. Kious and Tilling (1996)
13. Ibid.
14. http://eclipse.gsfc.nasa.gov/SEhelp/ApolloLaser.html Accessed 24 May 2008
15. Mojzsis, S J, et al. (2001) "Oxygen-Isotope Evidence from Ancient Zircons for Liquid Water at the Earth's Surface 4,300 Myr Ago" *Nature* 409, 178–181
16. Watson, E B and T M Harrison (2005) "Zircon Thermometer Reveals Minimum Melting Conditions on Earliest Earth" *Science* 308: 5723, 841–844
17. Nutman, Allen P (2006) "Comment on 'Zircon Thermometer Reveals Minimum Melting Conditions on Earliest Earth' Ii" *Science* 311: 5762, 779
18. Glikson, Andrew (2006) "Comment on 'Zircon Thermometer Reveals Minimum Melting Conditions on Earliest Earth' I" *Science* 311: 5762, 779
19. Lunine (1999) pp. 130–131
20. Morbidelli, A, et al. (2000) "Source Regions and Timescales for the Delivery of Water to the Earth" *Meteoritics & Planetary Science* 35: 6, 1309–1320
21. Lunine (1999) pp. 127–130
22. http://solarsystem.nasa.gov/scitech/display.cfm?ST_ID=446 Accessed 10 June 2014
23. Lunine (1999) p. 132
24. Kasting, James (2001) "Essay Review of Peter Ward and Don Brownlee's Rare Earth: Why Complex Life Is Uncommon in the Universe" *Perspectives in Biology and Medicine* 44, 117–131
25. Lunine (1999) p. 165
26. Ibid. pp. 165–176
27. Gribbin (2004) pp. 200–223

28. Kasting, J F and D Catling (2003) "Earth: Evolution of a Habitable Planet" *Ann Rev Astron Astrophys* 41, 429–463
29. http://www.seti.org/seti/seti-science Accessed 21 February 2008
30. Rowan-Robinson (2004) p. 63
31. Ward and Brownlee (2000)
32. Gonzalez, Guillermo, et al. (2001) "The Galactic Habitable Zone: Galactic Chemical Evolution" *Icarus* 152: 1, 185–200
33. Ward and Brownlee (2000)
34. Kasting, J F, et al. (1993) "Habitable Zones around Main Sequence Stars" *Icarus* 101: 1, 108–12
35. JPL Document D-34923 of 12 June 2006: http://planetquest.jpl.nasa.gov/TPF/STDT_Report_Final_Ex2FF86A.pdf
36. Wolszczan, A and D A Frail (1992) "A Planetary System around the Millisecond Pulsar Psr1257 + 12" *Nature* 355: 6356, 145–147
37. http://www.nasa.gov/mission_pages/kepler/main/index.html#.U_PC0Cj5hhK Accessed 19 August 1014
38. Finkbeiner, Ann (2014) "Astronomy: Planets in Chaos" *Nature* 511, 22–24
39. Ward and Brownlee (2000)
40. http://www.planetary.org/explore/topics/compare_the_planets/terrestrial.html Accessed 25 May 2008
41. http://earthobservatory.nasa.gov/Study/Paleoclimatology_Evidence/ Accessed 11 June 2008
42. http://www.newscientist.com/article/mg21228384.600-aliens-dont-need-a-moon-like-ours.html#.U5covij5hhI Accessed 10 June 2014
43. Morowitz (2004) pp. 58–62

Chapter 13: Life

1. Nagler (1987) p. 265
2. *The Upanishads* (1987) pp. 155–172
3. Nagler (1987) p. 265
4. Moira Yip, Professor of Phonetics and Linguistics at University College London, personal communication, 28 January 2008
5. Gottlieb (2001) pp. 13–14
6. Ibid. p. 311
7. Ibid. pp. 230–239; "Vitalism" *The Oxford Dictionary of Philosophy* Oxford University Press, 2005
8. Bechtel, William and Robert C Richardson (1998) "Vitalism" *Routledge Encyclopedia of Philosophy* edited by E Craig. London: Routledge
9. Krieger (1993) p. 7
10. Krieger, Dolores (1975) "Therapeutic Touch: The Imprimatur of Nursing" *The American Journal of Nursing* 75: 5, 784–787
11. Gordon, A, et al. (1998) "The Effects of Therapeutic Touch on Patients with Osteoarthritis of the Knee" *J Fam Pract* 47: 4, 271–277
12. Rosa, Linda, et al. (1998) "A Close Look at Therapeutic Touch" *JAMA* 279: 13, 1005–1010
13. Bohm (1980)
14. Laszlo (2006)

15. Sheldrake (2009)
16. Maddox, John (1981) "A Book for Burning?" *Nature* 293, 245–246
17. Stenger, Victor J (1991) "Bioeenergetic Fields" *The Scientific Review of Alternative Medicine* 3: 1
18. Wilson (1998) p. 58
19. Crick (1995) p. 11
20. Davies (1990) p. 61
21. Cited by ibid. pp. 61–62
22. Quoted in McFadden (2000) p. 13
23. Davies (1990) p. 59
24. Ibid. p. 65
25. Ball, Philip (2004) "What Is Life? Can We Make It?" *Prospect* August, pp. 50–54
26. See, for example definitions of virus given in *Gale Genetics Encyclopedia*, 2003; *McGraw-Hill Science and Technology Encyclopedia*, 2005; *Columbia Electronic Encyclopedia* Accessed 31 July 2008
27. Smolin (1998) p. 194
28. Capra (1997)
29. Ibid. p. 96
30. McFadden (2000) pp. 13–16

Chapter 14: The Emergence of Life 1: Evidence

1. Barghoorn, Elso S and Stanley A Tyler (1965) "Microorganisms from the Gunflint Chert" *Science* 147: 3658, 563–577
2. Schopf, J William (1993) "Microfossils of the Early Archean Apex Chert: New Evidence of the Antiquity of Life" *Science* 260: 5108, 640–646
3. Mojzsis, S J, et al. (1996) "Evidence for Life on Earth before 3,800 Million Years Ago" *Nature* 384: 6604, 55–59
4. Fedo, Christopher M and Martin J Whitehouse (2002) "Metasomatic Origin of Quartz-Pyroxene Rock, Akilia, Greenland, and Implications for Earth's Earliest Life" *Science* 296: 5572, 1448–1452
5. Mojzsis, S J and T M Harrison (2002) "Origin and Significance of Archean Quartzose Rocks at Akilia, Greenland" *Science* 298: 5595, 917–917
6. Brasier, Martin D, et al. (2002) "Questioning the Evidence for Earth's Oldest Fossils" *Nature* 416: 6876, 76–81
7. Dalton, Rex (2002) "Microfossils: Squaring up over Ancient Life" *Nature* 417: 6891, 782–784
8. McFadden (2000) pp. 26–27
9. Kashefi, Kazem and Derek R Lovley (2003) "Extending the Upper Temperature Limit for Life" *Science* 301: 5635, 934
10. Henbest, Nigel (2004) "The Day the Earth Was Born" *Origins* United Kingdom: Channel 4, 21 February
11. Lin, Li-Hung, et al. (2006) "Long-Term Sustainability of a High-Energy, Low-Diversity Crustal Biome" *Science* 314: 5798, 479–482
12. See, for example, Ouzounis, Christos A, et al. (2006) "A Minimal Estimate for the Gene Content of the Last Universal Common Ancestor: Exobiology from a Terrestrial Perspective" *Research in Microbiology* 157: 1, 57–68
13. Quoted in Doolittle (2000)

14. Ragan, et al. (2009)
15. Ibid.
16. Theobald, Douglas L (2010) "A Formal Test of the Theory of Universal Common Ancestry" *Nature* 465: 7295, 219–222
17. Cavalier-Smith, Thomas (2009) "Deep Phylogeny, Ancestral Groups and the Four Ages of Life" *Philosophical Transactions of the Royal Society B: Biological Sciences* 365: 1537, 111–132
18. Polanyi, Michael (1968) "Life's Irreducible Structure" *Science* 160: 3834, 1308–1312

Chapter 15: The Emergence of Life 2: Hypotheses

1. Miller, S L (1953) "A Production of Amino Acids under Possible Primitive Earth Conditions" *Science* 117: 3046, 528–529
2. McFadden (2000) pp. 85–88
3. Ibid. pp. 95–98
4. Lee, David H, et al. (1996) "A Self-Replicating Peptide" *Nature* 382: 6591, 525–528
5. Cairns-Smith (1990)
6. Ferris, James P, et al. (1996) "Synthesis of Long Prebiotic Oligomers on Mineral Surfaces" *Nature* 381: 6577, 59–61
7. Wächtershäuser, G (1988) "Before Enzymes and Templates: Theory of Surface Metabolism" *Microbiol Rev* 52, 452–484
8. McFadden (2000) pp. 91–92
9. Crick, F H C and L E Orgel (1973) "Directed Panspermia" *Icarus* 19, 341–346
10. Hoyle and Wickramasinghe (1978)
11. Wickramasinghe, Chandra, et al. (2003) "SARS—a Clue to Its Origins?" *The Lancet* 361: 9371, 1832–1832
12. See, for example, de Leon, Samuel Ponce and Antonio Lazcano (2003) "Panspermia—True or False?" *The Lancet* 362: 9381, 406–407
13. Napier, W M, et al. (2007) "The Origin of Life in Comets" *International Journal of Astrobiology* 6: 04, 321–323
14. http://www.astrobiology.cf.ac.uk/News3.html Accessed 12 August 2008
15. Bostrom, N (2003) "Are You Living in a Computer Simulation?" *Philosophical Quarterly* 53: 211, 243–255
16. Behe (1996)
17. Pallen, Mark J and Nicholas J Matzke (2006) "From the Origin of Species to the Origin of Bacterial Flagella" *Nat Rev Micro* 4: 10, 784–790
18. News (1981) "Hoyle on Evolution" *Nature* 294, 105
19. Hoyle, Fred (1982) *Evolution from Space: The Omni Lecture Delivered at the Royal Institution, London on 12 January 1982* Cardiff: University College of Cardiff Press
20. Carter, B (1974) "Large Number Coincidences and the Anthropic Principle in Cosmology" 291-298 in *Confrontation of Cosmological Theories with Observational Data* edited by M S Longair: Springer
21. Barrow and Tipler (1996)
22. Penrose (1989) p. 561
23. McFadden (2000) pp. 219–240
24. Kauffman (1996)
25. Wilson (1998) p. 97

26. McFadden (2000) p. 94
27. Morowitz (2004)

Chapter 16: Development of Scientific Ideas about Biological Evolution

1. "Carolus Linnaeus" *Encyclopædia Britannica Online* http://www.britannica.com/EBchecked/topic/342526/Carolus-Linnaeus Accessed 20 December 2008
2. Maillet (1968)
3. "Georges-Louis Leclerc, Comte de Buffon" *Gale Encyclopedia of Biography* 2006
4. Darwin, Erasmus (1796) and "Erasmus Darwin" *Encyclopædia Britannica Online* http://www.britannica.com/EBchecked/topic/151960/Erasmus-Darwin Accessed 16 February 2010
5. Darwin, Erasmus (1803)
6. Pearson, Paul N (2003) "In Retrospect" *Nature* 425: 6959, 665–665
7. Clifford, David "Jean-Baptiste Lamarck" (2004) http://www.victorianweb.org/science/lamarck1.html Accessed 16 February 2010; Shanahan (2004) pp. 14-23; Graur, Dan, et al. (2009) "In Retrospect: Lamarck's Treatise at 200" *Nature* 460: 7256, 688–689; http://www.ucmp.berkeley.edu/history/lamarck.html Accessed 16 February 2010
8. Étienne Geoffroy Saint Hillaire Collection, American Philosophical Society http://www.amphilsoc.org/mole/view?docId=ead/Mss.B.G287p-ead.xml;query=;brand=default Accessed 20 February 2010
9. Green, J H S (1957) "William Charles Wells, F.R.S. (1757–1817)" *Nature* 179, 997–999
10. Desmond, Adrian (1984) "Robert E. Grant: The Social Predicament of a Pre-Darwinian Transmutationist" *Journal of the History of Biology* 17, 189–223, plus Desmond, personal communication 2 April 2010
11. http://www.ucmp.berkeley.edu/history/matthew.html Accessed 3 December 2009
12. Smith (1998, 2000–14)
13. Darwin, Francis (1887) p. 68
14. Ibid. p. 116
15. Darwin's correspondence prior to the Linnean Society presentation is given in ibid. pp. 116–127
16. http://www.linnean.org/index.php?id=380 Accessed 20 February 2010
17. Pearson, Paul N (2003) "In Retrospect" *Nature* 425: 6959
18. Desmond, Adrian and Sarah E Parker (2006) "The Bibliography of Robert Edmond Grant (1793–1874): Illustrated with a Previously Unpublished Photograph" *Archives of Natural History* 33: 2, 202–213
19. Darwin, Francis (1887) Vol 2 pp. 206–207
20. Darwin, Charles (1861) p. iv
21. Ibid. pp. xiv–xv
22. http://anthro.palomar.edu/evolve/evolve_2.htm Accessed 16 December 2008
23. Steinheimer, Frank D (2004) "Charles Darwin's Bird Collection and Ornithological Knowledge During the Voyage of HMS 'Beagle', 1831–1836" *Journal of Ornithology* 145: 4, 300–320; Sulloway, Frank J (1982) "The Beagle Collections of Darwin's Finches (Geospizinae)" *Bulletin of the British Museum of Natural History (Zoology)* 43 (2): 49–94; Sulloway, Frank J (1982) "Darwin and His Finches: The Evolution of a Legend" *Journal of the History of Biology* 15: 1, 1–53
24. Lack (1947)

25. Desmond and Moore (1992) p. 209
26. Darwin, Charles (1872) pp. 70–71
27. Ibid. p. 49
28. Ibid. p. 106
29. Ibid. p. 49
30. Ibid. p. 103
31. See, for example, ibid. p. 42 and p. 47
32. Ibid. p. 137
33. Ibid. p. 50
34. Darwin, Charles (1882) p. 107
35. Darwin, Charles (1872) p. 59
36. Darwin, Charles (1868) pp. 5–6
37. Ibid. p. 6
38. Darwin, Charles (1882) p. 606
39. Darwin, Francis (1887) p. 215
40. Darwin, Charles (1859) p. 134
41. Darwin, Charles (1882) p. v
42. Darwin, Charles (1872) p. 429
43. Darwin, Charles (1958) p. 85
44. Darwin, Francis (1887) Vol 2
45. Ibid. Vol 2, pp. 243–244
46. Desmond (1989)
47. Lyons, Sherrie L (1995) "The Origins of T H Huxley's Saltationism: History in Darwin's Shadow" *Journal of the History of Biology* 28: 3, 463–494
48. Gould (2002) pp. 355–395
49. Kropotkin (1972) p. 18
50. Huxley, T H (1888) "The Struggle for Existence: A Programme" *Nineteenth Century* p. 165
51. Kropotkin (1972) p. 71
52. Ibid.
53. Ibid. p. 30
54. Sapp (2009) pp. 115–120
55. http://nobelprize.org/nobel_prizes/medicine/laureates/1933/morgan-bio.html Accessed 20 December 2008
56. Schrödinger (1992) based on lectures given in 1943
57. Watson and Stent (1980); Crick (1990)
58. Crick, Francis (1970) "Central Dogma of Molecular Biology" *Nature* 227: 5258, 561–563
59. Coyne (2006)
60. Woese, C R, et al. (1990) "Towards a Natural System of Organisms: Proposal for the Domains Archaea, Bacteria, and Eucarya" *Proceedings of the National Academy of Sciences of the United States of America* 87: 12, 4576–4579

Chapter 17: Evidence of Biological Evolution 1: Fossils

1. May, R M (1992) "How Many Species Inhabit the Earth?" *Scientific American* 267: 4, 42–48
2. UNEP (2007) p. 164

3. Torsvik, Vigdis, et al. (2002) "Prokaryotic Diversity—Magnitude, Dynamics, and Controlling Factors" *Science* 296: 5570, 1064–1066
4. Harwood and Buckley (2008)
5. Whitman, William B, et al. (1998) "Prokaryotes: The Unseen Majority" *Proceedings of the National Academy of Sciences* 95: 12, 6578–6583
6. Isaac, N J B, et al. (2004) "Taxonomic Inflation: Its Influence on Macroecology and Conservation" *Trends in Ecology & Evolution* 19: 9, 464–469
7. Mallet, J (2001) "The Speciation Revolution" *Journal of Evolutionary Biology* 14: 6, 887–888
8. Mallet, James (2008) "Hybridization, Ecological Races and the Nature of Species: Empirical Evidence for the Ease of Speciation" *Philosophical Transactions of the Royal Society B: Biological Sciences* 363: 1506, 2971–2986
9. Mayr (1982) p. 285
10. Mayr, Ernst (1996) "What Is a Species, and What Is Not?" *Philosophy of Science* 63: 2, 262–277
11. Coyne (2004) p. 30
12. Mace, Georgina, et al. (2005) "Biodiversity" 87-89 in *Current State & Trends*: Millennium Ecosystem Assessment
13. Mallet, James (2008)
14. Leakey and Lewin (1996) p. 39 and p. 45
15. Ibid. p. 45
16. Conway Morris (1998)
17. Leake, Jonathan and John Harloe (2009) "Origin of the Specious" *Sunday Times* London, 24 May, News 16; Henderson, Mark (2009) "Ida, the Fossil Hailed as Ancestor of Man, 'Wasn't Even a Close Relative'" *The Times* London, 22 October, News 25
18. Harding, Luke (2005) "History of Modern Man Unravels as German Scholar Is Exposed as Fraud" *The Guardian* London, 19 February, News 3
19. http://www.dmp.wa.gov.au/5257.aspx Accessed 24 March 2010
20. Han, Tsu-Ming and Bruce Runnegar (1992) "Megascopic Eukaryotic Algae from the 2.1-Billion-Year-Old Negaunee Iron-Formation, Michigan" *Science* 257: 5067, 232–235
21. Albani, Abderrazak El, et al. (2010) "Large Colonial Organisms with Coordinated Growth in Oxygenated Environments 2.1gyr Ago" *Nature* 466: 7302, 100–104
22. Donoghue, Philip C J and Jonathan B Antcliffe (2010) "Early Life: Origins of Multicellularity" *Nature* 466: 7302, 41–42
23. Knoll, A H, et al. (2006) "Eukaryotic Organisms in Proterozoic Oceans" *Philosophical Transactions of the Royal Society B: Biological Sciences* 361: 1470, 1023–1038
24. http://www.princeton.edu/main/news/archive/S28/14/71M11/index.xml?section= topstories#top 17 August 2010
25. "Ediacara fauna" *Encyclopædia Britannica Online* http://www.britannica.com/ EBchecked/topic/179126/Ediacara-fauna Accessed 12 June 2014
26. http://www.simonyi.ox.ac.uk/dawkins/WorldOfDawkins-archive/Dawkins/Work/ Articles/alabama/1996-04-01alabama.shtml Accessed 20 December 2008
27. Hans Thewissen, personal communication 22 July 2010
28. Thewissen, J G M, et al. (2007) "Whales Originated from Aquatic Artiodactyls in the Eocene Epoch of India" *Nature* 450: 7173, 1190–1194; Thewissen, J G M, et al. (2006) "Developmental Basis for Hind-Limb Loss in Dolphins and Origin of the Cetacean Bodyplan" *Proceedings of the National Academy of Sciences* 103. 22,

8414–8418; Thewissen, J G M, et al. (2001) "Eocene Mammal Faunas from Northern Indo-Pakistan" *Journal of Vertebrate Paleontology* 21(2), 347–366

29. Pallen (2009) p. 83
30. "Extinction" (2005) *McGraw-Hill Encyclopedia of Science and Technology,* 2005
31. "Dinosaurs" American Museum of Natural History http://www.amnh.org/exhibitions/dinosaurs/extinction/mass.php Accessed 29 October 2008
32. Alvarez (1997)
33. See, for example "Dinosaur extinction link to crater confirmed" http://news.bbc.co.uk/1/hi/sci/tech/8550504.stm 4 March 2010
34. Schulte, Peter, et al. (2010) "The Chicxulub Asteroid Impact and Mass Extinction at the Cretaceous-Paleogene Boundary" *Science* 327: 5970, 1214–1218
35. Courtillot, Vincent and Frederic Fluteau (2010) "Cretaceous Extinctions: The Volcanic Hypothesis" *Science* 328: 5981, 973–974
36. Keller, Gerta, et al. (2010) "Cretaceous Extinctions: Evidence Overlooked" *Science* 328: 5981, 974–975
37. Archibald, J David, et al. (2010) "Cretaceous Extinctions: Multiple Causes" *Science* 328: 5981, 973
38. Elliott (2000); Officer, David K (1993) "Victims of Volcanoes: Why Blame an Asteroid?" *New Scientist*: 1861, 34
39. Eldredge and Gould (1972)
40. Gould (1980) p. 182
41. Sheldon, Peter R (1987) "Parallel Gradualistic Evolution of Ordovician Trilobites" *Nature* 330: 6148, 561–563
42. Eldredge (1995) pp. 70–74
43. Cheetham, Alan H (1986) "Tempo of Evolution in a Neogene Bryozoan: Rates of Morphologic Change within and across Species Boundaries" *Paleobiology* 12: 2, 190–202
44. Cheetham, Alan H (1987) "Tempo of Evolution in a Neogene Bryozoan: Are Trends in Single Morphologic Characters Misleading?" *Paleobiology* 13: 3, 286–296
45. Eldredge pp. 69–70
46. Ayala (2014)
47. Eldredge and Tattersall (1982) pp. 45–46
48. Ayala (2014)
49. http://www.britannica.com/EBchecked/topic/360838/mammal Accessed 11 January 2015
50. Luo, Zhe-Xi, et al. (2011) "A Jurassic Eutherian Mammal and Divergence of Marsupials and Placentals" *Nature* 476, 442–445
51. Wible, J R, et al. (2007) "Cretaceous Eutherians and Laurasian Origin for Placental Mammals near the K/T Boundary" *Nature* 447: 7147, 1003–1006

Chapter 18: Evidence of Biological Evolution 2:
Analyses of Living Species

1. Darwin, Charles (1872) p. 386
2. Gehring (1998) pp. 207–216
3. Chouard, Tanguy (2010) "Evolution: Revenge of the Hopeful Monster" *Nature* 463, 864–867
4. Ayala (2014)
5. Thomas, Christopher M and Kaare M Nielsen (2005) "Mechanisms of, and Barriers to, Horizontal Gene Transfer between Bacteria" *Nat Rev Micro* 3: 9, 711–721

6. Boto, Luis (2010) "Horizontal Gene Transfer in Evolution: Facts and Challenges" *Proceedings of the Royal Society B: Biological Sciences* 277: 1683, 819–827

7. Soltis, P S (2005) "Ancient and Recent Polyploidy in Angiosperms" *New Phytologist* 166: 1, 5–8

8. Gregory, T Ryan and Barbara K Mable (2005) "Polyploidy in Animals" pp. 501–502 in Gregory (2005)

9. Gallardo, M H, et al. (2004) "Whole-Genome Duplications in South American Desert Rodents (Octodontidae)" *Biological Journal of the Linnean Society* 82: 4, 443–451

10. Coyne, Jerry A (1998) "Not Black and White" *Nature* 396: 6706, 35–36

11. http://www.gen.cam.ac.uk/research/personal/majerus/Darwiniandisciple.pdf [2004] Accessed 18 October 2010

12. Sargent, T D, et al. (1998) "The 'Classical' Explanation of Industrial Melanism: Assessing the Evidence" in *Evolutionary Biology:* Vol 23 edited by Max K Hecht and Bruce Wallace. New York: Plenum Press

13. Cunha, H A, et al. (2005) "Riverine and Marine Ecotypes of *Sotalia* Dolphins are Different Species" *Marine Biology* 148: 2, 449–457

14. Weiner (1994) p. 9

15. Grant, Peter R and B Rosemary Grant (1997) "Genetics and the Origin of Bird Species" *Proceedings of the National Academy of Sciences of the United States of America* 94: 15, 7768–7775

16. Pray, Leslie (2008) "Transposons, or Jumping Genes: Not Junk DNA?" *Nature Education* 1

17. Pennisi, Elizabeth (2012) "Encode Project Writes Eulogy for Junk DNA" *Science* 337: 6099, 1159–1161; http://www.genome.gov/10005107 Accessed 11 April 2014; ENCODE, Consortium (2007) "Identification and Analysis of Functional Elements in 1% of the Human Genome by the Encode Pilot Project" *Nature* 447: 7146, 799–816

18. http://www.ornl.gov/sci/techresources/Human_Genome/faq/compgen.shtml Accessed 17 August 2010

19. Ridley, Matt "The Humbling of Homo Sapiens" *The Spectator* 14 June 2003

20. See, for example Demuth, J P, et al. (2006) "The Evolution of Mammalian Gene Families" *PLoS One* 1: 1; Britten, Roy J (2002) "Divergence between Samples of Chimpanzee and Human DNA Sequences Is 5%, Counting Indels" *Proceedings of the National Academy of Sciences* 99: 21, 13633–13635

21. Schwartz, Jeffrey H and Bruno Maresca (2006) "Do Molecular Clocks Run at All? A Critique of Molecular Systematics" *Biological Theory* 1: 4, 357–371

22. Ragan, Mark A, et al. (2009) "The Network of Life: Genome Beginnings and Evolution" *Philosophical Transactions of the Royal Society B: Biological Sciences* 364: 1527, 2169–2175

23. Doolittle, W Ford (2009) "The Practice of Classification and the Theory of Evolution, and What the Demise of Charles Darwin's Tree of Life Hypothesis Means for Both of Them" *Philosophical Transactions of the Royal Society B: Biological Sciences* 364: 1527, 2221–2228

Chapter 19: Evidence of Biological Evolution 3: Behaviour of Living Species

1. Brown, Sam P, et al. (2009) "Social Evolution in Micro-Organisms and a Trojan Horse

Approach to Medical Intervention Strategies" *Philosophical Transactions of the Royal Society B: Biological Sciences* 364: 1533, 3157–3168

2. Crespi, B J (2001) "The Evolution of Social Behavior in Microorganisms" *Trends in Ecology & Evolution* 16: 4, 178–183

3. West, Stuart A, et al. (2007) "The Social Lives of Microbes" *Annual Review of Ecology, Evolution, and Systematics* 38: 1, 53–77

4. Shapiro, James A (1998) "Thinking About Bacterial Populations as Multicellular Organisms" *Annual Reviews in Microbiology* 52: 1, 81–104

5. Queller, David C and Joan E Strassmann (2009) "Beyond Society: The Evolution of Organismality" *Philosophical Transactions of the Royal Society B: Biological Sciences* 364: 1533, 3143–3155

6. Pearson, Joseph C, et al. (2005) "Modulating Hox Gene Functions During Animal Body Patterning" *Nat Rev Genet* 6: 12, 893–904

7. Leake, Jonathan (2010) "Check...Science Closes in on Intelligence Gene Test" *Sunday Times* London, 19 September; News 13

8. Most examples of insect social behaviour are taken from Ratnieks, Francis L W and Heikki Helanterä (2009) "The Evolution of Extreme Altruism and Inequality in Insect Societies" *Philosophical Transactions of the Royal Society B: Biological Sciences* 364: 1533, 3169–3179

9. See, for example, Laland, K N, et al. (2011) "From Fish to Fashion: Experimental and Theoretical Insights into the Evolution of Culture" *Philosophical Transactions of the Royal Society B: Biological Sciences* 366: 1567, 958–968

10. Laland, Kevin N (2008) "Animal Cultures" *Current Biology* 18: 9, R366–R370

11. Thornton, Alex and Aurore Malapert (2009) "Experimental Evidence for Social Transmission of Food Acquisition Techniques in Wild Meerkats" *Animal Behaviour* 78: 2, 255–264

12. Thornton, Alex and Katherine McAuliffe (2006) "Teaching in Wild Meerkats" *Science* 313: 5784, 227–229; Thornton, Alex and Tim Clutton-Brock (2011) "Social Learning and the Development of Individual and Group Behaviour in Mammal Societies" *Philosophical Transactions of the Royal Society B: Biological Sciences* 366: 1567, 978–987

13. van Schaik, Carel (2010) "Orangutan Culture and Its Cognitive Consequences" *Culture Evolves* Royal Society, London, 28 June

14. Laland, Kevin N (2008) "Animal Cultures" *Current Biology* 18: 9, R366–R370

15. Reader, Simon M, et al. (2011) "The Evolution of Primate General and Cultural Intelligence" *Philosophical Transactions of the Royal Society B: Biological Sciences* 366: 1567, 1017–1027

16. Kropotkin (1972) pp. 30–31

17. Ibid. p. 69

18. Taylor, Angela K "Living Wih Other Animals" 105–109 in Halliday (1994)

Chapter 20: The Human Lineage

1. Ayala (2014)

2. Schwabe, Christian (1986) "On the Validity of Molecular Evolution" *Trends in Biochemical Sciences* 11: 7, 280–283

3. Schwartz, Jeffrey H and Bruno Maresca (2006) "Do Molecular Clocks Run at All? A Critique of Molecular Systematics" *Biological Theory* 1: 4, 357–371. See

also Schwartz, Jeffrey H (in press) "Systematics and Evolution" in *Encyclopedia of Molecular Cell Biology and Molecular Medicine* edited by R A Meyer. Winheim: Wiley-VCH Verlag

4. Ragan, Mark A, et al. (2009) "The Network of Life: Genome Beginnings and Evolution" *Philosophical Transactions of the Royal Society B: Biological Sciences* 364: 1527, 2169–2175

5. Doolittle, W Ford (2009) "The Practice of Classification and the Theory of Evolution, and What the Demise of Charles Darwin's Tree of Life Hypothesis Means for Both of Them" *Philosophical Transactions of the Royal Society B: Biological Sciences* 364: 1527, 2221–2228

6. http://www.timetree.org/time_query.php?taxon_a=9606&taxon_b=9598 Accessed 18 August 2010

Chapter 21: Causes of Biological Evolution: the Current Orthodox Account

1. Barreiro, Luis B, et al. (2008) "Natural Selection Has Driven Population Differentiation in Modern Humans" *Nat Genet* 40: 3, 340–345

2. http://www.britannica.com/EBchecked/topic/406351/natural-selection Accessed 14 June 2014

3. "Ernst Mayr" *Gale Encyclopedia of Biography* 2006

4. Schwartz, Jeffrey H and Bruno Maresca (2006) "Do Molecular Clocks Run at All? A Critique of Molecular Systematics" *Biological Theory* 1: 4, 357–371

5. Coyne (2006)

6. Mayr (2001) p. 195

7. Williamson, Peter G (1981) "Morphological Stasis and Developmental Constraint: Real Problems for Neo-Darwinism" *Nature* 294, 214–215

8. Hotopp, Julie C Dunning, et al. (2007) "Widespread Lateral Gene Transfer from Intracellular Bacteria to Multicellular Eukaryotes" *Science* 317: 5845, 1753–1756

9. Boto, Luis (2010) "Horizontal Gene Transfer in Evolution: Facts and Challenges" *Proceedings of the Royal Society B: Biological Sciences* 277: 1683, 819–827

10. Lynch, Michael (2007) "The Evolution of Genetic Networks by Non-Adaptive Processes." *Nat Rev Genet* 8: 10, 803–813

11. Anway, Matthew D, et al. (2005) "Epigenetic Transgenerational Actions of Endocrine Disruptors and Male Fertility" *Science* 308: 5727, 1466–1469

12. Jablonka, Eva and Gal Raz (2009) "Transgenerational Epigenetic Inheritance: Prevalence, Mechanisms, and Implications for the Study of Heredity and Evolution" *The Quarterly Review of Biology* 84: 2, 131–176

13. West, Stuart A, et al. (2011) "Sixteen Common Misconceptions About the Evolution of Cooperation in Humans" *Evolution and Human Behavior* 32: 4, 231–262

14. Darwin, Charles (1872) p. 428

15. Heylighen (1999)

16. Valentine, J W, et al. (1994) "Morphological Complexity Increase in Metazoans" *Paleobiology* 20: 2, 131–142

17. Bonner (1988) p. 5

18. Gould (2004)

19. Ibid.

20. Ibid.

21. Borowsky, R and H Wilkens (2002) "Mapping a Cave Fish Genome: Polygenic Systems and Regressive Evolution" *J Hered* 93: 1, 19–21
22. http://www.bio.sci.osaka-u.ac.jp/~hfuruya/dicyemids.html Accessed 24 February 2011
23. Simpson (1949)
24. Bains, William (1987) "Evolutionary Paradoxes and Natural Non-Selection" *Trends in Biochemical Sciences* 12, 90–91
25. Nitecki (1988)
26. Gould, Stephen Jay (1988) "On Replacing the Idea of Progress with an Operational Notion of Directionality" in *Evolutionary Progress* edited by Matthew H Nitecki, Chicago; London: University of Chicago Press, p. 319
27. Bains (1987)
28. Mayr (1988) pp. 251–252
29. Simpson (1949) p. 262
30. Huxley (1923) p. 40
31. Dobzhansky (1956) p. 86

Chapter 22: Complementary and Competing Hypotheses 1: Complexification

1. Wells (2000)
2. Behe (2007)
3. Eldredge (1995) pp. 64–66
4. Eldredge, Niles and Stephen J Gould (1972) "Punctuated Equilibria: An Alternative to Phyletic Gradualism" 82-115 in *Models in Paleobiology* San Francisco: Freeman Cooper
5. Dawkins (1996) pp. 250–251
6. Maresca, B and J H Schwartz (2006) "Sudden Origins: A General Mechanism of Evolution Based on Stress Protein Concentration and Rapid Environmental Change" *The Anatomical Record Part B: The New Anatomist* 289B: 1, 38–46
7. Stebbins, G. Ledyard and Francisco J Ayala (1981) "Is a New Evolutionary Synthesis Necessary?" *Science* 213: 4511, 967–971
8. Williamson, Peter G (1981) "Morphological Stasis and Developmental Constraint: Real Problems for Neo-Darwinism" *Nature* 294, 214–215
9. Kimura (1983)
10. Orr, H Allen (2009) "Testing Natural Selection" *Scientific American* 300, 44–51
11. Kasahara, Masanori (2007) "The 2R Hypothesis: An Update" *Current Opinion in Immunology* 19: 5, 547–552
12. Shubin, Neil, et al. (2009) "Deep Homology and the Origins of Evolutionary Novelty" *Nature* 457: 7231, 818–823
13. Conway Morris (2005)
14. Kauffman, Stuart A (1991) "Antichaos and Adaptation" *Scientific American*, 78–84 and Kauffman (1996)
15. Koonin, Eugene V (2011) "Are There Laws of Genome Evolution?" *PLoS Comput Biol* 7: 8, e1002173
16. Shapiro (2011)
17. Noble (2006) p. 21
18. Brenner, Sydney (2010) "Sequences and Consequences" *Philosophical Transactions of the Royal Society B: Biological Sciences* 365: 1537, 207–212

19. Lovelock (1991) p. 99
20. Sheldrake (2009) pp. 222–229

Chapter 23: Complementary and Competing Hypotheses 2: Collaboration

1. Wilson (2000)
2. Darwin, Charles (1882) pp. 131–132
3. Williams (1996) p. 93
4. Ibid. p. vii
5. Haldane, J B S (1955) "Population Genetics" *New Biology* 18, 34–51
6. Hamilton, W D (1964) "The Genetical Evolution of Social Behaviour 1" *Journal of Theoretical Biology* 7: 1, 1–16
7. Harman (2010)
8. Wilson, David Sloan and Edward O Wilson (2007) "Survival of the Selfless" *New Scientist*: 3 November
9. Trivers, Robert L (1971) "The Evolution of Reciprocal Altruism" *The Quarterly Review of Biology* 46: 1, 35–57
10. Wilkinson, Gerald S (1984) "Reciprocal Food Sharing in the Vampire Bat" *Nature* 308: 5955, 181–184; Wilkinson, Gerald S (1985) "The Social Organization of the Common Vampire Bat" *Behavioral Ecology and Sociobiology* 17: 2, 123–134
11. See http://harvardmagazine.com/2010/08/harvard-dean-details-hauser-scientific-misconduct 20 August 2010; http://www.thecrimson.com/article/2010/9/14/hauser-lab-research-professor/?page=single 14 September 2010; and http://grants.nih.gov/grants/guide/notice-files/NOT-OD-12-149.html "Findings of Research Misconduct" 10 September 2012
12. Hauser, Marc, et al. (2009) "Evolving the Ingredients for Reciprocity and Spite" *Philosophical Transactions of the Royal Society B: Biological Sciences* 364: 1533, 3255–3266
13. West, Stuart A, et al. (2011) "Sixteen Common Misconceptions About the Evolution of Cooperation in Humans" *Evolution and Human Behavior* 32: 4, 231–262
14. Gardner, Andy, et al. (2007) "Spiteful Soldiers and Sex Ratio Conflict in Polyembryonic Parasitoid Wasps" *The American Naturalist* 169: 4, 519–533
15. Wilson, Edward O (2008) "One Giant Leap: How Insects Achieved Altruism and Colonial Life" *BioScience* 58: 1, 17–25
16. Dawkins (1989) p. 2
17. Ibid. p. 87
18. Ibid. p. 2
19. Ibid. pp. 265–266
20. Ibid. p. 248
21. Dawkins, Richard (2006) "It's All in the Genes" *Sunday Times* London, 19 March; Books 43–44
22. Dawkins (1989) p. 45
23. Ibid. p. 88
24. Ibid. p. 264–266
25. Koslowski (1999) p. 308
26. Gould (2002) p. 614
27. Dawkins (1989) p. 37

28. Ibid. p. 36
29. Ibid. p. 105
30. Ibid. p. 140
31. Ibid. p. 101
32. Dawkins, Richard (2009) "The Genius of Charles Darwin, Episode 2" *The Genius of Charles Darwin* United Kingdom: Channel 4 Television, 4 October
33. Roughgarden (2009)
34. Dawkins, Richard (2007) "Genes Still Central" *New Scientist* 15 December
35. Clutton-Brock, T, et al. (2009) "The Evolution of Society" *Philosophical Transactions of the Royal Society B: Biological Sciences* 364: 1533, 3127–3133
36. Kropotkin (1978) p. 69
37. Margulis (1970)
38. Margulis, Lynn, et al. (2005) "'Imperfections and Oddities' in the Origin of the Nucleus" *Paleobiology* 31 (sp5), 175–191 plus personal communications 8 August to 3 October 2011
39. Teresi, Dick (2011) "Lynne Margulis" *Discover Magazine* April, 66–71
40. http://whyevolutionistrue.wordpress.com/2011/04/12/lynn-margulis-disses-evolution-in-discover-magazine-embarrasses-both-herself-and-the-field/ 12 April 2011
41. Doolittle, W Ford (2000) "Uprooting the Tree of Life" *Scientific American* 282, 90–95

Chapter 24: The Evolution of Consciousness

1. Conway Morris pp. 197–200
2. Chomsky (2006)
3. Tomasello (2003)
4. Reader, Simon M and Kevin N Laland (2002) "Social Intelligence, Innovation, and Enhanced Brain Size in Primates" *Proceedings of the National Academy of Sciences of the United States of America* 99: 7, 4436–4441
5. Azevedo, Frederico A C, et al. (2009) "Equal Numbers of Neuronal and Nonneuronal Cells Make the Human Brain an Isometrically Scaled-up Primate Brain" *The Journal of Comparative Neurology* 513: 5, 532–541
6. Kendrick, Keith (2010) "Understanding the Brain: A Work in Progress" *Gresham College Lecture*. London, 22 November
7. Deaner, R O, et al. (2007) "Overall Brain Size, and Not Encephalization Quotient, Best Predicts Cognitive Ability across Non-Human Primates" *Brain, Behavior and Evolution* 70: 2, 115–124
8. Lahr, Marta Mirazón (2011) "African Origins – the Morphological and Behavioural Evidence of Early Humans in Africa" *Human Evolution, Migration and History Revealed by Genetics, Immunity and Infection* Royal Society, London, 6 June
9. Jolicoeur, Pierre, et al. (1984) "Brain Structure and Correlation Patterns in Insectivora, Chiroptera, and Primates" *Systematic Zoology* 33: 1, 14–29

Chapter 26: The Emergence of Humans

1. Stringer (2011) p. 28
2. Morris (1986)

3. Green, Richard E, et al. (2010) "A Draft Sequence of the Neandertal Genome" *Science* 328: 5979, 710–722
4. Jorde, Lynn B and Stephen P Wooding (2004) "Genetic Variation, Classification and 'Race'" *Nat Genetics* 36, 528–533
5. Derricourt, Robin (2005) "Getting 'out of Africa': Sea Crossings, Land Crossings and Culture in the Hominin Migrations" *Journal of World Prehistory* 19: 2, 119–132
6. Stoneking, Mark (2008) "Human Origins" *EMBO Reports* 9: S1, S46–S50
7. Stringer (2011) pp. 33–44
8. Cann, R L, et al. (1987) "Mitochondrial DNA and Human Evolution" *Nature* 325: 6099, 31–36
9. Stringer (2011) pp. 23–24
10. Cruciani, Fulvio, et al. (2011) "A Revised Root for the Human Y Chromosomal Phylogenetic Tree: The Origin of Patrilineal Diversity in Africa" *The American Journal of Human Genetics* 88: 6, 814–818
11. Underhill, P A, et al. (2001) "The Phylogeography of Y Chromosome Binary Haplotypes and the Origins of Modern Human Populations" *Annals of Human Genetics* 65: 1, 43–62
12. http://web.archive.org/web/20070706095314/http://www.janegoodall.com/chimp_central/default.asp Accessed 8 December 2011
13. "Bonobo" *The Columbia Electronic Encyclopedia*, Sixth Edition Columbia University Press, 2011 Accessed 8 December 2011; http://www.britannica.com/EBchecked/topic/73224/bonobo Accessed 16 January 2012
14. Brunet, Michel, et al. (2005) "New Material of the Earliest Hominid from the Upper Miocene of Chad" *Nature* 434: 7034, 752–755; Zollikofer, Christoph P E, et al. (2005) "Virtual Cranial Reconstruction of Sahelanthropus Tchadensis" *Nature* 434: 7034, 755–759
15. Henke, et al. (2007)
16. White, T D (2009) "Human Origins and Evolution: Cold Spring Harbor, Déjà Vu" *Cold Spring Harbor Symposia on Quantitative Biology*; Lovejoy, C Owen (2009) "Reexamining Human Origins in Light of Ardipithecus Ramidus." *Science* 326: 5949, 74, 74e1–74e8
17. Pickering, Robyn, et al. (2011) "Australopithecus Sediba at 1.977 Ma and Implications for the Origins of the Genus Homo" *Science* 333: 6048, 1421–1423
18. Spoor, Fred (2011) "Palaeoanthropology: Malapa and the Genus Homo" *Nature* 478: 7367, 44–45
19. Leakey, Meave G, et al. (2012) "New Fossils from Koobi Fora in Northern Kenya Confirm Taxonomic Diversity in Early Homo" *Nature* 488: 7410, 201–204
20. Reich, David, et al. (2010) "Genetic History of an Archaic Hominin Group from Denisova Cave in Siberia" *Nature* 468: 7327, 1053–1060
21. Green, Richard E, et al. (2010) "A Draft Sequence of the Neandertal Genome" *Science* 328: 5979, 710–722
22. "Mousterian" *The Concise Oxford Dictionary of Archaeology*. Oxford University Press, 2003
23. Zilhão, João, et al. (2010) "Symbolic Use of Marine Shells and Mineral Pigments by Iberian Neandertals" *Proceedings of the National Academy of Sciences* 107: 3, 1023–1028
24. Brown, P, et al. (2004) "A New Small-Bodied Hominin from the Late Pleistocene of Flores, Indonesia" *Nature* 431: 7012, 1055–1061

25. Brown, P and T Maeda (2009) "Liang Bua *Homo Floresiensis* Mandibles and Mandibular Teeth: A Contribution to the Comparative Morphology of a New Hominin Species" *Journal of Human Evolution* 57: 3, 571–596
26. Reich, David, et al. (2010) "Genetic History of an Archaic Hominin Group from Denisova Cave in Siberia" *Nature* 468: 7327, 1053–1060
27. White, Tim D, et al. (2003) "Pleistocene Homo Sapiens from Middle Awash, Ethiopia" *Nature* 423: 6941, 742–747
28. Fleagle, John G, et al. (2008) "Paleoanthropology of the Kibish Formation, Southern Ethiopia: Introduction" *Journal of Human Evolution* 55: 3, 360–365
29. "Paleolithic" *McGraw-Hill Encyclopedia of Science and Technology* 2005
30. Bahn, Paul Gerard "Stone Age" *Microsoft Encarta Online Encyclopedia* (2005) http://uk.encarta.msn.com/
31. Stringer (2011) pp. 125–126
32. Brown, Kyle S, et al. (2009) "Fire as an Engineering Tool of Early Modern Humans" *Science* 325: 5942, 859–862
33. James, Steven R (1989) "Hominid Use of Fire in the Lower and Middle Pleistocene" *Current Anthropology* 30: 1, 1–26
34. http://www.wits.ac.za/academic/research/ihe/archaeology/blombos/7106/blomboscave.html Accessed 1 February 2012
35. Botha and Knight (2009); Henshilwood, Christopher S, et al. (2002) "Emergence of Modern Human Behavior: Middle Stone Age Engravings from South Africa" *Science* 295: 5558, 1278–1280; d'Errico, Francesco, et al. (2009) "Additional Evidence on the Use of Personal Ornaments in the Middle Paleolithic of North Africa" *Proceedings of the National Academy of Sciences* 106: 38, 16051–16056
36. d'Erico at al. (2009)
37. Roberts, Richard G, et al. (1994) "The Human Colonisation of Australia: Optical Dates of 53,000 and 60,000 Years Bracket Human Arrival at Deaf Adder Gorge, Northern Territory" *Quaternary Science Reviews* 13: 5, 575–583
38. Bahn (2005)
39. Stringer (2011) p. 126
40. Bowler, James M, et al. (2003) "New Ages for Human Occupation and Climatic Change at Lake Mungo, Australia" *Nature* 421: 6925, 837–840
41. Zorich, Zach (2011) "A Chauvet Primer" *Archaeology* 64: 2
42. Balter, Michael (2008) "Going Deeper into the Grotte Chauvet" *Science* 321: 5891, 904–905
43. Balter, Michael (2000) "Paintings in Italian Cave May Be Oldest Yet" *Science* 290: 5491, 419–421
44. http://www.lascaux.culture.fr/#/en/04_00.xml Accessed 1 February 2012
45. Many of the sculptures were displayed at the British Museum exhibition "Ice Age Art: Arrival of the Modern Mind", visited 17 May 2013. See also Cook (2013) and Conard, Nicholas J. (2009) "A Female Figurine from the Basal Aurignacian of Hohle Fels Cave in Southwestern Germany" *Nature* 459: 7244, 248–252
46. Schneider, Achim (2004) "Ice-Age Musicians Fashioned Ivory Flute" *Nature News.* http://www.nature.com/news/2004/041217/full/news041213-14.html Accessed 28 January 2012; Stringer (2011) pp. 119–120
47. Renfrew, Colin, et al. (2008) "Introduction. The Sapient Mind: Archaeology Meets Neuroscience" *Philosophical Transactions of the Royal Society B: Biological Sciences* 363: 1499, 1935–1938

48. Department of Arts of Africa, Oceania, and the Americas, The Metropolitan Museum of Art, 2000 http://www.metmuseum.org/toah/hd/apol/hd_apol.htm Accessed 15 December 2011

49. http://www.abc.net.au/news/2010-05-31/megafauna-cave-painting-could-be-40000-years-old/847564 Accessed 15 December 2011

50. http://whc.unesco.org/en/list/925/ Accessed 15 December 2011

51. Enard, Wolfgang, et al. (2002) "Molecular Evolution of FOXP2, a Gene Involved in Speech and Language" *Nature* 418: 6900, 869–872

52. Shu, Weiguo, et al. (2007) "FOXP2 and FOXP1 Cooperatively Regulate Lung and Esophagus Development" *Development* 134: 10, 1991–2000

53. Sources include Stringer (2011); d'Errico, Francesco and Chris B Stringer (2011) "Evolution, Revolution or Saltation Scenario for the Emergence of Modern Cultures?" *Philosophical Transactions of the Royal Society B: Biological Sciences* 366: 1567, 1060–1069; Renfrew, Colin, et al. (2008); Green, Richard E, et al. (2010); Klein, Richard G, et al. (2004) "The Ysterfontein 1 Middle Stone Age Site, South Africa, and Early Human Exploitation of Coastal Resources" *Proceedings of the National Academy of Sciences of the United States of America* 101: 16, 5708–5715; McBrearty, Sally and Alison S Brooks (2000) "The Revolution That Wasn't: A New Interpretation of the Origin of Modern Human Behavior" *Journal of Human Evolution* 39: 5, 453–563; Tuttle (2014)

54. McBrearty, Sally and Nina G Jablonski (2005) "First Fossil Chimpanzee" *Nature* 437: 7055, 105–108

55. Maslin, Mark, et al. (2005) "A Changing Climate for Human Evolution" *Geotimes* September http://www.geotimes.org/sept05/feature_humanclimateevolution.html Accessed 17 June 2014

56. www.smithsonianmag.com/arts-culture/Q-and-A-Rick-Potts.html Accessed 31 January 2012

Chapter 27: Human Evolution 1: Primeval Thinking

1. Hawks, John, et al. (2007) "Recent Acceleration of Human Adaptive Evolution" *Proceedings of the National Academy of Sciences* 104: 52, 20753–20758

2. Jorde, Lynn B and Stephen P Wooding (2004) "Genetic Variation, Classification and 'Race'" *Nat Genetics* 36, 528–533

3. Jones, Steve (2008) "Is Human Evolution Over?" *UCL Lunchtime Lecture.* London, 25 October

4. Keeley (1996)

5. LeBlanc and Register (2003)

6. Hill and Hurtado (1996)

7. http://www.survivalinternational.org/tribes Accesssed 4 May 2014

8. http://www.britannica.com/EBchecked/topic/302707/Jericho Accessed 4 May 2014

9. *The Revised English Bible* Joshua 6: 1–27

10. Fagan (2004)

11. Ibid.; Kuhrt (1995)

12. Wilson, E Jan (1999) "Inside a Sumerian Temple: The Ekishnugal at Ur" in *The Temple in Time and Eternity* edited by Donald W Parry and Stephen David Ricks. Provo, Utah: Foundation for Ancient Research and Mormon Studies at Brigham Young University; Gavin White, personal communications 22 and 27 June 2012

13. Apart from specific citations, evidence from ancient Egypt is taken from http://www.digitalegypt.ucl.ac.uk/ Accessed 24 May 2012 and Romer (2012)
14. Van de Mieroop (2011) pp. 163–164
15. "Stonehenge" *The Concise Oxford Dictionary of Archaeology*, Oxford University Press 2003; "Stonehenge" *The Columbia Electronic Encyclopedia* Sixth Edition, Columbia University Press 2012
16. Oliver, Neil (2012) "Orkney's Stone Age Temple" *A History of Ancient Britain* United Kingdom: BBC HD TV, 1 January; Kinchen, Rosie (2012) "Temple Discovery Rewrites Stone Age" *Sunday Times* London, 1 January, News 10; http://www.scotsman.com/news/cathedral-as-old-as-stonehenge-unearthed-1-764826 13 August 2009
17. Eogan (1986); O'Kelly (1991); plus visit by author in 1993
18. Sources include Possehi (1996); Allchin and Allchin (1997); "Indus valley civilization" *The Columbia Electronic Encyclopedia* Sixth Edition, Columbia University Press 2012; http://www.infinityfoundation.com/mandala/history_overview_frameset.htm Accessed 27 May 2012
19. See, for example, http://www.hindunet.org/hindu_history/ancient/mahabharat/mahab_vartak.html Accessed 27 May 2012
20. Sources include Lagerwey and Kalinowski (2009); "Anyang" *The Concise Oxford Dictionary of Archaeology* Oxford University Press 2003; http://www.britannica.com/EBchecked/topic/114678/Zhou-dynasty Accessed 5 May 2014
21. Department of Asian Art. "Neolithic Period in China" In Heilbrunn *Timeline of Art History* New York: The Metropolitan Museum of Art, http://www.metmuseum.org/toah/hd/cneo/hd_cneo.htm October 2004
22. Sources include Alonzo (1995); Morales (1993); Davies (1990); plus visits by the author in 1996 and 1997
23. Crystal (1987)
24. del Carmen Rodríguez Martínez, Maria, et al. (2006) "Oldest Writing in the New World" *Science* 313: 5793, 1610–1614
25. Ruggles (2005) pp. 133–134
26. Hawkins and White (1971)
27. O'Kelly (1991)
28. Melville (2005); Nissen, et al. (1993); White (2008); Robson, Eleanor (2002) "Words and Pictures: New Light on Plimpton 322" *American Mathematical Monthly*, 105-120
29. http://www-gap.dcs.st-and.ac.uk/~history/HistTopics/Indian_sulbasutras.html Accessed 29 May 2012
30. Kak, Subhash C (1997) "Archaeoastronomy and Literature" *Current Science* 73: 7, 624–627; Kak, S C (1995) "The Astronomy of the Age of Geometric Altars" *Quarterly Journal of the Royal Astronomical Society* 36: 4, 385–395
31. "Chinese astronomy" *Dictionary of Astronomy*, John Wiley & Sons, Inc. Wiley-Blackwell 2004
32. Needham and Wang (1959); Needham and Ronan (1978)
33. Magli (2009) pp. 172–182
34. Vail, Gabrielle and Christine Hernández, 2011 *The Maya Codices Database, Version 3.0.* http://www.mayacodices.org Accessed 18 August 2012
35. "Akhenaten" *Gale Encyclopedia of Biography* 2006
36. *The Upanishads* (1987) pp. 7–11; Smart (1992) pp. 53–55
37. Lagerwey and Kalinowski (2009) pp. 4–34
38. Alonzo (1995) pp. 266–268

39. *The Revised English Bible* 2 Kings 17
40. Ibid. Exodus 7–12
41. Ibid. Ezekiel 1
42. Ibid. Isaiah 6: 1–7
43. Ibid. Daniel 8: 15–26 and 9: 20–27
44. Ibid. Daniel, 10: 1–21
45. Ibid. Job 1: 6–22 and 2: 1–7

Chapter 28: Human Evolution 2: Philosophical Thinking

1. Russell (1946) p. 21
2. Quinton, Anthony (1995) "History of Centres and Departments of Philosophy" 670–672 in *The Oxford Companion to Philosophy* edited by Ted Honderich. Oxford: Oxford University Press
3. Sources for this section include *The Upanishads* (1987); Nagler (1987); *The Upanishads* (1884); *The Ten Principal Upanishads* (1938); Honderich (1995); Smart (1992); plus two weeks at the Mandala Yoga Ashram (Director, Swami Nishchalananda Saraswati) in 2002
4. Quoted in Nagler (1987) p. 300
5. *The Upanishads* (1987) p. 21
6. Nagler (1987) p. 253
7. Mohanty, Jitendra Nath "Philosophy, Indian" *Microsoft Encarta Online Encyclopedia* (2005) http://uk.encarta.msn.com
8. Chakrabarti, Arindam (1995) "Indian Philosophy" 401–404 in *The Oxford Companion to Philosophy*
9. David Frawley, personal communication 1 March 2004
10. Quoted in Nagler (1987) p. 256
11. Sources for this section include Confucius (1893); Zhuangzi (1891); Riegel (2012); Shun, Kwong-loi (1995) "Taoism" 864–865 in *The Oxford Companion to Philosophy*; Fraser (2012); Hansen (2012); "Confucianism" *The Columbia Electronic Encyclopedia* Sixth Edition, Columbia University Press 2012; Smart (1992) pp. 103–114; Capra (2000) pp. 101–118
12. Confucius 15:24
13. Zhuangzi (1891) Chapter 22:5
14. Sources for this section include Russell (1946) pp. 10–101; Gottlieb (2001) pp. 1–108; O'Grady (2006); Couprie (2006); Curd (2012); Hussey, E L (1995) "Heraclitus of Ephesus" 351–352 in *The Oxford Companion to Philosophy*; Taylor, C C W (1995) "Sophists" 839–840 in *The Oxford Companion to Philosophy*
15. Russell (1946) p. 37
16. Sources for this section include *The Revised English Bible*; Armstrong (2006)
17. Sources for this section include Chakrabarti, Arindam (1995) "Indian Philosophy" 401–404 in *The Oxford Companion to Philosophy*; Frawley (1992); Batchelor (1998); Bronkhorst (2011); Keown (2003); Mohanty (2005); Smart (1992) pp. 55–102; Pauling (1997); Shun (1995); plus a nine-day study and meditation retreat at the Buddhist Gaia House in Devon led by Stephen and Martina Batchelor, several one-day guided meditation retreats at Gaia House London, and courses and guided meditation sessions at the North London Buddhist Centre.
18. Gombrich (2006) p. 8

19. Sources for this section include Hansen (2012); Fraser (2012); Smart (1992) pp. 103–129
20. Sources for this section include Gottlieb (2001) pp. 131–431; Russell (1946) pp. 102–510; Plato (1965); Shields (2012); Charles, David (1995) "Aristotelianism" 50-51 in *The Oxford Companion to Philosophy*
21. Quoted in Gottlieb (2001) p. 131
22. Marcus Aurelius *Meditations* VII, 28 quoted in Gottlieb (2001) p. 314
23. Plotinus (2010) VI, 9th Tractate, 10
24. Ibid. VI, 9th Tractate, 11
25. Quinton, Anthony (1995) "History of Centres and Departments of Philosophy" 670–672 in *The Oxford Companion to Philosophy*

Chapter 29: Human Evolution 3: Scientific Thinking

1. Roberts (1989) pp. 75–81
2. Quoted in Stachel (2002) p. 89
3. Ni (1995)
4. Ragep, F Jamil (2007) "Copernicus and His Islamic Predecessors: Some Historical Remarks" *History of Science* 45, 65–81; Saliba, George (1999) "Whose Science is Arabic Science in Renaissance Europe?" http://www.columbia.edu/~gas1/project/visions/case1/sci.1.html#t1 Accessed 30 November 2012
5. Gottlieb (2001) p. 386
6. Ibid. pp. 402–403
7. Van Helden, Albert (1995) http://galileo.rice.edu/sci/instruments/telescope.html#4 Accessed 11 May 2014
8. Gottlieb (2001) p. 414
9. http://www.eso.org/public/teles-instr/e-elt.html Accessed 8 March 2013
10. Al-Rawi, Munim M (2002) "The Contribution of Ibn Sina (Avicenna) to the Development of Earth Sciences" *Foundation for Science Technology and Civilisation* http://www.muslimheritage.com/uploads/ibnsina.pdf Accessed 11 May 2014
11. Nuland, Sherwin B (2007) "Bad Medicine" *New York Times* 8 July, Book Review
12. Crick (1995) p. 3
13. Ramachandran, V S and Colin Blakemore (2003) "Consciousness" in *The Oxford Companion to the Body* Oxford: Oxford University Press
14. Cosmides, Leda and John Tooby (1997) "Evolutionary Psychology: A Primer" http://www.cep.ucsb.edu/primer.html Accessed 30 January 2013
15. Cosmides and Tooby (1997); Wright (1996)
16. See, for example, MacIntyre, Ferren and Kenneth W Estep (1993) "Sperm Competition and the Persistence of Genes for Male Homosexuality" *Biosystems* 31: 2–3, 223–233
17. Price (1965)
18. Larsen, Peder Olesen and Markus Ins (2010) "The Rate of Growth in Scientific Publication and the Decline in Coverage Provided by Science Citation Index" *Scientometrics* 84: 3, 575–603
19. "Life Expectancy" *Gale Encyclopedia of American History* (2006)
20. Wilson (1998) pp. 40–41
21. Bohm (1980) p. 134
22. Schrödinger (1964)

23. Quoted in Braden (2008) p. 212 from a lecture "Das Wesen der Materie" given by Planck in 1944 in Florence.
24. Capra (2000) p. 25

Chapter 30: Uniqueness of Humans

1. Roth (2001) p. 555
2. Pollard, K (2009) "What Makes Us Human?" *Scientific American Magazine* 300: 5, 44–49
3. http://www.janegoodall.org/chimpanzees/tool-use-hunting-other-discoveries Accessed 11 May 2013
4. Pinker (2000) pp. 367–374
5. Guihard-Costa, Anne-Marie and Fernando Ramirez-Rozzi (2004) "Growth of the Human Brain and Skull Slows Down at About 2.5 Years Old" *Comptes Rendus Palevol* 3: 5, 397–402
6. Patton, Paul (2008) "One World, Many Minds: Intelligence in the Animal Kingdom" *Scientific American Mind* December, 72–79
7. Mercader, Julio, et al. (2007) "4,300-Year-Old Chimpanzee Sites and the Origins of Percussive Stone Technology" *Proceedings of the National Academy of Sciences* 104: 9, 3043–3048

Chapter 31: Conclusions and Reflections on the Emergence and Evolution of Humans

1. http://www.pepysdiary.com/diary/1660/10/ Accessed 31 March 2012
2. http://www.deathpenaltyinfo.org/states-and-without-death-penalty Accessed 13 May 2014
3. Pinker (2012) pp. 62–63
4. http://www.pugwash.org/about/manifesto.htm Accessed 1 April 2013
5. http://coursesa.matrix.msu.edu/~hst306/documents/indust.html Accessed 1 April 2013
6. http://www.nobelprize.org/nobel_prizes/peace/laureates/2012/ Accessed 31 March 2012
7. See, for example, http://www.iranhrdc.org/english/publications/reports/3401-surviving-rape-in-iran-s-prisons.html#.UVn_jHD5hhK Accessed 1 April 2013
8. Kropotkin (1972) p. 114
9. Cannadine (2013)
10. Kropotkin (1972) pp. 113–140
11. For an account of the origins and development of the cooperative movement, see Hands (1975) pp. 13–28
12. http://ica.coop/en Accessed 23 March 2013
13. http://ica.coop/en/whats-co-op/co-operative-facts-figures Accessed 25 March 2013
14. http://www.itu.int/ITU-D/ict/facts/2011/material/ICTFactsFigures2011.pdf Accessed 5 April 2013
15. http://www.ericsson.com/news/1775026 7 April 2014
16. See, for example, http://www.hrw.org/news/2013/04/25/bangladesh-tragedy-shows-urgency-worker-protections 25 April 2013
17. Teilhard de Chardin (1965); Huxley (1965)

Chapter 32: Limitations of Science

1.   Borjigin, Jimo, et al. (2013) "Surge of Neurophysiological Coherence and Connectivity in the Dying Brain" *Proceedings of the National Academy of Sciences* 12 August

2.   Parnia, Sam, et al. (2014) "AWARE—AWAreness During REsuscitation—a Prospective Study" *Resuscitation* 85: 12, 1799–1805

3.   Beauregard (2012) p. 10

4.   Persinger, M A, et al. (2002) "Remote Viewing with the Artist Ingo Swann: Neuropsychological Profile, Electroencephalographic Correlates, Magnetic Resonance Imaging (MRI), and Possible Mechanisms" *Perceptual and Motor Skills* 94: 3, 927–949

5.   Targ (2012)

6.   Radin (2009)

7.   Radin (2006)

8.   Flammer, Erich and Walter Bongartz (2003) "On the Efficacy of Hypnosis: A Meta-Analytic Study" *Contemporary Hypnosis* 20: 4, 179–197

9.   See, for example, the statement issued in 1995 by the National Institute for Health "Integration of Behavioral & Relaxation Approaches into the Treatment of Chronic Pain & Insomnia" http://consensus.nih.gov/1995/1995BehaviorRelaxPainInsomniata017PDF.pdf

10   Grau, Carles, et al. (2014) "Conscious Brain-to-Brain Communication in Humans Using Non-Invasive Technologies" *PLoS One* 9: 8, e105225

11   Fanelli, Daniele (2009) "How Many Scientists Fabricate and Falsify Research? A Systematic Review and Meta-Analysis of Survey Data" *PLoS One* 4: 5, e5738

12   Huth, Alexander G, et al. (2012) "A Continuous Semantic Space Describes the Representation of Thousands of Object and Action Categories across the Human Brain" *Neuron* 76: 6, 1210–1224

13   http://www.ucl.ac.uk/news/news-articles/1009/10091604 16 September 2010

14   Ellis (2002) Chapter 8

Bibliography

A date in brackets is the date of the edition consulted. A date at the end of a title is the date of the first edition of the work if this is different.

Adams, Fred and Greg Laughlin (1999) *The Five Ages of the Universe: Inside the Physics of Eternity* New York: Free Press

Ahmed, Akbar S (2007) *Journey into Islam: The Crisis of Globalization* Washington, DC: Brookings Institution Press

Allchin, Bridget and F Raymond Allchin (1997) *Origins of a Civilization: The Prehistory and Early Archaeology of South Asia* New Delhi: Viking

Alonzo, Gualberto Zapata (1995) *An Overview of the Mayan World* 1993 Mérida: Ediciones Alducin

Alvarez, Walter (1997) *T Rex and the Crater of Doom* Princeton, NJ: Princeton University Press

Armstrong, Karen (2006) *The Great Transformation: The World in the Time of Buddha, Socrates, Confucius and Jeremiah* London: Atlantic

Arp, Halton C (1998) *Seeing Red: Redshifts, Cosmology and Academic Science* Montreal: Apeiron

Ayala, Francisco J (2014) "Evolution" *Encyclopædia Britannica Online* http://www.britannica.com/EBchecked/topic/197367/evolution Accessed 7 April 2014

Barrow, John D and Frank J Tipler (1996) (paperback ed.) *The Anthropic Cosmological Principle* 1986 Oxford: Oxford University Press

Batchelor, Stephen (1998) *Buddhism Without Beliefs: A Contemporary Guide to Awakening* 1997 London: Bloomsbury

Beauregard, Mario (2012) *Brain Wars: The Scientific Battle Over the Existence of the Mind and the Proof That Will Change the Way We Live Our Lives* New York: HarperOne

Behe, Michael J (1996) *Darwin's Black Box: The Biochemical Challenge to Evolution* New York; London: The Free Press

— (2007) *The Edge of Evolution: The Search for the Limits of Darwinism* New York: Free Press

Bohm, David (1980) *Wholeness and the Implicate Order* London: Routledge and Kegan Paul

Bonner, John Tyler (1988) *The Evolution of Complexity by Means of Natural Selection* Princeton, NJ: Princeton University Press

Botha, Rudolf P and Chris Knight (2009) *The Cradle of Language* Oxford: Oxford University Press

Braden, Gregg (2008) *The Spontaneous Healing of Belief: Shattering the Paradigm of False Limits* London: Hay House

Braibant, Sylvie, et al. (2012) (2nd ed.) *Particles and Fundamental Interactions: An Introduction to Particle Physics* Dordrecht; London: Springer

Bronkhorst, Johannes (2011) *Buddhism in the Shadow of Brahmanism* Leiden: Brill

Bryson, Bill (2004) *A Short History of Nearly Everything* 2003 London: Black Swan

Buddha, The (1997) *Anguttara Nikaya* translated by Thanissaro Bhikku, Access to Insight http://www.accesstoinsight.org/tipitaka/an/index.html – an04.077.than

Cairns-Smith, A G (1990) *Seven Clues to the Origin of Life: A Scientific Detective Story* 1985 Cambridge: Cambridge University Press

Cannadine, David (2013) *The Undivided Past: Humanity Beyond Our Differences* London: Allen Lane

Capra, Fritjof (1997) *The Web of Life: A New Synthesis of Mind and Matter* 1996 London: Flamingo

— (2000) (4th ed.) *The Tao of Physics: An Exploration of the Parallels between Modern Physics and Eastern Mysticism* 1975 Boston: Shambhala

Chang, Raymond (2007) (9th ed.) *Chemistry* Boston, MA; London: McGraw-Hill Higher Education

Chomsky, Noam (2006) (3rd ed.) *Language and Mind* 1968 Cambridge; New York: Cambridge University Press

Confucius *Analects* (1893) translated by James Legge, http://www.sacred-texts.com/cfu/conf1.htm

Conway Morris, S (1998) *The Crucible of Creation: The Burgess Shale and the Rise of Animals* Oxford; New York: Oxford University Press

— (2005) *Life's Solution: Inevitable Humans in a Lonely Universe* 2003 Cambridge: Cambridge University Press

Cook, Jill (2013) *Ice Age Art: Arrival of the Modern Mind* London: British Museum Press

Couprie, Dirk L (2006) "Anaximander" *The Internet Encyclopedia of Philosophy* 2006 http://www.iep.utm.edu/anaximan/ Accessed 17 June 2014

Coyne, Jerry (2006) "Intelligent Design: The Faith That Dare Not Speak Its Name" in *Intelligent Thought* edited by John Brockman, New York: Random House

Coyne, Jerry A and H Allen Orr (2004) *Speciation* Sunderland, MA: Sinauer Associates

Crick, Francis (1990) *What Mad Pursuit: A Personal View of Scientific Discovery* London: Penguin

— (1995) *The Astonishing Hypothesis: The Scientific Search for the Soul* 1990 London: Touchstone

Crystal, David (1987) *The Cambridge Encyclopedia of Language* Cambridge: Cambridge University Press

Curd, Patricia (2012) "Presocratic Philosophy" *The Stanford Encyclopedia of Philosophy* edited by Edward N Zalta, http://plato.stanford.edu/entries/presocratics/ Accessed 17 June 2014

Darwin, Charles (1859) (1st ed.) *On the Origin of Species by Means of Natural Selection, or the Preservation of Favoured Races in the Struggle for Life* London: John Murray

— (1861) (3rd ed.) *On the Origin of Species by Means of Natural Selection, or the Preservation of Favoured Races in the Struggle for Life* 1859 London: John Murray

— (1868) (1st ed.) *The Variation of Animals and Plants under Domestication* London: John Murray

— (1872) (6th ed., with additions and corrections) *On the Origin of Species by Means of Natural Selection, or the Preservation of Favoured Races in the Struggle for Life* 1859 London: John Murray

— (1882) (2nd ed., revised and augmented) *The Descent of Man, and Selection in Relation to Sex* London: John Murray

— (1929) *Autobiography of Charles Darwin: With Two Appendices Comprising a Chapter of Reminiscences and a Statement of Charles Darwin's Religious Views* 1882 London: Watts and Co

— (1958) *The Autobiography of Charles Darwin, 1809–1882: With Original Omissions Restored* London: Collins

Darwin, Erasmus (1796) (2nd ed.) *Zoonomia; or, the Laws of Organic Life* 1794 London: J Johnson

— (1803) *The Temple of Nature; or, the Origin of Society a Poem, with Philosophical Notes* London: J Johnson

Darwin, Francis (editor) (1887) *The Life and Letters of Charles Darwin, Including an Autobiographical Chapter* London: John Murray

Davies, Nigel (1990) *The Ancient Kingdoms of Mexico* 1982 Harmondsworth: Penguin

Davies, Paul (1990) *God and the New Physics* 1983 London: Penguin

Dawkins, Richard (1989) (2nd ed.) *The Selfish Gene* 1976 Oxford: Oxford University Press

— (1996) (with a new introduction) *The Blind Watchmaker* 1987 New York; London: Norton

Desmond, Adrian J (1989) *The Politics of Evolution: Morphology, Medicine, and Reform in Radical London* Chicago: University of Chicago Press

Desmond, Adrian and James R Moore (1992) (new ed.) *Darwin* London: Penguin

Dobzhansky, Theodosius (1956) *The Biological Basis of Human Freedom* New York; London: Columbia University Press; OUP

Doolittle, W Ford (2000) "Uprooting the Tree of Life" *Scientific American* 282, 90–95

Einstein, Albert (1949) (abridged ed.) *The World As I See It* translated by Alan Harris, New York: Philosophical Library

Eldredge, Niles (1995) *Reinventing Darwin: The Great Evolutionary Debate* London: Weidenfeld and Nicolson

Eldredge, Niles and Stephen J Gould (1972) "Punctuated Equilibria: An Alternative to Phyletic Gradualism" 82–115 in *Models in Paleobiology* San Francisco: Freeman Cooper

Eldredge, Niles and Ian Tattersall (1982) *The Myths of Human Evolution* New York; Guildford: Columbia University Press

Elgin, Duane (1993) *Awakening Earth: Exploring the Dimensions of Human Evolution* New York: Morrow

Elliott, David K (2000) "Extinctions and Mass Extinctions" in *Oxford Companion to the Earth* edited by Paul L Hancock and Brian J Skinner, Oxford; New York: Oxford University Press

Ellis, George (2002) "The Universe around Us: An Integrative View of Science and Cosmology" http://www.mth.uct.ac.za/~ellis/cos8.html Accessed 13 August 2013

— (2007) "Issues in the Philosophy of Cosmology" in *The Philosophy of Physics* edited by Jeremy Butterfield and John Earman, Amsterdam; New Holland: Elsevier

Eogan, George (1986) *Knowth and the Passage Tombs of Ireland* London: Thames and Hudson

Fagan, Brian M (2004) *The Long Summer: How Climate Changed Civilization* London: Granta

Finkelstein, Israel and Neil Asher Silberman (2001) *The Bible Unearthed: Archaeology's New Vision of Ancient Israel and the Origin of Its Sacred Texts* New York: Free Press

Fraser, Chris (2012) "Mohism" *Stanford Encyclopedia of Philosophy* edited by Edward N Zalta, http://plato.stanford.edu/entries/mohism/ Accessed 31 August 2012

Frawley, David (1992) *From the River of Heaven: Hindu and Vedic Knowledge for the Modern Age* 1990 Dehli: Motilal Banarsidass

Futuyma, Douglas J (1983) *Science on Trial: The Case for Evolution* New York: Pantheon Books

Gehring, W J (1998) *Master Control Genes in Development and Evolution: The Homeobox Story* New Haven, CN; London: Yale University Press

Gombrich, Richard F (2006) (2nd ed.) *Theravada Buddhism: A Social History from Ancient Benares to Modern Colombo* 1988 London: Routledge

Gorst, Martin (2001) *Measuring Eternity: The Search for the Beginning of Time* New York: Broadway Books

Gottlieb, Anthony (2001) *The Dream of Reason: A History of Western Philosophy from the Greeks to the Renaissance* 2000 London: Penguin

Gould, Stephen Jay (1980) *The Panda's Thumb: More Reflections in Natural History* New York; London: Norton

— (2002) *The Structure of Evolutionary Theory* Cambridge, MA; London: Belknap

— (2004) "The Evolution of Life on Earth" 1994 *Scientific American* 14, 92–100

Graves, Robert (1955) *The Greek Myths* Baltimore: Penguin Books

Gregory, T Ryan (editor) (2005) *The Evolution of the Genome* Burlington, MA; San Diego, CA; London: Elsevier Academic Press

Gribbin, John R (2004) *Deep Simplicity: Chaos, Complexity and the Emergence of Life* London: Allen Lane

Griffiths, David J (1987) *Introduction to Elementary Particles* New York; London: Harper & Row

Guth, Alan H (1997) *The Inflationary Universe: The Quest for a New Theory of Cosmic Origins* London: Jonathan Cape

Halliday, Tim (editor) (1994) *Animal Behavior* Norman: University of Oklahoma Press

Hands, John (1975) *Housing Co-operatives* London: Society for Co-operative Dwellings

Hansen, Chad (2012) "Taoism" *Stanford Encyclopedia of Philosophy* edited by Edward N Zalta, http://plato.stanford.edu/entries/taoism/ Accessed 3 August 2012

Harman, Oren Solomon (2010) *The Price of Altruism: George Price and the Search for the Origins of Kindness* London: Bodley Head

Harwood, Caroline and Merry Buckley (2008) *The Uncharted Microbial World: Microbes and Their Activities in the Environment* Washington, DC: The American Academy of Microbiology

Hawking, Stephen W (1988) *A Brief History of Time: From the Big Bang to Black Holes* London: Bantam

Hawkins, Gerald Stanley and John Baker White (1971) *Stonehenge Decoded* 1966 London: Collins

Henke, Winfried, et al. (2007) *Handbook of Paleoanthropology* Berlin: Springer-Verlag

Heylighen, Francis (1999) "The Growth of Structural and Functional Complexity During Evolution" 17–44 in *The Evolution of Complexity* edited by F Heylighen, et al., Dordrecht: Kluwer Academic Publishers

Hill, Kim and A Magdalena Hurtado (1996) *Aché Life History: The Ecology and Demography of a Foraging People* New York: Aldine de Gruyter

The Holy Qur'an (1938) translated by Yusuf Ali, http://www.sacred-texts.com/isl/quran/index.htm

Honderich, Ted (editor) (1995) *The Oxford Companion to Philosophy* Oxford: Oxford University Press

Hoyle, Fred and Nalin Chandra Wickramasinghe (1978) *Lifecloud: The Origin of Life in the Universe* London: Dent

Huxley, Julian (1923) *Essays of a Biologist* London: Chatto & Windus

— (1965) "Introduction" 11–28 in *The Phenomenon of Man*, 1959 London: Collins

Kauffman, Stuart A (1996) *At Home in the Universe: The Search for Laws of Self-Organization and Complexity* 1995 London: Penguin

Keeley, Lawrence H (1996) *War before Civilization* New York; Oxford: Oxford University Press

Keown, Damien (2003) *A Dictionary of Buddhism* Oxford: Oxford University Press, 2004

Kimura, Motoo (1983) *The Neutral Theory of Molecular Evolution* Cambridge: Cambridge University Press

Kious, W Jacquelyne and Robert I Tilling (1996) "This Dynamic Earth: The Story of Plate Tectonics" Version 1.12 http://pubs.usgs.gov/gip/dynamic/dynamic. html – anchor19309449 Accessed 25 April 2008

Kitcher, Philip (1982) *Abusing Science: The Case against Creationism* Cambridge, MA; London: MIT Press

Koslowski, Peter (1999) "The Theory of Evolution as Sociobiology and Bioeconomics: A Critique of Its Claim to Totality" 301–326 in *Sociobiology and Bioeconomics: The Theory of Evolution in Biological and Economic Theory* edited by Peter Koslowski, Berlin: Springer-Verlag

Krieger, Dolores (1993) *Accepting Your Power to Heal: The Personal Practice of Therapeutic Touch* Santa Fe, NM: Bear & Co.

Krishnamurti, J and David Bohm (1985) *The Ending of Time* London: Gollancz

— (1986) *The Future of Humanity: A Conversation* San Francisco: Harper & Row

— (1999) *The Limits of Thought: Discussions* London: Routledge

Kropotkin, Peter (1972) *Mutual Aid: A Factor of Evolution* 1914 London: Allen Lane

Kuhn, Thomas S and Ian Hacking (2012) (4th edition) *The Structure of Scientific Revolutions* 1962 Chicago; London: The University of Chicago Press

Kuhrt, Amélie (1995) *The Ancient Near East: c.3000–330 BC* London: Routledge

Lack, David Lambert (1947) *Darwin's Finches* Cambridge: Cambridge University Press

Lagerwey, John and Marc Kalinowski (2009) *Early Chinese Religion* Leiden: Brill

Laszlo, Ervin (2006) *Science and the Reenchantment of the Cosmos: The Rise of the Integral Vision of Reality* Rochester, VT: Inner Traditions

Leakey, Richard E and Roger Lewin (1996) *The Sixth Extinction: Biodiversity and Its Survival* 1995 London: Weidenfeld & Nicolson

LeBlanc, Steven A and Katherine E Register (2003) *Constant Battles: The Myth of the Peaceful, Noble Savage* New York: St Martin's Press

Lerner, Eric J (1992) (paperback ed.) *The Big Bang Never Happened* 1991 New York, NY: Vintage

Linde, Andrei (2001) "The Self-Reproducing Inflationary Universe" *Scientific American* Special Issue 26 November http://www.sciam.com/specialissues/039 8cosmos/0398linde.html Accessed 17 August 2006

Lochner, James C, et al. (2005) "What Is Your Cosmic Connection to the

Elements?" *Imagine the Universe* http://imagine.gsfc.nasa.gov/docs/teachers/
elements/imagine/contents.html Accessed 24 June 2007

Lovelock, James (1991) *Healing Gaia: Practical Medicine for the Planet* New York:
Harmony Books

Lunine, Jonathan Irving (1999) *Earth: Evolution of a Habitable World* Cambridge:
Cambridge University Press

Maddox, John (1998) *What Remains To Be Discovered: Mapping the Universe, the
Origins of Life, and the Future of the Human Race* London: Macmillan

Magli, Giulio (2009) *Mysteries and Discoveries of Archaeoastronomy: From Giza to
Easter Island* New York; London: Copernicus Books

Magueijo, João (2003) *Faster Than the Speed of Light: The Story of a Scientific
Speculation* Cambridge, MA: Perseus Book Group

Maillet, Benoit de (1968) *Telliamed; or, Conversations between an Indian
Philosopher and a French Missionary on the Diminution of the Sea* translated by
Albert V Carozz, Urbana: University of Illinois Press

Margulis, Lynn (1970) *Origin of Eukaryotic Cells: Evidence and Research
Implications for a Theory of the Origin and Evolution of Microbial, Plant, and
Animal Cells on the Precambrian Earth* New Haven; London: Yale University
Press

Mayr, Ernst (1982) *The Growth of Biological Thought: Diversity, Evolution, and
Inheritance* Cambridge, MA: Belknap Press

— (1988) *Towards a New Philosophy of Biology* Cambridge, MA: Belknap

— (2001) *What Evolution Is* New York: Basic Books

McFadden, Johnjoe (2000) *Quantum Evolution* London: Flamingo

Melville, Duncan J "Mesopotamian Mathematics" (2005) *St Lawrence University*
http://it.stlawu.edu/~dmelvill/mesomath Accessed 25 February 2006

Morales, Demetrio Sodi (1993) *The Maya World* 1989 Mexico: Minutiae Mexicana

Morowitz, Harold J (2004) *The Emergence of Everything: How the World Became
Complex* 2002 New York: Oxford University Press

Morris, Desmond (1986) (revised ed.) *The Illustrated Naked Ape: A Zoologist's
Study of the Human Animal* 1967 London: Cape

Nagler, Michael N (1987) "Reading the Upanishads" 251–301 in *The Upanishads*
translated by Eknath Easwaran, Petaluma, CA: Nilgiri Press

Narlikar, Jayant and Geoffrey Burbidge (2008) *Facts and Speculations in Cosmology*
Cambridge: Cambridge University Press

Needham, Joseph and Colin A Ronan (1978) *The Shorter Science and Civilisation
in China: An Abridgement of Joseph Needham's Original Text* Cambridge; New
York: Cambridge University Press

Needham, Joseph and Ling Wang (1959) *Science and Civilisation in China Vol.
3: Mathematics and the Sciences of the Heavens and the Earth* Cambridge:
University Press

Ni, Maoshing (translator) (1995) *The Yellow Emperor's Classic of Medicine (the
Neijing Suwen)* Boston and London: Shambala Publications

Nissen, H J, et al. (1993) *Archaic Bookkeeping: Early Writing and Techniques of the Economic Administration in the Ancient Near East* 1990 Chicago: University of Chicago Press

Nitecki, Matthew H (editor) (1988) *Evolutionary Progress* Chicago; London: University of Chicago Press

Noble, Denis (2006) *The Music of Life: Biology Beyond the Genome* Oxford; New York: Oxford University Press

O'Grady, Patricia (2006) "Thales of Miletus" *The Internet Encyclopedia of Philosophy* 2006 http://www.iep.utm.edu/thales/ Accessed 17 June 2014

O'Kelly, Claire (1991) *Concise Guide to Newgrange* Blackrock: C O'Kelly

Pallen, M. (2009) *The Rough Guide to Evolution* London: Rough Guides

Pauling, Chris (1997) (3rd ed.) *Introducing Buddhism* 1990 Birmingham: Windhorse

Penrose, Roger (1989) *The Emperor's New Mind: Concerning Computers, Minds, and the Laws of Physics* Oxford: Oxford University Press

— (2004) *The Road to Reality: A Complete Guide to the Laws of the Universe* London: Jonathan Cape

Pinker, Steven (2000) *The Language Instinct: The New Science of Language and Mind* 1994 London: Penguin

— (2012) *The Better Angels of Our Nature: A History of Violence and Humanity* 2011 London: Penguin

Plato (1965) *The Republic* 1955 translated by H D P Lee, London: Penguin

Plotinus *The Six Enneads* (2010) translated by Stephen MacKenna and B S Page, Christian Classics Ethereal Library http://www.ccel.org/ccel/plotinus/enneads.toc.html

Possehi, Gregory L (1996) "Mehrgarh" in *The Oxford Companion to Archaeology* edited by Brian M Fagan, New York; Oxford: Oxford University Press

Price, Derek J de Solla (1965) *Little Science, Big Science* 1963: New York: Columbia University Press

Radin, Dean I (2006) *Entangled Minds: Extrasensory Experiences in a Quantum Reality* New York: Paraview Pocket

— (2009) *The Noetic Universe: The Scientific Evidence for Psychic Phenomena* 1997 London: Corgi

Ragan, Mark A, et al. (2009) "The Network of Life: Genome Beginnings and Evolution" *Philosophical Transactions of the Royal Society B: Biological Sciences* 364: 1527, 2169-2175

Reanney, Darryl (1995) *The Death of Forever: A New Future for Human Consciousness* 1991 London: Souvenir Press

Rees, Martin (1997) *Before the Beginning: Our Universe and Others* Reading, MA: Addison-Wesley

Rees, Martin J (2000) *Just Six Numbers: The Deep Forces That Shape the Universe* 1999: London: Weidenfeld & Nicolson

The Revised English Bible (1989) Oxford University Press and Cambridge University Press

Riegel, Jeffrey (2012) "Confucius" *The Stanford Encyclopedia of Philosophy* edited by Edward N Zalta, http://plato.stanford.edu/archives/spr2012/entries/confucius/ Accessed 3 August 2012

The Rig Veda (1896) translated by Ralph Griffith, Internet Sacred Text Archive http://www.sacred-texts.com/hin/rigveda/rvi10.htm

Roberts, Royston M (1989) *Serendipity: Accidental Discoveries in Science* New York: Wiley

Romer, John (2012) *A History of Ancient Egypt: From the First Farmers to the Great Pyramid* London: Allen Lane

Roth, Gerhard (2001) "The Evolution of Consciousness" 554–582 in *Brain Evolution and Cognition* edited by Gerhard Roth and Mario F Wullimann, New York: Wiley

Roughgarden, Joan (2009) *The Genial Gene: Deconstructing Darwinian Selfishness* Berkeley, CA; London: University of California Press

Rowan-Robinson, Michael (2004) (4th ed.) *Cosmology* Oxford: Clarendon

Ruggles, C L N (2005) *Ancient Astronomy: An Encyclopedia of Cosmologies and Myth* Santa Barbara, CA; Oxford: ABC Clio

Russell, Bertrand (1946) *History of Western Philosophy and Its Connection with Political and Social Circumstances from the Earliest Times to the Present Day* London: G Allen and Unwin Ltd

Sapp, Jan (2009) *The New Foundations of Evolution: On the Tree of Life* New York; Oxford: Oxford University Press

Schrödinger, Erwin (1964) *My View of the World* 1961 translated by Cecily Hastings, Cambridge: Cambridge University Press

— (1992) *What Is Life? with Mind and Matter and Autobiographical Sketches* 1944 Cambridge; New York: Cambridge University Press

Shanahan, Timothy (2004) *The Evolution of Darwinism: Selection, Adaptation, and Progress in Evolutionary Biology* Cambridge, UK; New York: Cambridge University Press

Shapiro, James Alan (2011) *Evolution: A View from the 21st Century* Upper Saddle River, NJ: FT Press

Sheldrake, Rupert (2009) (3rd ed.) *A New Science of Life: The Hypothesis of Formative Causation* 1981 London: Icon

Shields, Christopher "Aristotle" *The Stanford Encyclopedia of Philosophy* (2012) edited by Edward N Zalta, http://plato.stanford.edu/cgi-bin/encyclopedia/archinfo.cgi?entry=aristotle Accessed 3 August 2012

Simpson, George Gaylord (1949) *The Meaning of Evolution: A Study of the History of Life and of Its Significance for Man* New Haven, CN; London: Yale University

Singh, Simon (2005) *Big Bang: The Most Important Scientific Discovery of All Time and Why You Need To Know About It* 2004 London: Harper Perennial

Smart, Ninian (1992) *The World's Religions: Old Traditions and Modern Transformations* 1989 Cambridge: Cambridge University Press

Smith, Charles H "The Alfred Russel Wallace Page" (1998, 2000–14) *Western*

Kentucky University http://people.wku.edu/charles.smith/index1.htm Accessed 19 February 2015

Smolin, Lee (1998) *The Life of the Cosmos* 1997 London: Phoenix

— (2007) *The Trouble with Physics: The Rise of String Theory, the Fall of a Science, and What Comes Next* 2006 London: Allen Lane

Sproul, Barbara C (1991) *Primal Myths: Creation Myths Around the World* San Francisco: Harper & Row

Stachel, John (2002) *Einstein from 'B' to 'Z'* Boston: Birkhäuser

Steinhardt, Paul J and Neil Turok (2004) "The Cyclic Model Simplified" Physics Department, Princeton University http://www.phy.princeton.edu/~steinh/dm2004.pdf Accessed 11 March 2007

Steinhardt, Paul J and Neil Turok (2007) *Endless Universe: Beyond the Big Bang* London: Weidenfeld & Nicolson

Stringer, Chris (2011) *The Origin of Our Species* London: Allen Lane

Susskind, Leonard (2005) *Cosmic Landscape: String Theory and the Illusion of Intelligent Design* New York: Little, Brown and Co

Targ, Russell (2012) *The Reality of ESP: A Physicist's Proof of Psychic Phenomena* Wheaton, IL: Quest Books

Teilhard de Chardin, Pierre (1965) *The Phenomenon of Man* 1959 (English edition), 1955 (French original) translated by Bernard Wall, London: Collins

The Ten Principal Upanishads (1938) 1937 translated by Swami Purohit Shree and W B Yeats, London: Faber and Faber

Tolman, Richard Chace (1987) *Relativity, Thermodynamics, and Cosmology* Oxford: Clarendon Press, 1934. New York: Dover Publications

Tomasello, Michael (2003) *Constructing a Language: A Usage-Based Theory of Language Acquisition* Cambridge, MA; London: Harvard University Press

Tuttle, Russell Howard "Human Evolution" *Encyclopædia Britannica Online* (2014) http://www.britannica.com/EBchecked/topic/275670/human-evolution Accessed 1 May 2014

UNEP (2007) "Global Environment Outlook 4" http://www.unep.org/geo/geo4/report/GEO-4_Report_Full_en.pdf Accessed 15 June 2010

The Upanishads (1884) translated by Max Müller, Internet Sacred Text Archive http://www.sacred-texts.com/hin/sbe15

The Upanishads (1987) translated by Eknath Easwaran, Petaluma, CA: Nilgiri Press

Van de Mieroop, Marc (2011) *A History of Ancient Egypt* Oxford: Wiley-Blackwell

Ward, Peter Douglas and Donald Brownlee (2000) *Rare Earth: Why Complex Life Is Uncommon in the Universe* New York: Copernicus

Watson, James D and Gunther S Stent (1980) *The Double Helix: A Personal Account of the Discovery of the Structure of DNA* New York: Norton

Weinberg, Steven (1994) *Dreams of a Final Theory* 1992 New York: Vintage Books

Weiner, Jonathan (1994) *The Beak of the Finch: The Story of Evolution in Our Time* London: Jonathan Cape

Wells, Jonathan (2000) *Icons of Evolution: Science or Myth? Why Much of What We Teach About Evolution Is Wrong* Washington, DC: Regnery Publishing

White, Gavin (2008) (2nd revised ed.) *Babylonian Star-Lore: An Illustrated Guide to the Star-Lore and Constellations of Ancient Babylonia* London: Solaris Publications

Williams, George C (1996) *Adaptation and Natural Selection: A Critique of Some Current Evolutionary Thought* 1966 Princeton, N J; Chichester: Princeton University Press

Wilson, Edward O (1998) *Consilience: The Unity of Knowledge* London: Little, Brown

Wilson, Edward O (2000) (25th anniversary ed.) *Sociobiology: The New Synthesis* 1975 Cambridge, MA; London: Belknap Press of Harvard University Press

Woit, Peter (2006) *Not Even Wrong: The Failure of String Theory and the Continuing Challenge to Unify the Laws of Physics* London: Jonathan Cape

Wright, Robert (1996) *The Moral Animal: Why We Are the Way We Are* 1994 London: Abacus

Zhuangzi (1891) *The Book of Zhuangzi* translated by James Legge, http://ctext.org/zhuangzi/knowledge-rambling-in-the-north

Glossary

The following definitions are not the only meanings of the words listed. They are the precise meanings used in this book and are given here in order to minimize misunderstanding what I am saying. Words in italics are defined elsewhere in this Glossary.

abstraction A general concept formed by extracting common features from concrete realities, actual instances, or specific examples.

aesthetics The branch of *reasoning* that that attempts to understand and communicate the essence of beauty in nature and in human creations.

altruism Behaviour characterized by unselfish concern for the welfare of others; selflessness.

amino acid A *molecule* consisting of a carbon atom bonded to an amino group (–NH2), a carboxyl group (–COOH), a hydrogen atom, and a fourth group that differs from one amino acid to another and often is referred to as the –R group or the side chain. The –R group, which can vary widely, is responsible for the differences in chemical properties of the molecule.

archaea *Prokaryotes* that differ from *bacteria* in their genetic make-up and the composition of their plasma membranes and cell walls. They include most *extremophiles*. Although structurally similar to bacteria, their chromosomal *DNA* and their cell machinery more closely resemble those found in *eukaryotes*.

artistic insight Direct understanding that results in the creation of beautiful or thought-provoking visual, musical, or written works.

asteroid A rocky or metallic object, smaller than a *planet* but bigger than a *meteoroid*, that orbits the Sun or another star; also known as a minor planet. The orbits of most, but certainly not all, of the ones in the solar system are between the orbits of Mars and Jupiter.

astronomy The observational study of moons, planets, stars, galaxies, and other matter beyond the Earth's atmosphere together with their motions.

atom The fundamental unit of a chemical *element*. It comprises a dense central nucleus consisting of positively charged *protons* and uncharged *neutrons* surrounded by orbiting clouds of negatively charged *electrons* equal in number and charge to the protons, thus making the atom electrically neutral. The number of protons uniquely determines what that element is.

atomic number   The number of protons in an *atom*, denoted by Z; it is the number that fundamentally determines what the *element* is and distinguishes it from all other elements.

bacteria   Extremely small, single-celled organisms whose genetic information, encoded in a folded loop of double-stranded *DNA*, is not enclosed in a membrane-bound nucleus (and hence they are *prokaryotes*). In addition to this nucleoid, the cell may include one or more plasmids, separate circular strands of DNA that can replicate independently and are not responsible for the reproduction of the organism. Most reproduce by splitting in two and producing identical copies of themselves. They occur in a variety of shapes, including spheres, rods, spirals, and commas.

baryon   *Protons* and *neutrons*, which comprise most of the mass of ordinary matter, plus a number of short-lived particles such as Sigmas, Deltas, and Xis; according to the Standard Model of Particle Physics, each is composed of three *quarks*.

BCE   Before Common Era; by convention this counts back from the notional year of birth of Jesus in the Christian calendar.

Big Crunch   The result of an expanding universe having a mass density exceeding a critical value, when gravity will eventually reverse the expansion and cause the universe to collapse into the state conjectured for its beginning.

binary fission   The splitting of a cell into two, whereby each cell produced is identical to, and usually grows to the size of, the original cell.

biological evolution   A process of change in organisms that results in new species.

Biological Evolution, First Law of   Competition and rapid environmental change cause the extinction of species.

Biological Evolution, Second Law of   Collaboration causes the evolution of species.

Biological Evolution, Third Law of   Living things evolve by progressive *complexification* and centration along fusing and diverging lineages that lead to stasis in all but one lineage.

Biological Evolution, Fourth Law of   A rise in *consciousness* correlates with increasing *collaboration*, *complexification*, and centration.

black hole   An object with such a strong gravitational field that nothing, including light, can escape because the velocity required to do so exceeds that of light. Black holes are hypothesized to form from the collapse of stars with masses of several Suns or more when their nuclear fuel runs out. A black hole can grow in mass by its powerful gravitational field sucking in surrounding objects. Quantum theory suggests that black holes emit blackbody radiation, but the effect is significant only for very small black holes.

boson   A subatomic particle with integral spin (0, 1, and a hypothesized 2) that does not obey the *Pauli Exclusion Principle*, which means that there is no limit to the number of bosons that can occupy the same quantum state.

BP   Before Present.

Brahma   The creator god in the Vedic trinity whose other members are Vishnu the preserver and Shiva the destroyer, he was later eclipsed by these two. Not to be confused with *Brahman*. (Sanskrit)

Brahman   Ultimate reality, which exists out of space and time, from which everything springs, and of which everything consists; often interpreted as the Cosmic Consciousness or Spirit or Supreme Godhead beyond all form. (Sanskrit)

catalyst   A substance, usually used in small amounts relative to the reactants, that changes the rate of a chemical reaction without itself being changed in the process.

CE   Common Era; by convention this dates from the notional year of birth of Jesus in the Christian calendar.

chemistry   The branch of *science* that investigates the properties, composition, and structure of substances and the changes they undergo when they combine or react under specified conditions.

chromosome   A structure that contains the genetic information of a cell. In a eukaryotic cell this structure consists of threadlike strands of *DNA* wrapped in a double helix around a core of proteins within the cell nucleus; in addition to this nuclear chromosome, the cell may contain other small chromosomes within, for example, a *mitochondrion*. In a prokaryotic cell it consists of a single tightly coiled loop of DNA; the cell may also contain one or more smaller circular DNA molecules called *plasmids*.

cladistics   A system for classifying living things by only those characteristics shared by all descendants of a common ancestor. It is based on the assumption of Darwinian descent with modification; such genealogical grouping into clades (from the Greek for "branch") produces a cladogram, or branching family tree.

cognition   The mental faculty of knowing—through such processes as perceiving, recognizing, insight, reasoning, memorizing, or conceiving—as distinct from the experiences of feeling or of willing.

collaboration   All forms of working together, including *cooperation* and *collectivization*.

collectivization   Working together involuntarily, whether by instinct or conditioned learning or coercion (see *cooperation*).

comet   A small body, usually a few kilometres across, containing icy chunks and frozen gases with bits of embedded rock and dust, and possibly a rocky core, often described as a dirty snowball. Comets have elongated and eccentric orbits, typically with a much greater inclination to the plane of the Earth's orbit. When they pass close to the Sun they develop diffuse gaseous envelopes and often long glowing tails.

complex   A whole composed of distinct, interrelated parts.

complexification   The process of becoming more *complex*.

complexity   The state of being *complex*.

comprehension   The ability to grasp the meaning of something.

conjecture   An opinion or conclusion based on incomplete or inconclusive evidence; a *speculation*.

consciousness   Awareness of the environment, other organisms, and self that can lead to action; a property shared by all organisms in differing degrees, from rudimentary levels in very simple organisms to more sophisticated levels in organisms with complex cerebral systems. (Defined as such to differentiate this property from *reflective consciousness*.)

Conservation of Energy, Principle of   Energy can neither be created nor destroyed. Thus the total energy of an isolated system remains constant, although it may be transformed from one form into another.

cooperation   Working together voluntarily for mutual benefit or to achieve commonly agreed aims.

cosmology   The study of the origin, nature, and large-scale structure of the physical *universe*, which includes the distribution and interrelation of all the galaxies, clusters of galaxies, and quasi-stellar objects.

cosmos   All that exists, which includes various speculations of dimensions other than the three dimensions of space and one of time that we perceive and also other universes with which we have no physical contact and about which we cannot obtain observable or experimental information.

creativity   The ability to use *imagination* to produce new things such as ideas, solutions to problems, images, sounds, smells, tastes, and artefacts.

cytoplasm   Everything outside the cell nucleus and inside the cell membrane. It consists of a gelatinous, water-based fluid called a cytosol, which contains salts, organic molecules and *enzymes*, and in which are suspended organelles, the metabolic machinery of the cell.

Dao   Ultimate, ineffable reality; the Way everything is brought into existence and is manifest in the Way the *cosmos* functions. (Chinese)

Darwinism   The hypothesis that all species of the same genus have evolved from a common ancestor. The principal cause of this biological evolution is *natural selection*, or the survival of the fittest, whereby offspring whose variations make them better fitted to compete with others of their species for survival in a particular environment will live longer and produce more offspring than those less fitted. These advantageous variations are somehow inherited and, over successive generations, will gradually come to dominate the increasing population in that environment while less well-adapted variants will be killed, starved, or driven to extinction. Sexual selection of traits favourable to mating, and use and disuse of organs, are also heritable and cause biological evolution.

Data Interpretation, Law of   The degree to which a scientist departs from an objective interpretation of the data from his investigation is a function of four factors: his determination to validate a *hypothesis* or confirm a *theory*, the time the investigation has occupied his life, the degree of his emotional investment in the project, and his career need to publish a significant paper or safeguard his reputation.

deduction   A process of reasoning in which the conclusion must be true if the premises, or proposition, are true; *inference* from the general to the specific.

dharma   A Sanskrit word originating in the Upanishads of ancient India that means the natural law which regulates and coordinates the operation of the universe and everything within it, and also the conduct of individuals in conformity with this law. The concept is used in Hinduism, Buddhism, and Jainism.

DNA   Deoxyribonucleic acid, located in cells, contains the genetic instructions used in the development and functioning of all known independent organisms and some viruses.

Each DNA *molecule* normally consists of two long chains of four nucleotides in a characteristic sequence; the chains (usually referred to as strands) are twisted into a double helix and joined by hydrogen bonds between the complementary bases adenine (A) and thymine (T) or cytosine (C) and guanine (G) so that its structure resembles a twisted ladder.

When DNA is copied in a cell the strands separate and each serves as a template for assembling a new complementary chain from molecules in the cell.

DNA strands also act as templates for the synthesis of proteins in a cell through a mechanism that makes another nucleic acid, *RNA*, as an intermediary.

dualism   The speculation or belief that there are two fundamental constituents to the universe: matter and mind, or consciousness.

Earth sciences   The branches of *science* that study the origin, nature, and behaviour of the Earth and its parts, including their interactions.

electromagnetic interaction   The force associated with electric and magnetic fields, which are manifestations of a single electromagnetic field. It governs the interaction between two electrically charged particles like a *proton* and an *electron*, and is responsible for chemical interactions and the propagation of light. Like the *gravitational interaction*, its range is infinite and its strength is inversely proportional to the square of the distance between the particles; unlike gravitational interaction, it can either be attractive, when the two charges are unlike (positive and negative), or else repulsive, when the two charges are like (both positive or both negative). The electromagnetic interaction between atoms is 10^{36} times stronger than their gravitational interaction.

According to the Standard Model of Particle Physics, it operates by the exchange of a messenger, or intermediary, particle, the massless *photon*, for which there is experimental evidence.

electron   A fundamental particle that is a constituent of every *atom*. It carries a unit charge of negative electricity (approx. 1.6×10^{-19} Coulombs) and its mass is approximately $1/1836$ that of the *proton*, which carries a unit charge of positive electricity.

element   A substance that cannot be decomposed into simpler substances by chemical means. All *atoms* of the same element are characterized by having the same *atomic number*.

emergence   The appearance of one or more new properties of a complex whole that none of its constituent parts possesses.

"Weak emergence" is where novel properties at the higher level are explained solely by the interaction of the constituent parts.

"Strong emergence" is where novel higher-level properties can neither be reduced to, nor predicted from, the interaction of the constituent parts.

"Systems emergence" is where novel higher-level properties causally interact with lower level properties; this top-down as well as bottom-up causality frequently forms part of a systems approach that, in contrast to the reductionist approach, sees each component as an interdependent part of the whole.

endosymbiosis   An association in which a smaller organism lives inside a larger one, usually collaboratively with each organism feeding on the other's metabolic excretion (see *symbiosis* and *symbiogenesis*).

engineering   The application of knowledge about the natural world, usually obtained through *science*, in order to design means of achieving desired objectives.

entropy   A measure of the disorder or disorganization of the constituent parts of a closed system; a measure of the energy that is not available for use. The lower the entropy the greater the organization of its constituent parts and, consequently, the more energy that is available for use and the more information that can be gained by observing its configuration. At maximum entropy the configuration is random and uniform, with no structure and no energy available for use; this occurs when the system has reached a state of equilibrium. Statistically this is expressed as:

$$S = k \, ln \, \Omega$$

where S represents entropy

k is a constant, called Boltzmann's constant after the scientist who formulated the equation

ln is the natural logarithm

Ω is the number of different ways in which the equilibrium state can occur.

enzyme   A biological *catalyst*, or chemical that speeds up the rate of a chemical reaction without being consumed by the reaction. Such catalysts are essential for the functioning of an organism because they make possible processes that would otherwise occur far too slowly without the input of energy (measured by an increase in temperature) to activate the reaction but which would damage or destroy the organism.

epigenetic inheritance   The transmission from parent cell to offspring cell, in either asexual replication or sexual reproduction, of variations that give rise to variations in the characteristics of the organism but do not involve any variations in *DNA* base sequences.

epigenetics   The study of mechanisms of *gene* regulation that cause changes in the *phenotype* of an organism but involve no change in the *DNA* sequences of the genes themselves.

epistemology   The branch of *reasoning* that investigates the nature, sources, validity, limits, and methods of human knowledge.

equinox   The two days of the year when the Sun is exactly above the equator and day and night are of approximately equal length all over the Earth. The vernal equinox occurs around 21 March and the autumnal equinox around 22 September.

ethical insight   Direct understanding of how, and often why, humans should behave either as individuals or as a group towards other individuals and other groups.

ethics   The branch of *reasoning* that evaluates human behaviour and often attempts to produce codes governing good conduct between individuals, between an individual and a group of individuals (like a society or state), and between groups of individuals.

eukaryotes   Organisms whose cells incorporate a membrane-bound nucleus, which contains the genetic information of the cell, plus organelles, which are discrete structures that perform specific functions. Larger and structurally and functionally more complex than *prokaryotes*, they comprise single-celled organisms, like amoeba, and all multicellular organisms, like plants, animals, and humans.

evolution   A process of change, occurring in something, especially from a simple to a more complex state (see also *biological evolution*).

extremophile   An organism that lives in extreme conditions of heat, pressure, or chemical environment such as high acidity or salt concentration.

gene   The fundamental unit of inheritance, which normally comprises segments of *DNA* (in some viruses they are segments of *RNA* rather than DNA); the sequence of the bases in each gene determines individual hereditary characteristics, typically by encoding for *protein* synthesis. The segments are usually split, with some parts located in distant regions of the *chromosome* and overlapping with other genes.

genetic drift   The variation in the frequencies of alleles (gene pairs) in a small population that takes place by chance events rather than through *natural selection*. It can result in genetic traits being lost from, or becoming widespread in, a population irrespective of the survival or reproductive value of those genetic traits.

genome   The entire genetic content of an organism consisting of long strands, called *chromosomes*, of *DNA* molecules or, in certain viruses, *RNA*. It includes the *genes*, the coding regions of DNA that are translated into *protein* and RNA molecules, and also regulatory and non-coding regions.

genotype   The genetic makeup of an organism, as distinct from its physical characteristics (see *phenotype*).

gravitational interaction   In Newtonian physics it is an instantaneous force of interaction between all particles of mass. Uniquely among the four fundamental forces (see *electromagnetic interaction*, *strong interaction*, and *weak interaction*) it is the only universal interaction. Its range is infinite, it is always attractive, and its strength is given by multiplying the masses and dividing the product by the square of the distance between the centres of mass of the particles, and then multiplying the result by a universal constant, G, called Newton's gravitational constant. Mathematically:

$$F = \frac{Gm_1m_2}{r^2}$$

where F is the gravitational force, m_1 and m_2 are the masses, r is the distance between the centres of mass, and the constant, G, is an incredibly small number, 6.67×10^{-11} metre^3 (kilogram-second^2)$^{-1}$.

In relativity theory, however, it is not a force but a warp in the space-time fabric caused by mass and is not instantaneous.

gravity   A natural phenomenon by which physical bodies appear to attract each other by a force proportional to their masses, described by Newton but modified by relativity theory. See *gravitational interaction.*

homogeneous   Having a uniform composition or structure. A homogeneous universe is the same at every point.

human   The only species known to possess *reflective consciousness.*

human culture   A society's knowledge, beliefs, values, organization, customs, *creativity* expressed in its arts, and innovation expressed in its *science* and *technology*, that are learned and developed by its members and transmitted to each other and to succeeding generations.

hypothesis   A provisional *theory* put forward to explain a phenomenon or set of phenomena and used as a basis for further investigation; it is usually arrived at either by *insight* or by inductive reasoning after examining incomplete evidence and must be capable of being proven false.

idealism   The speculation or belief that material things do not exist independently but exist only as constructions of the *mind* or *consciousness.*

imagination   The ability of the *mind* to form images, sensations, and ideas not seen or otherwise experienced at the time, including things that can never be experienced by the senses such as a man with the head of a lion.

Increasing Entropy, Principle of   During any process in an isolated system the *entropy* either stays the same or, more usually, increases, i.e. disorder increases, available energy decreases, and information is lost over time as the system moves towards a state of equilibrium.

induction   The method of assembling facts, detecting a pattern, and making a generalized conclusion, or law, based on that pattern; the conclusion may be true, but is not necessarily true as is a conclusion reached by *deduction* from valid premises.

inference   A process of reasoning from premises or known facts that derive a conclusion which may be either necessarily or else probably true (see *deduction* and *induction*).

insight   Seeing clearly the essence of a thing, usually suddenly after disciplined meditation or following an unsuccessful attempt to arrive at an understanding through reasoning.

(For types of insight, see *mystical insight, spiritual insight, scientific insight, mathematical insight, psychological insight, ethical insight,* and *artistic insight.* For contrast, see *reasoning.*)

instinct   An innate, impulsive response to stimuli, usually determined by biological necessities such as survival and reproduction.

intellect   The ability to learn, reason, and understand.

intelligence   The capacity to acquire and successfully apply knowledge for a purpose, especially in new or challenging situations.

invention   The ability to make something new, either by creative use of the *imagination* or by trial and error.

ion   An *atom* that has lost or gained one or more electrons and thus has a positive or negative charge.

isotope   Atoms with the same *atomic number* but a different number of neutrons—and hence different mass—are called isotopes of the *element.*

isotropic   Having physical properties that do not vary in any direction. An isotropic universe is one that appears to be the same in every direction; if the universe is isotropic when observed from every point, then it is necessarily *homogeneous.*

knowledge   Information about a thing acquired by experience, inference, insight, or education.

language   The communication of feelings, perceptions, narratives, explanations, or ideas by a complex structure of learned spoken or written or signed symbols that convey meaning in the culture within which it is employed.

law, scientific or natural   A succinct, general statement capable of being tested by observation or experiment and for which no repeatable contrary results have been reported that a set of natural phenomena invariably behaves in an identical manner within specified limits. Typically it is expressed by a single mathematical equation. The result of applying such a law may be predicted from knowing the values of those variables that specify the particular phenomenon under consideration. See also *principle* and *theory.*

lepton   A group of fundamental particles that do not take part in the *strong interaction.* A lepton can carry one unit of electrical charge, like an *electron,* or be neutral, like the neutrino.

life   The ability of an enclosed entity to respond to changes within itself and in its environment, to extract energy and matter from its environment, and to convert that energy and matter into internally directed activity that includes maintaining its own existence.

life sciences   The branches of *science* that study features of living organisms (such as plants, animals, and humans) and also relationships between such features.

logic   The branch of *reasoning* that aims to systematically distinguish valid from invalid inferences by devising rules principally for deductive and inductive reasoning.

MACHOs   Massive Compact Halo Objects are forms of dense matter, like black holes, brown dwarfs and other dim stars, speculated by astrophysicists to explain *dark matter* in the universe. See also *WIMPs.*

magnetic monopole   A hypothetical particle that has only one pole of magnetic charge (a north pole without a south pole or vice-versa) instead of the usual two.

materialism   The speculation or belief that only material things or phenomena are real and that all things, such as mind or consciousness or thoughts, will eventually be explained as material things or their interactions. (See *physicalism.*)

mathematical insight   Direct understanding of the properties of, or relationships between, numbers, real and abstract shapes, and often the rules governing any such relationships.

medical sciences   The branches of science that are applied to maintain health, prevent and treat disease, and treat injuries.

megaverse   A speculated higher dimensional *universe* in which our universe of three spatial dimensions is embedded. Some speculations have many megaverses comprising the *cosmos.*

metaphysics   The branch of *reasoning* that investigates and attempts to understand ultimate reality or the essence and causes of all things whether material or immaterial.

meteor   A streak of light in the night sky caused by a natural solid body, called a *meteoroid*, heated to incandescence by friction as it plunges into the atmosphere of the Earth or another planet.

meteorite   A natural solid body that reaches the surface of the Earth or another planet after plunging through the atmosphere.

meteoroid   A natural solid body that plunges into the atmosphere of the Earth or another planet.

mind   That which conceives, perceives, reasons, wills, and remembers.

mitochondrion   A membrane-bound organelle located in the *cytoplasm* of almost all eukaryotic cells; its primary function is to generate energy.

model   A simplified version of a *theory* constructed to facilitate calculation or visual representation.

molecule   The smallest physical unit of a substance that can exist independently; it consists of one or more *atoms* bonded by sharing *electrons* and is electrically neutral.

monism   The speculation or belief that all existing things are formed from, or are reducible to, the same ultimate reality or principle of being. It may be divided into *materialism* (or *physicalism*), *idealism*, and *neutral monism*. It contrasts with *dualism* and *pluralism.*

morality   The custom of right or good conduct.

morphology   The size, shape, and structure of an organism; the study thereof.

multiverse   A speculated *cosmos* that contains our *universe* plus multiple if not infinite other universes with which we have no physical contact and about which we cannot obtain observable or experimental information. Several different kinds of multiverse, each having different properties, have been proposed.

mystical insight   Direct understanding of ultimate reality: the essence and cause of all things.

myth   A traditional story, either wholly or partly fictitious, especially one involving supernatural beings or ancestors or heroes, that explains some natural or social phenomenon or some cultural or religious practice.

natural philosophy   The branch of *reasoning* that investigates and attempts to understand the natural world perceived by our five senses and how it operates.

natural selection (Darwinian and NeoDarwinian)   The cumulative effect of small, randomly produced variants inherited over very many generations that enables organisms to survive longer and reproduce more in a particular environment than organisms lacking those variants; it results in an increase in the number of such favourable, or best adapted, variants for that environment and the elimination of unfavourable variants. See also *Darwinism*, *NeoDarwinism*, and *UltraDarwinism*.

NeoDarwinism   The synthesis of Darwinian *natural selection* with Mendelian and population genetics in which randomly produced genetic variations responsible for characteristics that make individuals in a species population better adapted to compete for resources in their environment survive longer and produce more offspring. These favourable genes are thus inherited in increasingly greater numbers causing the gene pool—the total of all genes in the population—gradually to change over very many generations until a new *species* emerges. Members of the population lacking genetic variations responsible for such adaptive characteristics are killed, starved, or gradually become extinct in that environment.

nervous system   An organized group of cells, called *neurons*, specialized for the conduction of electrochemical stimuli from a sensory receptor through a nerve network to an effector, the site at which a response occurs.

neuron   A eukaryotic cell specialized for responding to stimulation and for conducting electrochemical impulses.

neutral monism   The speculation or belief that both the mental and physical can be reduced to some sort of third entity.

neutron   An electrically neutral subatomic particle found in the nucleus of all *atoms* except ordinary hydrogen. In light nuclei the stable configuration consists of neutrons and *protons* in almost equal numbers, but as the elements become heavier neutrons outnumber protons. Although a neutron is stable within a nucleus, a free neutron has a half-life of about 15 minutes and decays to produce a *proton*, an *electron*, and an antineutrino.

noetic   Of, pertaining to, or originating in, the *mind*.

observable universe   That part of the *universe* containing matter capable of being detected by astronomical observation. According to current orthodox cosmology it is circumscribed by the speed of light and the time since matter and radiation decoupled some 380,000 years after the universe came into existence in the Big Bang.

ontogenesis, or ontogeny   The origin and development of an individual organism from embryo to adult.

ontology   The sub-branch of *metaphysics* that investigates what it is that exists.

orthogenesis   The hypothesis that *biological evolution* has a direction caused by intrinsic forces; versions range from those that hold adaptation also plays a significant role in the evolution of *species*, through adaptation only influences variations within species, to the view that direction demonstrates a purpose to biological evolution.

paradigm   The prevailing pattern of largely unquestioned thought and assumptions in a scientific discipline within which research is carried out and according to which results are interpreted.

parameter   In science generally, one of a set of measurable factors, such as temperature and pressure, that define a system and determine its behaviour; in an experiment one parameter is often varied while others are kept constant. In mathematics, which is the favoured instrument of theoretical physicists, it is a constant in an equation that varies in other equations of the same general form.

parthenogenesis   The development of an egg into an offspring without fertilization by a male.

particle horizon   We cannot be causally influenced by, or obtain information on—and therefore detect—any particles, whether of positive or zero mass, further from us than the distance travelled at light speed since time began.

Pauli Exclusion Principle   No two *electrons* in an *atom* or *molecule* can have the same four quantum numbers. More generally, no two types of fermion (a class of particles including *electrons*, *protons*, and *neutrons*) in a given system can be in states characterized by the same quantum numbers at the same time.

peptide   A chain of two or more *amino acids* formed by chemically linking the carboxyl group of one amino acid to the amino group of another amino acid.

phase change   A significant change in the behaviour of a substance. For example, ice heated to 0° Celsius undergoes a phase change from solid to liquid water; liquid water heated to 100° Celsius changes to gaseous steam.

phenomenon   Something that can be perceived or experienced by the senses.

phenotype   The observable characteristics of an organism, such as shape, size, colour, and behaviour.

philosophical thinking   The second phase of *reflective consciousness* when reflection on self and its relationship with the rest of the universe branches from *superstition* into *philosophy*.

philosophy   Love of wisdom; thinking about ultimate reality, the essence and causes of things and of existence, the natural world, behaviour, and thinking itself by means of *insight* or *reasoning*.

photon   A quantum of light or other form of electromagnetic energy that possesses both particle and wave properties and has zero mass, no electric charge, and an indefinitely long lifetime.

phyletic, or pseudo, extinction   A *species* evolves into one or more new species; the first species has become extinct, but the evolutionary lineage continues.

phylogenetics   The study of evolutionary relationships between groups of organisms, particularly the patterns of lineage branching.

physical sciences   The branches of *science* that study inanimate phenomena; they include *astronomy, physics, chemistry,* and the *Earth sciences.*

physicalism   The speculation or belief that only physical things are real and that all other things, such as mind or consciousness or thoughts, will eventually be explained as physical things or their interactions; also called *materialism,* although it incorporates a wider view of physicality than matter, e.g. non-material forces like gravity that arise from matter.

physics   The branch of *science* that investigates matter, energy, force, and motion, and how they relate to each other.

Planck length, l_p   The smallest unit of length in any quantum theory of gravity: below this length quantum fluctuations are incompatible with relativity theory's continuous space-time.
Mathematically it is expressed
$l_p = \surd\,(\hbar G/c^3)$ and is approximately 10^{-35} metres
(see *Planck units* for explanation of symbols)

Planck mass, m_p   The mass of the hypothesized most massive elementary particle: with a greater mass the elementary particle would be overwhelmed by its own gravitational force and collapse into a black hole.
Mathematically it is expressed
$m_p = \surd\,(\hbar c/G)$ and is approximately 10^{19} proton masses or 10^{-5} grams.
(see *Planck units* for explanation of symbols)

Planck time, t_p   The time it takes light to travel a Planck length.
Mathematically it is expressed
$t_p = \surd\,(\hbar G/c^5)$ and is approximately 10^{-43} seconds.
(see *Planck units* for explanation of symbols)

Planck units   A system of absolute scales of measurement chosen so that universal physical constants all equal one, specifically
$G = c = k = \hbar = 1$
where
G is Newton's gravitational constant, which measures the strength of the gravitational force
c is the constant speed of light
k is Boltzmann's constant, which measures the entropy, or degree of disorganization, of a closed system

h̵ is Planck's constant, *h*, which measures the scale of quantum phenomena, divided by 2π

Consequently these constants disappear from equations of physical laws that use these scales.

planet An object of approximately constant mass and volume that usually orbits a star or stellar remnant, has sufficient mass for its gravitational field to pull it into an approximately spherical shape but insufficient mass to cause thermonuclear fusion in its core, and is not a satellite of another planet; it has cleared its orbital zone of planetesimals and other debris and may hold some of these in orbit round itself either as moons or dust rings.

plasma A phase of matter comprising an ionized gas of positively charged nuclei of *atoms* (*ions*) and free negatively charged *electrons* plus neutral particles, with no net overall charge; it is electrically conductive and is affected by magnetic fields.

plasmid A circular molecule of *DNA*, sometimes *RNA*, located in the *cytoplasm* of most *prokaryotes* and in the *mitochondria* of some *eukaryotes*. They replicate independently of a cell's *chromosome*.

pluralism The speculation or belief that reality is made up of many kinds of being or substance.

polyploidy The possession of more than two sets of chromosomes in a cell.

prana Vital energy, the power of life; the essential substrate of all forms of energy. (Sanskrit)

primeval thinking The first phase of *reflective consciousness* when reflection on self and its relationship with the rest of the universe is rooted principally in survival and superstition.

principle, scientific or natural A *law* considered to be fundamental and universally true. For example, the First Law of Thermodynamics applies to work and heat energy, whereas the Principle of Conservation of Energy applies to all forms of energy.

prokaryote A cell that doesn't have its genetic material enclosed by a membrane within the cell enclosure.

protein A *molecule* consisting of a chain of between fifty to several thousand *amino acids* that provides structure, or controls reactions, in all cells. A particular protein is characterized by the sequence of up to twenty different amino acids that comprise the chain plus the three-dimensional configuration of the chain.

proton A stable subatomic particle that carries a unit charge of positive electricity and constitutes the nucleus of a hydrogen *atom*. Protons, together with slightly more massive *neutrons*, are found in the nucleus of all atoms, and the number of protons defines the chemical element.

psychological insight Direct understanding of why and how individuals, or groups of individuals, think and behave as they do.

psychology The branch of *science* that investigates the mental processes and behaviour of individuals and groups.

quantum gravity   The hoped-for quantum theory of gravity that would enable gravitational energy to be unified with other forms of energy in a single quantum theoretical framework.

quantum mechanics   The theory that explains the behaviour of matter at the scale of an atom or smaller based on *quantum theory* and incorporates Heisenberg's *Uncertainty Principle* and the *Pauli Exclusion Principle*.

quantum theory   The theory that energy is emitted and absorbed by matter in tiny, discrete amounts, each of which is called a quantum that is related to the radiation frequency of the energy, and thus possesses properties of both a particle and a wave. It gave rise to *quantum mechanics*. The term is now used in a general sense to refer to all subsequent theoretical developments.

quark   A group of fundamental particles that make up *protons*, *neutrons*, and other particles that feel the *strong interaction*.

reasoning   An attempt to understand the essence of a thing by a logical process, based either on evidence or on assumptions taken as self-evident.

reductionism   The method of breaking down a thing into its constituent parts in order to understand or explain it; the belief that anything can be understood and explained by studying its constituent parts and the way in which they interact. (By contrast see *emergence*.)

reflective consciousness   The property of an organism by which it is conscious of its own *consciousness*, that is, not only does it know but also it knows that it knows; the ability of an organism to think about itself and its relationship with the rest of the universe of which it knows it is a part.

religion   An organization established to preserve, interpret, apply, teach and, usually, propagate the insights, beliefs, and resultant exhortations of the person claimed as the founder. Members of the organization accept the truths of such insights and interpretations as matters of faith. The organisation typically secures loyalty from, and cohesion of, its members by three principal means: inculcation, especially of the young; rules, the breaking of which incurs penalties; and ritual, the practice of which generates emotional satisfaction.

retrodiction   A result that has occurred in the past and is deduced or predicted from a later scientific law or theory.

ribosome   A round particle composed of *RNA* and *protein* in the *cytoplasm* of cells that serves as an assembly site for proteins by translating the linear genetic code carried by messenger RNA into a linear sequence of *amino acids*.

RNA   Ribonucleic acid resembles *DNA* in that it consists of a chain of four nucleotides in a characteristic sequence, but uracil (U) replaces thymine (T) alongside adenine (A), cytosine (C), and guanine (G), and the strands are single, except in certain viruses.

rule   A direction for scientists to conduct a procedure or solve a problem, as distinct from a *law* that natural phenomena follow (nature obeys laws, scientists employ rules).

scalar   A quantity, such as mass, length, or speed, that is completely specified by its magnitude and has no direction.

scalar field   In mathematics and physics a scalar field associates a scalar value to every point in a space. The scalar may either be a mathematical number or a physical quantity.

science   The attempt to understand and explain natural phenomena by using systematic, preferably measurable, observation or experiment, and to apply reason to the knowledge thereby obtained in order to infer testable laws and make predictions or *retrodictions*.

scientific insight   Direct understanding of the essence or causes of natural phenomena, their interactions or other relationships, and often the rules governing such interactions or relationships.

scientific method (notional)

1. Data is collected by systematic observation of, or experiment on, the phenomenon being studied;
2. a provisional conclusion, or *hypothesis*, is inferred from this data;
3. predictions deduced from this hypothesis are tested by further observations or experiments;
4. if these tests confirm the predictions, and the confirmations are reproduced by independent testers, then the hypothesis is accepted as a scientific *theory* until such time as new data conflict with the theory;
5. if new data conflict with the theory, then the theory is either modified or discarded in favour of a new hypothesis that is consistent with all the data.

scientific thinking   The third phase of *reflective consciousness* when reflection on self and its relationship with the rest of the universe branches into empiricism.

scientism   The belief that science, incorporating the methods of the natural sciences, is the only means of arriving at true knowledge and understanding.

singularity   A hypothetical region in space-time where gravitational forces cause a finite mass to be compressed into an infinitely small volume and therefore to have infinite density, and where space-time becomes infinitely distorted.

solstice   The time (21 June or 22 December) at which the Sun is furthest north or south from the Earth's celestial equator and appears to stand still before returning towards the equator. The longest day occurs at the summer solstice; the shortest day at the winter solstice.

species   A population of organisms whose defining adult heritable characteristics have undergone irreversible change from those of the population, or populations, from which it evolved.

speculation   An idea or conclusion based on incomplete or inconclusive evidence; a *conjecture*.

spiritual insight   Claimed revelation from a god or God, or his or her messenger, that usually exhorts the recipient to advocate a course of action for believers in the deity.

Standard Model of Particle Physics   Aims to explain the existence of, and interactions between, everything except gravity we observe in the universe in terms of fundamental particles and their movements. Currently it describes 17 such types of fundamental particles, grouped as *quarks*, *leptons*, or *bosons*. When corresponding antiparticles and boson variations are taken into account, the total of fundamental particles is 61.

strong interaction   One of the four fundamental interactions between elementary particles of matter (see *electromagnetic interaction*, *weak interaction*, and *gravitational interaction*). It is thought to be the force holding quarks together to form *protons*, *neutrons*, and other hadrons, and for binding protons and neutrons together to form the nucleus of an atom, thereby overcoming the electrical repulsion of the positively charged protons. Its range is approximately that of an atomic nucleus and at such distances its strength is about 100 times that of the electromagnetic force.

superstition   A belief that conflicts with evidence or for which there is no reasonable basis and which usually arises from a lack of understanding of natural phenomena or fear of the unknown.

symbiogenesis   The merging of separate organisms to form a single new organism (see *endosymbiosis*).

symbiosis   The physical association of two or more organisms through most of the life history of one of them.

taxonomy   The hierarchical classification of organisms into named groups according to common characteristics, from the most general characteristics to particular ones.

technology   The invention, making, and use of tools or machines to solve a problem.

terminal extinction   A *species* ceases to exist without leaving any evolved descendants.

theory   An explanation of a set of phenomena that has been confirmed by a number of independent experiments or observations and is used to make accurate predictions or *retrodictions* about such phenomena. See also *hypothesis*, *model*, *law*, and *principle*.

Thermodynamics, First Law of   The increase in the internal energy of a system using or producing heat is equal to the amount of heat energy added to the system minus the work done by the system on its surroundings. It is a specific application of the *Principle of Conservation of Energy*.

Thermodynamics, Second Law of   Heat never passes spontaneously from a cold body to a hot body; energy always goes from more usable to less usable forms. It is a specific application of the *Principle of Increasing Entropy*.

Thermodynamics, Third Law of   The entropy of a perfectly ordered crystal at a temperature of absolute zero is zero.

think   To apply the mind, whether by reasoning or by insight, to something, as distinct from responding by instinct.

thought   The process of applying the mind to dwell upon something; the results of this process.

UltraDarwinism   Any hypothesis that employs the concept of the evolution of things other than organisms by means of natural selection in which the cumulative effect of small random variations in the characteristics of those things, or else characteristics caused by those things, over numerous generations makes them increasingly better fitted to compete for survival and reproduction in their environment.

Uncertainty Principle (Heisenberg)   A principle in quantum mechanics that says the more certain we are in measuring an object's position the less certain we are in measuring its velocity at the same time. This also applies to measuring the energy of an object at a specific time. For visible objects the product of the two uncertainties in each case is so small as to be negligible. However, for objects with the mass of an *atom* or less, like an *electron*, the uncertainty is significant.

universe   All the matter and energy that exists in the one dimension of time and three dimensions of space perceived by our senses, as distinct from the *observable universe* and from the *cosmos*.

Upanishad   An account of the teachings of a seer in ancient India arising from his mystical insights. Traditionally they are attached to the end of one of the *Vedas*, but they are concerned with aspects of ultimate reality rather than worship of gods. (Sanskrit)

Veda   Revealed wisdom; specifically, one of the four collections that comprise the Hindu scriptures. It is often used to refer to the first, and most ancient, part of each collection, the samhita, a collection of hymns to the gods. (Sanskrit)

visual horizon   According to the Big Bang model, we can only see back to the time of decoupling of matter and electromagnetic radiation—currently estimated at 380,000 years after the Big Bang—because before that photons were scattered by continual interaction with the initial plasma, thereby making the universe opaque.

weak interaction   One of the four fundamental interactions between elementary particles of matter (see *electromagnetic interaction*, *strong interaction*, and *gravitational interaction*). It plays an important role in transforming particles into others, e.g. through radioactive decay. It is responsible for transforming an electron and a proton into a neutron and a neutrino, an essential stage of nuclear reactions. It is several orders of magnitude weaker than the electromagnetic interaction and much weaker still than the strong interaction, while its range is about one thousandth the diameter of an atomic nucleus.

will   Use of the *mind* to make decisions about things.

WIMPs   Weakly Interacting Massive Particles are particles left over from the Big Bang, like neutrinos with a hundred times the mass of a proton, and so on, speculated by particle physicists to explain *dark matter* in the universe. See also *MACHOs*.

Illustration credits

Figures 3.2 and 3.3 From *The Inflationary Universe* © 1997 Alan Guth by kind permission of the author

Figure 4.1 NASA

Figure 5.1 From *A Brief History of Time* © 1988 Stephen Hawking by kind permission of Writers House LLC on behalf of the author

Figure 5.2 From *Facts and Speculations in Cosmology* © Jayant Narlikar and Geoffrey Burbidge 2008 by kind permission of Jayant Narlikar

Figure 5.3 By kind permission of Paul Steinhardt

Figure 8.1 From *The Inflationary Universe* © 1997 Alan Guth by kind permission of the author

Figure 8.2 NASA

Figure 8.3 European Southern Observatory

Figures 8.4, 9.1, and 9.2 NASA

Figure 9.3 By kind permission of the International Union of Pure & Applied Chemistry

Figure 10.1 NASA

Figure 12.2 Dna-Dennis

Figure 12.3 From *Environmental Science: Earth as a Living Planet* by Daniel D Botkin and Edward A Keller (Media Support Disk) © 1999 by John Wiley and Sons Inc. by kind permission of the publisher

Figure 12.4 NASA

Figure 12.5 Redrawn by permission from an original by Stephen Nelson

Figures 12.6, 12.7, and 12.8 The United States Geological Survey

Figure 14.1 By kind permission of Norman Pace

Figure 14.2 Mariana Ruiz Villarreal

Figure 14.3 The National Human Genome Research Institute

Figure 14.4 Adapted from an original by Yassine Mrabet

Figure 14.5 Yassine Mrabet

Figures 15.1 and 15.2 From *At Home in the Universe* © 1995 Stuart Kauffman; figures © 1993 Oxford University Press, reproduced by kind permission of Oxford University Press

Figure 16.1 John Gould

Figures 17.1 and 17.2 From *The Crucible of Creation* © 1998 Simon Conway Morris, by kind permission of the author and Oxford University Press

Figure 17.3 © John Hands, digitized by Kevin Mansfield

Figure 17.5 From http://web.neomed.edu/web/anatomy/Pakicetid.html

Figure 17.6 From http://web.neomed.edu/web/anatomy/Pakicetid.html illustration by Carl Buell

Figure 17.7 By kind permission of Encyclopædia Britannica Inc © 2011

Figure 18.1 Trevor Bounford

Figure 19.1 Mediran

Figure 20.1 © John Hands

Figure 20.2 By kind permission of Douglas L Theobald

Figure 23.1 © John Hands

Figure 23.2 Kathryn Delisle, reproduced by kind permission of Lynn Margulis

Figure 24.1 Adapted from the drawing of Quasar Jarosz

Figure 24.2 © John Hands, redrawn by Trevor Bounford

Figure 24.3 Redrawn by Trevor Bounford from an original by William Tietien.

Figure 24.5 Adapted by kind permission from the HOPES website at Stanford University https://www.stanford.edu/group/hopes/cgi-bin/wordpress/?p=3787

Figure 24.6 By kind permission from http://brainmuseum.org/ and https://www.msu.edu supported by the US National Science Foundation

Figure 26.1 © John Hands

Figure 26.2 By kind permission of Encyclopædia Britannica Inc © 2013

Figure 27.1 © John Hands

Figure 28.1 Klem

Figures 28.2, 29.1, 29.2, 31.1, and 31.2 © John Hands. Figure 31.1 digitally redrawn by Kevin Mansfield. Figure 31.2 digitally redrawn by Fakenham Prepress Solutions

Where permission is not listed, the illustration is not copyright. Every effort has been made to trace all copyright holders, and the author and publisher will gladly rectify in future editions any errors or omissions brought to their attention.

Index

Page numbers in *italics* refer to tables and illustrations.